RESIDENTIAL INTEGRATION SERIES

Residential Integrator's Basics

SAMUEL DiPAOLA
TECHNICAL EDITOR

THOMSON
DELMAR LEARNING

Australia Canada Mexico Singapore Spain United Kingdom United States

THOMSON
DELMAR LEARNING

Residential Integrator's Basics
Samuel DiPaola, Technical Editor

Vice President, Technology and Trades ABU:
David Garza

Director of Learning Solutions:
Sandy Clark

Executive Editor:
Stephen Helba

Product Manager:
Sharon Chambliss

Senior Editorial Assistant:
Dawn Daugherty

Marketing Director:
Deborah S. Yarnell

Channel Manager:
Dennis Williams

Marketing Coordinator:
Stacey Wiktorek

Curriculum Manager:
Elizabeth Sugg

Director of Production:
Patty Stephan

Production Manager:
Larry Main

Content Project Manager:
Nicole Stagg

Technology Project Manager:
Kevin Smith

Technology Project Specialist:
Linda Verde

COPYRIGHT © 2007 Thomson Delmar Learning, a division of Thomson Learning Inc. All rights reserved. The Thomson Learning Inc. logo is a registered trademark used herein under license.

Printed in the United States of America
1 2 3 4 5 XX 08 07 06

For more information contact Thomson Delmar Learning Executive Woods
5 Maxwell Drive, PO Box 8007, Clifton Park, NY 12065-8007

Or find us on the World Wide Web at
www.delmarlearning.com

ALL RIGHTS RESERVED. No part of this work covered by the copyright hereon may be reproduced in any form or by any means—graphic, electronic, or mechanical, including photocopying, recording, taping, Web distribution, or information storage and retrieval systems—without the written permission of the publisher.

For permission to use material from the text or product, contact us by
Tel. (800) 730-2214
Fax (800) 730-2215
www.thomsonrights.com

Library of Congress Cataloging-in-Publication Data:

ISBN: 1-4180-1407-9

NOTICE TO THE READER

Publisher does not warrant or guarantee any of the products described herein or perform any independent analysis in connection with any of the product information contained herein. Publisher does not assume, and expressly disclaims, any obligation to obtain and include information other than that provided to it by the manufacturer.

The reader is expressly warned to consider and adopt all safety precautions that might be indicated by the activities herein and to avoid all potential hazards. By following the instructions contained herein, the reader willingly assumes all risks in connection with such instructions.

The publisher makes no representation or warranties of any kind, including but not limited to, the warranties of fitness for particular purpose or merchantability, nor are any such representations implied with respect to the material set forth herein, and the publisher takes no responsibility with respect to such material. The publisher shall not be liable for any special, consequential, or exemplary damages resulting, in whole or part, from the readers' use of, or reliance upon, this material.

Contents

Series Preface *xiii*
Preface *xv*

1
Introduction 1

What Is a Residential Network? 1
The Evolution of Residential Networks 2
Summary 8
Key Terms 8
Review Questions 9

2
Networking Fundamentals 10

Basic Networking, Architecture, and Topology 10
Network Protocols and Their Functions 41
TCP/IP and the Internet 41
Computer Network Addressing 46

Other Concerns 47
Low Voltage Residential Network Applications 48
Summary 49
Key Terms 50
Review Questions 51

3

Wireless Communications 53

Introduction 53
Cellular Mobile Telephone Systems 54
Analog Versus Digital Access 56
Wireless Applications and Products 63
Wireless LANs (WLANs) 67
Satellite Communications 73
International Wireless Communications 83
Summary 86
Key Terms 86
Review Questions 87

4

The Technology of Communications 88

Introduction 88
The Operation of Telephones 89
Video Transmission 90
CATV Networks 92
Summary 93
Key Terms 93
Review Questions 93

5

Voice and Data Applications 95

Voice Applications 95
Data Applications 100
Voice Over IP 103

Summary   109
Key Terms   110
Review Questions   110

6
Entertainment Applications   112

Introduction   112
Audio and Music   115
Video   122
Home Theater   133
Gaming   136
Entertainment Server   137
Future Applications   138
Summary   144
Key Terms   145
Review Questions   146

7
Home Automation   147

Introduction   147
Climate Control   148
Lighting Control   150
Security Systems   152
Fire Alarm Systems   154
Smart Appliances   155
Integrated Systems   157
Summary   161
Key Terms   161
Review Questions   162

8
The Construction Process   163

Introduction   164
The Construction Process   164

Information and Instruction 165
Permits and Inspections 172
Regional Differences 173
Introduction to Job Site Safety 174
Summary 191
Key Terms 191
Review Questions 193

9

Boxes 194

Introduction 194
Functions of Boxes 194
Types of Boxes 195
Sizing Boxes 205
Installing Boxes 209
Conduit Bodies 219
Summary 219
Key Terms 219
Review Questions 220

10

Residential Electrical Cabling Installation 221

Introduction 222
NEC® Requirements for Drilling or Notching Studs, Rafters, and Joists 222
Requirements for Drilling or Cutting Studs, Joists, and Rafters 222
Environmental Airspaces 226
Bundling Cables 228
Cables Run Parallel to Framing Members 230
Connecting Conductors with Wire Nuts or Splice Caps 230
Pigtail Connections 233
Identification of Conductors 235
Organizing the Box 239

Summary   242
Key Terms   242
Review Questions   242

11
Cabling Standards 244
Introduction   244
TIA-568 Series   245
TIA-570-B   253
Environmental and Safety Issues   260
International Cabling Standards   264
Summary   265
Key Terms   265
Review Questions   266

12
Optical Fiber and Cable 268
Introduction   268
Fiber Attenuation   271
Fiber Bandwidth   272
Bending Losses   273
Fiber-Optic Cable   273
Choosing a Cable   274
Fiber-Optic Cable Installation   276
Summary   278
Key Terms   278
Review Questions   279

13
Classification of Circuits, *Article 725* of the *NEC*® 280
Types of Electrical Circuits   281
Article 725 of the *NEC*®   282

Classification of Circuits and Class 1   283
Class 2 and Class 3 Circuits   287
Wiring Methods for Class 2 and Class 3 Circuits, Supply-Side and Load-Side Applications, *Sections 725.51* and *725.52*   288
Power-Limited Tray Cable and Instrumentation Tray Cable   289
Reclassification of Class 2 and Class 3 Circuits, Markings, and Separation Requirements   290
The Installation Requirements for Multiple Class 2 and Class 3 Circuits and Communications Circuits, *Section 725.56*   295
Support of Conductors and Cables   296
Calculate the Number and Size of Conductors in a Raceway   297
Summary   299
Key Terms   299
Review Questions   301

14
Telephone Wiring   302

Residential or Commercial?   303
Inside Wiring   303
Installation and Code Requirements   305
Circuit Protection   306
Interior Communications Conductors   307
Telephone Wiring Components   307
Second Line Installations   309
Small Office Installations   310
Planning the Installation   310
Separation and Physical Protection for Premises Wiring   312
Installation Safety   313
Installation Steps   314
Testing   315
Installation Checklist   316
Residential Networking   316
Summary   318
Key Terms   319
Review Questions   319

15
Video System Installations   321

Introduction   321
Video Cabling   321
Installation   322
Termination   322
Code Requirements   324
Summary   327
Key Terms   327
Review Questions   328

16
Network Cabling   329

Introduction   330
Cabling Requirements   330
Network Cabling Types   332
Wireless Transmission   333
Infrared Transmission   334
Power Line Carrier Networks   334
Other Transmission Means   335
Network Cable Handling   335
Typical Installations   336
Pulling Cables   337
Mounting Hardware in Closets or Equipment Rooms   343
Cabling in the Closet or Equipment Room   343
Summary   345
Key Terms   346
Review Questions   346

17
Cabling for Wireless Networks   348

Introduction   348
Wireless Is Not Wireless   349

Installation:The Site Survey 350
Summary 352
Key Terms 352
Review Questions 352

18
Testing Voice, Data, and Video Wiring 353

Introduction 354
Coax Cable Testing 354
Unshielded Twisted-Pair (UTP) Testing 354
Wire Mapping 355
Impedance, Resistance, and Return Loss 360
Cable Length 360
Attenuation 361
Near End Cross-talk (NEXT) 362
Power Sum NEXT 362
Attenuation to Cross-talk Ratio (ACR) 363
FEXT and ELFEXT 364
Propagation Delay and Delay Skew 364
Cable Plant "Certification" to Standards 365
Summary 365
Key Terms 365
Review Questions 365

19
Communications Infrastructure Design 367

Introduction 367
The Infrastructure Design Process 368
Future Flexibility 375
Retrofit Installations 376
Example with Floor Plan 378
Summary 382

Key Terms 383
Review Questions 383

20
Television 384
Installing the Wiring for Home Television 384
Satellite Antennas 389
Code Rules for the Installation of Antennas and Lead-In Wires (*Article 810*) 391
Summary 395
Key Terms 396
Review Questions 396

21
Smoke, Heat, and Carbon Monoxide Alarms, and Security Systems 397
National Fire Alarm Code 398
Smoke, Heat, and Carbon Monoxide Alarms 399
Detector Codes 400
Types of Smoke Alarms 401
Types of Heat Alarms 402
Installation Requirements 403
Maintenance and Testing 408
Carbon Monoxide Alarms 410
Fire Alarm Systems 412
Security Systems 413
Summary 416
Key Terms 416
Review Questions 416

22
Home of the Future 419
Introduction 419
Universal Plug-and-Play 420

Home Automation   420
Video   423
Home Game Consoles   423
Wireless Roaming   424
The Car   424
Out There   424
Pen Tablets: Notebooks of the Future?   427
Summary   427
Key Terms   427
Review Questions   428

Glossary 429
Index 439

RESIDENTIAL INTEGRATION SERIES

Series Preface

You've heard these catch phrases many times. Home theaters. High-Definition Television (HDTV). Distributed audio and digital video. Lighting and home control systems. Broadband Internet access and wireless networking. What do all of these terms mean? Simply stated, it's cutting-edge technology that is changing how we live—and will live—in our homes.

Why is this technology emerging? For one thing, the need for technical content and information is growing. But perhaps more importantly, people want the convenience and time- and cost-saving benefits of controlling their homes electronically as a way to spend more quality time with their families.

The residential integration industry is enjoying unprecedented growth in the installation of these new technologies. Due to the high demand, residential integration companies are seeking qualified individuals who understand the traditional fields such as new home construction, along with newer disciplines such as structured low-voltage wiring and data systems. Now is the time to join the bandwagon!

Recognizing the need for educational resources in this field, Thomson Delmar Learning is excited to present a four textbook suite, called the *Residential Integration Series,* which addresses this exploding industry. These four texts encompass many of the aspects of the residential integration industry to include the basics of the business and how to get started in it, customer service skills, project management, and finally a text on the information required to prepare for the various certifications that are currently offered within the industry.

More specifically, these textbooks are:

Residential Integrator's Basics is the foundational text for this industry. Here you'll find comprehensive information on computer networks, communications, home automation, cabling, wiring, and other topics. The final chapter addresses the Home of the Future.

Residential Integrator's Customer Relations pinpoints the types of customer service skills a residential integrator needs to be successful in this industry. These include working with both internal and external clients, working in teams (very important in this field), handling difficult client relationships, communications skills, training the client on both technology and equipment, and ensuring client satisfaction.

Residential Integrator's Project Management is based on *A Guide to the Project Management Body of Knowledge* (PMBOK), which is a widely accepted work published by the Project Management Institute (**www.pmi.org**). This text is divided into four sections, each covering a major phase of a residential integration project: The Foundation, Defining the Project, Planning the Project, and Executing, Monitoring/Controlling, and Closing the Project. This text is accompanied by a CD-ROM that contains project management templates specific to the industry that can be adapted to the needs of a particular company.

Residential Integrator's Certification provides in-depth coverage of the information required to prepare for the CompTIA HTI+ exam and the CEDIA Designer Classification Series Exam. This includes low-voltage cabling, high-voltage wiring, computer networking, audio/video, security systems, home controls, and other industry-related topics. Two appendices contain exam objectives for both exams.

The textbooks are pedagogically rich with chapter objectives, critical thinking questions, study tips, and other suggested activities, along with chapter summaries and review questions to help you learn and retain the material.

Build Your Perfect Course Solution

It's your course, so why compromise? Now you can create a text that exactly matches your syllabus using **Thomson Custom Solutions** online book-building application, **TextChoice.** TextChoice allows you to easily browse and select content from leading Thomson textbooks and custom collections—even include your own content—to create a text that is tailor-fit to the way you teach. Visit TextChoice at **www.textchoice.com** and learn how Thomson Custom Solutions can help you teach your course, your way.

Preface

The current age of digital electronics has brought about a virtual tidal wave of consumer-driven expectations and needs, all aimed at moving society toward a singular goal, the desire to be connected. As manufacturers continue to develop new products, the interoperability and connectivity of systems becomes the main issue. Convergence is about the linking of separate systems into a unified functional network, one that is assessable from multiple locations, and easy to use.

Residential Integrator's Basics has been developed as an all-inclusive source, bringing together the theory of different applications, such as voice, data, video, security, and home automation, as well as the basics of residential construction and cabling.

Intended Audience

Residential Integrator's Basics is intended for use at the technical or community college level. It may be used by residential electrical or electronics programs wishing to add a low-voltage component to their curriculum, or by any low-voltage design programs already teaching residential systems.

Code References

Code references listed in *Residential Integrator's Basics* are taken from the 2005 edition of the *National Electrical Code*® (NFPA-70), and the 2002 edition of the *National Fire Alarm Code Handbook* (NFPA-72).

Chapter Descriptions

Chapter 1 asks the question, "What is a residential network?" The chapter then goes on to describe the evolution of a residential network, from the early days of a simple telephone or cable television installation to the more modern high-speed data or wireless networks involving residential gateways and computer automation.

Chapter 2 details the basics of computer networking. The chapter starts with a brief history of computer networking, leading then into a discussion of the basic components that make up various types of computer networks. Topics include, network architecture, topology, network access, methods of communication, and collision detection. From there, chapter 2 will discuss the seven layers of the OSI model, TCP/IP protocols and their functions, network addressing, network hardware (network interface cards (NIC), hubs, routers, switches, bridges), cabling and connecting hardware, and some basic troubleshooting techniques.

Chapter 3 introduces the reader to the theory and study of wireless communications. Topics include cellular technology, wireless applications and standards, wireless LANS, and satellite communication.

Chapter 4 describes the theory and operation of telephone circuits, as well as the process of transmitting video signals, the use and purpose of coaxial cable, and the distribution model of a cable television system.

Chapter 5 introduces the reader to voice and data applications within the home. Topics include analog telephony, the distribution of analog telephony signals, structured wiring, connectivity, the private branch exchange (PBX), and voice over IP.

Chapter 6 discusses the architecture of a home entertainment network. Topics include digital entertainment devices, digital audio and digital video standards, home theatre systems, and the various options for cabling and installing such systems.

Chapter 7 details the four categories of home automation: climate control, lighting control, security systems, and fire systems. The connectivity of each system is discussed, as well as the pros and cons of having smart appliances and an integrated home automation system.

Chapter 8 details the construction process. Topics include work safety rules, the hierarchy of a job site, licensing and inspection, the organization and use of the *National Electrical Code®*, units of measurement, listing and labeling, and the use of construction drawings, specifications, scope-of-work documents, and wiring diagrams.

Chapter 9 describes the use and purpose of electrical boxes. Topics include types of boxes, reason for installing boxes, box fill, conduit bodies, and the differences between lighting outlet boxes and device boxes.

Chapter 10 describes the *NEC®* and International Residential Code (IRC) requirements for installing residential cable. Topics include notching and drilling studs, joists and rafters, the installation of conductors in environmental airspaces, cable bundling, splicing and pigtailing conductors, conductor identification, and the importance of organizing conductors in junction or device boxes.

Chapter 11 explains the standards, terminology, and architecture of structured cabling for the residential environment. Topics include the use and purpose of recognized standards such as TIA/EIA, or ISO/IEC; a comparison between the Commercial Building Telecommunications Cabling Standard TIA/EIA-568-B and the Residential Telecommunications Infrastructure Standard TIA-570-B; the evolution of UTP cable from CAT3 to CAT6; important cable specifications, such as NEXT, FEXT, ACR, and Delay Skew; optical fiber specifications; NFPA ratings for copper and fiber cables; specifications for connecting hardware; and grounding and bonding requirements.

Chapter 12 describes the theory and use of fiber-optic cable. Topics include glass vs. plastic, the physical properties of fiber-optic cable, specifications and *NEC®* requirements, signal loss, and the purpose and use of cable varieties such as simplex cable, distribution cable, breakout cable, and loose tube cable.

Chapter 13 takes an in-depth look at *NEC® Article 725*. *NEC® Article 725* details the classification of remote control and signaling circuits, including Class 1, Class 2, and Class 3 circuits. The chapter also discusses the comparison and use of power-limited tray cable as well as the use of instrumentation tray cable, as referenced in *NEC® Article 727*. Additional topics include the installation requirements of Class 1, Class 2, and Class 3 circuits, installation options with communication circuits, support of conductors, and the number of conductors to be placed in a raceway.

Chapter 14 describes the installation of residential telephone wiring. Topics include the network interface, modular plugs, surge protectors and arrestors, telephone grounding, *NEC® Article* 800, color codes, installation requirements, safety issues, testing, and the ANSI/TIA/EIA-570-Residential Telecommunications Cabling Standard for residential networking.

Chapter 15 describes the installation of residential video systems. Topics include the types of video application, cabling, and connectivity; the process for terminating coaxial cable; *NEC® Article* 820 requirements; grounding; and the designated classifications of coaxial cable, including substitution hierarchy.

Chapter 16 describes the installation of network cabling. Topics include network cabling specifications and performance, network cable installation requirements, the standards for wireless transmission, and the parameters and specifications of power line carrier transmission.

Chapter 17 covers the cabling and installation requirements of wireless networks. Topics include the fundamentals of IEEE 802.11, the differences between 802.11a, 802.11b, and 802.11g, and the design issues relating to the installation and troubleshooting of wireless systems.

Chapter 18 covers the testing of voice, data, and video cables. Topics include the testing of coaxial cable with a digital multimeter (DMM), the testing of UTP cables, wire mapping, common cable errors, troubleshooting, structural return loss, and the definition and purpose of advanced tests such as NEXT, ACR, and FEXT.

Chapter 19 details the step-by-step process for designing a communication infrastructure. Topics include components such as the DD, NID, and ADO; the requirements and locations of TOs; wireless design considerations; design requirements of entertainment networks and home automation systems; design documentation; and the requirements of retrofit installations.

Chapter 20 involves the installation and wiring of home television systems, satellite antennas, and the requirements of *NEC® Article 810*.

Chapter 21 describes the basics of fire and security systems. Topics include basic smoke and heat detectors, carbon monoxide alarms, installation requirements of devices, minimum levels of protection, UL standards of fire warning equipment, and a basic understanding of the National Fire Alarm Code (NFPA 72), and *NEC® Article* 760.

Chapter 22 describes the home of the future. Topics include the purpose and use of UPNP devices, a listing of various types of devices that will be networked and automated in the home of the future, a description of the Xanboo product line, the future of home game consoles, and an explanation of how GPS and VPN are changing the way individuals access the Internet.

Features

Chapter Objectives. Each chapter begins with a detailed list of the concepts to be accomplished within that chapter. This list provides the student with a quick reference to the chapter's contents and is a useful study aid.

Illustrations and Tables. Illustrations are provided to help the student visualize important concepts and procedures presented in the chapter.

Chapter Summaries. Each chapter is followed by a summary of the concepts presented in the chapter. These summaries provide a helpful recap and review of the chapter.

Key Terms. All key terms introduced in bold face type in the chapter are listed at the end of the chapter along with the definitions. This provides students with an opportunity to check their understanding of all the terms presented.

Review Questions. End-of-chapter review questions reinforce the ideas introduced in each chapter. These questions ensure that the student has mastered the concepts presented in the chapter.

Supplemental Material

The following supplemental materials are available with the text.

Faculty Guide with CD-ROM. The printed faculty guide offers a complete teaching package. Components for each chapter include chapter objectives, key terms, classroom discussion questions, learning activities, teaching tips, chapter summaries, answers to the end-of-chapter review questions, a resource correlation grid, and an outline of the PowerPoint® presentation.

CD-ROM. Each faculty guide has an accompanying CD-ROM that includes a PowerPoint® presentation, a computerized test bank, sample syllabi, and suggestions for additional resources.

HTI+ Home Networking Videos, The Training Dept.
The HTI+ Home Networking Videos provide step-by-step instructions on the how-tos of home networking, and are suitable for anyone interested in learning about this exciting new industry.

CD One contains ten video segments related to planning, installation, and safety requirements.

CD Two focuses on audio systems, the pre-wiring and trim-out procedures, and the distribution planning. Safety is also a focus, with segments devoted to job safety and OSHA residential regulations. In total, there are two full hours of video segments.

Lab Manual. A valuable resource for students, the lab manual provides exercises to further enhance the learning experience and to give students hands-on practice for both basic and complex concepts presented in the text.

Acknowledgments

The publisher wishes to acknowledge the following people for their suggestions and feedback in putting the manuscript together:

Sam DiPaola, Technical Editor
Chuck Dale, Eastfield College
A. Edward Kuehner, Luzerne County Community College
Chuck Jennings, Mid South-IEC

The contents for this text were taken from various Thomson Delmar Learning textbooks. For further information on these or other Thomson Delmar Learning titles please contact your local Thomson Sales Representative.

Titles

Residential Networks, Baxter (1-4018-6267-5) (Chapters 1, 5, 6, 7, 11, 19)

Introduction to Low Voltage Systems, DiPaola/DiPaola (1-4018-5656-X) (Chapters 2, 13)

Introduction to Telecommunications, 2E, Gokhale (1-4018-5648-9) (Chapter 3)

Data, Voice and Video Cabling, 2E, Hayes/Rosenberg (1-4018-2761-6) (Chapters 4, 12, 14, 15, 16, 17, 18)

Residential Wiring: An Introductory Approach, 2E, Sorge (1-4018-7866-0) (Chapters 8, 9, 10)

Electrical Wiring Residential: Based on the 2005 National Electrical Code®, Mullin. 15E, 1-4018-5019-7) (Chapters 20, 21)

Basic Home Networking, DeLeon/Coombs (0-7668-6180-5) (Chapter 22)

About the Technical Editor

Sam DiPaola holds a Power Limited Technician license with the Minnesota Board of Electricity. He is also a registered instructor with the state of Minnesota and the Minnesota Board of Electricity to teach low voltage continuing education courses and electronics. He is an associate of the North Central Electrical League through The Minnesota Statewide Limited Energy JATC, and holds a General Radiotelephone Operator License with the Federal Communications Commission. Mr. DiPaola worked for ten years as an audio technician and an audio system installer and later began working in manufacturing with Rosemount Engineering as an electronics design technician. Mr. DiPaola is currently the Training Director for the Minnesota Statewide Limited Energy JATC. For the past twenty years, Mr. DiPaola has been involved in the design and installation of low-voltage systems and automated process controls.

Introduction

After studying this chapter, you should be able to:

- Define residential networks.
- Understand the evolution of residential networking.

OBJECTIVES

What Is a Residential Network?

The Evolution of Residential Networks

OUTLINE

What Is a Residential Network?

There are a number of definitions of residential (or home) networks. The Consumer Electronic Association's (CEA) Home Networking and Information Technology (HNIT) division defines residential networks as follows:

> A home network interconnects electronic products and systems, enabling remote access to and control of those products and systems, and any available content such as music, video, or data.

According to the International Engineering Consortium:

> Home networking is the collection of elements that process, manage, transport, and store information, enabling the connection and integration of multiple computing, control, monitoring, and communication devices in the home.

Audio Advisors (**http://www.audioadvisors.com**) has a definition that illustrates how home networking has changed in recent years:

> The working definition of Home Networking has involved connecting multiple computers in a home in order to share files, printers, and Internet access. Today, Home Networking has become a term that covers practically anything in your home that can be connected to anything else in your home.

The most important aspects of the network are: first, it interconnects electronic devices so they can communicate (that is, exchange information) with each other; second, the network enables the user to control all the devices on the network, not just the one he is directly interacting with; and third, the user can access *content* (movies, photographs, music, data files, etc.) on any device in the network. A residential network can, for example, allow two or more personal computers (PCs) to share a high-speed Internet-access line or to access a remote printer. More exotic applications are possible, although not necessarily provided by today's networks. For example, if you have video clips or digital photographs stored on your PC, it would be nice to play them back on the big-screen television in the family room. Residential networks facilitate this type of application.

The Evolution of Residential Networks

Although residential networks are a relatively recent innovation, Arno Penzias has remarked that "[f]uture households will define themselves as much by their home networks as they now do by walls and fences."[1]

As recently as a couple of decades ago, the concept of residential networks would not even have made sense. At that time, the communication needs of most homes were served by a single telephone line that terminated in a hardwired phone. The cabling to the phone was owned and maintained by the local phone company. Television was received by an antenna on the roof or a "rabbit-ear" antenna on the television set. Typically, two or three channels were available with reasonably good quality reception.

More recently, additional phone lines became commonplace and modular jack connections allowed the phones to be moved if desired. With the breakup of the Bell System in 1984, the inside wiring became the responsibility of the home or building owner. In the commercial arena, the field of structured cabling developed rapidly, allowing building owners to build, modify, and maintain flexible, multipurpose networks at a reasonable cost.

Structured cabling systems (SCSs) are general-purpose cabling systems that are designed to handle a large number of different networks and applications. An SCS is the opposite of an application-specific cabling system, which is designed for one particular network or application.

Structured cabling did not take off as quickly in the residence as in commercial buildings. Due to the lack of applications in the home, there was not a clear value proposition that benefited either the home owner or builder. But communication-based applications in the home continued to grow and develop. In the 1980s, when PCs started to appear in homes, they were initially used as stand-alone devices. Eventually, the PC was often connected to the phone line with a voiceband modem. Prior to this, communication, computation, and entertainment were distinctly different activities. Attaching the PC to the phone line was the first step in blurring the boundaries between these activities—a trend that has continued unabated to the present day.

Also in the 1980s, cable television became increasingly popular and widely deployed. **Coaxial (coax) cable** was used to connect the television set to the cable television feed. In the mid-1990s, satellite television (such as DBS) became available and started growing rapidly. The total number of pay television

---

[1] Arno Penzias, "The Next Fifty Years: Some Likely Impacts of Solid-State Technology," *Bell Labs Technical Journal* (Autumn 1997).

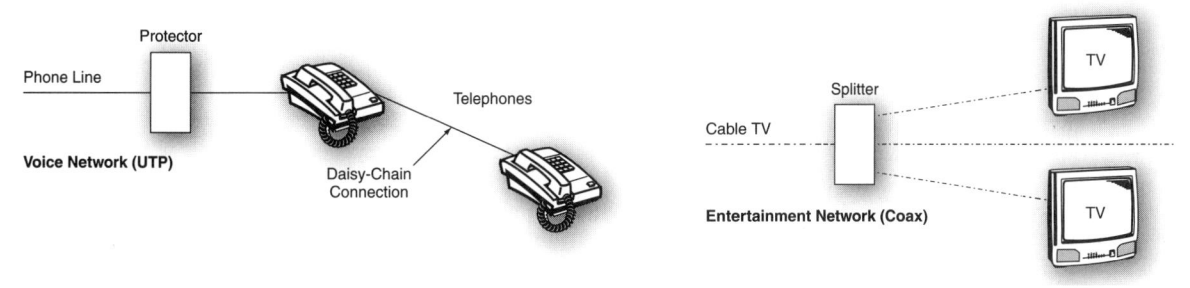

FIGURE 1–1 Simple Residential Network, circa 1990

subscribers (cable and satellite) grew from just over 8 million in 1980 to more than 115 million in 2001. By the 1990s, a typical well-equipped home would have had two distinct networks in parallel, as illustrated in Figure 1–1. These networks would be:

- A voice network consisting of **unshielded twisted-pair (UTP) cable**, which was used for one or more voice telephones and for dial-up service on PCs using voiceband modems
- An entertainment network consisting of coaxial cable that connected one or more television sets to the cable television service entrance.

The cabling from this era was so simple that, at the time, it was not generally considered a network but it clearly foreshadowed future developments.

In the mid-1990s, several factors converged to create an explosion of demand for communication and communication capabilities within the home. These factors included the general availability of the Internet, rapidly falling prices for PCs and associated equipment such as printers and scanners, and the availability of broadband access at a reasonable price (via either **digital subscriber line (DSL)** or **cable modem**). Applications such as e-mail, instant messaging, and online shopping and banking gained rapid acceptance. As a result, by the turn of the century, most homes in America had a PC with Internet access and many homes had two or more PCs. Table 1–1 shows the rapid growth in PC ownership and Internet access in U.S. homes.

The rapid rise of multi-PC households, coupled with the rapid development (and falling prices) of **Ethernet local area networks (LANs)** in the commercial market, led to another type of network in the home. A data network, typically some variant of Ethernet, connected PCs and related equipment, including Internet-access

TABLE 1–1
Growth of Personal Computers and Internet Access

Percentage of U.S. Households	*1995*	*1998*	*2000*	*2001*
With computer	37.0	42.1	51.0	56.4
With Internet access		26.2	41.5	50.5

Source: U.S. Census Bureau, Statistical Abstract of the United States

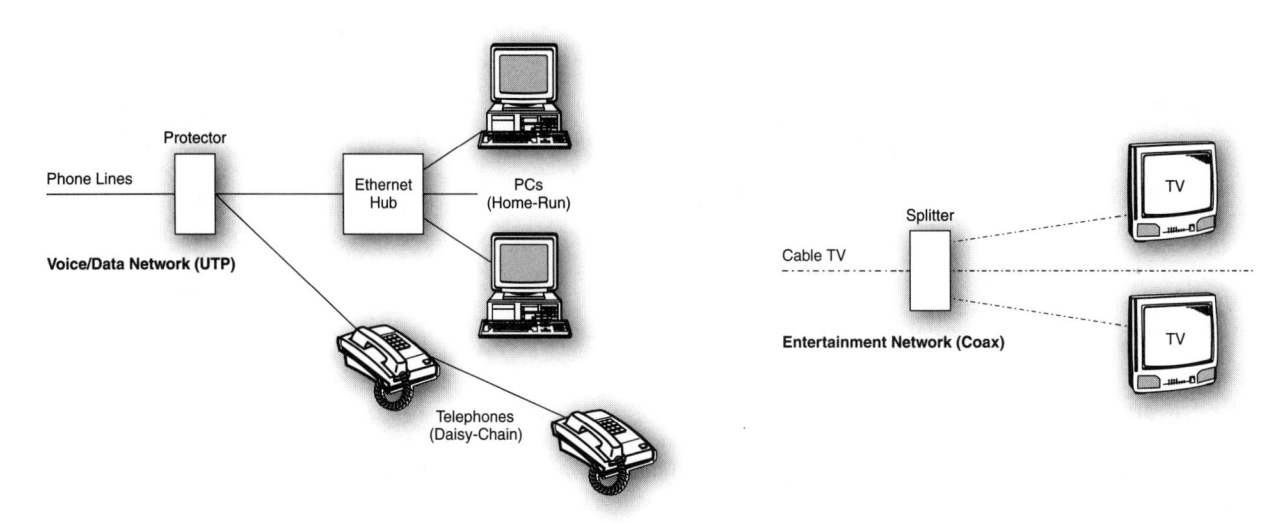

FIGURE 1–2 Residential Network, circa 2000

devices. Figure 1–2 shows a typical home network in the 2000 time frame. Note that the three networks are largely separate. Each has its own wiring—coax for the entertainment network, **daisy-chain** UTP for the voice network, and a separate **home-run** UTP network for data.

The trend toward **teleworking**[2] and home-based businesses has also encouraged the installation of home networks. According to the National Association of Home-Based Businesses (**http://www.usahomebusiness.com**), in 2001 there were more than 60 million people working from home. Home-based businesses account for about a third of that group.

The entertainment network within the home has also been expanding in scope. Starting from just a single television connected to cable service, the entertainment network in many households has grown to include, for example, multiple televisions, VCRs, DVD players, home theaters, video games, camcorders, and digital cameras. Increasingly, digital video is being used in consumer electronics to provide higher quality, lower cost, and compatibility with networks. More sophisticated networks to serve the needs of home entertainment devices, such as IEEE 1394 (also known as **FireWire**® and iLink®)[3] have been developed and are being deployed.

Another major addition to residential communications has been the cell phone. Cell phones have proliferated very rapidly, from a little over 5 million subscribers in 1990 to more than 159 million in 2003. This has siphoned some of the voice traffic away from the traditional telephone network, but most households still maintain at least one wired telephone line in addition to one or more cell phones.

When you mention home networks today, most people assume that you are talking about the data network. Currently, we are seeing a gradual merging of the

---

[2] Teleworking is a more general term than the more familiar *telecommuting*, which implies that the employee normally commutes to a particular place.
[3] FireWire is a trademark of Apple Computer. iLink is a trademark of Sony.

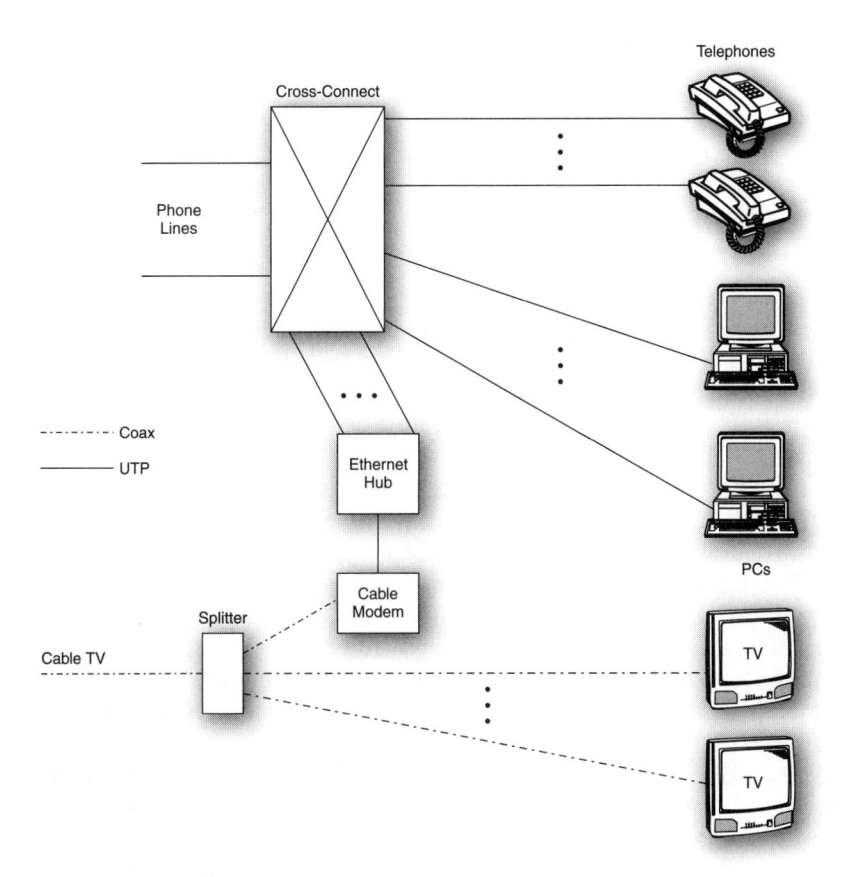

FIGURE 1–3 Modern Residential Network

three networks, as data, entertainment, and, in some cases, voice telephony are carried over the same network or over networks connected by a gateway. This merging of voice, data, and entertainment is often referred to as **convergence.** Figure 1–3 shows a modern home network supporting all of the aforementioned functions. Compared to the previous figures, note that the cabling has much more structure. All of the cables (coax and UTP) for all the applications are home-run to a common point called the **distribution device (DD).** This allows much greater flexibility in modifying and expanding the network.

Many other types of devices could be added to the home network in the future. Some examples include:

- **Residential gateways (RGs).** The use of RGs is currently getting a lot of attention. The primary purpose of an RG is to link the residential network to the **access network** outside the house. This involves providing interfaces to different transmission media, adaptation of different speeds, and translation of different message formats.
- Home automation devices, such as heating, air conditioning, and lighting. A **home automation system (HAS)** allows you to control all these functions from a common point, which could be outside the home. For example, you could log into your residential network from the office and adjust the air conditioning before going home.

- Home security devices, such as security cameras, for example. Networking these devices has the same benefits as home automation—you can control and monitor all the devices from a single point.
- Home appliances can be networked as well. This has been the subject of a lot of futuristic speculation, but it has been difficult to come up with a value proposition that consumers are willing to pay for. Is there really any advantage, for example, to being able to control your washing machine while you are away from home?

Home networking has been the subject of quite a bit of hype over the last few years. Following a lot of optimism in the late 1990s, home networking suffered from the same malaise as most of the communications industry during the "tech wreck" period. For a couple of years, headlines such as "High-End Networking on Hold at Cisco, 3Com" [CNET News.com (January 9, 2001)] and "Consumers Snub Costs, Complexity of Home Networking" [Reuters (June 7, 2002)] were commonplace. Currently, however, there seems to be renewed interest in the field and the news is more likely to be of a positive nature, such as "Big Names in Tech Join Forces to Sync Consumer Electronics" [*San Francisco Chronicle* (June 25, 2003)] and "Converging on the Couch" [*Scientific American* (August 2003)].

A number of vendors have announced home networking products recently, for example, the BridgeCo system illustrated in Figure 1–4. This network allows data and entertainment devices throughout the house to be connected on a common network.

At the CEDIA conference in September 2003, the **1394 Trade Association** demonstrated a home network that delivers high-definition video throughout a model house, as shown in Figure 1–5. A diagram of the network is given in Figure 1–6. This network demonstrates the interconnection of PCs, video equipment (HDTV, VCR, set-top box, DVD player), and audio equipment on an IEEE 1394 network that runs throughout the house. This is indicative of the start of the art in this time frame.

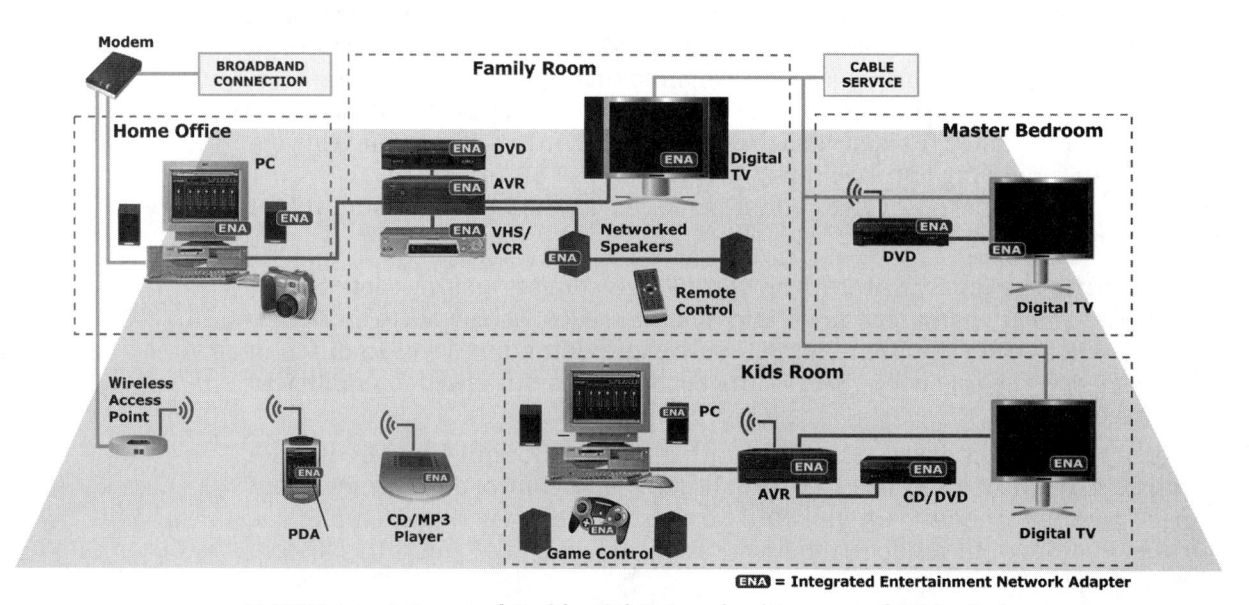

FIGURE 1–4 Integrated Residential Network (Courtesy of BridgeCo.)

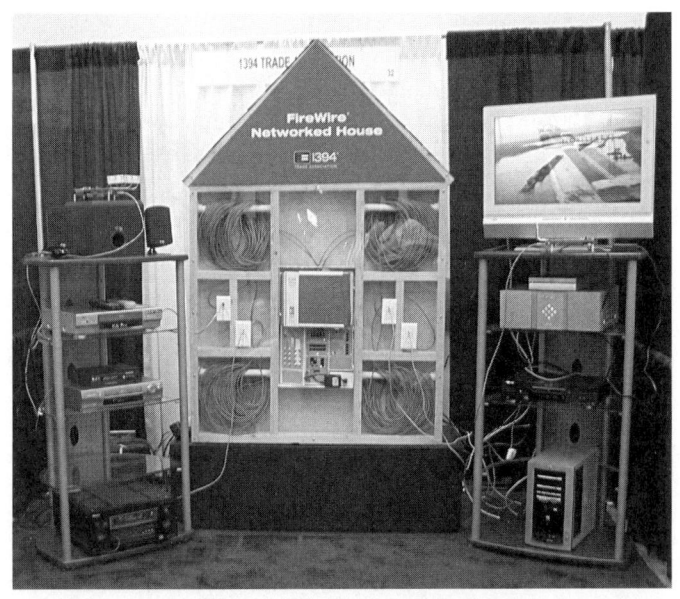

FIGURE 1–5 CEDIA Demo System (Courtesy of 1394 Trade Association)

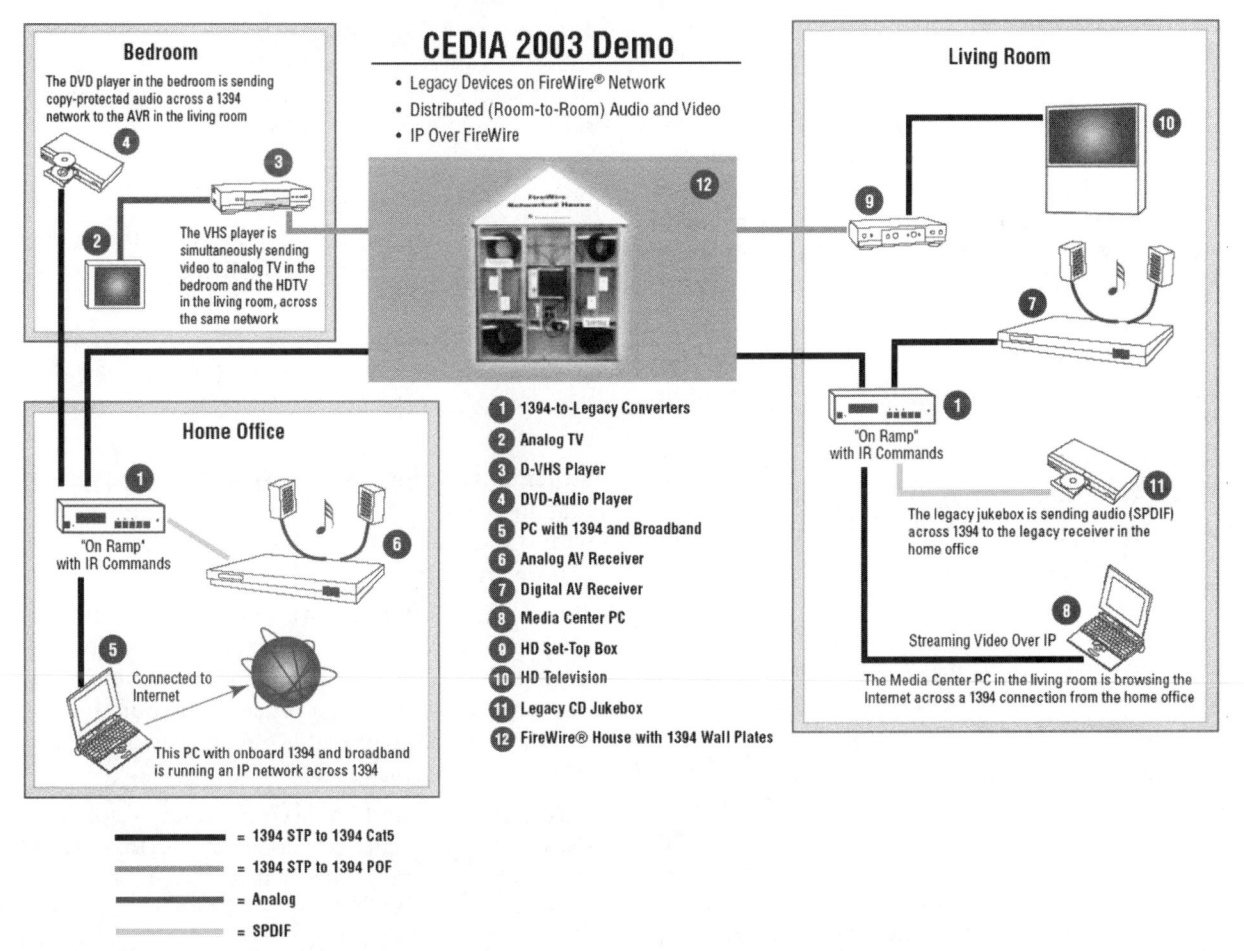

FIGURE 1–6 Diagram of FireWire® Home Network (Courtesy of 1394 Trade Association)

Summary

- Residential networks provide three primary functions: (1) communication among electronic devices and systems, (2) control of all devices on the network, and (3) sharing of information and content (movies, music, data files) among devices.

- Residential networks currently emphasize voice, data (PC), and entertainment applications.

- In the future, home control and automation applications will become more important.

Key Terms

1394 Trade Association   A nonprofit industry association dedicated to promoting the proliferation of IEEE 1394 technology in the computer, consumer, peripheral, and industrial markets. The 1394 TA supports several working groups that issue technical specifications and bulletins relating to applications of IEEE 1394.

Access network   The facilities that connect the residence to the long-haul networks (telephone and Internet).

Coaxial (coax) cable   A cable in which a single center conductor is surrounded by a dielectric material, and a cylindrical shield that is often composed of layers of foil and metallic braid. Coax has excellent high-frequency characteristics and is most commonly used for cable television signals.

Convergence   The merging of telephony, computing, and television into a single digital network.

Daisy-chain   A method of cabling where end points are connected directly together without returning to a common point.

Digital subscriber line (DSL) or cable modem   A technology that uses the existing copper loop plant to provide broadband access by using frequencies above the voice band.

Distribution device (DD)   A facility within the residence that contains the main cross-connect or interconnect where one end of each of the outlet cables terminates.

Ethernet   The first computer networks, known as Ethernets, were developed in the 1970s by Xerox, which ran at maximum speeds of 2.94 Mb/sec on coaxial cable. Xerox soon joined with Intel and Digital Equipment Corporation to develop a standard for a 10-Mb/sec Ethernet, known as DIX (Digital, Intel, Xerox). Later, the development of the 10Base-T Ethernet in 1987 communicated at speeds of 10 Mbps on a twisted pair line; such networks could reliably transmit over a 100-m run.

FireWire®   FireWire®, also known as iLink®, is a high-bandwidth, serial communication standard developed by IEEE (IEEE 1394).

Home automation system (HAS)   Equipment and infrastructure that support automatic control of the home services, such as climate and lighting control, security and fire alarms, and so on.

Home-run   A method of cabling where each end point is connected back to a central point.

Local area networks (LANs)   Refers to the linking of computers and users in a small geographical area, such as an office complex, school, or college campus.

Residential gateways (RGs)   Devices that enable communication among networks in the residence and between residential networks and service providers' networks.

Structured cabling systems (SCSs)   Cabling systems that are designed to support a range of applications with standard interfaces and specified transmission performance.

Teleworking   Doing work outside the traditional workplace. This could be a traditional telecommuting environment where the employee works from home instead of traveling to an office, or it could involve working on the road while traveling, working at a customer's site, and so on.

Unshielded twisted-pair (UTP) cable   Cable consisting of eight insulated copper conductors twisted together into four pairs without any screen or shields.

Review Questions

1. Define *residential network*.
2. What are the three main types of applications that are most commonly served by residential networks?
3. What are the three main functions that residential networks provide?
4. What is a structured cabling system?
5. Why did the market for structured cabling systems develop more slowly in the residence than in commercial buildings?
6. State a few trends that are currently encouraging the installation of home networks.
7. Name three common home networking technologies.
8. What is the purpose of a residential gateway?

Networking Fundamentals

OBJECTIVES *After studying this chapter, you should be able to:*

- Compare a Wide Area Network (WAN) and a Local Area Network (LAN) and explain their differences.
- Describe some popular network architectures.
- Describe the purpose of the OSI model and list the various levels.
- List various TCP/IP Protocols and their functions.
- Explain the process of addressing computers on a network.
- Categorize the cabling used to send data between multiple points of a network.
- List and describe the various hardware used in building a network.
- Explain a variety of installation techniques that will ensure a clean install.
- Describe the purpose and use of various operating systems and device drivers.
- List the various methods and type of connections needed to access the Internet.

OUTLINE Basic Networking, Architecture, and Topology

Network Protocols and Their Functions

TCP/IP Protocols and the Internet

Computer Network Addressing

Other Concerns

Low Voltage Residential Network Applications

Basic Networking, Architecture, and Topology

Networking involves the intercommunication between computers. Network topology involves the manner in which multiple computers are physically connected. Network media refers to the method of transporting communication signals. Three common methods are copper wire, optical fiber, and wireless. The desired speed of communication will be determined based on which method is used. In this chapter, we will discuss the basics of computer networking and how systems are interconnected locally and over long distances.

A Brief History of the Network

A **network** is made up from a collection of individual computers, computer processing equipment, printers, or communication devices, which are linked together by interconnected routes or pathways, for common use by all participants. The first networks, known as Ethernets, were developed in the 1970s by Xerox; they ran at maximum speeds of 2.94 Mb/s. Xerox soon joined with Intel and Digital Equipment Corporation to develop a standard for a 10-Mb/s Ethernet, known as DIX (Digital, Intel, Xerox). During the same period of time, The Institute of Electrical and Electronic Engineers (IEEE) was developing the 802 task force that was working to establish technical standards for **local area networks (LANs)** and metropolitan-area networks (MANs). Figure 2–1 illustrates a local area network.

Local area networks evolved out of the personal computer revolution. The LAN succeeded in linking the communications of multiple users for file exchanges, exchange messaging (e-mail), and the sharing of resources such as printers, information databases, and file servers. A local area network or LAN refers to the linking of users in a small geographical area, such as an office complex, school or college campus.

Metropolitan area networks, or now more commonly known as **Wide area networks (WANs),** interconnect multiple local area networks over a wide geographical area, such as a city or state. Wide area networks will achieve a communication link over long distance through a network line brought in by a service provider. The service provider can provide access to wide area networks and the Internet by a variety of methods. Examples include, a dial-up telephone **modem, Integrated Services Digital Network (ISDN),** high-speed digital coaxial cable, **Digital Subscriber Line (DSL),** satellite, optical fiber, and wireless. We will further discuss connection to the Internet later in the chapter. Figure 2–2 illustrates a wide area network.

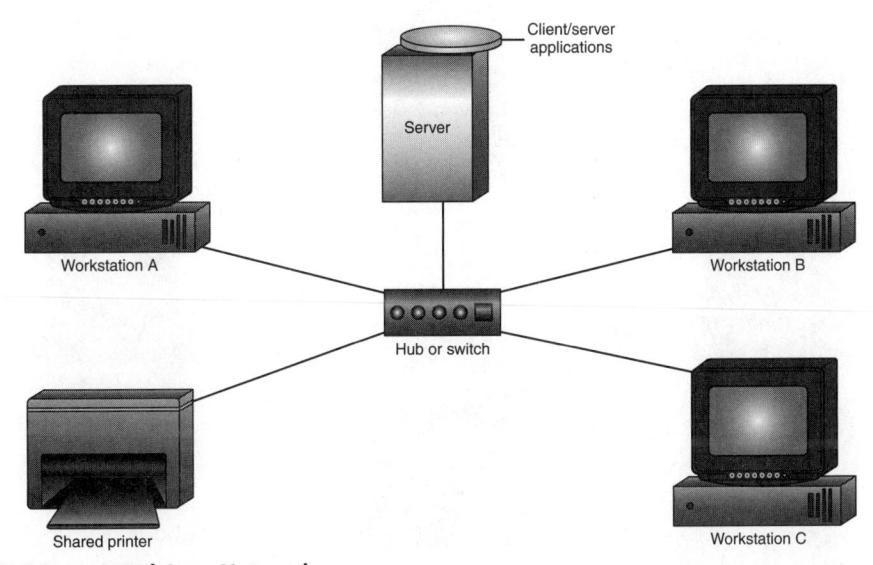

FIGURE 2–1 Local Area Network

12 • Chapter 2

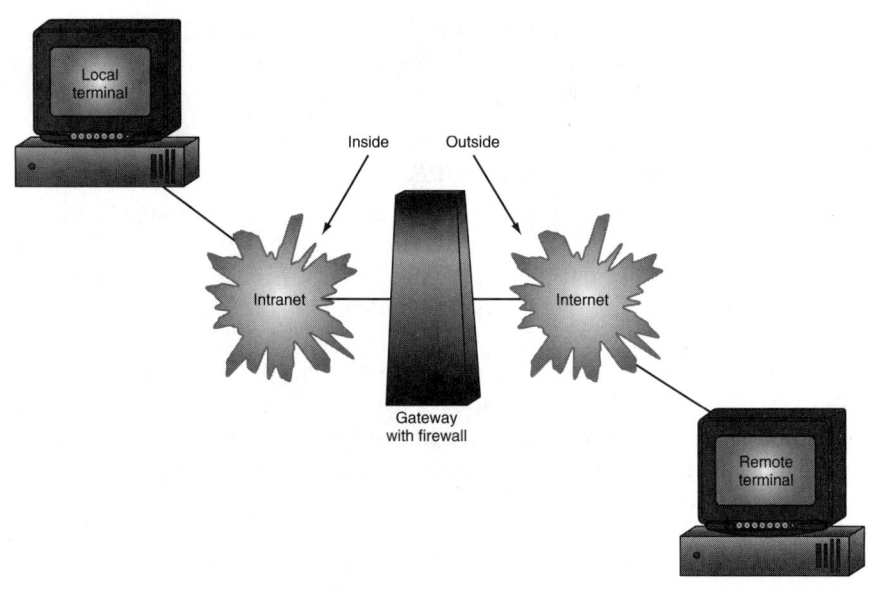

FIGURE 2–2 Wide Area Network

By 1985, IBM developed the **Token Ring Network** (TRN) that linked computers through a dedicated system of data relay. The ring functioned by passing a bit stream, referred to as the token, from station to station. Every station in the loop was required to pass on the circulating information until the token returned back to the source, thereby ensuring that all participants in the network had equal access to the circulating data. Figure 2–3 illustrates a Token Ring network. Addressable data stream packets (to be discussed later in the chapter) were then developed as a security measure to ensure that individual stations around the ring only had access to data packets addressed to them; the token, containing the remainder of unopened data packets, was then passed on to the next designated station in the loop.

The token ring process is similar to a mail carrier delivering mail to only one house on the block; each house would then only be responsible for opening their

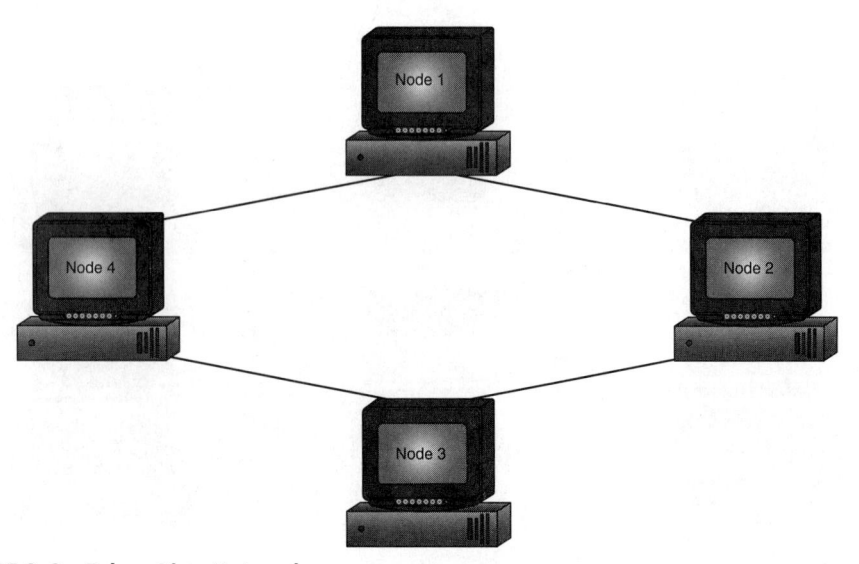

FIGURE 2–3 Token Ring Network

individual mail, as the bag moved from house to house. Eventually, all the letters would make it around the block to the required destinations. The bag would then be empty of any outgoing letters and refilled with new incoming letters as soon as it returned to the originating mail-stop.

The IBM Token Ring network was so successful that it was quickly adopted by the IEEE 802 task force and standardized. It remained a popular network design through the mid-1990s.

As the physical topologies of network connections were improved and adapted to permit higher speeds of communication, ease of connectivity and more reliable cabling was needed. Customers also insisted on a more telephone-based cabling structure over the bulky and more expensive coaxial cable, which was typically being used on early **Ethernet** and Token Ring networks. Ultimately, **unshielded twisted pair (UTP)** was adopted as an industry standard since it was less expensive and easier to install.

In 1987, **Fiber Distributed Data Interface (FDDI)** networks were developed that used optical fiber to transport token ring data at speeds of up to 100 Mbps. Nevertheless, within a few short years of its release and due to the high cost of optical fiber, shielded and unshielded, copper, twisted pair became the desired industry cable of choice. As a result, the Copper Distributed Data Interface (CDDI) was organized to develop standards and specifications for the more commonly used UTP and **Shielded Twisted-Pair (STP)** cables.

Fiber Distributed Data Interface did not, however, disappear altogether. It still exists as the main backbone for many installations requiring long-distance runs; examples include universities and large corporate complexes. FDDI is also commonly used to interconnect two or more local area networks, since optical fiber inherently transmits a higher quality signal over greater distance than does copper.

The expansion of Token Ring networks was ultimately slowed by the development of the 10Base-T Ethernet in 1987. 10Base-T refers to a communication speed of 10 Mbps (bits per second) on a twisted-pair line; such networks can reliably transmit over a 100-meter run. The downside of Token Ring networks was their inability to provide quality control as network traffic increased because the addition of more stations on the ring would ultimately slow down the entire network. A proprietary system of permissions to specific files was also not in the nature of Token Ring networks; data was simply passed on from one computer to the next, with no concern as to whom the recipient may be, or whether or not they had an actual right to the data. Ethernet solved the problem by only transmitting out to desired stations within a network via a hub, digital switch, or router. As a result, data moved faster, and in a more direct manner, without having to visit each individual station along the way. Even though 4, 16, and even 100 Mbps were possible through Token Ring designs, the eventual 10Base-T and later 100Base-T and 1000Base-T Ethernets began to dominate the industry. This dominance was cost-related in many respects, since token ring was a proprietary system and expensive to implement.

Access to the Internet

As stated earlier, an individual needing to gain access to a wide area network or to the Internet can do so through a variety of means: dial-up modem, DSL, ISDN, cable modem, satellite, optical fiber, or wireless. In most cases, the choice will be determined by price, speed, and availability of service.

Dial-up Modem

A dial-up modem is a device that allows a computer to communicate over an analog telephone line. The word modem is a conjunction of *mod*ulation and *dem*odulation. Modulation places an intelligent signal onto a communication carrier for transmission. An intelligent signal may be voice, data, or video. The carrier then carries the signal through a designated channel to the recipient of the communication. In a wired system, the channel is the connecting wire; in a wireless system, the channel is the air. Demodulation is the opposite of modulation; it removes the intelligent signal from the carrier so that the recipient can interpret the message. A dial-up modem also provides a digital-to-analog conversion so that the analog phone line can use and pass the signal.

The maximum speed of most dial-up modems is 56K. Limitations on connection speed are also introduced into the system by the telephone line. Very often, Internet connection speed over a telephone line is limited due to low bandwidth and traffic on the line. The type of information being transmitted will also affect the rate of data flow because signals heavy in video and graphics will ultimately take up more space and slow down the system. In most cases, 56K is not even possible, depending on how far the connection is from the telephone company switchgear. The closer an individual is to the telephone company, the better. Longer transmission distances add unwanted capacitance, resistance, and inductive loading to the telephone line that inhibits the passage of high-speed communications. Figure 2–4 illustrates a dial-up modem network connection.

Modems are also limited by the type of data they can transmit. Some have the ability to handle voice and fax, as well as data. Another option may include data compression to help optimize and increase data-transfer rates. When purchasing a modem, be sure to read all the specifications carefully, as well as the available options, otherwise issues of hardware incompatibility may arise.

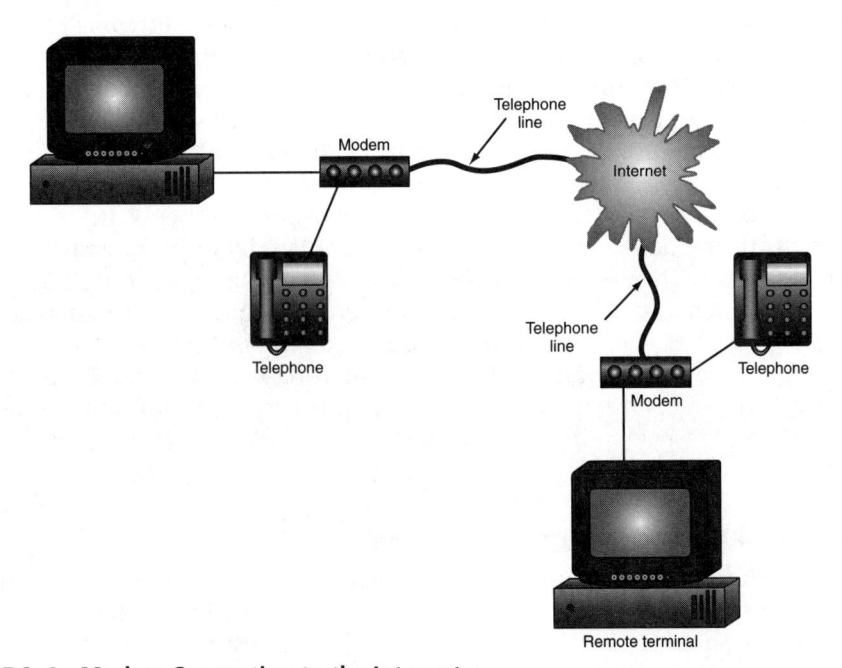

FIGURE 2–4 Modem Connection to the Internet

Digital Subscriber Line (DSL)

Digital Subscriber Line (DSL) offers the next alternative to a standard dial-up modem because it allows high-speed computer data to share the telephone line with the analog voice communication. The service provider of a DSL system will install a special DSL modem that filters and splits the analog voice signal from the high-speed digital data. A passive low-pass filter referred to as a plain old telephone system (POTS) splitter is used to separate the voice signal from the digital data at frequencies just below 20 kHz; 20 kHz represents the upper threshold of human hearing. The DSL uses all frequencies above 25 kHz for computer communication and connection to the Internet. Splitting the signals also ensures that if the digital section of the DSL modem ever fails then the voice section will continue to operate.

The speed of DSL can operate up to 50 times faster than dial-up modems at nearly 1.5 Mbps. However, just as with the dial-up modem, maximum speeds will not be realized if the connection is located very far from the telephone company, or telco. Line loading and line-conditioning equipment installed by the telephone company to optimize the voice signal will often inhibit high-frequency transmission at great distances, not to mention the added capacitance and resistance of the cable.

DSL may also introduce noise or hum to the voice section of the telephone line. To help prevent such interference, low-pass filters are added to phone jacks that do not use the DSL connection. While the telephone must still dial out to make a voice call, the DSL is always connected and requires no dial-up to access the service provider. The DSL connection is on 24 hours a day, 7 days a week.

DSL comes in six varieties, all varying in speed, upstream and downstream symmetry, and maximum distance. The chart in Figure 2–5 details the six varieties of DSL.

1. **ADSL.** ADSL, or asymmetrical DSL, provides an upstream rate of 1.5 Mbps and a downstream rate of 640 kbps. As a result, uploads from an individual's computer will be slower than downloads. ADSL is also limited to a distance of 18,000 feet from the telephone company.

2. **SDSL.** SDSL, or symmetric DSL, allows upstream and downstream rates to be equal, to a maximum of 2.3 Mbps. SDSL does not, however, allow simultaneous access to the telephone line. Maximum distance of SDSL is limited to 22,000 feet from the telephone company.

Type of DSL	Upstream Rate	Downstream Rate	Distance
ADSL	640 Kbps	1.5 Mbps	18,000 feet
SDSL	2.3 Mbps	2.3 Mbps	22,000 feet
RADSL	1.0 Mbps	7.0 Mbps	18,000 feet
HDSL	1.5 Mbps	1.5 Mbps	12,000 feet
VHDSL	16 Mbps	52 Mbps	4,000 feet
ISDL	144 Kbps	144 Kbps	35,000 feet

FIGURE 2–5 Digital Subscriber Line (DSL) Specifications

3. ***RADSL.*** RADSL, or rate-adaptive DSL, is a variation of ADSL that allows the modem to vary the speed of transmission based on the length and the quality of the line. RADSL is asymmetrical, allowing maximum downloads rates of 7 Mbps and maximum upload rates of 1 Mbps. The maximum distance from the phone company is limited to 18,000 feet.

4. ***HDSL.*** HDSL, or high-bit-rate DSL, is symmetrical, offering a maximum data rate of 1.5 Mbps in each direction. The down side is that it requires three phone lines: two for the DSL connection, one being uplink and the other downlink, and the third line for the regular telephone. HDSL has a maximum transmission distance of 12,000 feet from the telephone company.

5. ***VHDSL.*** VHDSL, or very-high-bit-rate DSL, provides the fastest data rates—52 Mbps for download and 16 Mpbs for upload. But to achieve such speeds, the maximum distance from the phone company can not be greater than 4,000 feet.

6. ***ISDSL.*** ISDSL, or ISDN DSL, is for existing users of Integrated Service Digital Network (ISDN) who choose to use their existing equipment. While the rates of ISDL are the slowest of all DSL options, coming in at 144 kbps each direction, it does offer the greatest transmission distance—35,000 feet from the phone company.

Termination of a DSL to a computer will be to either the USP port or the RJ-45 port, depending on the type of DSL transceiver being used. The service provider will often refer to the transceiver as an ADSL Termination Unit-Remote (ATU-R).

Integrated Services Digital Network (ISDN)

Integrated Services Digital Network (ISDN) is a dial-up service to the Internet. It does not offer a dedicated or always-on connection to a network like DSL. In some respects, ISDN is very similar to a standard dial-up modem connection, except that the ISDN has the ability to share the line with the telephone, fax, and computer simultaneously. A standard dial-up modem connection can only provide access to one user at a time. The service provider of ISDN will provide a network terminal adapter, which is nothing more than a special type of modem that allows the integration of voice, fax, and computer on a single telephone line. ISDN allows all devices to share the phone line simultaneously.

To accomplish a shared line, the ISDN terminal adapter provides the use of a bearer channel (B channel) that automatically scales back the bandwidth usage of the line to allow the fax or telephone access to 64 kbps on incoming calls. In this way, the computer does not actually disconnect but instead slows down to achieve the integration of services on the line.

Two levels of service exist through ISDN. These services are:

- Basic Rate Interface (BRI), consisting of two 64-kbps B channels, and one 16-kbps D channel. The Basic Rate Interface can provide up to 128 kbps when other devices are not using the B channels.
- Primary Rate Interface (PRI), consisting of 23 B channels, and one 64-kbps D channel, in the United States, or 30 B channels and one D channel in Europe.

The B channels carry voice, data, and special services, while the D channel, or Delta channel, carries control and signaling information.

The local telephone company in most urban areas can provide ISDN service. The cost, however, is quite high when compared to that of DSL, and is based on usage. ISDN is simply not a competitive option to DSL due to the lower transmission rates

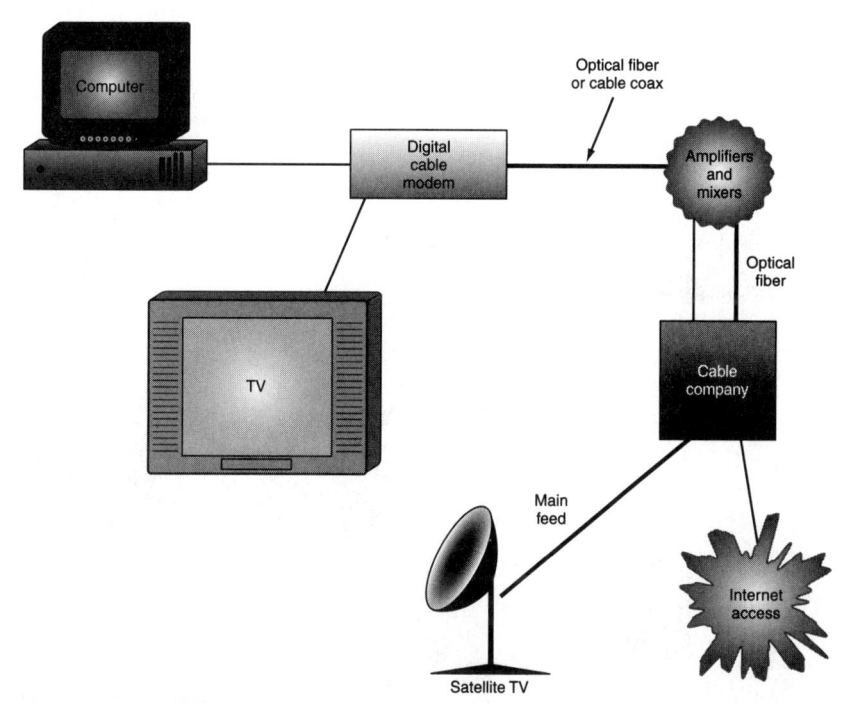

FIGURE 2–6 Cable Modem

and higher costs, and also because it cannot provide an always-on, dedicated line to the user.

Cable Modem

The cable television (CATV) service provider in most regional areas provides cable modem Internet service. The modem signals travel on a two-way hybrid fiber or coaxial cable, often at speeds ranging from 320 kbps to 10 Mbps. Subscribers, however, typically achieve maximum download speeds of only 1.5 Mbps as a result of having to share bandwidth with other subscribers on the system. Figure 2–6 illustrates a cable modem network connection.

The cable modem provides two outputs, one for the television and one for the PC. The modem itself is capable of receiving and processing information at rates of up to 30 Mbps, thousands of times faster than a standard dial-up telephone modem. The obvious advantages to using a cable modem service are higher speeds and greater bandwidth. Cable service is also often more readily available than DSL. DSL must frequently be brought into a location at additional charge, whereas television cable service already has an existing infrastructure in most locations.

Satellite

From a computer networking point of view, a Direct Broadcast Satellite (DBS) connection is capable of downloading Internet service at speeds of up to 45 Mbps. An uplink to a DBS satellite is not possible though, since the system is capable of only providing a one-way signal from the satellite in the sky to a ground base station. To achieve a network uplink, the system will still need a secondary telephone connection, which, as described earlier, is considerably slower. Figure 2–7 illustrates a satellite network connection with a secondary uplink connection. Satellite

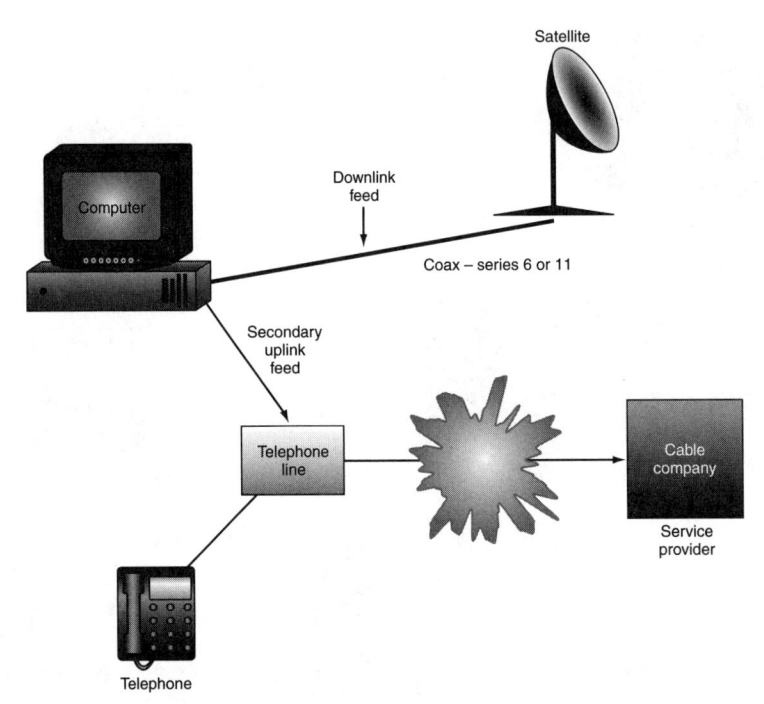

FIGURE 2–7 Satellite Downlink with Secondary Uplink Via the Telephone Line to Service Provider

uplinks can only be made by the service provider. The data follows a point-to-multipoint path, meaning that multiple recipients have the ability to download signals originating from a single ground station.

Satellite systems do exist, however, that offers the ability to uplink and downlink data to and from an Internet Service Provider. In most cases, such systems are quite expensive, often running as high as $150.00 per month.

The main advantage to using satellite is that it provides a wireless high-speed connection from just about anywhere in the world. All that's needed are a dish, a receiver, and an unobstructed, line-of-site view to the orbiting transceiver in the sky. For rural areas, it is often the only available choice if high-speed network connections are required.

Optical Fiber

Optical fiber remains the best choice when considering a hard-wired connection to the Internet. Fiber can transmit data at rates in the gigabits and over very long distances without having to worry about electromagnetic interference (EMI) or signal attenuation. The cost of optical fiber can be somewhat prohibitive, which is why most people opt for DSL or cable modems. Many new construction projects are now bringing optical fiber into the home. While business and industry have been using optical fiber for years, it has only recently become more affordable for residential use.

Network Interface Card (NIC)

The network interface card (NIC), as shown in Figure 2–8, is the link between the computer and the network. The card slides into the expansion slot of the computer. The main function of the card is to interpret data passing between the

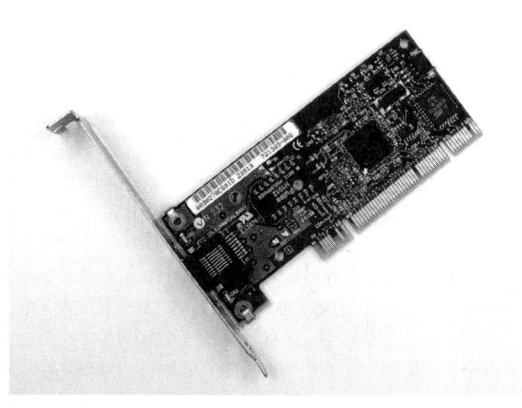

FIGURE 2–8 Network Interface Card

microprocessor and the network. The card also manages the flow of data to ensure that bidirectional communication is logically taking place on both sides of the link.

The network interface card also provides a unique identifier number known as the Media Access Control (MAC) address. The number is built into the electronics of the card. A secondary address known as the **Internet Protocol (IP)** address (discussed later in the chapter), combines with the MAC address to form a logical address, thereby enabling the computer to communicate to any device or computer on the network.

Digital Computers

Phones and television are basically analog transmission networks. They begin as electrical signals of continuously varying voltages or currents traveling over the cables (Figure 2–9). Computers, however, are digital. Computer data is represented by "ones" and "zeroes" that can be stored and manipulated by digital processing. Computer programs express all data, graphics, and programs as digital data.

Analog signals are subject to distortion and degradation by noise and attenuation. Digital signals are immune to noise and tolerate greater attenuation simply because the receiver only needs to distinguish between a one or a zero, not a continuously varying signal. Besides computers, most phone signals are digitized as soon as they reach the phone company central office. This not only allows for better signal quality over long distances, but also for the multiplexing or compressing of digital data into high-bandwidth signals carrying numerous phone calls.

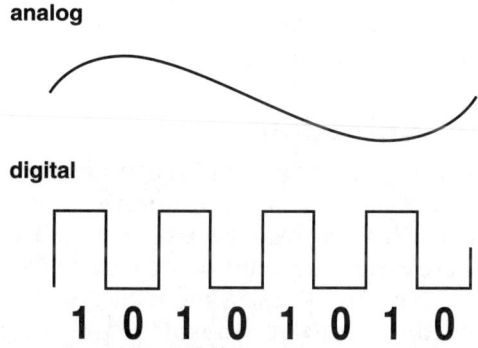

FIGURE 2–9 Analog systems transmit data as a continually varying signal, whereas digital systems transmit data as a stream of ones and zeros.

A Byte

A byte refers to a digital word consisting of an 8-bit, binary count ranging anywhere from 0000 0000, to 1111 1111. Computers talk to each other through binary switching. A '0' refers to a logic low, off, or no, while a '1' refers to a logic high, on, or yes. The entire word of 8 binary bits can also represent the decimal equivalent of a base 10 number. Since computers can only deal with binary math, any numbers must be ultimately reconverted back to base-10 decimal for the operator. As an example, 0010 1001 will represent a decimal equivalent of 41. Bytes are read from right to left. The farthest 1 to the right, bit 1, is the least significant digit (LSD), while the farthest 0 to the left, bit 8, represents the most significant digit (MSD). The base-10 decimal equivalent of each bit in a binary word would look like this: 128(bit 8), 64(bit 7), 32(bit 6), 16(bit 5) 8(bit 4), 4(bit 3), 2(bit 2), 1(bit 1). To obtain the decimal total of a binary word, simply add the decimal equivalent of each bit indicating an "on" position, or a value of 1. In our previous example, 0010 1001, add bits 1, 4, and 6 to obtain a decimal equivalent of 41; 1 + 8 + 32 = 41.

Computer programmers use digital, binary logic to write programs and make computers work. How fast a binary word can clock or switch its way through the computer, or the network, will determine how fast the computer can complete a desired task or even send and receive information on the Internet.

Networking Challenges

The ultimate goal of any network is to have sufficient bandwidth, allowing for fast communication, high performance, and reliability. The original Ethernets of 10 Mbps have mostly been upgraded and improved to the more common levels of 100 Mbps, and now even 10s of Gbps, for very high-speed digital communication, translating to well over 10 billion bytes per second.

Today, the challenge of most networks has more to do with the management, support, and security of many disparate technologies. Different types of equipment operating at varying speeds, through different styles of communication and across a variety of media types, makes for a continuously changing and demanding marketplace. Networks must therefore be reliable, relatively problem-free, and yet remain flexible enough to adapt to the continual changes and challenges of the industry. Security cannot be overstated. As more and more networks are interconnected, the protection and security of information systems from outside attack and hacking becomes a central concern to the overall connectivity and reliability of the system. Internal attacks must also not be overlooked; most security breaches result from insiders, who have access to security passes and codes. These breaches can very effectively and efficiently undermine the normal everyday operations of any networked system, naively assumed to be secure.

Methods of Communication

Multiple possibilities exist for sending and receiving data bits or binary words between any two points of a communication network. Possible communication methods may include a **serial communication** link, a **universal serial bus (USB)**, a high-speed **FireWire®**, or a multilane parallel connection. The chosen method will typically be based on the size and width of the data highway and the style and type of connecting hardware available to the devices. The data highway refers to the number of physical connections between two points of the communication link, along with the total distance and maximum travel time. Possible communication methods are explained and compared below.

FIGURE 2–10 Serial Connector, RS-232

Serial Communication

Serial communication sends data bits, one bit at a time, between two points in a computer network or system. The process is slow, since binary words cannot be decoded or processed by the system until the entire word has arrived at the opposite end. The process is similar to catching balls out of an automatic pitching machine. Imagine that each ball has an individual letter written on it, and you are told to catch the next eight balls before you can read the intended message. This is how serial communication works. A serial port and serial cable are shown in Figure 2–10.

At the receiver end of the system, data bits are stored into memory locations called registers and reassembled to their maximum readable size. When the entire binary word has been received, it is sent to another registry location for processing. The previous register will then be cleared and made ready for the next available bit stream to be received and assembled. The process will ultimately continue until all arriving data bits have been reassembled into complete binary words and processed for meaning.

Universal Serial Bus (USB)

Universal serial bus (USB) is a high-bandwidth data communication standard, which can support multiple devices within an entire system. Figure 2–11 shows a USB port and cable. The maximum number of devices is limited to 127, with a

FIGURE 2–11 Universal Serial Bus Connector

FIGURE 2–12 FireWire® Cable

maximum data rate of 12 Mbps. A revision to the USB standard, known as USB 2.0, can now support data transfer speeds of 480 Mbps.

FireWire®

FireWire®, also known as iLink®, is a high-bandwidth, serial communication standard developed by IEEE (IEEE 1394). A FireWire® cable is shown in Figure 2–12. The connection allows a maximum of 63 devices, with a data transfer rate of 400 Mbps. While FireWire® is ideal for the transfer of video and audio information, it is also commonly used to connect hard drives and computer storage devices because of its high bandwidth and speed. Maximum cable lengths for FireWire® are limited to 14 feet. Any attempt to use increased lengths will significantly lower the overall transfer speeds and increase the number of data errors on the bus.

Parallel Communication

Parallel communication sends multiple binary bits, all at one time, between two points in a communication system. In most cases, entire binary words can be sent simultaneously, helping to reduce the total transfer time and making parallel communication much faster than that of serial—eight times faster in actuality. A parallel port and cable are shown in Figure 2–13.

Parallel communication, however, requires a separate pathway or wire for each binary bit, which means that multiple connections need to be made between the send and receive points of the system. For instance, if the system requires 16 bits to travel at one time, then there must be 16 physical lines connected between

FIGURE 2–13 Parallel Port Connection

two points in order to get the data to transfer. The maximum number of data bits that can be transferred at any one time will then be limited to how many wires have been connected. For this reason, parallel cables tend to be large and bulky, as compared to those of serial connections.

Network Cabling

Once you have a network interface in place, you need some method of getting the data signals from one computer on the network to another. This can be done in a number of ways. Although any of the following methods can be used, the first item on the list (twisted-pair cables) is far and away the most common:

1. UTP cables
2. Shielded twisted-pair cables
3. Coaxial cables
4. Optical fiber cables
5. Radio waves
6. Infrared light
7. Electronic signals sent through power lines

The choice of network cabling (or communications medium, as it is sometimes called) is rather important because of the extremely high frequencies of the signals. Sending 60-cycle utility power through a wire rarely presents a difficulty; but sending a 100 million bits per second signal can be a little more tricky. For this reason, the method of sending signals, and the materials they are sent through, can be important.

The types of signals that are sent through the network and the speed at which they are sent are extremely important details of a network. All parts of the system must be coordinated to send, carry, and receive the same types of signals. Usually, these details are not something that you have to consider as long as all parts of your network come from, or at least are specified by, the same vendor or were designed to the same industry standard.

Because networking evolved over several decades, many different cabling solutions have been used. Today, virtually all cabling has moved to UTP or fiber-optic cable specified in the TIA/EIA 568 standard. Because older equipment is sometimes used on newer cable plants, adapters called "baluns" have become available to make that adaptation.

UTP cable has been developed to support most networks up to 200 Mb/s (megabits per second). The cable has very tightly controlled physical characteristics, especially the twist on the wire pairs and characteristics of the insulation on the wire that control the characteristics of the cable at high frequencies (Figure 2–14).

Normally, wire pair cables cannot carry high-bandwidth signals fast enough for LAN communications. The wires also radiate like antennas, so they interfere with other electronic devices. However, several developments allowed for the use of simple wire for high-speed signals.

The first development is using twisted-pair cables. The two conductors are tightly twisted (one to three twists per inch) to couple the signals into the pair of wires. Each pair of wires is twisted at different rates to minimize cross-coupling. Then, "balanced transmission" is used to minimize electromagnetic emissions (Figure 2–15).

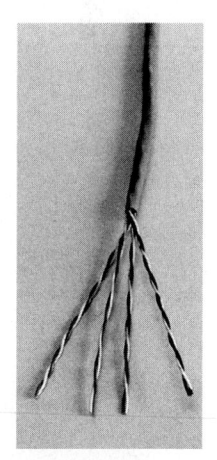

FIGURE 2–14 Unshielded twisted-pair (UTP) cable is the most common medium for networking today.

Balanced Pair Transmission

Equal but opposite signals on a pair of wires

Output is the sum of both signals

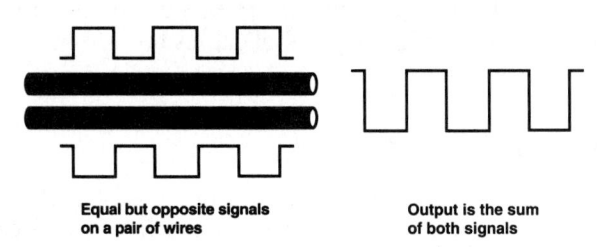

FIGURE 2–15 Balanced pair transmission minimizes emission of radiation from and cross-talk among pairs.

Balanced transmission works by sending equal but opposite signals down each wire. The receiving end looks at each wire and sees a signal of twice the amplitude carried by either wire. Although each wire radiates energy because of the signal being transmitted, the wires carry opposite signals so the two wires cancel out the radiated signals and reduce interference.

By using these techniques and having two pairs of wires sending signals in opposite directions, it has been possible to adapt UTP cables to work with Ethernet and Token Ring networks. At higher speeds, signals are compressed and encoded and multiple pairs are used for transmission in each direction to allow operation with networks of over 1,000 Mb/s per second.

UTP has four twisted pairs in the cable, two of which are used to simultaneously transmit signals in opposite directions, creating a full-duplex link. Some of the higher-speed networks now use all four pairs to reduce the total bandwidth requirement of any single pair.

Some cables have been offered with high performance guaranteed only on two of the four pairs, but these cables can be troublesome if the cable plant is ever upgraded to high-speed networks. Although most networks only use two pairs of the four pairs available, it is not possible to use the other pair for telephone or other LAN connections because of possible cross-talk problems.

Today, all network wiring has migrated to a "physical star" network with all cables going out from a central communications room. This network topology is specified by the TIA/EIA 568 standards (Figure 2–16).

These are the basic considerations of network systems. All of the confusing terms associated with networks simply describe things associated with these basics. In reality, the basic operations of networks are fairly simple. The real problem exists in the methods used to accomplish parts of these operations. As we continue

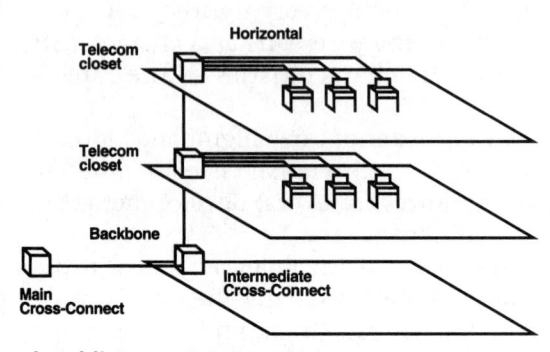

FIGURE 2–16 Network Cabling per TIA/EIA 568

through our discussion of networks, do not lose track of the fact that all the other details are merely methods of accomplishing these basics, and nothing more—no matter how terrifying they sound.

Seven Layers of the OSI

For a network to be functional, information from one software package on one computer must move through a physical medium in order to communicate with a software package on a secondary computer. As an example, if a Microsoft Word document exists on a file server, and a remote computer connected to the network would like to open the existing document, the two computers ultimately need to communicate with each other in order to transfer the data. Early on, when the computer industry was attempting to iron out details such as how the interconnection and communication would or should take place, many computer manufacturers were attempting to come up with their own proprietary systems, ultimately raising issues of incompatibility. It became obvious early on that, whatever level of communication was adopted, there needed to be some form of standardization to make the physical interconnectivity of different devices plausible. As a result, a conceptual model, or framework, was developed by the International Organization for Standardization (ISO) in 1984. Subsequently, the **Open System Interconnection (OSI)** reference model was created; it is now considered the primary architectural model for intercomputer communications. The Open System Interconnection (OSI) reference model divides the movement of communications between computers into seven tasks or layers. A group of tasks is then assigned to each layer, which makes them independently self-contained operations of the overall communication system. Conceptually dividing the model into seven self-contained tasks helps to reduce the complexity of the network. Smaller tasks are ultimately easier to understand and manage.

The seven layers are listed as follows:

Layer 7	Application
Layer 6	Presentation
Layer 5	Session
Layer 4	Transport
Layer 3	Network
Layer 2	Data Link
Layer 1	Physical

Keep in mind that the OSI model is only a framework of how communication should take place; it is not a working system in and of itself. Ultimately, manufacturers do not need to conform to such standards. But having the OSI model makes the conceptualization of the network simpler to understand. Engineers can work more efficiently on modular functions, and provided their layer effectively communicates with adjacent layers, the overall interoperability of the system will function reliably even between inherently dissimilar systems.

Layers 5, 6, and 7 are called the upper layers of the OSI and deal with application and software-related communication issues. Application layer 7, the highest layer, is closest to the user end of the system. The computer operator interacts directly with layer 7.

```
       7  Application  ⎫
                       ⎬ Upper
       6  Presentation ⎪ Layers
                       ⎭
       5  Session

       4  Transport    ⎫
                       ⎪
       3  Network      ⎬ Lower
                       ⎪ Layers
       2  Data-Link    ⎪
                       ⎭
       1  Physical
```

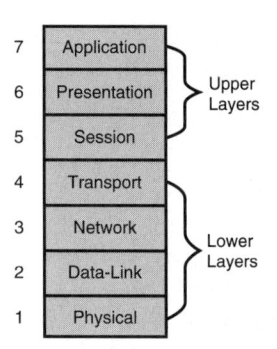

FIGURE 2–17 Open Systems Interconnection (OSI) Model

Layers 1 through 4 are considered the lower layers of the OSI and deal with data transport, the physical network, cabling, and hardware. The actual physical connection to the network medium is closest to layer 1, the physical layer. The implementation of communication is accomplished through a combination of software and hardware as the communication works its way through the individual layers. Figure 2–17 illustrates the layers of the OSI model.

Communication between computers is made possible through a set of **protocols,** which are a formal set of rules governing the transfer and exchange of data through the network medium and individual layers of the OSI. A wide variety of protocols exist, such as LAN or WAN protocols, network protocols, or routing protocols. As an example, **router** protocols exist in layer 3, the network layer of the OSI. Their purpose is to provide an organized exchange between routers in order to select the best available pathway for network traffic. Since the actual network appears more like a spiderweb than a road, there are always multiple pathways that can accomplish the same destination outcome. The router protocols attempt to simplify the data traffic as it moves through the network, helping to prevent crashes or redundancy of information.

Communication and the OSI Model

Communication starts at layer 7, the Application Layer, as the user interacts with a software application such as e-mail, to communicate with other systems resource files, printers, network resources, or simply to share resources between network locations. The application layer provides direct interaction of the software package necessary to implement network communications. Its functions typically include the identification of communication partners, the determination of available resources, and the synchronization of data flowing back and forth on the network.

For any type of communication to take place, the identity of partners must be clearly known; resources must be made available; and the logical synchronization of data must occur in an organized manner for it to be clearly and intelligibly transferred between desired locations. Let's use a telephone conversation as an example. To make a call you need to know who you are calling: *identification of partners*. To be able to make the call, the phone must not already be in use by someone else: *available resources*. And finally, a clear protocol of speaking and passing information back and forth across the phone line must be implemented so as not to cause confusion; I talk, you listen, you talk, I listen: *synchronization of data*. Protocols are implemented at each level of the OSI to help implement these three crucial steps in the communication process.

The exchange of information along the OSI is implemented from one layer to the next in a top-down, bottom-up method, as information winds its way between communicating parties. If System A wishes to communicate with System B, data will transfer from the application layer of system A, to the presentation layer, layer 6.

The Presentation Layer

The Presentation Layer converts the data from a user format to a computer or Internet format. Common data formats are the American Standard Code for Information Interchange (ASCII) and the Extended Binary Coded Decimal Information Code (EBCDIC). As an example, in the ASCII format each key symbol on a standard keyboard will have binary computer code associated with it for communicating with the computer. From layer 6, the data is transferred to layer 5, the session layer. This layer is also important for encryption because both systems need to be on the same page when encoding and decoding the data. Go back to the phone example: are we speaking English, German, Spanish, or French?

The Session Layer

The Session Layer establishes, manages, and terminates the communication session between prospective parties. It can be thought of as the local operator of the telephone system. The session layer is the last level of the upper OSI. From there the data is transferred to the transport layer, which represents the lower level of the OSI and the physical connection of the computer to the network.

The Transport Layer

The Transport Layer accepts data from a session and transports it across the network. The transport layer is responsible for dividing the data stream into smaller addressable chunks, making sure the data is delivered error free. The transport layer also manages flow control so that multiple parties are not communicating simultaneously across the network. **Transport Control Protocols (TCP)** will be discussed in more detail in the next section of this chapter. From the transport layer, data is then sent to layer 3, the network layer. This is where the operator sets up and tears down the actual connection. Think of it as the old style plug-and-pull operator switch stations at the telephone company.

The Network Layer

The Network Layer of the OSI controls the address, destination, and routing of messages between nodes or locations on the network. Layer 3 also determines the way that the data will be sent to a recipient device, such as over an Ethernet or phone line through a token ring or FDDI, and provides the Internet protocols necessary to achieve such transmissions.

Internet Protocol (IP). IP or Internet Protocol is a routing protocol which contains the addresses of the sender and the intended network destination. IP addresses can be permanently or temporarily assigned, depending on how the computer is connected to the network. For most dial-up connections through telephone modems, a new IP address is assigned to the user by the Internet provider at the start of each session. Corporate connections, however, assign each individual computer on the network a permanent address that is unique to each workstation. The network administrator managing the local network is responsible for setting up the IP addresses for each individual user or station. To illustrate this point, think about addressing an envelope when you write a letter. The envelope will have the name and address of the recipient as well as the name and address of the sender written on it. IP uses this type of addressing style on every packet sent from a device. The source and destination addressing use either the IP address or network address. The next stage of communication along the OSI is the data-link layer, layer 2.

The Data-Link Layer

The Data-Link Layer provides the reliable transfer of data across the physical network. Depending on the data-link layer, the protocols can include physical addressing, network topology, error notification, sequencing and framing of data, and flow control. **Physical addressing,** which is different from network addressing, defines how devices are addressed at the data-link layer. This is similar to a zip code as opposed to an actual house address. Network **topology** refers to the type of physical connection between computers, such as a bus, ring, or star pattern. Topology will be further discussed in the next part of this chapter. Error notification alerts the upper layers of the ISO that transmission errors have occurred. Flow control manages the flow of data on the network so as not to overwhelm the network with more data than it can handle. The data-link layer communicates next with the physical layer, which represents the last stage between the computer and the network.

The Physical Layer

The Physical Layer defines the electrical and mechanical functions and protocol specifications for activating, maintaining, and deactivating the physical link between computers. The physical layer is responsible for the following characteristics: voltage levels and the timing of transitions, data rates, distance of transmission lines, and the physical connection. Each bit (0 or 1) is represented by a specific signal value.

Encapsulations

From the physical layer, the information from System A is transferred across the network to System B. Once arriving at System B, the information works its way up the OSI model, starting at physical layer 1 and progressing through successive layers until reaching application layer 7 and the subsequent user of System B. The user of system B can now respond, causing the process to repeat in reverse order—from System B, application layer 7, back down to physical layer 1, and across the network to System A, into physical layer 1 and up the OSI chain to application layer 7, where it originated. This entire process is called encapsulation. Figure 2–18 illustrates the encapsulation process. Each layer of the OSI model corresponds to its counterpart on the receiving device. Therefore, the physical layer talks to the physical layer and the data-link layer talks to the data-link layer and so on up the line. Information that is encapsulated at layer 7 is not truly exposed or open until it gets up to the receiver's layer 7. A toy many toddlers play with can illustrate the process. Remember from childhood the assortment of five or six different colored cups that fit inside of each other? Starting with the big cup, you have to place the cups inside of each other; you put one inside the big cup, then another inside of that one, and another inside of that one until you are out of cups. Simply put, this is what encapsulation does. The littlest cup represents the data sent by the application.

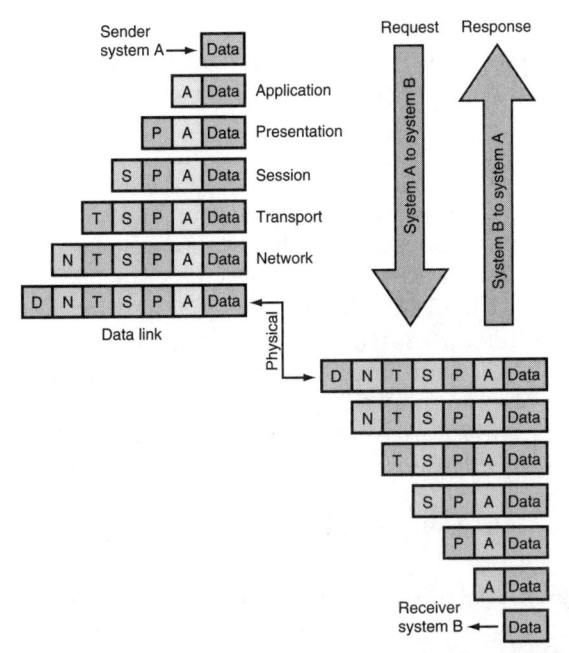

FIGURE 2–18 Encapsulation of the Open Systems Interconnection (OSI)

Servers and Hosts

Most networks are client-server networks where a central computer (server) acts as the main depository of all the files that may be shared by all users (clients).

When we are discussing file servers as being the center of most networks, it is important to talk about what types of computers these servers are. Typically, when the word server is used, it refers to a high-quality personal computer or workstation, usually a new, state-of-the-art unit. When, however, you hear the word **host,** this refers to a **mainframe** or **minicomputer** being used as the center of the network.

Host computers are usually used in large networks at large companies. These companies frequently have large centralized operations that require huge amounts of storage and processing capabilities available only in mainframe computers. For these companies, it simply makes sense to connect their networks to this central information source that is already running most of their company's operations.

Small companies, with their different needs, rarely require a mainframe computer. They nearly always use file servers, rather than host computers or peer-to-peer networks.

Networking Software

Just as a computer has an operating system that controls all its hardware and allows applications software to run on the computer, networks have operating systems that control the flow of information on the networks. The network operating system establishes what computers are on the network, what peripherals they share, and how data is formatted for transfer. It even detects and corrects errors in data caused during transmission.

Because networks send various routing commands through their system, software that is not written for networks (where it will be exposed to these strange commands) often will not work properly. Many of the more popular types of software come in a "network version." It is very important to verify that your software is compatible with your network before you install it. Copyright laws may also limit the use of software on networks, depending on the licensing agreement of the manufacturer.

Network Topology

Local area networks are designed to connect a group of computers within a building or group of buildings. Individual workstations are usually connected through a central computer system called an NOS (Network Operating System), which allows multiple users to access the same software, the same printer, or even the same document. Network administrators manage the individual user rights, allowing access to certain documents by some individuals, while preventing similar access by others. The right to change or alter a document must also be administered, thus ensuring the security and safety of the system and stored data. As an example, an instructor will have rights to post and change student grades on an Internet Web site, but the students will have read-only access rights, allowing them to see the grades but not giving them the necessary permissions to alter any of the posted documentation.

The physical geometry of the network structure can take on many forms.

Star Network

A Star network connects individual workstations or computers through a central device that acts as a hub to the surrounding satellites. Each of the network connections is considered a node to the central hub. The star topology works well for organizations needing a centralized system where multiple workstations must have access to a common communication line. The downside of the star topology is that if the central device were to fail, the entire network would be disabled, thus preventing the free access and exchange of any data on the system.

Ring Network

The Ring network connects each node of the network in series with the next node through a connecting cable. Communication is achieved by passing a short message in the form of binary data packets, called a token. The token passes continually around the ring from node to node in the form of a message relay. When a station wishes to send data, it changes the token to busy, and alters the data packets prior to passing on the message. Each station receiving the token examines it to see if they are the intended destination; if so, the message is opened and processed. The processed message, now altered, will continue along the path, passing from node to node until eventually reaching the original destination. If the intended destination of the message is for a node other than the receiving node, the token is simply passed on to the next station until it eventually makes its way back to the originator. The originator, once receiving the reply, releases the token from busy to free, and then passes it on.

Token passing ensures that only one message is on the network at one time. It is a very effective way of managing network traffic, helping to prevent congestion or multiple messages in the loop. The downside of token ring networks is that the failure of any node on the network will crash the entire system, preventing the return of the tokens. Also, every additional node added to the network has the effect of slowing down the entire network. Since each node must examine the token prior to passing it on, the time it takes for the token to make a complete loop around the

Networking Fundamentals • 31

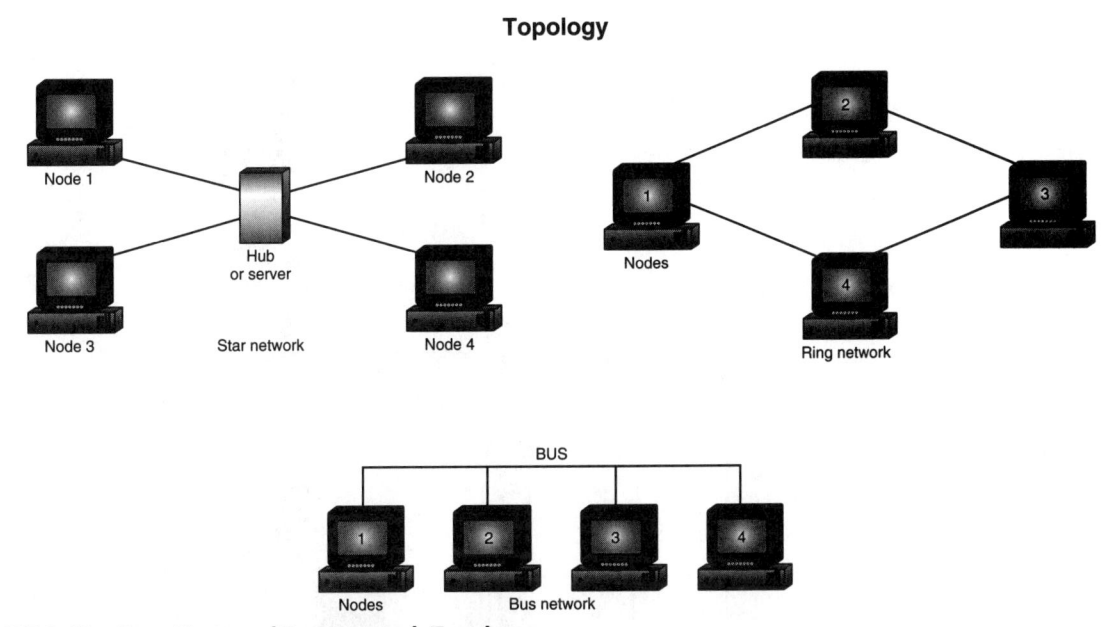

Topology

FIGURE 2–19 Star, Ring, and Bus Network Topology

ring increases. The originator of the busy token will likely see reply delays as more and more participants are added to the loop.

Bus Network

The Bus network connects each node of the network to a common line. It is a broadcast environment where all devices on the bus hear all transmissions simultaneously. There is no central device such as in the star design. A token can still be passed on a bus, but instead of traveling in a ring, the token would be passed up and down the line from computer to computer. Special timing and addressing provisions must also be put into effect to prevent multiple messages from being sent along the bus simultaneously. At any time there should only be one message appearing on the bus. Figure 2–19 illustrates the star, ring, and bus configurations of network topology.

Combinational Topology

Combinational network topologies such as the Star-Wired Ring, Star-Wired Bus, or Tree exist. The star-wired ring connects all nodes of the ring through a central **hub** or Multi-Station Access Unit (MSAU), which ensures the continual use of the network by all participants in the event of failure of a specific node. In the traditional ring design, a single failure can take down the entire network. With the star-wired ring design, all failures are bypassed by the central hub or MSAU so as to continue the passing of the token on to the next destination. Figure 2–20 illustrates the star-wired ring topology configuration.

Another type of combinational topology is the star-wired bus. The star-wired bus does not use a host computer as the master central connection; instead, all nodes are terminated to a central hub or multiport repeater. The distance to or from any node is then considered only one hop away through the central repeater. **Tree topology** is nearly identical to **bus topology,** except multiple branches from nodal points are possible.

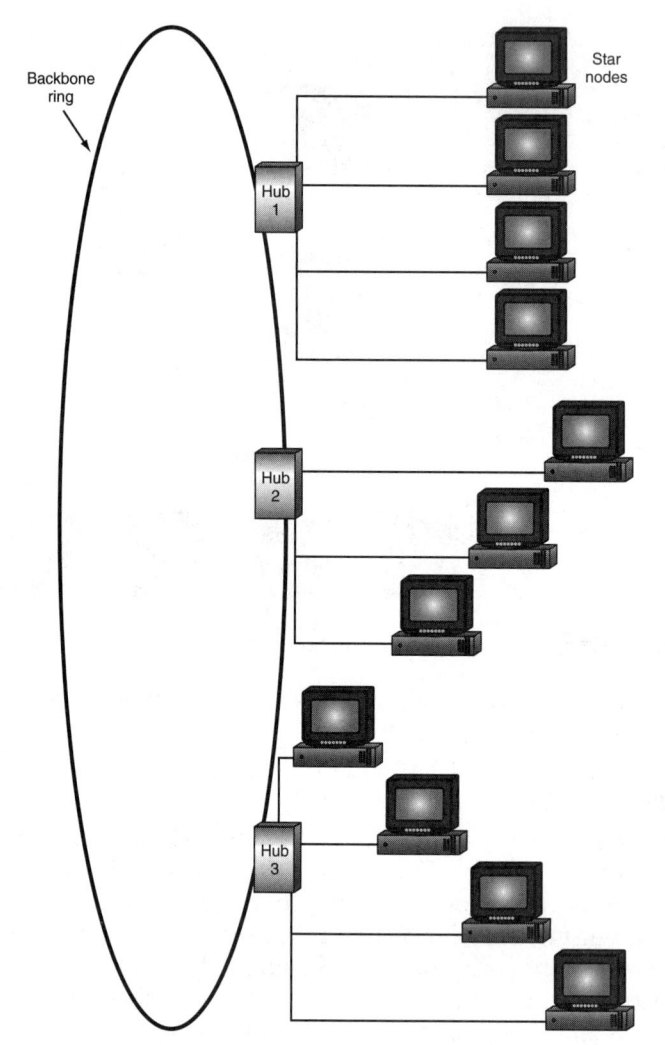

FIGURE 2–20 Ring-Wired Star Network

Collision Detection

One method for preventing multiple transmissions is a process called carrier sense multiple access with collision detection (CSMA/CD). By using CSMA/CD, network terminals monitor the line for data. If the line is not busy, a data packet will be sent out on to the bus having a specified address or destination, and only a node having the correct address can access the data. Problems can occur, however, if one node begins a transmission after another node has already started a transmission. Since it takes a specific amount of time for data to transfer down the line, there is always the possibility for one computer to think there is no data on the line, when in fact it is, but has not yet reached the listener. Two or more nodes transmitting simultaneously will result in what is termed a data collision. A collision will immediately corrupt the data on the line and cause errors. At such a point, all participants in the collision must stop and wait a determined amount of time before retransmitting their data. CSMA/CD sets up the required protocols for managing collisions and for the transmission and retransmission of data on the bus.

Token ring systems inherently avoid collisions, however, they can take longer since each participant must wait for the token in order to access the network. Bus designs using CSMA/CD tend to move data faster because there is no token to pass and participants can simply transmit at will provided the line is not busy or already transmitting. Collision rates will increase as more nodes are added and bus traffic is heavy, which can ultimately slow down the network as the number of retransmissions due to errors begins to rise.

Transmission Methods

Communication along a LAN or network can take place in one of two common ways: baseband or broadband. (See *National Electrical Code® [NEC®] Definition 830.2 Network Interface Unit*).

Baseband

The baseband method is used by most LANs and Ethernet systems. Transmissions usually consist of high-speed, digital data along one shared channel, meaning that the combinational transmitter/receiver signals of the transceiver share the same cable connection. Protocols must be in place to ensure that the transmitter and receiver, residing on the network interface card (NIC) of the computer, are not operating simultaneously, otherwise data errors will occur. Baseband is cheaper and easier to connect and commonly uses either coaxial or twisted-pair cables. Typical uses for baseband are digital voice or data transfers from one computer to the next.

Broadband

Broadband differs from baseband in that it uses multiple channels to communicate. The signals are analog in nature, they transmit and receive on separate radio frequencies, and they must also pass through a specially designed modem to properly negotiate the protocols for the sending and receiving of data packets. A cable modem is a perfect example of a broadband modem. Inside a cable modem, frequencies are allocated to different data streams. One frequency is for the digital video, while another is for Internet traffic.

National Electrical Code® Article 830.1

Broadband communications are considered asymmetrical since they transmit and receive along separate channels, at different frequency rates; whereas, baseband is considered symmetrical since the communication remains at a constant rate for both the transmitter and receiver along the same pair of wires or length of coaxial cable. Typical uses for broadband are data, voice, video, or CCTV. Cable television systems inside of hotels are examples of broadband networks, which supply customers not only with in-room movies, but also with a variety of services and games.

National Electrical Code® Article 830.179 as Related to Broadband Equipment and Cables

Cable to be used for broadband application must be listed as being suitable for the purpose for which they are being used by a listing lab such at Underwriters Laboratories. Cable used for medium-power network broadband applications must be rated for a minimum of 300 volts and should be marked as type BMU, BMR, or BM with type BMU being suitable for outdoor underground use. Cable used for low-power network broadband applications must be rated for a minimum of 300 volts and should be marked as type BLU, BLX or BLP with type BLU being suitable for outdoor underground use.

National Electrical Code® Article 830.90 as Related to Protection

Primary electrical protection is to be provided on all network broadband communication conductors that are not grounded. Primary protectors must be located within the city block where the service is being provided. Fuseless protectors may be used where the current on the protected cable is safely limited to its current-carrying capacity. Fused protectors can be used where requirements cannot be met by the means explained above. Fused-type protectors must provide a fuse in series with each conductor to be protected. The location should be one approved in article 830.90 (B) and should be within a practical distance from the grounding point.

Unicast, Broadcast, and Multicast

Transmissions along an Ethernet can be of the unicast, broadcast, or multicast variety. Unicast transmissions involve the transmission of data from one individual to another. The connection involves a single transmitting station and a single receiving station. Broadcast transmissions are sent out by a single station to all participants on the network. Broadcasts do not discriminate as to destination or recipient. Multicast transmissions involve a list of select recipients within the broadcast domain. While there may be a variety of multicast address lists to choose from, multicast transmissions do not have to be limited to a single multicast address. Compared to unicast, broadcast and multicast provide a more efficient means of data transmission across the network.

Network Devices

Networks can be connected through a variety of devices on the physical layer as a way to either interconnect multiple networks, which would otherwise not communicate, or to extend cable distances beyond maximum limitations. The following is a list of commonly used devices and their definitions:

Repeater

A **repeater** is a network device that amplifies and retimes incoming data signals prior to sending them out to their final destination. As cable lengths increase, signal quality can become degraded, causing voltages to attenuate or fall off. Data bits may also lose shape, become rounded, or even become overly skewed, often taking on a more triangular shape than the ideal square. Excessive skewing will cause data bits to become blurred, making it more and more difficult for the microprocessor to distinguish between a 0 pulse and a 1 pulse. Electrical or atmospheric conditions such as lightning and thunderstorms may also influence the quality of the signal. Multiple repeaters may be used at various points on long cable runs, as necessary, to help reshape, re-amplify, and retime signals along the network to ensure error-free transmissions. System timing requirements will dictate how many repeaters may be used on a specific segment of the network.

Hub

A hub connects multiple user stations through multiple ports. The hub does not make any decision or determination on the data. It simply acts as a multiport repeater. Since the hub functions as a central relay station, having a low level of intelligence, collisions will result as more and more ports are connected. This device fits into layer 1 of the OSI model.

Bridge

A **bridge** is a network device that is used to connect and disconnect computer clusters operating in two separate domains. As an example, consider a situation where there are 15 computers operating in the accounting department of a college or university, and another 25 computers operating in the chemistry department across the street. The basic problem is that the accounting staff does not really need to know what the chemistry department is up to; and likewise, the chemistry staff does not need to know about the daily operations of the accounting department. For all practical purposes they can probably exist as separate networks since they rarely need to communicate with each other. The question remains, why should 40 computers exist as a single network when they would essentially be much better off operating separately? By operating as smaller, separate networks, they will produce lower levels of network congestion, have fewer collisions and errors, and ultimately be faster and more efficient. But what happens when accounting needs to consult with the chemistry department about an overdue charge for 500 test tubes? At some point, they will need to communicate with each other, and a network bridge can solve the problem. Figure 2–21 illustrates the concept of a network bridge.

A bridge has the ability to separate computer domains from each other, and then reconnect as necessary, based on communication requests from either side. As an example, if a computer in domain A would like to communicate with any other computer in domain A, the bridge will remain off. As a result, domain B will not be aware of the additional network traffic flowing in domain A. However, when a computer in domain A wishes to communicate with a computer in domain B, the bridge will respond by turning the necessary connection on, thus allowing the transfer of data to flow between the two sides. By having the ability

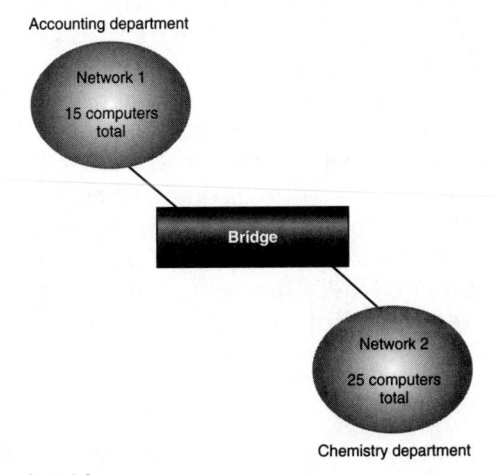

FIGURE 2–21 A Network Bridge

to switch the interconnection of domains on or off, a bridge can directly control and manage the amount of network traffic and ultimately prevent excessive collisions and possible network errors. For such switching to take place, however, the bridge must be intelligent enough to recognize addresses. Addresses are typically either programmed into a bridge manually by a network administrator, or designed to be updated and stored in their internal memory automatically. The type of bridge being used will determine the setup process. As a result, bridges are considered smart devices, as compared to repeaters or hubs that lack the overall intelligence to make any kind of network decision about where data should flow and to whom. These devices are defined at the data-link layer of the OSI model. The actual connection switching occurs at this level.

Router

A router is essentially a network bridge that is used to link entirely separate networks. A bridge connects clusters of computers within the same network, whereas a router is used to connect isolated networks. As an example, a router would be used to link separate entities, such as the University of California to the University of Minnesota. The router still operates from a list of programmed addresses; however, the address is targeting a location inside the network layer of the OSI by using Internet Protocol (IP) addressing instead of from the data-link layer where CSMA/DC manages network traffic.

Switch

A switch is an Ethernet or network device that can send incoming data packets to one of several address destinations. A switch can be thought of as a multiport bridge. Some switches have higher level functions that allow a network administrator to turn ports on and off, or separate the LANs so that they act like they are on their own switch, known as a Virtual LAN (VLAN). A virtual LAN has the ability to separate computers into separate networks as a way to impose security restrictions on specific terminals or to help reduce contention between nodes attempting to transmit simultaneously.

There are three varieties of switches available: Fragment-Free, Cut-Through, and Store-And-Forward. The fragment-free style will use the calculation of propagation over the Ethernet to determine which section of a transmitted data frame may be corrupted. This works out to be the first 64 bytes of any Ethernet frame. The cut-through switches operate faster because they forward data on to their destination as soon as the destination address has been determined. The store-and-forward variety stores the entire frame of data in internal memory prior to sending it on to the required destination. This allows the frame to be checked for errors using a mathematical process. At the end of each frame there is a Frame Check Sequence (FCS) that is an answer to the same mathematical process performed when the frame was sent. By comparing the two, it can be determined if the frame has been corrupted during transmission. As a side note, this process is done by all devices within an Ethernet network when a frame enters layer 2 of the OSI model. A switch can operate from data-link layer 2. Since a bridge operates on layer 2, essentially a layer 2 switch is a multiport bridge

LAN Extender

LAN extenders are remote access, multilayer switches that are used to filter out and forward network traffic coming from a host router. While they can be used to help filter out unwanted traffic by address, they are unable to segment or create security **firewalls.** Firewalls require passwords and user IDs in order to gain access to the network.

Gateway

Gateways are used to interconnect incompatible computer systems, electronic mail systems, or any network devices that would otherwise be unable to communicate directly with each other. An example would include Macintosh to IBM PC, or even different varieties of mainframe computers. Another example would be when different LANs have different protocols, say a TCP/IP network talking to a IPX/SPX network.

Network Software Management

Once the network had been designed and the topology constructed, the management of network software becomes a secondary issue along with the availability and user rights of shared resources on the network. There are two basic designs for managing network software: peer-to-peer and **client/server.**

Peer-to-Peer Networks

Peer-to-peer networks are meant for smaller systems where each node of the system has equal rights and access to any shared software, resource, or device on the network. Individual nodes are able to control their network availability with respect to what resources may be shared or used by other nodes. The construction of peer-to-peer networks is relatively simple and inexpensive to create. A typical setup will include fewer than ten computers.

Client/Server

A Client/Server network designates one node of the system as the main file server and system resource for all other connecting nodes of the network. The main server is typically dedicated as a network resource, and normally would not be used as a separate individual workstation. This is the only downside of the client/server model. There are, however, many benefits in using a clients/server network since they inherently tend to be more orderly and organized when managing system resources and shared files.

A client/server network also has the ability to monitor system usage and software licensure. Software packages are typically sold based on quantity of maximum users. As an example, if specific software requires a maximum of 25 users at any one time, the centralized control of the main server can more easily keep track of and manage client requests, ensuring the proper enforcement of licensure agreements. As a result, users may occasionally find themselves locked out of certain software resources on the network until they are released for use and made available by the main server. Connecting nodes or workstations of the client/server model may also have the ability to share personal resources among themselves, from workstation to workstation, if necessary.

Multiple software packages residing on the main server must also be able to run simultaneously to various nodes in the network, all based on client usage and demand. The main server must therefore have multitasking abilities in order to accommodate the various requests. As a result, **operating systems** such as Unix or Microsoft Windows NT and XP were specifically designed for the management and daily operations of the client/server network, thus allowing multiple programs to run simultaneously on various nodes without the risk of interference. These operating systems are known as network operating systems (NOS).

The client/server network also has the ability to manage and keep track of alterations to files, thus helping to reduce the possibility of error and confusion. As an example, if a specific file is being updated or changed by a client while other network participants are requesting use of the document, the requesting parties will be given read-only rights by the server and told that the document is currently locked and unavailable for alteration. Since complete user rights to a document can be given to only one network participant at a time, any requesting participants must ultimately wait for the document to be released back to the main server by the current user before any additional changes can be made. The main server can also permanently lock out rights to certain clients, such as students requesting class grades. In such circumstances, the main server will only allow students to read current grade data; they will not have the ability to make changes or alterations. Students could, however, have access to personal files where they do have rights to update user information such as address, phone number, etc. In all such cases, network participants must provide a log-in and password to the main server in order to identify and verify user rights and privileges to available documents and software packages residing in the system.

As stated earlier, client/server networks are inherently more organized and able to handle and manage system resources more effectively than peer-to-peer networks, due to the centralized control of the main server.

Point-to-Point Links

Point-to-point links consist of leased lines, often from the telephone company, which are used by network carriers or service providers to connect customers across a WAN to remote networks. Leased lines are quite expensive and are often priced based on transmission length and bandwidth requirements. The benefit of having a point-to-point network link is that the line is not shared. The customer is paying for a dedicated line, which will be made available by the network provider, 24 hours a day, for the continuous transfer of data, as needed.

Circuit Switching

The alternative to expensive, dedicated, leased lines is to connect to network circuitry only as needed, and then disconnect when not in use. The concept of circuit switching is very similar to making a standard telephone call, but over a digital line called an Integrated Services Digital Network (ISDN). A system router makes the connections available, as required by the customer, and then disconnects or terminates the connection once the data transfer has been completed. A customer will pay less since the network connection is never tied up for any long period of time. The line is therefore made available to many customers throughout the day, provided that they do not intend to monopolize large amounts of time transmitting or receiving data. The downside of circuit switching is that the connection must be redone every time the address of the destination changes. Another alternative to dedicated, leased lines or circuit switching would be to use packet switching.

Packet Switching

Packet switching involves the transfer of datagrams across the network by best available means, meaning that the network itself determines the delivery pathway without directly connecting or involving the original sender. Back in the mid-1970s, packet switching protocols were being developed by an organization known as the Defense Advanced Research Project Agency (DARPA). The main goal of the organization was to develop a system where incompatible computer systems would not only be able to send and receive data, but also be able to make the transfer irregardless of downed lines or intermittent connection across the network or WAN. The U.S. Department of Defense realized early on that they needed to develop a highly redundant and secure data delivery model that would be able to negotiate and quickly sidestep any sporadic outages of power or telephone service, possibly the result of a terrorist or nuclear attack. Such an event of unthinkable magnitude could have ultimately had a crippling effect on the nationwide communication infrastructure, which at the time was largely dependent on the circuit switching protocols of the analog telephone system. Defense planners therefore envisioned their ideal delivery model as one that would be able to utilize multiple pathways and connections at will through the linking of intelligent hardware, thus ensuring the safe and guaranteed transfer of information, especially in times of greatest risk or need. As a result, a connectionless data transfer system was soon established.

Connectionless implies that the original sender of a data packet does not ever actually connect to its destination node address. Instead, the originator simply relays a data packet to a network router. The network router is then responsible for negotiating the transfer of data by first storing the data packet, which includes the destination address, in its internal memory and then sending it on only after having chosen the best available route to the final destination. As a result, packet switching networks are often called store-and-forward networks. Once the network router has taken control of the data packet, the original sender is released from the connection and remains uninvolved with the rest of the transmission.

Packet switching was ultimately designed as a way to route and re-route datagrams across the network, by best available means, in order to guarantee their reliable and safe delivery. The odds are very remote that a terrorist group could ever effectively disable all the interconnected and cross-connected pathways of a WAN. By utilizing intelligent routing hardware and packet switching, a datagram is virtually guaranteed to reach its final destination—unless of course the primary connection to the access server or router is severed, which would effectively isolate data packets and permanently prevent them from gaining access to the Internet or WAN. In most cases the primary access to the Internet is either available through a dial-up, telephone connection or through a secondary cable provider. As long as the primary Internet connection remains available, any data placed on the network will appear to have an almost limitless variety of pathways to choose from while negotiating its route between nodes A and B. In fact, this process is very similar to finding a route from Chicago to California through the network of national roads and highways. Imagine a car intelligent enough to decide, of its own free will, how to get from point A to B—and the *how* doesn't matter, just as long as it gets there. Packet switching accomplishes such an outcome. So what are datagrams?

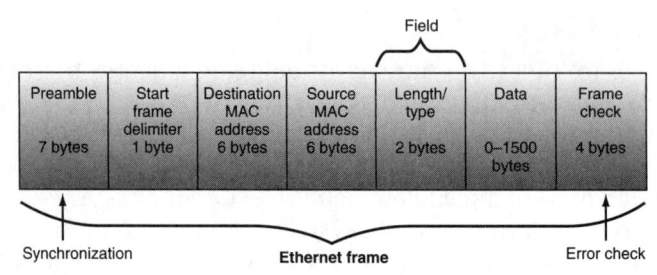

FIGURE 2–22 Datagram

Datagrams

Datagrams are data packets that have been organized into frames, representing compartmentalized sections of binary code, each designed to carry specific information about the transmission along with the actual data. Each compartmentalized section of a frame is referred to as a field. Figure 2–22 illustrates the concept of a datagram. The example shown is an Ethernet frame. Depending on the type of network protocol being used, the individual fields within the frame may represent all or some of the following transmission details: preamble, used to synchronize the system; start and stop delimiters, indicating when to begin and end a transmission; destination address; source address; data type and length; control information; error check; and the actual data. Each field is limited to a specific size based on its coded purpose. In most cases, the data field will occupy the largest part of the packet.

Types of Data Errors

The occurrence of data errors on a digital network may result from one or all of the following: latency, jitter, bandwidth, or lost packets.

Latency

Latency involves the time delays between a transmitting node and a receiving node—the time it takes for data to be processed and transferred across the network.

Jitter

Jitter involves time shifts between data packets, which can ultimately create a shaky or jerky quality, especially in video streams where the necessity for smooth, high-speed data flow is required to recreate the desired image.

Low Bandwidth

Bandwidth represents the size of the pipeline. Low bandwidth may cause data to be delayed, increase network traffic, and ultimately produce higher rates of collision and error. This is similar to what would happen if a three-lane highway were reduced to a single lane.

Lost Packets

Lost packets will occur when a device such as a hub, router, or switch is unable to keep up with the incoming network traffic. Losses will occur when data arrives too fast or too often causing the network device to drop or throw away data packets, especially as internal memory becomes overloaded and unable to store or buffer the incoming stream. The conceptualization of lost packets can be compared to overfilling a glass of milk. Eventually the vessel will become full and any further attempts to force more into the container will result in overflow and spillage. As network memory becomes overloaded or full, data spillage will occur and packets will be lost.

The increasing need for most organizations to achieve higher network bandwidths, higher transmission speeds, and lower error rates is primarily due to current high demand for streaming video and real-time video conferencing, which has now become a daily routine and necessary business tool for many networked corporations around the world. The convergence of voice, data, and video signals across a single network connection has significantly increased traffic density for all participants, making the topics of traffic management and network performance a primary concern of most network providers and analysts.

Network Protocols and Their Functions

Network protocols provide three basic functions. They determine the size of the data packet, they organize the frame layers, and they determine the requirements for data control helping to provide error-free transmissions. TCP/IP, which stands for Transmission Control Protocol and Internet Protocol, is an example of an Internet protocol for data packets.

TCP/IP and the Internet

Two of the most commonly used protocols for accessing the Internet and World Wide Web (WWW) are known as Transmission Control Protocol (TCP) and Internet Protocol (IP), often referred to as TCP/IP. TCP/IP works as a suite of protocols providing connectivity between communicating systems across the Internet. The protocols were originally developed by the U.S. military in 1969 for use on a 4-node network referred to as ARPANET (Advanced Research Projects Agency Network). Since then, they have grown significantly beyond their initial inception into a much broader range of protocols that currently have the ability and flexibility to handle a wide variety of applications, ultimately providing the communication link between any similar or dissimilar mediums and data structures, making it extremely reliable.

TCP/IP

TCP/IP forms a hierarchy of protocols; it is a protocol suite, very similar to the OSI, but centered on a 4-layer model instead of 7 layers. The four layers of TCP/IP are listed as:

Layer 4	Application/Process
Layer 3	Transport
Layer 2	Internet
Layer 1	Network Access/Interface

Note that the top layer (Application/Process) has combined the session, presentation, and application layers into a single entity entitled Application/Process. The

Transport layer and Internet layer of the TCP/IP suite map directly to the Transport and Network layers of the OSI model, respectively. The last layer of the TCIP/IP suite maps to the bottom two layers of the OSI model (Data-Link and Physical).

The functionality of TCP/IP protocols are split between layers 3 and 4 of the OSI. The Internet Protocol (IP) starts working on network layer 3, while the Transmission Control Protocol (TCP) resides in transport layer 4. Since TCP/IP operates above the physical and data-link layers, any compatibility issues between dissimilar mediums or data systems become a non-issue. As a result, TCP/IP has the ability to provide linkage to any communication system as long as the physical and data-link layers are properly designed and working appropriately. Therefore, the type of system or data becomes irrelevant since TCP/IP does not concern itself with layers 1 or 2, and ultimately it can provide linkage to the Internet through a variety of network connections and over any style LAN or WAN.

Internet Protocol (IP)

Internet Protocol (IP) provides two main functions to the Internet: it sends out continuous streams of datagrams to the required network nodes, and it manages the removal of lost or recirculation packets in the system. IP is considered connectionless since it does not concern itself with the results of a transmission, nor does it care about the order of error checking of data packets. Datagrams are simply moved through the system without keeping track of their final outcome as they propagate from node to node; any error checking, sorting, or data rearranging will eventually take place in the transport layer by the Transmission Control Protocol (TCP).

The management of lost packets becomes a secondary function of IP. Lost packets, if not properly managed, could potentially migrate around the network forever, without ever arriving at their final destination. Remember that data packets do not necessarily take a direct route across the network, nor are they required to take the same route each time. For this reason there can be a great number of packets, broken apart and out of order, continuing to endlessly circulate the network. Lost packets, if not properly controlled, can ultimately increase the level of network traffic, resulting in additional congestion, loss of bandwidth, higher rates of collision, and error. In order to solve this problem, IP has been given the ability to control a data field inside the packet frame referred to as the "time to live" count. As a packet attempts to "hop" from one node to the next, the IP will reduce the count by one. Once the "time to live" count has reached zero, the system will automatically discard and terminate the packet.

Transmission Control Protocol (TCP)

Transmission Control Protocol (TCP) operates from the transport layer at both ends of the communication link. Its capabilities include data stream transfer management, flow control, full-duplex operation, and communication multiplexing. TCP is mainly responsible for the sorting and reassembling of data packets, error correction, and the detecting, requesting and resending of lost packets. Full-duplex operation means that TCP can send and receive packets simultaneously, while multiplexing refers to the ability to communicate multiple conversations simultaneously over a single connection. Since TCP is not operating at the network level, it does not concern itself with the destination routes of the various packets; instead, it simply operates from the end points of the communication link, spending the majority of its time sorting and reorganizing packets as they arrive. TCP ensures that all packets have been received and are in order before handing them up to the next layer. If data errors or lost packets are detected, then the retransmission of specific bytes is requested through means of acknowledgements. The sending computer waits for confirmation of receipt; if the confirmation does not arrive, then the transmission is resent. This process does not happen forever. There is a time associated with retransmission called a "time out" that tells the sender when to stop trying. If a sending computer does not receive a confirmation known as an acknowledgement and the time out value has expired, an error code is reported to the upper layers of the sending computer. Typically, the error is related to "connection lost." TCP is the connection-oriented protocol in the TCP/IP suite, which means that it is responsible for ensuring a connection is established between hosts. The connection is maintained by the "acknowledgements" during a transmission. When paired with the ability of IP, the communication can traverse many different networks to a destination.

User Datagram Protocol (UDP)

User Datagram Protocol (UDP) is a connectionless protocol that does not send back any type of acknowledgement to the originating transmitter regarding the status of data packets. A UDP transmission must therefore rely upon upper-layer protocols for reliability since data checks are not built into the communication.

Port numbers are assigned to TCP and UDP segments to distinguish the presence of multiple applications running from a single device. The port number contained within the TCP or UDP data segment is used to identify the application for which the data packets belong. In order to keep the level of confusion to a minimum, standardized port numbers have been assigned to applications so that differing and multiple implementations of the TCP/IP protocol suite can interact and operate together.

Application and Process Layer

The Application and Process layer provides all the necessary protocols needed for a Web browser, such as Internet Explorer, to communicate with a Web server. The following items are examples of application and process layer protocols: **Hypertext Transport Protocol (HTTP), file transfer protocols (FTP),** Telnet, **Simple Mail Transport Protocol (SMTP),** and **Post Office Protocol 3 (POP3).**

Hypertext Transport Protocol (HTTP)

Network Web browsers such as Internet Explorer, AOL, and Netscape are all examples of application programs that utilize Hypertext Transport Protocol (HTTP) to communicate across the Internet. HTTP communicates through a programming language known as HTML or hypertext markup language. HTML has all the functionality of a word processor while also allowing links to other Web pages residing on the same or even different servers. HTML essentially allows an Internet Web browsing program the ability to read and display text files and or graphics.

While HTML may be suitable for reading or displaying Internet text or graphics on a Web browser, there are often times when a Web user needs to interact with a specific sight rather than just be a passive observer. Two additional programming languages exist, known as Java and Flash, each of which offers its own level of specialization and functionality to Web browsing through HTTP.

Java. Java was developed as an interactive Web browsing programming language. Since HTML can only display the contents of a Web page, Java allows the user the ability to fully interact with or even make changes to the contents of the site. A common example would be an on-line shopping page where a buyer would be required to enter in their name, address, and payment option in order to purchase a desired item. Without the ability to interact with the Web site, on-line ordering would not be possible.

Flash. Flash takes Java to another level; it provides the functionality of interactive Web browsing and also allows for the design and presentation of multimedia programming across the Internet. Flash has the ability to produce very high quality graphics and/or animations, including audio and video, which can then be displayed and/or played by visitors to a Web site. Although Java has the ability to create some graphics and simple animations, the overall process would be quite daunting and cumbersome to the programmer if they attempted to design a Web page as fully interactive as Flash. Flash, being more highly adaptable and functional than Java, is simply a better programming tool, especially when a multimedia experience on the Internet is desired.

File Transfer Protocol (FTP)

File Transfer Protocol (FTP) is often used to upload or download files to or from network servers. FTP is built into most Unix- and Windows-based operating systems and Web browsers. When files need to be transferred between servers or network nodes, the FTP is engaged to perform the desired task and is often transparent from the user's point view. FTP is how files move around the network or Internet.

Telnet

Telnet was designed as a communication link between a server or mainframe computer and a remote terminal, primarily to perform administrative tasks such as turning switches on or off, configuring routers and hubs, or setting user permissions. By having the ability to log-in and alter network parameters from a remote location, the network administrator has the freedom to control the system from anywhere through the use of a single keyboard connection and monitor. The process, however, is not very user friendly and demands a higher level of command line knowledge and insight into computer code to be able to communicate effectively with the main server. Telnet looks very much like the old DOS operating systems or present day Unix system. Each command line code must be typed in separately, and the user must then wait for a response from the main server or device before attempting to transmit any additional commands.

Although Telnet is still available, allowing individuals the ability to access, change, and move data from a remote location, it has been losing popularity over recent years to the more graphically based and secure protocols on the present day Internet. Keep in mind that Telnet was originally developed when network connections were more text-based, connection-oriented systems. The use of flashy graphical interfaces and connectionless packet switching was nearly nonexistent since the main priority of the day was simple two-way communication and information transfer modeled after the public telephone system. The term "telnet" is actually a shortened form of "telephone network." It would have also been considered a great luxury of the time to use higher-level graphical displays, which would have required far more memory to operate, gobbling up precious network bandwidth in the process, and ultimately bogging down the entire system. As computer speeds increased substantially from the low megahertz of the past to present day levels of a gigahertz or more, the availability of reliable, more affordable memory chips became a reality, thus allowing the higher-level graphical interfaces of HTTP to flourish. The landscape of computer networking was permanently changed once users were able to choose between the high learning curve and burdensome code aspects of DOS as compared to the simplicity of a point-and-click, graphical, Windows-based operating systems, and later an Internet Web browser such as Microsoft's Internet Explorer.

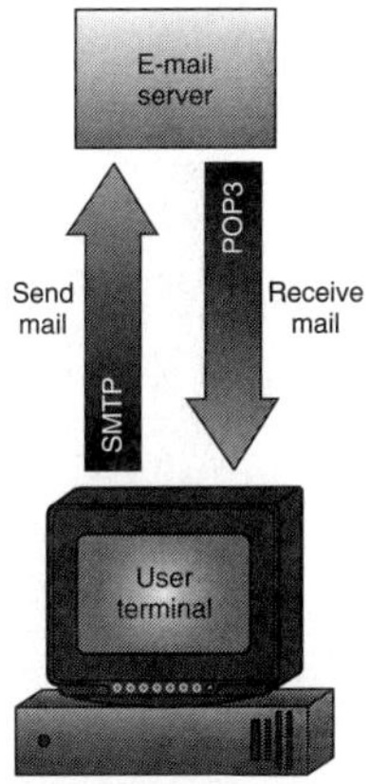

Simple Mail Transport Protocol (SMTP)

Simple Mail Transport Protocol (SMTP) allows for the sending of electronic mail and messages from a user station to an e-mail server. The process is one directional. As a result, SMTP cannot request mail, it can only send. To retrieve mail the user must use Post Office Protocol 3 (POP3).

FIGURE 2–23
E-mail Protocols

Post Office Protocol 3 (POP3)

Post Office Protocol 3 (POP3) defines the method and protocol for retrieving e-mail from an e-mail server; both protocols, POP3 and SMTP, work together to send and receive electronic messages across the Internet. Figure 2–23 illustrates the process.

There are also a variety of other protocols, briefly explained below:

Simple Network Transfer Protocol (SNTP)

Simple Network Transfer Protocol (SNTP) exists as a means of monitoring computers, routers, switches, and network usage.

Internet Control Message Protocol (ICMP)

Internet Control Message Protocol (ICMP) is used for IP errors and control messages.

Address Resolution Protocol (ARP)

Address Resolution Protocol (ARP) is used to map IP addresses to MAC addresses in the physical layer (layer 1 of the OSI).

Internet Mail Access Protocol (IMAP)

An alternate replacement to Post Office Protocol 3 (POP3) is Internet Mail Access Protocol (IMAP). IMAP adds features and abilities, allowing a user to manipulate and search messages that are still on the server rather than simply to retrieve them.

Applications utilizing TCP/IP are constantly being developed as the growth and emergence of the Internet continues. Teleconferencing, voice-over IP, and streaming audio and video have all become commonplace in the network environment, and new applications are appearing daily to take the place of the older, outdated modes of communication. The list of protocols shown here for TCP/IP simply represents the current, most commonly used variety. Remember that the strength of the Internet resides in its versatility and the power of those connectionless data packets, which ultimately can adapt to any change or unknown as the growth of Internet usage moves forward into uncharted territories.

Computer Network Addressing

Internet Addressing

All available resources or files on the Internet must have a unique address known as a **URL** or **Uniform Resource Locator.** The URL provides the necessary protocols for finding and accessing Web sites or information files residing on the Internet. URLs are registered through network providers to ensure that they are unique and available for use, since two individual Web sites are not allowed to have the same URL.

An Internet computer address presently consists of a 32-bit binary number, broken into 4 groups of 8 bits called octets, which specify the network it belongs to and its host address on that network. The 4 groups of 8 bits are displayed in base-10 format and expressed as dotted-decimal. The following number represents an example of a possible internet address: 145.22.78.82. Mathematically, the maximum number of possible combinations for a 32-bit address is 4,294,967,296, which is equal to 2^{32} (32 binary characters, two states each). This may seem like a lot, but as the size of the Internet continues to grow at a rapid pace, available address space will eventually start to fall short of demand. To help alleviate this problem, version 6 of IP protocol has increased the allocated size of the address field to 128 bits, thus ensuring the continued growth of the Internet for many years to come.

To make Internet addressing easier to remember, the URL is written in word format rather than a meaningless series of numbers. As an example, let's say the above Internet address 145.22.78.82 represents the location on the Web for a local pet shop. Instead of having to remember the four segments of numbers, the URL can be coded as: **http://www.petshop.com/newpets/blacklabs.html.** However, for this process to work there needs to be a conversion between the words and the numbers. The translation is accomplished by a network computer called a **Domain Name Server (DNS).** The domain name server is regularly updated with a current list of computer addresses presently connected to its network. Since a domain server cannot possibly have the address list for every computer connected to the Internet, there must be a process for finding or searching out unknown computers. This process is defined by the Domain Name System, both the Domain Name Service and Domain Name System are referred to as DNS. They really are the same, they are just different names for the same function. The overall system is defined to convert or resolve friendly computer names to IP addresses. Each DNS server has the ability, if desired, to communicate with other DNS servers. All organizational LANs will contain a DNS server for outside access to the local names. A request for resolving the name of an IP address from an outside computer will be answered by the DNS server. If the name to IP mapping is unknown, then a process of recursion can be used, where the DNS server asks other DNS servers if they have the mapping. All this is done on behalf of the computer that is asking. Most organizations point their DNS servers to the Internet's root DNS servers that can resolve names for the entire Internet space.

The fictitious address for the pet store, **http://www.petshop.com/newpets/blacklabs.html,** breaks down in the following manner:

- http:// (or sometimes possibly ftp://) represents the protocol resource to be used when communicating

- www represents the pointer to a computer or a resource on the World Wide Web

- petshop.com represents the domain name of the desired resource
- The remainder of the address, /newpets/blacklabs.html, represents the hierarchical name or pointer to a specific file (in this case blacklabs.html) located on the pet store server.

The URL could also point to CGI programs, Java or Flash programs, graphic files, streaming media, or any other available Internet resources.

Firewall

Often, private organizations need to be connected to the Internet, but they do not want to have to worry about outside networks or users accessing confidential files or restricted parts of their internal network, or Intranet. The Intranet represents an internal network of a private company or organization that can only be accessed by individuals internal to the organization or by those having recognizable user names, log-ins, and passwords. This is usually accomplished through a special type of software referred to as a firewall. A firewall will only provide an individual access to a private Intranet as long as they know the desired authentification codes. Firewalls are also necessary to help prevent computer hackers from gaining access and control of a network or a computer for malicious or destructive purposes, such as the stealing of valuable information or planting of viruses. The firewall creates a two-way controlled environment where individuals can communicate safely, inside and outside of the organization, while also preventing access to unauthorized individuals.

Other Concerns

Device Drivers

Almost all new hardware will require you to install a "device driver." A device driver is a program that tells the operating system how to control, or "drive" the device or hardware. For example, if you physically install a new CD-ROM drive in your computer, chances are it will not work once you boot up the computer. This is because you have not installed the information to tell the computer to look for this particular device. The information that is needed should be packaged along with the CD-ROM drive and needs to be installed once the drive is physically in place. The driver software will tell the computer to look for the CD-ROM drive and identify it for use. Presently most operating systems are what we call "plug and play." This means that you can pretty much plug in any device and the computer will recognize it. The latest version of Windows comes with thousands of device drivers built into the operating system software so hardware installation has become much quicker and easier.

Ping

"**Ping**" is a standard troubleshooting utility available on most network operating systems. Besides determining whether a remote computer is "alive," ping also indicates something about the speed and reliability of a network connection. Ping is like the game of ping-pong (table tennis); you hit the ball to the other side and it should be hit back. When you use ping you are actually sending a request directly to another computer simply to respond.

Operating systems

The operating system (OS) is the focus of a computing experience. It is the first software you will see when you boot up the computer, the software that will guide you through your applications, and the software that will see that the computer is safely shut down when you are ready to end your session. The operating system is referred to as a Graphical User Interface (GUI), meaning it uses pictures and icons to interact with the user. It is the software that allows programs to be launched and used along with organizing and controlling the hardware.

Of course, not all computers have operating systems. The computers that control operations in your car do not need a user interface such as an OS. The most common of all operating systems is the Windows generation of interfaces.

Operating systems can be grouped generally into four different groups based on the types of computers they are working for and the types of software application they support.

The first group is the real-time operating system (RTOS). These operating systems are used specifically in industrial settings to run controls or for instrumentation purposes. This operating system has very little interaction with the user since it is running a specific task with a specific outcome in mind. An RTOS executes an operation in a given amount of time over and over again, making it almost application-like in itself.

The second group is the single-user, single task operating system. This operating system will only allow one user to do one thing at a time. A good example of this operating system is what is usually found on a handheld computer.

Moving on, the third group is the single-user, multi-tasking operating system. This OS allows one user to do many different tasks virtually at the same time. This operating system is typically found on desktop or laptop computers where more than one process may need to be done at one time. For example, being able to print a word processing document, while surfing the Web, qualifies as a multitasking operating system.

The fourth group is the multi-user, multitasking operating system. It allows multiple users to access a computer's resources at the same time. To ensure proper operation, this operating system come with a series of checks and balances that automatically "balances" out the system so that a problem with one user will not affect the entire network of users. Examples of this style of operating system are Unix, Windows 95, Windows 98, Windows 2000, Windows XP, and Windows server 2003.

Most operating systems running network applications today require multitasking and networking capabilities.

Low Voltage Residential Network Applications

Applications for networked low voltage systems in a residential location include PCs, stereo gear, telephones, automated lighting, intercoms, surveillance cameras, and security and alarm systems. This is obviously not a complete list but illustrates many of the most commonly applied applications.

The term "structured wiring" now exists. It refers to the integration and interconnection of various low voltage, networked devices, all of which may ultimately communicate through a centrally-located household computer. By providing special (multimedia and control) outlets and terminations that consolidate telephone jacks and data jacks along with video, audio, and control cables into a single unified housing, the future management, maintenance, troubleshooting, and upgrade of such complex systems can be more easily accomplished. The inherent logic of terminating all convergent systems to a centrally-located junction box should be obvious.

Summary

- Networking involves the intercommunication between computers over copper wires or optical fiber, or by wireless means.
- Network topology involves the manner in which computers are physically connected.
- A LAN refers to the linking of users in a small geographical area.
- A WAN interconnects multiple LANs over a large geographical area through the use of modems, ISDN, cable, DSL, satellite, or wireless means.
- The link between the computer and the network is accomplished by the network interface card (NIC).
- The NIC manages the MAC address, flow of data, and bidirectional communication.
- The OSI reference model was created as a conceptual framework to organize the layers of intercommunication between computers.
- The seven layers of the OSI model include physical, data-link, network, transport, session, presentation, and application.
- Dividing the model into seven self-contained tasks helps reduce the complexity of the network.
- The types of network topology include star, ring, bus, and combinational.
- A method of preventing multiple transmissions on a network line is a process called carrier sense multiple access with collision detection (CSMA/CD).
- Collision detection monitors the line for data and determines the amount of time before retransmission can occur in the event of a collision.
- Communication across a LAN network can take place as baseband or broadband.
- Baseband allows only a single channel of transmission, whereas broadband communicates over multiple channels through a specially designed modem.
- Networks can connect through a variety of devices on the physical layer. Such devices include a repeater, hub, bridge, router, switches, LAN extenders, and gateways.
- Network software management can be designed as peer-to-peer or client/server.
- Client/server networks are inherently more organized and able to handle and manage system resources more effectively due to the centralized control of the main server.
- Network protocols determine the size of the data packet organization of the frame layer and the requirements and control of transmission.
- Examples of protocols include X.25, frame relay, ATM, and TCP/IP.
- Application and process protocols include HTTP, FTP, SMTP, and POP3.
- Such protocols provide the necessary communication link between the Internet and a Web server.
- The URL provides the necessary protocols for finding and accessing Web sites and information files residing on the Internet.
- The domain name server (DNS) provides the necessary translation between the numerical URL address and the coded word format.
- Firewalls are used to prevent unwanted intrusion into a private network.
- A device driver is a program that tells the operating system how to control a specific piece of hardware such as a CD-ROM or printer.

Key Terms

Bridge A network device that is used to connect and disconnect computer clusters operating in two separate domains. The two domains are normally not networked and are only temporarily connected through the bridge when communication is required.

Bus topology A method of computer networking that connects each mode of the network to a common line (bus). The bus represents a broadcast environment where all devices on the bus hear all transmissions simultaneously.

Client/Server A client/server network designates one node of the system as the main file server and system resource for other connecting nodes of the network.

Digital subscriber line (DSL) or cable modem A technology that uses the existing copper loop plant to provide broadband access by using frequencies above the voice band.

Domain Names Server (DNS) Converts the numbered Internet address code of a URL to a series of recognizable words.

Ethernet The first computer networks, known as Ethernets, were developed in the 1970s by Xerox, which ran at maximum speeds of 2.94 Mb/sec on coaxial cable. Xerox soon joined with Intel and Digital Equipment Corporation to develop a standard for a 10-Mb/sec Ethernet, known as DIX (Digital, Intel, Xerox). Later, the development of the 10Base-T Ethernet in 1987 communicated at speeds of 10 Mbps on a twisted pair line; such networks could reliably transmit over a 100-m run.

Fiber Distributed Data Interface (FDDI) A fiber distributed data interface (FDDI) uses optical fiber to transport high-speed data over a computer network.

File Transfer Protocols (FTP) Used to upload or download files to or from network servers.

FireWire® FireWire®, also known as iLink®, is a high-bandwidth, serial communication standard developed by IEEE (IEEE 1394).

Gateways Used to interconnect incompatible computer systems, e-mail systems, or any network devices that would otherwise be unable to communicate directly with each other.

Host Large computer used with terminals, usually a mainframe.

Hub A hub connects multiple user stations of a computer network through multiple ports. The hub does not make any decision or determination on the data and merely acts as a multiport repeater. This device fits into layer one of the OSI model.

Hypertext Transport Protocol (HTTP) The language that computers use when talking to each other for moving hypertext files across the Internet.

Integrated Services Digital Network (ISDN) Digital circuit for carrying voice, video, and data simultaneously.

Internet Protocol (IP) A routing protocol that contains the addresses of the sender and the intended network destination.

LAN extenders Remote-access, multiplayer switches that are used to filter out and forward network traffic coming from a host router.

Local-area Networks (LAN) Refers to the linking of computers and users in a small geographical area, such as an office complex, school, or college campus.

Mainframe A large computer used to store and process massive amounts of data.

Minicomputer A mid-sized computer, usually fitting within a single cabinet about the size of a refrigerator, that has less memory than a mainframe.

Modem A device that allows a computer to communicate over an analog telephone line.

Network Refers to the interconnection of computers for the purpose of intercommunication and the sharing of information.

Open System Interconnection (OSI) The reference model that represents the primary architectural model for intercomputer communications.

Operating System The OS is the focus of a computing experience. It is the first software you will see when you boot up a computer, the software that will guide you through your applications, and the software that will safely shut down the computer when you are ready to end a session. The OS is referred to as a graphic user interface (GUI), meaning it uses pictures and icons to interact with the user. It is the software that allows programs to be launched and used, together with organizing and controlling the hardware.

Parallel communication Sends multiple binary bits, all at one time, between two points in a communication system. In most cases, entire binary words can be sent simultaneously, helping to reduce the total transfer time and making parallel communications much faster than that of serial communication.

Physical addressing Defines how devices are addressed at the data link layer; it is similar to a zip code as opposed to an actual house address.

Ping A standard troubleshooting utility available on most network OSs. Besides determining whether a remote computer is "alive," ping also indicates something about the speed and reliability of a network connection.

Post Office Protocol 3 (POP3) Defines the method and protocol for retrieving e-mail from an e-mail server.

Protocol A formal set of rules governing the transfer and exchange of data through the network medium and individual layers of the OSI.

Repeater Provides the function of passing data or radio transmissions from one system or antenna cell to another.

Router A network bridge that is used to link entirely separate networks. A bridge connects clusters of computers within the same network, whereas a router is used to connect isolated networks.

Serial communication The process of sending data one bit at one time, sequentially, over a communications channel or computer bus.

Shielded Twisted-Pair Cable (STP) Character impedance of STP is typically 150 Ω. Although STP can provide more protection against noise and EMF, it is not widely used because of its incompatibility with most of the new modular connectors and terminators currently used in the industry.

Simple Mail Transport Protocol (SMTP) A network protocol that allows for the sending of e-mail and messages from a user station to an e-mail server.

Simple Network Transfer Protocol (SNTP) Exists as a means of monitoring computers, routers, switches, and network usage.

Token Ring Network (TRN) A type of network connection where each node of the network is connected in series with the next node through a connecting cable.

Topology Defines how computer networks are connected. Examples include token ring, bus, star, and tree.

Transport Control Protocols (TCP) One of the core protocols of the Internet protocol suite. Using TCP, applications on networked hosts can create *connections* to one another, over which they can exchange data or packets. The protocol guarantees reliable and in-order delivery of sender-to-receiver data. TCP also distinguishes data for multiple, concurrent applications (e.g., Web server and email server) running on the same host.

Tree topology A type of bus topology that connects nodal points of a computer network to multiple branches.

Uniform Resource Locator (URL) Provides the necessary protocols for finding and accessing Web sites or information files residing on the Internet.

Universal Serial Bus (USB) A high-bandwidth data communication standard that can support multiple devices within an entire system.

Unshielded Twisted Pair (UTP) Cable consisting of eight insulated copper conductors twisted together into four pairs without any screen or shields.

User Datagram Protocol (UDP) A connectionless network protocol that does not send back any type of acknowledgment to the originating transmitter regarding the status of data packets.

Wide area network (WAN) A computer network covering a wide geographical area, involving a vast array of computers. This is different from personal area networks (PANs), metropolitan area networks (MANs), or local area networks (LANs) that are usually limited to a room, building, or campus. The most well-known example of a WAN is the Internet.

Review Questions

1. List the seven layers of the OSI model.
2. Define the purpose of a protocol.
3. Which network topology will cause the entire network to fail if the central computer were to go down?
4. With respect to network topology, what does the term *node* mean?
5. When using Ring topology, does a central computer control the network?

6. Name one benefit to using a Bus topology when designing a computer network.
7. Which is faster, serial communication or parallel? Why?
8. What does 10base5 mean?
9. What is the purpose of using a hub in a computer network?
10. What type of device allows a large network to divide into a collection of smaller, independent networks?
11. How does a network bridge differ from a network router?
12. Describe the only drawback of using a client/server network. How does it affect the overall performance of the network?
13. Name four conditions that can cause a loss of data in a computer network.
14. Is it practical to expect data packets to arrive at their destination in the original sending order? Why or why not?
15. What are the most commonly used protocols for accessing the World Wide Web and the Internet?
16. Why is Internet Protocol (IP) considered connectionless?
17. The central computer in a network is usually called the _____.

Wireless Communications

OBJECTIVES

After studying this chapter, you should be able to:

- Develop an understanding of the evolution of the wireless industry.
- Describe the term *cellular* and discuss the advantage of this technology.
- Explain analog cellular technologies.
- Distinguish between different digital cellular technologies.
- Identify different wireless applications and standards.
- Discuss the characteristics and applications of different wireless LAN technologies.
- Assess the role of satellite communications in the wireless industry.
- Discuss the current status of international wireless communications.
- Determine appropriate wireless technologies for different applications.

OUTLINE

Introduction
Cellular Mobile Telephone Systems
Analog Versus Digital Access
Wireless Applications and Products
Wireless LANs (WLANS)
Satellite Communications
International Wireless Communications

Introduction

The **wireless** industry is growing at an exponential rate to keep pace with the rapidly increasing demand for "information at your fingertips." Wireless refers to a communications, monitoring, or control system in which electromagnetic waves carry a signal through atmospheric space rather than along a wire. The major factors that affect the design and performance of wireless networks are the characteristics of radio or electromagnetic wave propagation over the geographical area. *Propagation* refers to the various ways by which an electromagnetic wave travels from the transmitter to the receiver. The major regions of the earth's atmosphere that affect radio wave propagation are the troposphere and the ionosphere. For most communication links the troposphere, which extends up to about 15 km and includes atmospheric precipitation such as fog, raindrops, snow, and hail, makes the most difference.

Most wireless systems use radio frequency (RF) or infrared (IR) waves. RF includes any of the electromagnetic wave frequencies that lie in the range extending from 3 kHz to about 300 GHz (GHz is same as 10^9 Hz), which include the frequencies used for radio and television transmission. IR includes frequencies from 3 THz to 430 THz (THz is same as 10^{12} Hz). IR wireless products have appeal because they do not require any form of licensing by the FCC; in contrast, the FCC regulates and licenses the use of the radio spectrum in the United States. The ITU plays the same role internationally. The FCC and the ITU have played a large role in the development and deployment of wireless technologies. The wireless industry is evolving quickly and rather unpredictably as the products and services are changing. There are competing standards for communication protocols, interfaces, and networks. However, standardization efforts to assist the convergence of these essentially competing wireless access technologies are underway.

Cellular Mobile Telephone Systems

Mobile communication was mainly applied to military and public safety services until the end of World War II, but after the war it began to be applied to public telephone services. The first public mobile phone service was the Mobile Telephone System (MTS) introduced in the United Sates in 1946. In this system, operation was simplex, and call placement was handled by a manual operation. The MTS was then followed by the development of a full-duplex automatic switching system—the **Improved Mobile Telephone System (IMTS).** The IMTS was introduced in 1969 using a 450-MHz band. Although IMTS was widely introduced in the United States as a standard mobile phone system, it was not able to cope with the rapidly increasing demand for two reasons: it was a large-zone system, and its assigned bandwidth was insufficient. Therefore, a more advanced and high-capacity land mobile communication system called **Advanced Mobile Phone Service (AMPS)** was introduced in the United States in 1983 in Chicago. The most important feature of AMPS is that it employs a *cellular* concept to achieve high capacity.

Cellular Networks

A **cellular network** is any mobile communications network with a series of overlapping hexagonal cells in a honeycomb pattern, as shown in Figure 3–1. Cellular technology is a type of shortwave analog or digital transmission in which a user has a wireless connection from a mobile device to a relatively nearby base station. The **base station** consists of a transmitter, receiver, controller, and antenna system, and has a wireless link to the **Mobile Switching Center (MSC).** The base station's span of coverage is called a **cell.** Cells can be as small as an individual building (e.g., an airport or arena), or as big as 20 miles across, or any size in between. The base station controller routes the circuit-switched calls to mobile phones. The MSC contains all of the control and switching elements to

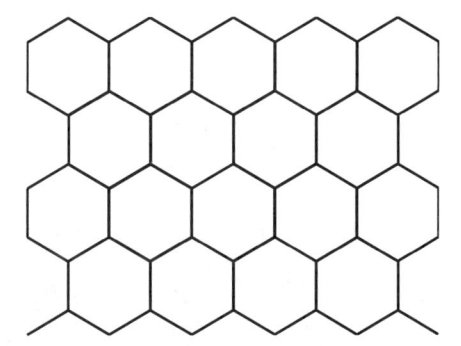

FIGURE 3–1 The cellular network is a series of overlapping hexagonal cells in a honeycomb pattern.

connect the caller to the receiver and is connected to the local exchange or central office, which is a switching center for wired telephones.

As the cellular device user moves from one cell or area of coverage to another, the MSC senses that the signal is becoming weak and automatically hands off the call to the base station in the cell into which the user is traveling. This is depicted in Figure 3–2. The purpose of this division of the geographic region into cells is to make the most out of a limited number of transmission frequencies. Cellular systems allocate a set number of frequencies to each cell. Two cells can use the same frequency for different conversations as long as the cells are not adjacent to each other. Typically, within a cellular network, every seventh cell uses the same set of channels or frequencies.

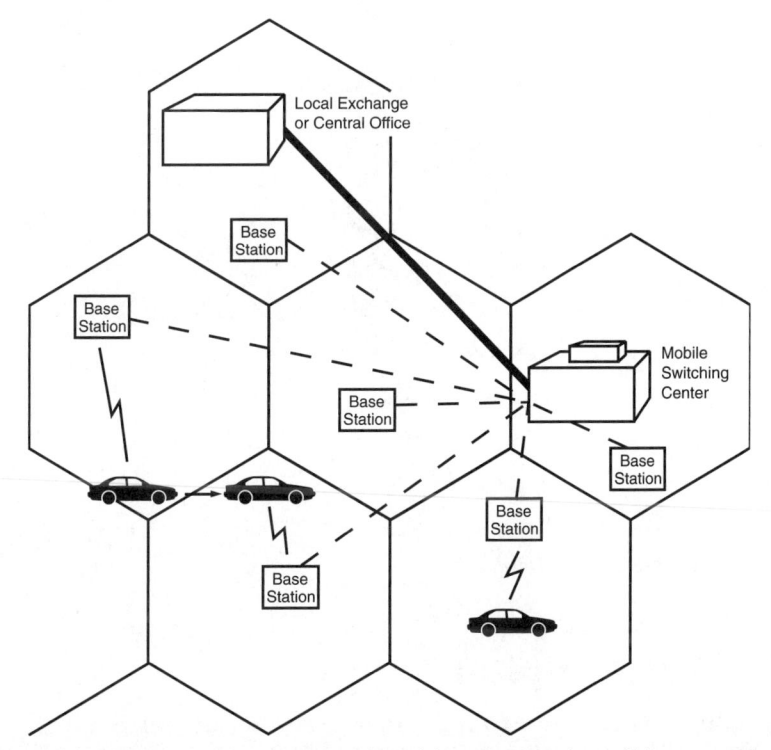

FIGURE 3–2 The Mobile Switching Center (MSC) hands over the call from one base station to another as the caller moves from one cell to the next. The MSC is also connected to the local exchange.

Personal Communications Systems (PCS)

Personal Communications Systems (PCS) has expanded the horizon of wireless communications beyond the limitations of cellular systems. Also called a *Personal Communications Network (PCN)*, its goal is to provide integrated communications (such as voice, data, and video) and near-universal access irrespective of time, location, and mobility patterns. At this time, PCS is not a single standard but a mosaic consisting of several incompatible versions coexisting rather uneasily with one another.

There are three categories of PCS: broadband, narrowband, and unlicensed. Broadband (1900 MHz) addresses both cellular and cordless handset services, while narrowband (900 MHz) focuses on enhanced paging functions. Unlicensed service is allocated from 1910 to 1930 MHz and is designed to allow unlicensed short-distance operation. PCS networks and the existing cellular networks should be regarded as complementary rather than competitive. PCS architecture resembles that of a cellular network with enhancements such as speech quality, flexibility of radio-link architecture, economics of serving high-user-density areas, and lower power consumption of the handsets.

Key features of PCS are variable cell size and hierarchical cell structure: picocell for low-power indoor applications; microcell for lower-power outdoor pedestrian applications; macrocell for high-power vehicular applications; and supermacro cells for satellites. The combined coverage of different types of cells is called hierarchical cell structure. For instance, the microcells create a second level of coverage under the existing level of macrocells, as shown in Figure 3–3. A PCS microcell has a radius of 1 to 300 meters, and a group of microcells are superimposed by one macrocell, as represented in Figure 3–4 on page 57.

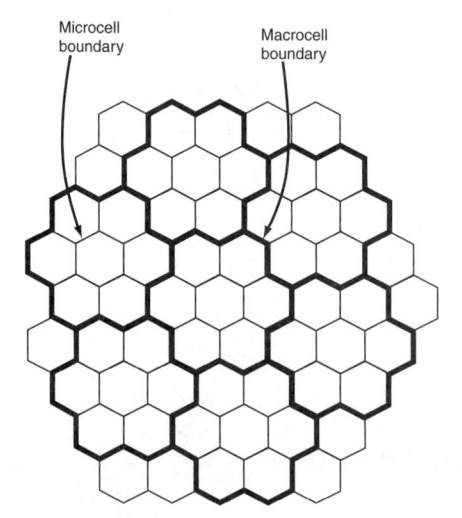

FIGURE 3–3 Hierarchical Cell Structure

Analog Versus Digital Access

Fundamentally, there are three modes of wireless access: (1) Frequency-Division Multiple Access (FDMA), (2) Time-Division Multiple Access (TDMA), and (3) Code-Division Multiple Access (CDMA).

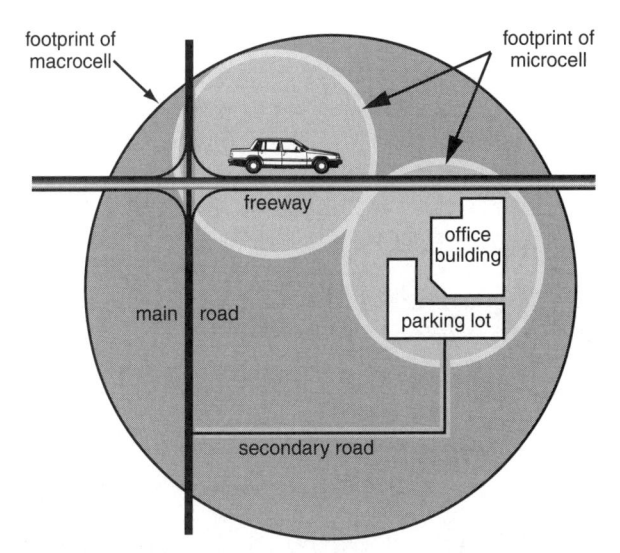

FIGURE 3–4 A PCS Environment with a Macrocell Superimposing a Group of Microcells

Until recently, most wireless data transmitted through radio communications have been analog. In analog systems, the actual sound of the caller's voice pattern is transmitted over the airwaves by means of a continuous wavelike signal. Next-generation wireless systems have turned toward the use of digital signals. Digital systems have several advantages, including better coverage, more calls per channel, broadband communications, less noise interference, and the ability to add new features and functions. Analog and digital technologies and their salient features are identified in Figure 3–5.

Analog Access

The analog cellular systems are referred to as first generation cellular technologies. The North and South American analog cellular systems conform to the AMPS

Cellular System Generation	Technology	Operating Frequency	Advantages	Disadvantages
First Generation	AMPS based on FDMA	800 MHz or 1800 MHz	✦ Widest coverage including rural areas	✦ Poor security ✦ Not optimized for data ✦ Limited capacity
Second Generation	TDMA	800 MHz or 1900 MHz	✦ Better security ✦ Higher capacity	✦ May experience an interruption during handoff
Third Generation	CDMA	800 MHz or 1900 MHz	✦ Very high security ✦ Improved capacity ✦ Greater immunity from interference ✦ Soft handoff with no interuption	✦ Limited coverage at this time

FIGURE 3–5 Cellular Technologies and Their Salient Features

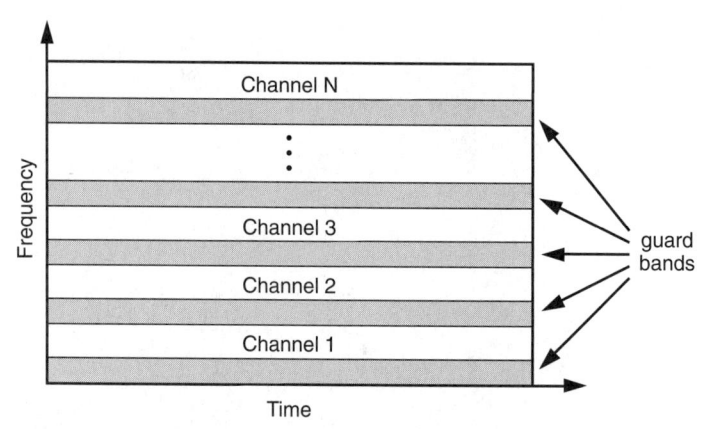

FIGURE 3–6 FDMA is the division of the frequency band into a number of channels separated by guard bands.

standard, which operates on the 800 MHz or 1800 MHz frequency band. But there are several other types of analog system standards in the rest of the world. In Europe and Asia, these include Total Access Communications System (TACS), Nordic Mobile Telephone (NMT), and others. The AMPS and TACS are both based on FDMA. **Frequency Division Multiple Access** is the division of the frequency band allocated for wireless cellular communication into a number of channels separated by guard bands, as shown in Figure 3–6. Each channel can carry only one voice conversation at a time.

By the end of 1995, the total number of subscribers to the AMPS system had reached around 30 million. The analog cellular system was inadequate to satisfy the rapidly increasing demand for cellular services. Also, analog cellular architecture has very poor or no security and is not optimized for data. Therefore, the United States, Europe, and Japan have independently developed second-generation and third-generation digital technologies.

Analog access is inherently less optimal than digital for transmitting data. However, it should be considered as a backup solution to digital technologies. The analog cellular system has the widest coverage, with service available in almost any city or town and on most major highways in the United States. About half of the world's cellular subscribers are still using a mobile telephone system based on AMPS. For this reason, analog cellular will remain the only wireless communications option in rural areas for some time to come.

Digital Access

Until now, digital wireless technologies have developed mainly because of the need for increased system capacity for voice transmission. However, this trend is now changing. In recent years, there has been a rapid growth in multimedia communications via the Internet, which has resulted in an increased demand for high-bandwidth transmissions rather than dedicated voice transmissions, even in the wireless world. Another important factor is the increasing need for global coverage.

North America currently has a multitude of digital cellular technologies for wireless radio communication. Collectively referred to as Personal Communications System (PCS), they operate at 1900 MHz, each with different coverage and capacities. By definition, analog cellular technology is not included as a PCS technology because PCS only refers to digital technologies that were designed specifically to

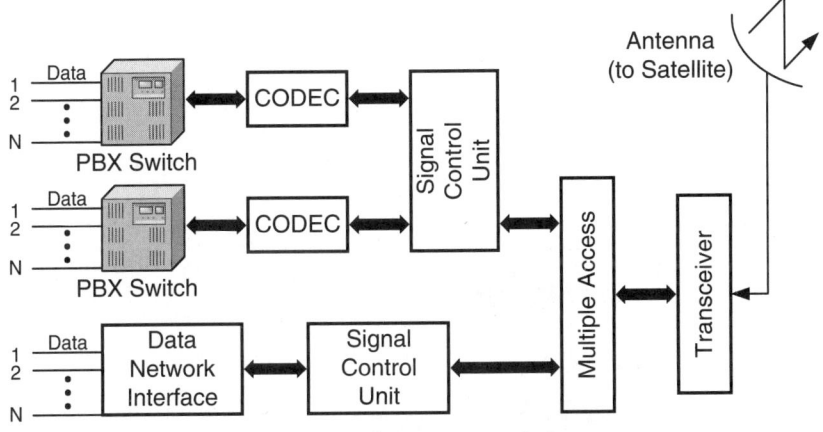

FIGURE 3–7 TDMA System for Voice and Data Transmission

provide improvements over analog. Spectrum is a scarce and limited resource, so multiple access schemes are designed to share the resource among a large number of wireless users.

Time Division Multiple Access (TDMA)

Time Division Multiple Access (TDMA)-based second-generation digital cellular technology is designed to coexist with the analog AMPS system. The first implementation of digital AMPS, often known as D-AMPS, used narrowband digital TDMA over the same frequencies as AMPS. The specifications for its operation were provided by the ANSI-136 and the TIA IS-54 standards. D-AMPS uses FDMA but adds TDMA to get three channels for each FDMA channel, tripling the number of calls that can be handled on a channel. In addition to voice conversations, it can carry digital data. Figure 3–7 represents a TDMA system for voice and data transmission. This requires digitizing voice, compressing it, and transmitting it in a regular series of bursts that are interspersed with other users' conversations.

TDMA enables users to access the assigned bandwidth on a time basis. D-AMPS divides the available radio spectrum into narrow channels, each of which carries one or more conversations using TDMA. It is designed to allocate three timeslots per 30-kHz channel, yielding a 3-to-1 capacity increase over AMPS. Each user is assigned his or her own time slot on a particular carrier frequency that cannot be accessed by any other subscriber until the original call is finished or handed off. Within each channel, conversations are separated by slight differences in time. Thus, each user occupies the whole channel bandwidth, but only for a fraction of the time, called a *slot*, on a periodic basis. TDMA thereby makes more efficient use of available bandwidth than the previous-generation analog technology, in addition to being optimized for both voice and data. The D-AMPS is much more secure than AMPS since encryption is built into the system. It would require sophisticated hackers to modify the scanner for different digitization techniques.

TDMA was first investigated for satellite communication systems in the late 1960s. When the TDMA mode of access is utilized, a single carrier is used by all earth stations in a time-sharing mode of operation, as shown in Figure 3–8.

When an earth station transmits a burst of information at a preassigned time to a satellite, the entire bandwidth can be utilized by that earth station. The satellite receives the transmitted burst, and amplifies, down-converts, and retransmits it back to earth. The next burst, perhaps originating from another earth station,

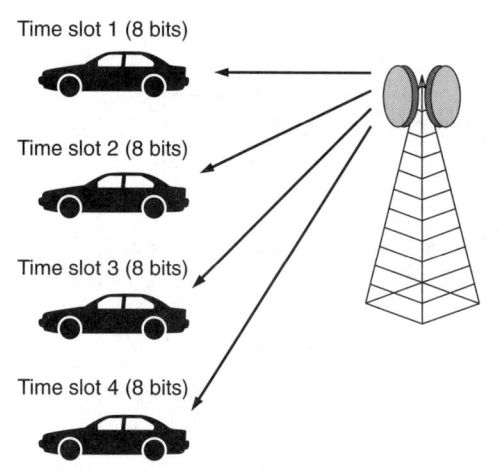

FIGURE 3–8 Time Division Multiple Access (TDMA) Applications for Satellite Communications

follows exactly the same pattern. The fact that bursts are transmitted in sequence separated by narrow guard bands necessitates optimum synchronization techniques so that burst overlap can be avoided.

The first commercial TDMA satellite system was operated in Canada in 1976. Investigation into the application of TDMA to land mobile communication systems started around the 1980s. At this time, unfortunately, TDMA was not considered to be suitable for land mobile communications because of its complicated timing controls. However, in 1982, the Conference of European Posts and Telecommunications (CEPT) started to make specifications for the pan-European digital cellular system, in which TDMA was included as an access scheme. Finally, TDMA was selected as the access scheme for the Global System for Mobile Communications (GSM) system in 1988.

As a result of the extensive studies on TDMA, this technology is now a very popular access scheme for land mobile communication systems. It is applied to many second-generation cellular systems and cordless phone systems all over the world. The TIA published the revised TDMA standard IS-136 in 1994. At the core of the IS-136 specification is the DCCH (Digital Control Channel), which provides a platform for PCS, introducing new functionalities and supporting enhanced features that make PCS a powerful digital system.

Like IS-54, IS-136 co-exists with analog channels on the same network. One advantage of this dual-mode technology is that users can benefit from the broad coverage of established analog networks and at the same time take advantage of the more advanced technology of IS-136 where it exists. TDMA IS-136 is available in North America at both the 800 MHz and 1900 MHz bands.

Code Division Multiple Access (CDMA)

Code Division Multiple Access (CDMA), in layman's terms, is like a room full of people talking in different languages at the same time. However, you can pick out the conversation in the language you understand and ignore all the others. CDMA is a method of sharing a single frequency among users by encrypting each signal with a different code. As a result, it supports many callers along the same carrier. Transmission signals are broken up into coded *packets* of information that hop available frequencies and are reassembled at the receiving end. Each earth station transmits coded information to the satellite, regardless of any overlap with other stations that may be transmitting simultaneously. At the receiver end, the separation of the transmitted information by each station is achieved through the detection of the individual earth station's transmitted identification code.

Like TDMA, CDMA operates in the 1900-MHz band as well as the 800-MHz band. CDMA has three times the capacity of TDMA and ten times that of FDMA. In 1992, the TIA published the first CDMA standard, IS-95. The third-generation CDMA systems provide both operators and subscribers with significant advantages over first (analog) and second (TDMA-based) generation systems. The advanced methods utilized in CDMA technology improve capacity, coverage, voice quality, and immunity from interference by other signals introducing a new generation of wireless networks.

Compared with conventional systems, the CDMA system makes frequency assignment easy because the same frequency can be used for each cell. Moreover, this system allows high-quality communications, since no frequency switching is required when moving from one cell to another. CDMA is the first technology to use a technique called **soft handoff,** which allows a handset to communicate with multiple base stations simultaneously. The system chooses the best signal in order to provide the user with the best audio at all times. In contrast to CDMA users, TDMA users can experience an interruption in the audio when the signal is handed off from one base station to another, resulting in higher interference during handoff and increased dropped calls.

CDMA employs a technique originally created by the military called a **spread spectrum,** which is significantly different from AMPS and TDMA technologies. Rather than dividing RF spectrum into separate user channels by frequency slices as in FDMA or time slots as in TDMA, spread-spectrum technology separates users by assigning them digital codes within a broad range of the radio frequency, as shown in Figure 3–9.

The system chops up the signal into data packets and spreads it over a wide band of frequencies. The data packets are then reassembled at the receiving end. This makes it very difficult for hackers to grab orderly, meaningful data, and ensures secure communications. In addition, the signal appears like low-level noise to conventional radio receivers. Since the receivers are designed to eliminate noise, the signal goes undetected. Also, this technology is ideal for densely populated areas because it prevents signal interference among mobile units such as cellular phones and portable computers. The spread-spectrum techniques can be divided into two families: Frequency Hopping Spread Spectrum (FHSS) and Direct Sequence Spread Spectrum (DSSS).

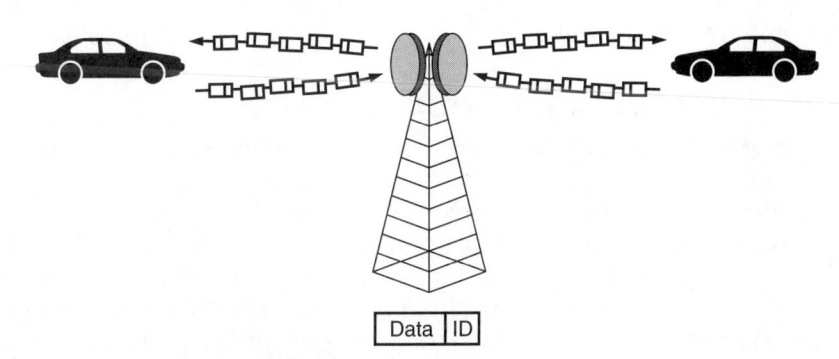

FIGURE 3–9 Code Division Multiple Access (CDMA)

Frequency-Hopping Spread Spectrum (FHSS)

Frequency-Hopping Spread Spectrum (FHSS) resists interference by jumping rapidly from frequency to frequency in a pseudo-random way. The receiving system has the same pseudo-random algorithm as the sender and jumps simultaneously. In the frequency-hopping pattern, a very long sequence code is used before the sequence is repeated, over 65,000 hops, making it appear random. Thus, it is very difficult to predict the next frequency at which such a system will transmit/receive data. The system appears to be a noise source to an unauthorized listener, which makes FHSS very secure against interference and interception. Another advantage of FHSS systems is that multiple hopping sequences can typically be assigned within the same physical area, thereby increasing the total amount of available bandwidth. In large facilities, especially those with multiple floors, it is necessary to space antennas, or access points, in an overlapping array. This ensures adequate coverage for wireless devices that are moving from cell to cell. With FHSS, system users can roam between access points on different channels. In this respect, this makes FHSS technology more flexible than DSSS.

Direct-Sequence Spread Spectrum (DSSS)

Direct-Sequence Spread Spectrum (DSSS) resists interference by mixing in a series of pseudo-random bits with the actual data. The receiver, using the same pseudo-random algorithms, strips out the extra bits. A redundant bit pattern (also called a chip) is produced for each bit of data to be transmitted. If one or more bits are damaged in transmission, the original data can be recovered as opposed to having to be retransmitted. This built-in safeguard significantly increases the efficiency of the data transmission process. DSSS can be used as a substitute for leased lines or fiber-optic cables to bridge LAN segments in point-to-point or multipoint connectivity between buildings. In DSSS, roaming can only be done between access points on the same channel, which makes the roaming capabilities of DSSS less robust than those of FHSS.

DSSS systems can transmit at bandwidths of about 11 Mbps for point-to-point connections up to 25 miles. In multipoint applications where the signal has to be broadcast over a wider arc, the distance the signal can actually travel may be much less than 25 miles, depending on the number of sites and the distances between them. The selection of microwave antennas at the distribution and receiving points therefore becomes crucial. If the network components are not properly balanced, locations close to the antenna could end up monopolizing bandwidth, while locations farther away could suffer performance degradation. Another issue of concern with DSSS is signal loss, which can be alleviated by devices such as high-gain antennas and amplifiers.

Cellular Digital Packet Data (CDPD)

Cellular Digital Packet Data (CDPD) has been combined with spread-spectrum radio transmission in several wireless communications products. CDPD allows a packet of information to be transmitted in between voice telephone calls. Even if all the subscribers in a particular area are using their cellular telephones, 30 percent of the frequency is still available for data transfer. For signal transmission, the system starts channel hopping. Anytime it finds an opening, it transmits the information in packets. This approach not only spreads the packets in a predetermined manner, but also causes them to hop frequencies.

CDPD enables data specific technology to be tacked onto the existing cellular telephone infrastructure. It supports wireless access to the Internet and other public packet-switched networks. CDPD is an open specification that adheres to the layered structure of the OSI model and, therefore, has the ability to be extended in the future. CDPD supports the Internet's IP protocols for broadcasting and multicasting, including IPv6, and the ISO Connectionless Network Protocol (CLNP).

For the mobile user, CDPD's support for packet-switching means that a persistent link is no longer a requirement. The same broadcast channel can be shared among a number of users at the same time. The user's modem recognizes the packets intended for its user. As data such as e-mail arrives, it is forwarded immediately to the user without a circuit connection having to be established. There is a circuit-switched version, called CS CDPD, which can be used where traffic is expected to be heavy enough to warrant a dedicated connection. Cellular telephone and modem providers that offer CDPD support make it possible for mobile users to get access to the Internet.

Wireless Applications and Products

There are two distinct, yet complementary, system approaches for addressing the demand for mobility: an enhanced cellular network, and a noncentralized wireless LAN. The two system approaches are complementary because the wide-area cellular network permits high mobility and globalizes communications through handoff and roaming, while wireless LANs can offer orders of magnitude higher data rates through coverage area restriction, which results in substantially reduced signal attenuation and multipath delay spread. Both system approaches have been the focus of extensive research and standards activities.

The features that are the most different in regard to the cellular products, cordless devices, and wireless LANs are terminal mobility, coverage, and bandwidth. Basically, mobility of the cordless phone and wireless LAN is very low compared with the cellular phone. Therefore, these systems do not require any location restriction or handoff functions, which means their system cost is much lower than that for cellular systems. Another feature difference among these three systems is the signal bandwidth. In the case of cellular phone or cordless phone systems, the signal bandwidth is very narrow—approximately 20 to 30 kHz—because they are used only for voice transmission. On the other hand, a desirable signal bandwidth for a wireless LAN system may be 10 MHz. A minimum bit rate of 1 Mbps is required to satisfy intersystem matching between wireless and wired LANs. Regarding Internet access, most vendors are moving toward unifying voice mail and data communications. There are several companies that provide a truly wireless solution by escaping the phone infrastructure.

Wireless Application Protocol (WAP) is a standard for wireless data delivery, loading, and navigation. Before WAP was developed, loading standard HTML pages on mobile phones and other handheld devices could be a problem because HTML pages are designed to be viewed on a computer monitor, not on a 1.5-inch LCD. WAP applications are designed to be viewed on tiny screens found on average mobile phones, but the micro-browser also allows for expanding the viewable area of an application if the application is being accessed from a machine with a larger screen, such as a Personal Digital Assistant (PDA). More important, WAP micro-browsers are configured to use less memory and CPU power, thus extending the battery life of a mobile device. They also allow for easier resumption of an

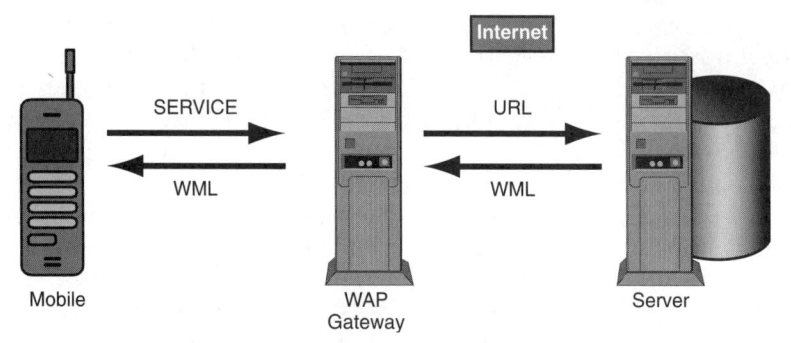

FIGURE 3–10 A Mobile Device Accessing the Internet Through a WAP Gateway

Internet session if a mobile device loses its signal. According to the WAP Forum, a standards-setting body, the WAP protocol eats up a lot less bandwidth than do standard HTTP/TCP/IP pages. An example of a mobile device accessing the Internet through WAP is shown in Figure 3–10.

WAP removes much of the overhead associated with IP, including large message headers and long session startup times. WAP solutions contain a client-side browser for the mobile device and a transmission gateway that uses WAP-over-IP to send and receive data and to gain access to the Internet or corporate Intranets. The browser reaches Web sites using the Wireless Markup Language (WML), which excludes large files such as Java scripts and picture files and sends only plain text information in response to a query. Thus, the information is suited for small display, keyboardless, low-restoration devices used to access information while on the road.

Bluetooth®

Bluetooth® is a uniting technology that allows any sort of electronic equipment from computers and cell phones to keyboards and headphones to make its own connections, without any wires, cables, or user intervention. A Bluetooth module block diagram is shown in Figure 3–11; the module is built on a 9 × 9 mm microchip incorporated into the mobile device or other electronic devices.

The idea originated in 1994 when mobile phone manufacturer Ericsson began working on a short-distance radio system. With the intent of achieving an open solution, Ericsson formed the Bluetooth Consortium with technology providers IBM, Intel, Nokia, and Toshiba. Each member of the consortium brought a different expertise to the solution, including radio technology and/or software. Bluetooth operates in the unlicensed 2.4-GHz ISM bands, an open frequency band in most countries, ensuring communication compatibility worldwide, although local regulations in Japan, France, and Spain reduce the bandwidth. Bluetooth is intended to be an open standard that works at the two lower layers of the OSI Model, the physical and data-link layers. The standard provides agreement on when bits are sent, how many will be sent at a time, and how the devices in a conversation can be sure that the message received is the same as the message sent.

Each Bluetooth device has a unique 48-bit address and uses built-in authentication and encryption at the device level and not the user level. When a manufacturer wants to make a product based on Bluetooth wireless technology, it contacts the IEEE registration authority and requests a single or a specific number of 24-bit organizationally unique identifiers (OUIs). The manufacturer then uniquely assigns the other 24 bits to make up the 48-bit Bluetooth device address.

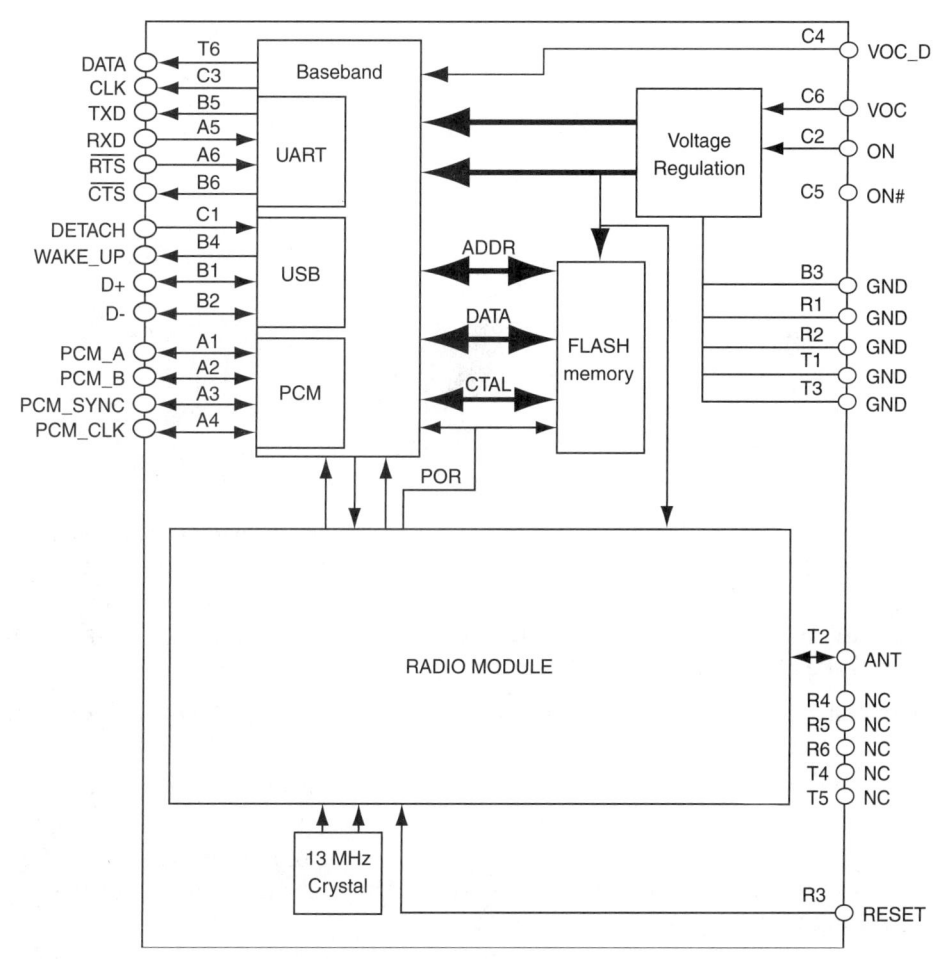

FIGURE 3–11 Bluetooth Module Block Diagram

When Bluetooth connects devices to each other, they become paired; one unit acts as a master for synchronization purposes, and other units act as slaves for the duration of the connection, forming a piconet, represented in Figure 3–12. Unlike many other wireless standards, Bluetooth includes application layer definitions for product developers to support data-, voice-, and content-centric applications.

A number of devices such as baby monitors, garagedoor openers, microwave ovens, and cordless phones make use of frequencies in the ISM band. Making sure that Bluetooth and these other devices do not interfere with one another has been a crucial part of the design process. Bluetooth radios use a spread-spectrum, frequency-hopping, full-duplex signal that changes frequencies 1,600 times every second. The signal hops among 79 individual, randomly chosen frequencies at 1-MHz intervals to give a high degree of interference immunity. Bluetooth radios consume very little power—between 30 to 100 mW while transmitting, and between 0.01 to 0.1 mW when not transmitting. The low power limits the range of a Bluetooth device to about 10 meters. But even with the low power, ceilings and walls do not stop the signal, making the standard useful for controlling several devices in different rooms.

Bluetooth wireless technology is supported by product and application development in a wide range of market segments. The specification allows up to seven

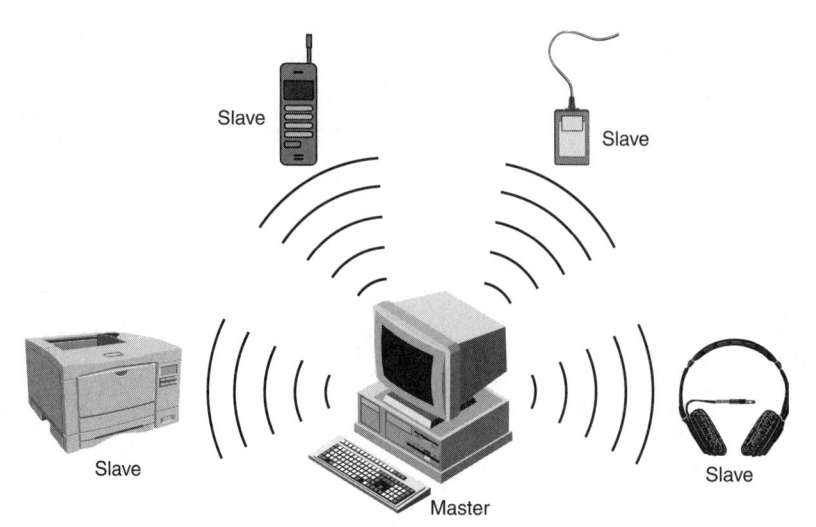

FIGURE 3–12 Bluetooth Piconet with One Master and Multiple Slaves (maximum seven slaves on one master)

simultaneous connections to be established and maintained by a single radio. Bluetooth has three generic applications:

- Personal area networks, where two or more Bluetooth products can communicate directly
- Local area networks (LANs), where products communicate with a company's broader network via a Bluetooth LAN access point
- Wide area networks (WANs), where a product with Bluetooth-enabled technology can communicate with a wireless WAN device to allow global connectivity, including Internet access.

The Bluetooth specification, a continuing process, allows member companies to enhance and extend the technology into new usage models and markets. However, Bluetooth on its own is not a complete solution. A mobile worker wishes to have all the information synchronized between the mobile devices and his networked applications. A common synchronization protocol is also needed to ensure that meaningful connectivity is not limited to devices from only one manufacturer. SyncML is an initiative to develop and promote an open data synchronization protocol. SyncML is an extensible and transport-independent technology that allows a device to support a single synchronization standard for both local synchronization over Infrared, Bluetooth, and USB, as well as remote synchronization over Internet and WAP.

The rapidly evolving wireless technology is resulting in two major benefits for the users: lower prices and better functionality. However, there are two major shortcomings of wireless products: speed and limited coverage for certain services. Vendors cite two measures of speed. *Data Rate* is raw radio transmission speed, and it is heavily affected by *overheads,* which consist of the transmission rules and protocols that govern a network. As shown in Example 3–1, by subtracting overhead from data rate, one arrives at the user rate or *throughput,* which is the capacity available to the user. Bandwidth is critical in maintaining high throughput for each user, especially when there are a large number of users. Mostly, wireless data rates are painfully slow when compared with wired options.

EXAMPLE 3–1

Overhead = Data Rate – Throughput

Problem

A user realizes 8 Mbps throughput on a 10-Mbps network. Find the overhead.

Solution

Overhead = Data Rate – Throughput
= (10 – 8) Mbps
= 2 Mbps

Wireless LANs (WLANs)

Wireless LANs (WLANs), which are increasingly being used by companies and organizations, are flexible communication systems implemented as extensions to or as alternatives for the wired LANs in buildings or on campuses. In the past, established vertical applications such as those in warehousing and retail enterprises have dominated the demand for WLANs. The real-time aspects of wireless networks improved the efficiency and productivity of factories, distribution centers, warehouses, and transportation hubs. More and more companies implemented radio-frequency data communications technology to provide real-time, on-line inventory and process control.

In recent years, however, the most dramatic demand for wireless applications has been driven by mobile professionals. Through the use of WLANs, company owners, managers, and other executives who spend at least 20 percent of their time away from their desk while at the office are able to maintain constant network connection to key business applications irrespective of where they roam. WLANs have evolved into a secure, high-performance, decision-support tool.

A WLAN transmits and receives data over the air, minimizing the need for wired connections, as depicted in Figure 3–13. It is recommended highly for

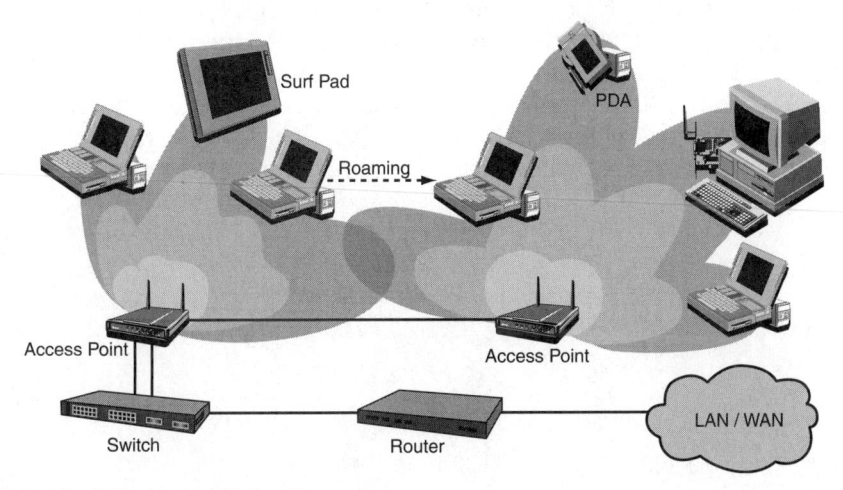

FIGURE 3–13 Wireless LAN Configuration

hard-to-wire sites. A wireless bridge works well for multisite organizations to connect branch offices to corporate LANs in a local geographic area, as well as to connect LANs within a college or corporate campus. A WLAN adapter is a *Personal Computer Memory Card Industry Association* (*PCMCIA*) card (commonly called PC card) for a laptop or notebook computer. It combines data connectivity with user mobility and, through simplified configuration, enables movable LANs.

WLANs solve several problems like cabling restrictions, frequent reorganizations, and networking of highly mobile employees. While wireless connections are currently more expensive than most types of physical cabling, often their use may be cost-effective. However, it is true that most existing wireless services are less functional and offer limited coverage when compared with their wireline counterpart. But the lifetime operational costs of WLANs can be substantially less than those of wired LANs, particularly in environments requiring frequent modification.

To be able to offer broadband communication services and provide universal connectivity to mobile users, it is necessary to have a suitable standard for WLANs in place. Similarly, an approach to connecting them to the existing wired LANs and broadband networks is needed. A key design requirement for the wireless LANs is that mobile hosts must be able to communicate with other mobile and wired hosts transparently. This mobility is handled at the *Medium Access Control* (*MAC*) layer. In addition, it is important that the performance available to mobile users be comparable to that available to wired hosts.

To satisfy these wireless data networking needs, study group 802.11 was formed under IEEE project 802 to recommend international standards for WLANs (Table 3–1). The mission of the study group is to develop MAC layer and physical layer standards for wireless connectivity of fixed, portable, and mobile stations within the local area. The 802.11-compliant solutions consist of access points—wireless transceivers that connect directly to a wired Ethernet LAN via a built-in Ethernet port—and wireless PC cards, which allow a mobile user equipped with a notebook to connect to an access point. The access point then admits the user to the wired LAN.

The IEEE 802.11b standard for 11-Mbps WLAN connectivity on the 2.4-GHz unlicensed radio band addresses a critical issue for many organizations: wireless access to a wired LAN. Proprietary solutions for WLANs have been available for years, but the 802.11b LAN standard provides for interoperability between different manufacturers' equipment. Corporations that need faster speed migrate to an 802.11a network, which uses the less crowded 5-GHz band, as opposed to the 2.4-GHz band for 802.11b. The 802.11a has a maximum throughput of 54 Mbps and can accommodate more users due to the increase in radio-frequency channels and increased operating bandwidth. However, because of a higher absorption rate at the 5-GHz spectrum, 802.11a devices have a shorter operating range of about 150 feet,

TABLE 3–1
Wireless LAN Standards

	IEEE 802.11a	*IEEE 802.11b*	*IEEE 802.11g*	*Bluetooth*
Frequency	5 GHz	2.4 GHz	2.4 GHz	2.45 GHz
Data Rate (max)	54 Mb/s	11 Mb/s	54 Mb/s	2 Mb/s
Range (ft)	150	300	300	33

compared with the 300 feet achievable by 802.11b. The range greatly depends on line-of-sight between the workstation and access point. Transmission speeds decrease as the distance between the portable workstation and the access point increases. The 802.11g, ratified in 2003, is the same high speed as 802.11a and is fully backward compatible with 802.11b since it too uses the 2.4 GHz band. Because the IEEE only sets specifications but does not test them, a trade group called the Wireless Ethernet Compatibility Alliance runs a certification program that guarantees interoperability. Wireless Fidelity, or Wi-Fi, is a certification issued by this group, which assures compliance among manufacturers. Ongoing work is being done on the standard to accommodate higher speeds and better performance.

WLANs typically use microwave signals, radio waves, or infrared light to transmit messages between computing devices. Subsequently, each of these technologies comes with its set of advantages and disadvantages. To select the appropriate technology, hardware, and software, one needs to consider several operational factors: data rate; number of users; interactive versus data transfer; expected response time; permissible error rate; regulatory issues; and most important of all, training, support, and maintenance of equipment and network.

Microwave LANs

Microwave technology is not really a LAN technology because it is not used to replace wired LANs within buildings. But it is classified in this category because its most important application is interconnecting LANs—bypassing T-1 circuits, providing back up against failure of the primary circuit route, connecting PBXs in a metropolitan network, and crossing obstacles such as highways and rivers. A microwave relay system is shown in Figure 3–14. Although analog and digital microwave systems are available, digital predominates in current products. A digital microwave system consists of three major components on both sides of the link: a modem, an RF unit, and an antenna. The RF unit is typically connected directly to the antenna, which is mounted on a tower or tall building to transmit and collect the microwave signals, as depicted in Figure 3–15 and Figure 3–16. They must be in line-of-sight and not more than about 30 miles apart. In installations where line-of-sight cannot be obtained, passive repeaters, which redirect the signal from one path to another, are used.

Typical microwave systems utilize signals above 30 MHz. Generally, frequencies in the range of 300 kHz to 30 MHz are not used for communications since there are many services located in that part of the spectrum. Additionally, these lower frequencies result in lower bandwidth supporting fewer channels. As a general rule, the higher the frequency, the greater the amount of bandwidth that can be carried. Therefore, the GHz bands offer the greatest value. Unfortunately, the

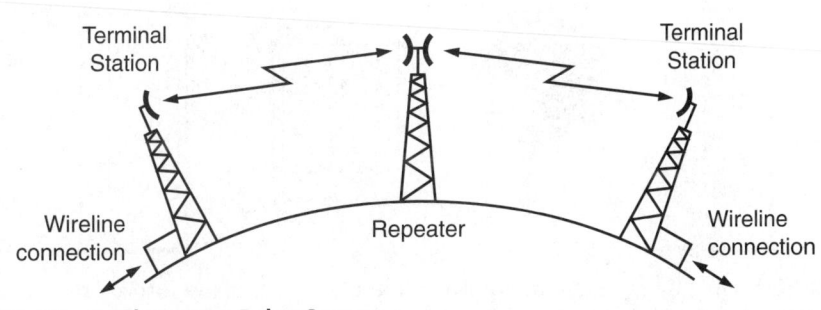

FIGURE 3–14 A Microwave Relay System

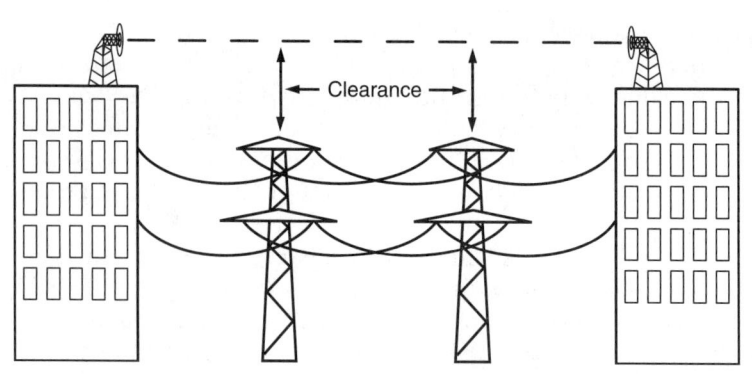

FIGURE 3–15 A microwave antenna must be mounted on a tower or tall building to keep the path above electric or telephone wires.

GHz bands that are shared by both the satellite and terrestrial microwave frequencies require line-of-sight and are affected by atmospheric conditions such as rain and humidity. One major drawback to the use of microwave technology is that the frequency band used requires licensing by the FCC. Once a license is granted for a particular location, that frequency band cannot be licensed to anyone else, for any purpose, within a 17.5-mile radius.

Radio LANs

Radio LANs basically come in two forms: narrow band or spread-spectrum. Narrow band has a cost advantage but a lower data throughput rate when compared with spread-spectrum technology. However, it can still provide response within seconds to several hundred terminals, making it a preferred choice for warehousing, distribution environments, and other industrial applications. In a narrow-band radio system, user information is transmitted and received on a specific radio frequency. Undesirable cross-talk between communications channels is avoided by carefully coordinating different users on different channel frequencies, while privacy and non-interference are accomplished through the use of separate radio frequencies. In addition, the radio receiver filters out all radio signals except the ones on its designated frequency. While wood, plaster, and glass are not serious barriers to radio LANs, brick and concrete walls can attenuate RF signals quite a

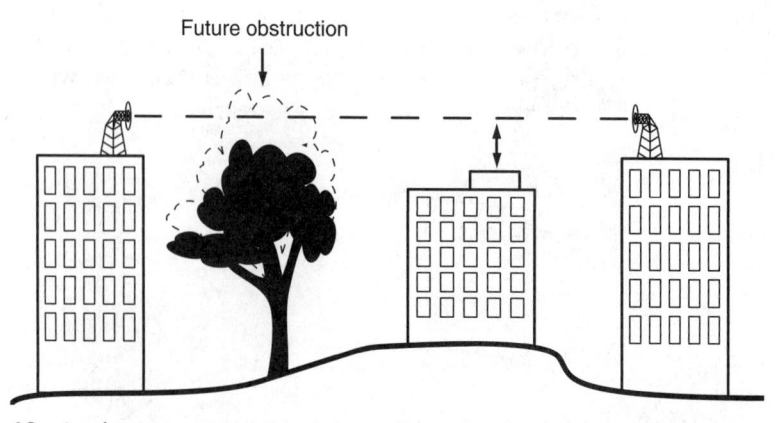

FIGURE 3–16 A microwave system requires a line-of-sight and any future obstructions must be taken into consideration before mounting antennas.

bit. Not surprisingly, the greatest obstacles to radio transmissions commonly found in office environments are metal objects such as desks, filing cabinets, elevator shafts, and even reinforced concrete.

In business applications, however, most wireless LANs use spread-spectrum technology. It is designed to trade off bandwidth efficiency for reliability, integrity, and security. Each cell has an access point or a base station that is hardwired to the media switch. An access point is a local wireless extension that bridges a cordless LAN to existing wired LANs. A worker can senselessly move or roam from one access point to another without ever losing the network connection. This technology is called roaming-enabled access points. Of the spread-spectrum technologies, FHSS is less expensive, easier to install, and uses less power, but DSSS has wider coverage and higher throughput.

Current WLANs are a combination of basically two technologies: PC card adapters and roaming-enabled access points. A credit-card-sized PC card plugs into the portable unit and lets users create their own wireless network. The PC card has a 32- or 64-bit address and data bus and provides support for the latest 3.3-volt platform in addition to the 5-volt technology. This enables the PC card products to use less power in battery-powered PCs and PDAs.

Infrared LANs

Currently, the Infrared Data Association's (IrDA) standard for wireless data communications is being adopted by leading computer and printer manufacturers. This standard, which uses high-speed infrared (IR) light for network connections, is the same basic technology used in remote control systems. Figure 3–17 provides a block diagram of an IR transmission system. IR systems consume very little power, and IrDA has thus become a natural standard for battery-operated devices. In addition, IR products have a cost advantage over spread-spectrum products and are not subject to any regulations or restrictions, worldwide. However, IR technology has not been widely used because of distance limitations. Also, unlike spread-spectrum signal, IR can only work in line-of-sight and the signal will not go through a wall.

The three types of IR systems on the market are line-of-sight, reflective, and scatter. *Line-of-sight* systems offer point-to-point, high-speed connectivity between stations located within 100 feet. They take advantage of the built-in infrared port on the computer. Increasingly, notebook and laptop computers and other devices (such as printers) come with IrDA ports that support roughly the same transmission rates as traditional parallel ports. The only restriction on their use is that there must be a clear line-of-sight between them. Any interruption such as an object or a person walking through will disrupt the signal. Usually, the two devices are within a few feet of each other. Reflective infrared systems are getting around the line-of-sight problem by bouncing the signal off walls, ceilings, and floors. Each workstation's transceiver is aimed at a spot on the wall or ceiling off of which

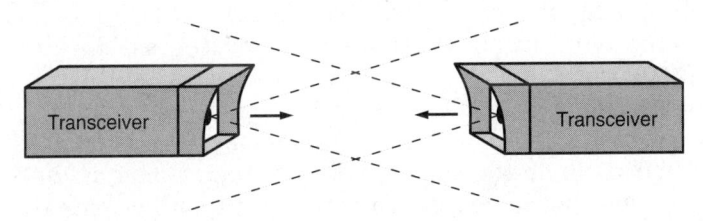

FIGURE 3–17 **Infrared Transmission System**

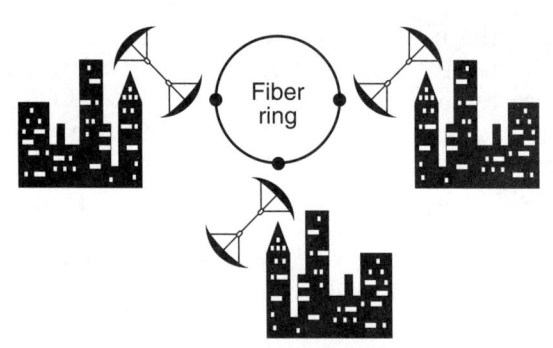

FIGURE 3–18 Wireless Local Loop (WLL) Configurations

the IR signal is reflected. *Reflective* systems are best suited for areas with high ceilings where people are unlikely to disrupt the IR signal. *Scatter* systems use diffused signals similar to the manner in which light scatters. The diffused signals bounce off walls and ceilings to cover an area of up to 100 square feet. Compared with line-of-sight, reflective and scatter are low-speed systems.

Broadband Wireless Systems

Wireless Local Loop (*WLL*) is a broadband (voice, video, and data) system that involves a low-power digital transceiver capable of supporting bidirectional communications in a small geographic area. WLL configurations include centralized antennae that support RF connectivity to matching antennae at the customer premises, as illustrated in Figure 3–18.

The WLL is being used in place of conventional wire-line connections between the local telephone exchange and customer premises. Conventionally, the local loop consists of a pair of copper wires. This is difficult to maintain and also increasingly expensive and time-consuming to deploy in view of the rising cost of copper and digging. Although WLL is essentially fixed and not mobile, many WLL systems are extended to cellular or PCS systems and enable mobile wireless communications. *Local Multipoint Distribution System* (*LMDS*) and *Multichannel Multipoint Distribution System* (*MMDS*) are two popular WLL technologies.

LMDS supports signal transmission over short distances of about two miles using line-of-sight transmissions. This option uses millimeter-wave radio at frequencies of about 30 GHz. A typical installation has a central base station with an omnidirectional antenna serving many residences, each of which has a directional dish aimed at the base station. Unlike many other microwave systems, LMDS can carry two-way signals from a single point to multiple points. Since LMDS systems send very high-frequency signals over short line-of-sight distances they are, by default, cellular. The radius of the cell ranges from 2 km to 5 km with a single hub transceiver in the center, which communicates with residences and businesses in the cell at blazingly fast data rates. Using Quaternary Phase Shift Keying (QPSK) modulation, LMDS supports data transfer rates of over 1 Gbps, which makes it desirable for high-speed data networks. With the use of low-powered residential transceivers, LMDS can be used for two-way communication of data, voice, and video signals.

LMDS cell layout has proven to be a very complex issue. The main factors that play into the cell size decision are line-of-sight, analog versus digital signals, overlapping cells versus single-transmitter cells, rainfall, transmission and receiver antenna height, and foliage density. Direct line-of-sight between the transmitter and receiver is essential. Locations without direct line-of-sight can occasionally fake it by using reflectors and, in some cases, amplifiers to bounce a strong signal

Technology	Frequency Allocation in the U.S.	Maximum Range	Characteristics
Radio LANs	902–928 MHz 2.4–2.483 GHz (most common) 5.725–5.875 GHz	25 miles	✦ Low deployment cost ✦ Low capacity ✦ Well suited for ad hoc networks
LMDS	27.5–28.35 GHz 29.1–29.25 GHz	2 miles	✦ Highest depoymet cost ✦ High capacity ✦ Can be used to serve many customers
MMDS	2.5–2.69 GHz	35 miles	✦ Lower depoymet cost then LMDS ✦ Lower capacity

FIGURE 3–19 Broadband Wireless Technologies and Their Characteristics

into shadow areas. Regarding analog vs. digital, analog signals degrade faster than digital signals because they have less tolerance for noise. Additionally, many more digital signals can be squeezed onto the same amount of RF spectrum by using advanced signal modulation techniques. Also, appropriately engineered digital signals are much more robust and less susceptible to rain and foliage. So, despite the fact that analog set-tops are less expensive than digital ones, most LMDS products in the United States are considering going digital. The LMDS technology is geared more toward dense, urban environments or multitenant buildings.

The characteristics of LMDS and MMDS are identified in Figure 3–19. A MMDS system transmits microwave signals over the 2-GHz band of the radio spectrum. Unlicensed spectrum in the 2.4-GHz band shows the most promise as the expense and difficulty of obtaining licenses necessary to provision MMDS service is eliminated. Without huge up-front license costs, there is incentive to build infrastructure even in low-population areas. The low-power nature of the unlicensed spread-spectrum band makes the equipment and installation inherently inexpensive as well. In addition, MMDS is not as range-limited as LMDS, allowing it to operate cost effectively in lower-density markets. But MMDS has lower capacity when compared with LMDS. The MMDS system was initially designed as a one-way broadcast (downlink) system for TV transmission. Various two-way systems are being deployed now, but most still offer asymmetrical bandwidth with high downlink data rates and low uplink data rates. While this system may suit a residential user, it does not completely fulfill the business user's needs.

Satellite Communications

Several attractive services can be offered by utilizing satellite technology, including PCS on a global scale, digital audio broadcasting, environmental data collection and distribution, remote sensing/earth observation, and several military applications. The International Telecommunication Satellite Organization (INTELSAT) series began with the launch of the INTELSAT-I satellite, also known as the Early Bird, in 1965. INTELSAT arose from a United Nations resolution to develop worldwide satellite communications on a nondiscriminatory basis. In the early phases of

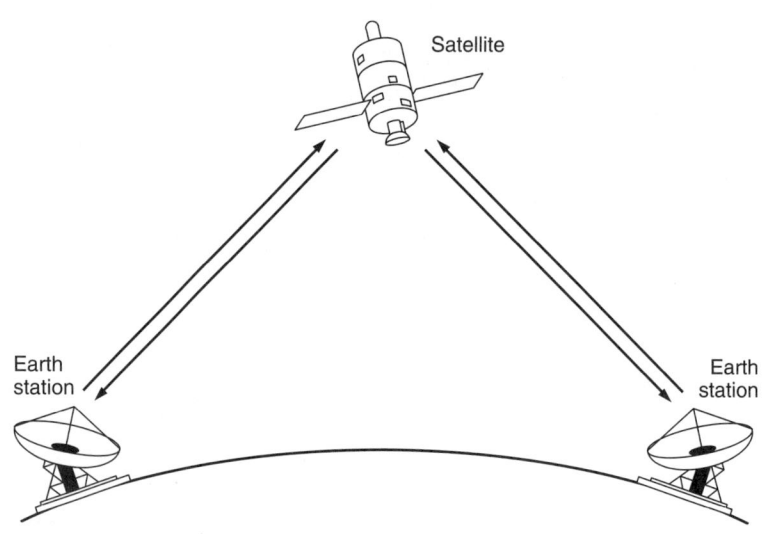

FIGURE 3–20 Single-Access Satellite System Used in INTELSAT-I

satellite communications, only two earth stations were capable of establishing communications links. INTELSAT-I system incorporated two earth antennas, four on-board antennas, two transponders, and four frequencies. The first transponder utilized the 6.30102-GHz uplink and the 4.081-GHz downlink frequencies; the second utilized the 6.3899-GHz uplink and the 4.16075-GHz downlink frequencies. Since a single station was able to access and utilize the entire transponder bandwidth, INTELSAT-I was referred to as a *single-access* system, as depicted in Figure 3–20. The ever-increasing demand for more earth stations generated the need for the design of wide-band repeaters involving *multiple-access* techniques. The concept of multiple access was first introduced in the INTELSAT-II satellite systems.

To facilitate satellite communications and eliminate interference between different systems, the ITU has divided the entire world into three regions, as shown in Figure 3–21. The same parts of the spectrum are reassigned to many nations throughout the world. The frequency spectrum allocations for satellite services are given in Figure 3–22. Each frequency band has different applications. For example, the Ku band is used for broadcasting and certain fixed satellite services. The C band

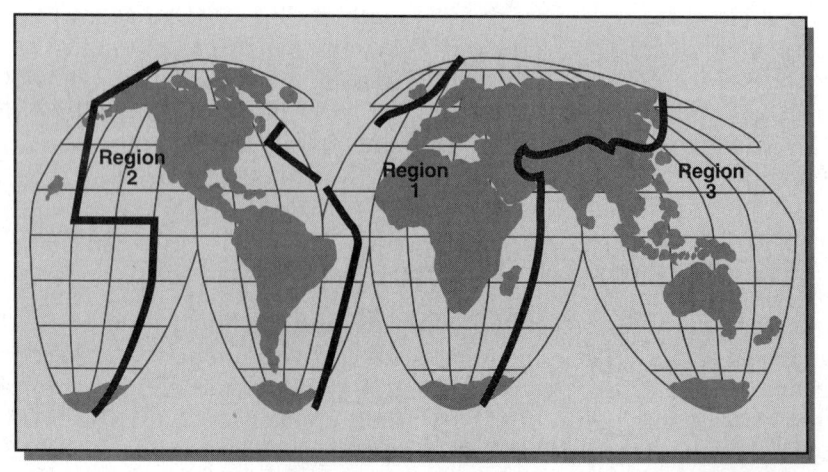

FIGURE 3–21 The ITU has divided the world into three regions for efficient satellite communications.

Frequency Band	Nominal Frequency Range (GHz)
L	1–2
S	1–4
C	4–8
X	8–12
Ku	12–18
K	18–27
Ka	27–40

FIGURE 3–22 Satellite Frequency Allocations

is exclusively for fixed satellite services. The L band is employed by mobile satellite services and navigation systems.

Each satellite band is divided into separate portions: one for earth-to-space links (the uplink) and one for space-to-earth links (the downlink). Figure 3–23 provides the general frequency assignments for uplink and downlink satellite frequencies. The uplink frequency bands are slightly higher than the corresponding downlink frequency bands in order to take advantage of the fact that it is easier to generate RF power within an earth station than it is on a satellite. The uplink transmission is highly focused to a specific satellite, while the downlink transmission is focused on a particular *footprint* or area of coverage.

A communications satellite is basically a microwave station placed in outer space. A *satellite* is a specialized wireless receiver/transmitter that is launched by a rocket and placed in orbit around the earth. There are hundreds of satellites currently in operation. They are used for such diverse purposes as weather forecasting, television broadcast, amateur radio communications, Internet communications, GPS, and the GSM. Signals are sent to the satellite by large ground station dishes and are received on the satellite by individual *transponders*. Each satellite contains about 24 transponders, which are roughly analogous to channels. These transponders then transmit the information back to earth.

The entire satellite system operates in a manner similar to that of a cellular telephone network. The main difference is that the transponders, or wireless transceivers, are in space rather than on the earth. Satellites bring affordable access to

Uplink Frequencies (GHz)	Downlink Frequencies (GHz)
5.9–6.4	3.7–4.2
7.9–8.4	7.25–7.75
14–14.5	11.7–12.2
27.5–30	17.7–20.2

FIGURE 3–23 Typical Uplink and Downlink Satellite Frequencies

interactive broadband communication to all areas of the Earth, including those areas that cannot be served economically by any other means. The satellite network combines the advantages of a circuit-switched network (low delay), and a packet-switched network (efficient handling of multi-rate and bursty data). The most feasible use of satellite connections is in locations where high-speed wire connections are not an option for geographic or financial reasons. Very Small Aperture Terminal (VSAT) networks, which are essentially fixed satellite systems, have become mainstream networking solutions for long-distance, low-density voice and data communications because they are affordable to both small and large companies due to their lower operating costs, ease of installation and maintenance, and ability to manage multiple protocols. For instance, VSAT systems are used for batch and transaction processing, including airline reservations and credit card purchases.

Satellite Earth Station

The main mission of a **satellite earth station** is to establish and maintain continuous communication links with all other earth stations in the system through the satellite repeater. It must also provide and maintain the necessary command and control links with the spacecraft. The main system components of a satellite earth station are the antenna subsystem, transmitter section, receiver section, and command and control section. Figure 3–24 illustrates a block diagram of a satellite earth station.

The principal function of the earth station transmitter is to generate the composite baseband signal by multiplexing all the incoming transponder signals (usually digital), then modulating the result and up-converting it to a carrier frequency through a frequency translation circuit. The carrier frequency is then amplified through a high-power amplifier and finally transmitted to satellite transponder via the antenna subsystem.

The receiver section of a satellite earth station performs the exact opposite function of the transmitter. The microwave beam intercepted by the receiver antenna is collected at the focal point of the secondary antenna and fed to the input of a low-noise amplifier (LNA). Both antennas must exhibit a very high gain and a narrow bandwidth. Although the antenna gain is in the order of 50 dB, the output signal from the antenna is extremely weak and very compatible with the input noise power. Therefore, the presence of an LNA is absolutely necessary.

In satellite communications systems, highly directional antennas are used to provide the much-needed antenna gain both for the ground segment and on board the spacecraft. This very high signal amplification is required in order to compensate for the substantial losses the signal suffers while propagating through space. The ability of an earth station to receive satellite signals is measured by the ratio of its receiver antenna gain (G_r) to the system noise temperature (T_{sys}). This ratio is referred to as **figure of merit,** which is given in Equation 3–1.

Standards for INTELSAT systems have set the figure of merit to be equal to or higher than 40.7 dB, as shown in Equation 3–2 on page 77. In order for a figure of merit to be equal to or better than 40.7, an excellent combination of receiver antenna gain and low system noise temperature must be achieved.

Geosynchronous Satellite (GEO)

In 1963, NASA (National Aeronautics and Space Administration) set out with its Synchronous Communications (Syncom) satellite concept. The first two Syncom satellites achieved an orbit that was geosynchronous but not geostationary. In

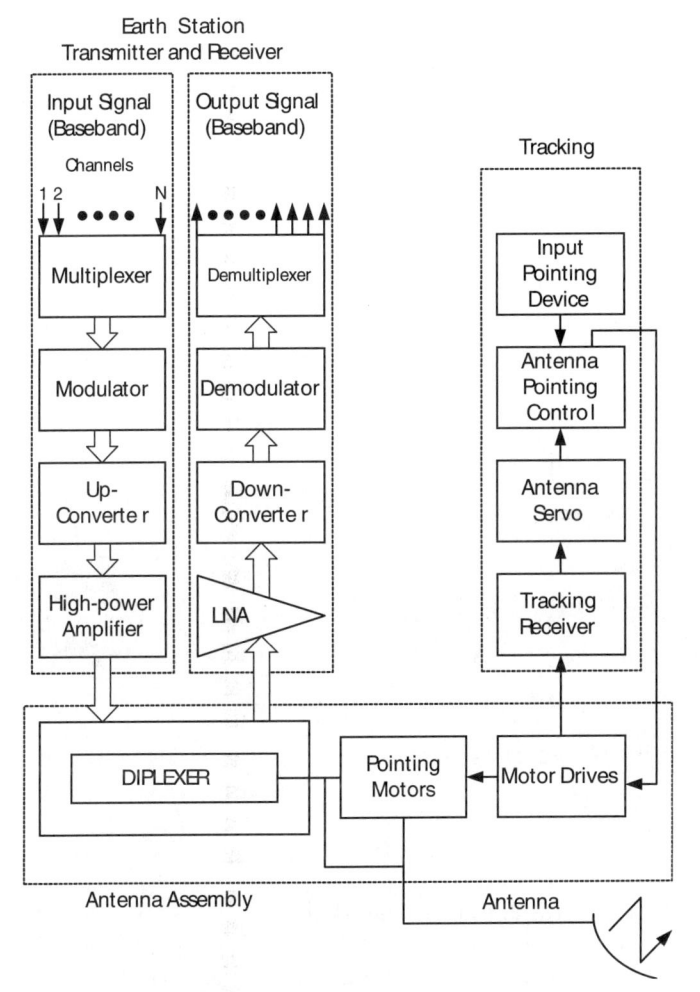

FIGURE 3–24 Satellite Earth Station Block Diagram

Figure of Merit — **EQUATION 3–1**

$$\text{Figure of Merit} = \frac{G_r}{T_{sys}}$$

where G_r = receiver antenna gain (dB)
T_{sys} = system noise temperature at the input of the LNA (dB)

For INTELSAT systems,

EQUATION 3–2

$$\frac{G_r}{T_{sys}} = 40.7 \text{ dB}$$

EXAMPLE 3–2

Problem

Determine the system noise temperature (T_{sys}) of a satellite receiver station in order to maintain a constant figure of merit equal to 40.7 dB with a receiver antenna gain of 55 dB.

Solution

In decibels,

$$\frac{G_r}{T_{sys}} = 40.7 \text{ dB}$$

$$G_r \text{ dB} - T_{sys} \text{ dB} = 40.7$$

$$-T_{sys} \text{ dB} = 40.7 - G_r \text{ dB}$$

$$T_{sys} \text{ dB} = G_r \text{ dB} - 40.7 \text{ dB}$$

Since $G_r = 55$ dB,

$$T_{sys} = 55 - 40.7 = 14.3 \text{ dB}$$

In degrees, Kelvin,

$$T_{sys} = \text{antilog } (1.43) \cong 27$$

$$T_{sys} = 27° \text{ K}$$

other words, their rotational period matched the Earth's own, but their orbits were inclined and eccentric. Syncom 3, launched in August 1964, circled the equator without inclination and successfully became the first geosynchronous or **geostationary orbit (GEO) satellite.**

GEOs must orbit the equator at an altitude of 22,237 miles, mostly using Ku-band (12 to 14 GHz) frequencies for television transmission. However, they were not used much for data transmission for a variety of reasons. First, most GEOs relied on a passive architecture. Essentially, these were electronic mirrors in the sky. They received signals from transceivers on Earth, amplified them, and sent them back down across their entire footprint. There were two major problems with GEOs. First, a GEO's footprint is very large since a single GEO can see about 40 percent of Earth's surface, as shown in Figure 3–25.

This was useful for a broadcast medium like television but was hardly optimal for data no matter how it was multiplexed. Secondly, GEOs must be spaced far apart, enough so that neither uplink nor downlink transmissions interfere with one another.

These problems were solved by the Advanced Communications Technology Satellite (ACTS) launch from the space shuttle Discovery in September 1993. It demonstrated that Ka-band (27 to 40 GHz) transmission was practical; these frequencies were previously considered unusable because of cloud cover. By unlocking a wide range of new frequencies, it made interference much less likely. The high frequencies used in the Ka-band also reduced power consumption and made smaller antennas more practical. Better still, ACTS featured antennas that could electronically divide the satellite's footprint into spot beam cells approximately

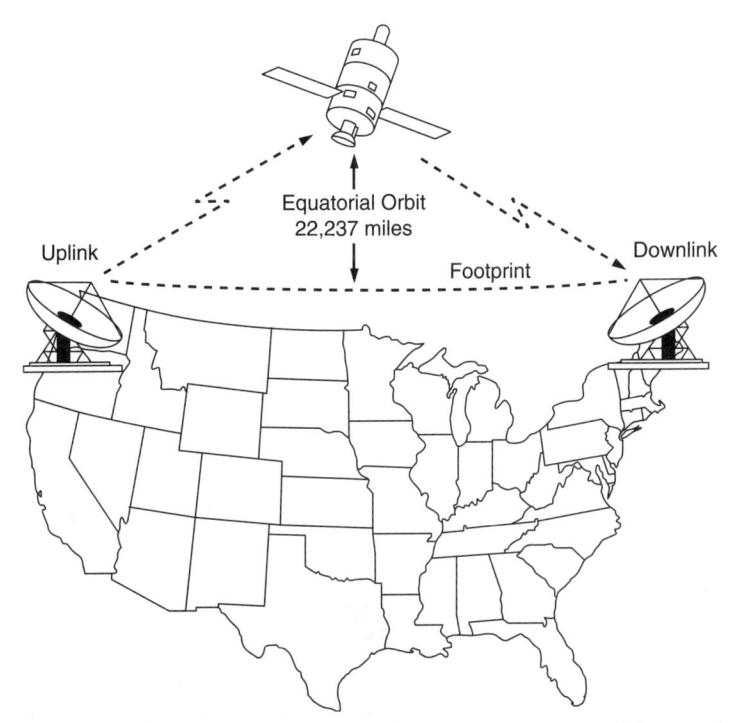

FIGURE 3–25 A GEO satellite's footprint covers about 40 percent of the Earth's surface.

150 miles in diameter. Used in conjunction with a satellite's onboard switching, spot beam technology sends traffic to and from its destination more efficiently. Its advanced antennas proved capable of transmitting data at up to 622 Mbps to 3.5-m dishes, or 45 Mbps to 60-cm dishes. ACTS and its successors brought satellites much closer to the goal of full-duplex, broadband data transmission.

However, none of the satellites could solve the problem of latency, which was a difficulty that developed as a result of the high elevation at which GEOs must orbit the equator. While electromagnetic energy travels faster in the vacuum of space than in earthbound copper or glass, it still cannot exceed the speed of light. Therefore, it takes nearly 240 ms (milliseconds) for a radio signal to make a round trip between a ground station and a GEO directly overhead. If the ground station is at the edge of the satellite's footprint, the delay might be as long as 270 ms.

GEOs are mainly used for international and regional communications. The INTELSAT family of satellites is utilized for international communications, whereas the SATCOM, ANIK, WESTAR, COMSTAR, GALAXY, and MOLNIYA systems are used for domestic service like local television distribution and regional communications. The National Oceanic and Atmospheric Administration, (NOAA) developed and operates a system of polar orbiting satellites, as well as GEO satellites, for meteorological, oceanic, and space environmental studies. Both systems are designed to perform specific tasks; for example, the GEOs monitor weather patterns as they develop in the tropics, while the polar satellites orbiting at higher altitudes are used for data collection.

Global Positioning System (GPS)

Global Positioning System (GPS), a worldwide radio-navigation system, is funded and constantly monitored by the U.S. Department of Defense (DoD). GPS is formed from a constellation of 24 satellites at 11,000 mile altitude and their ground stations.

80 • Chapter 3

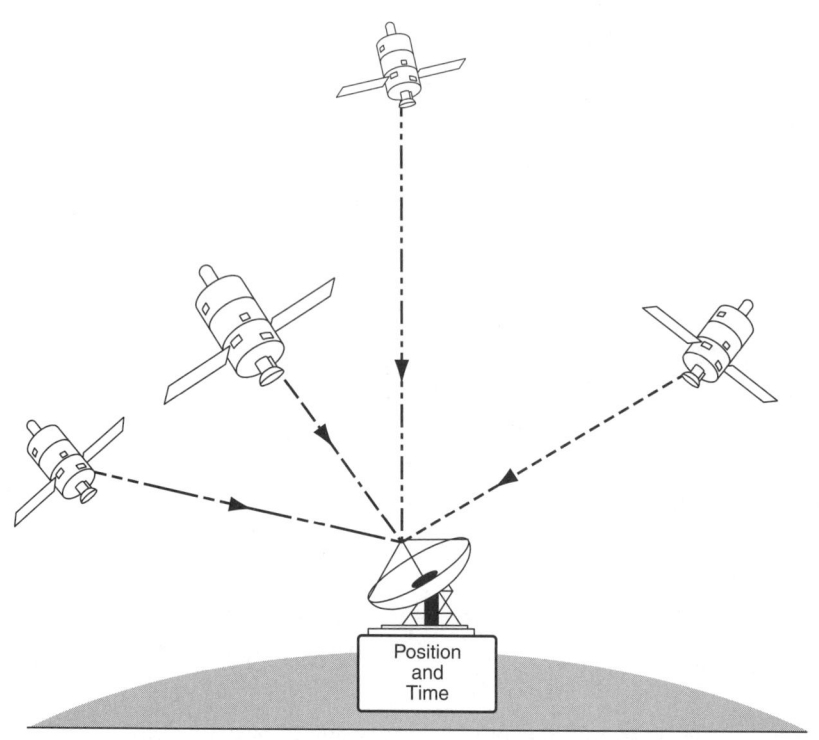

FIGURE 3–26 GPS Navigation System

The orbit altitude is such that the satellites repeat the same track and configuration over any point approximately each 24 hours (4 minutes earlier each day). There are six orbital planes equally spaced and inclined at about fifty-five degrees with respect to the equatorial plane. This constellation provides the user with between five and eight satellites visible from any point on the earth. The whole idea behind GPS is to use satellites in space as reference points for locations here on earth.

A GPS receiver measures distance by timing how long it takes for the radio signal sent from the satellite to arrive at the receiver; the signal speed is the same as the speed of light. Since Distance = Velocity × Time, accurate timing is the key to measuring distance to satellites. Satellites are accurate because they have atomic clocks on board, but GPS receivers with imperfect clocks must make four measurements simultaneously, as shown in Figure 3–26, and compute a correction factor for any delays. Each satellite has its own unique pseudo-random code, represented in Figure 3–27, which guarantees that a GPS receiver will not

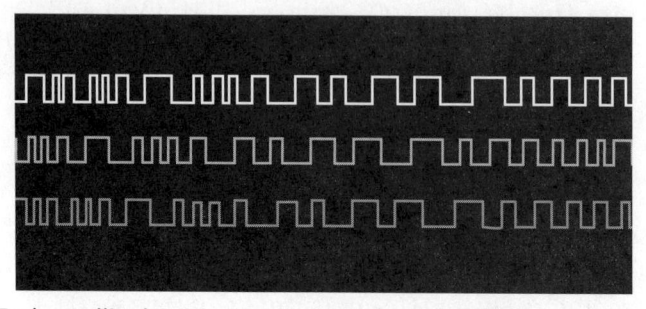

FIGURE 3–27 Each satellite has a unique pseudo-random code.

Wireless Communications • 81

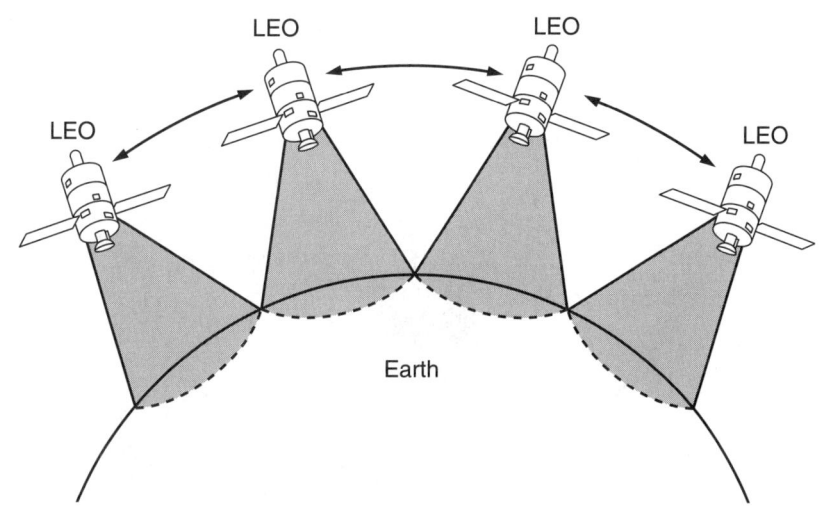

FIGURE 3–28 LEO Satellite Coverage

accidentally pick up another satellite's signal. Therefore, all satellites can use the same frequency without jamming each other. GPS receivers are becoming very economical and have found applications in cars, boats, planes, construction equipment, movie-making gear, farm machinery, and wireless devices including laptop computers.

Low Earth Orbit (LEO) and Medium Earth Orbit (MEO) Satellites

In order to overcome the delay and to complement current terrestrial cellular networks, companies that want to provide satellite-based data communications are mostly deploying low earth orbit (LEO) or medium earth orbit (MEO) satellite constellations, as depicted in Figure 3–28 and Figure 3–29. A well-designed LEO system makes it possible for anyone to access the Internet via wireless from any point on the planet using an antenna. The system consists of a large fleet of satellites, each in a circular orbit at a constant altitude of 500 to 1,000 miles, depicted

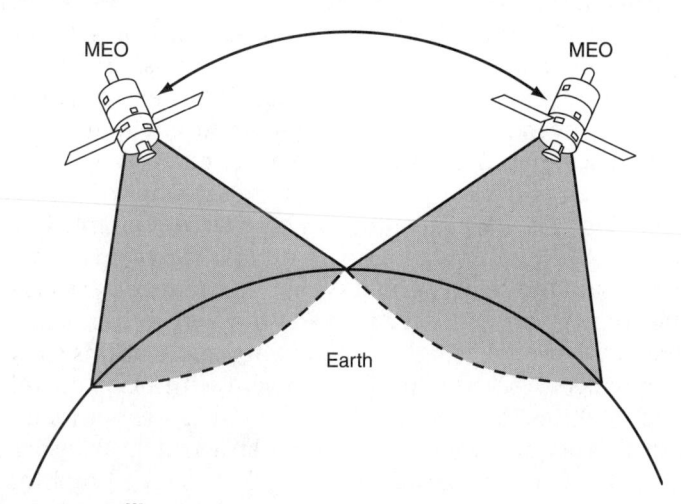

FIGURE 3–29 MEO Satellite Coverage

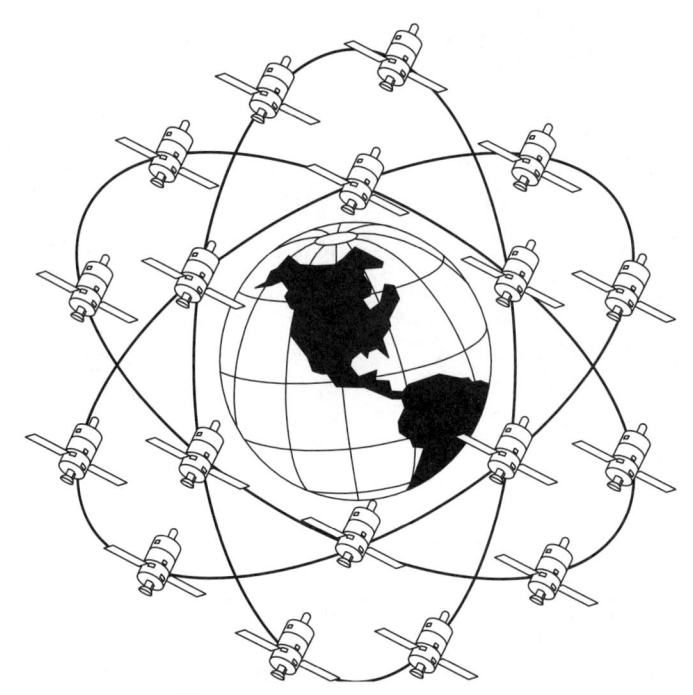

FIGURE 3–30 Low Earth Orbiting (LEO) Satellite Constellation

in Figure 3–30. This makes them capable of providing smaller more energy-efficient spot beams and delivering latency potentially equal to wired (transcontinental fiber-optic cable) communications. The closer a satellite is to Earth, the narrower its angle of view. Therefore, instead of the eight-satellite constellation that is ample when GEOs are used, at least 48 to 288 satellites are required to provide global coverage with LEOs. Each satellite takes approximately 90 minutes to a few hours to complete one revolution.

Because of their low orbit, LEOs are not geostationary. Instead, they are constantly in motion with respect to any point on the Earth. Therefore, in order to create a LEO constellation, one must provide enough satellites to circle the Earth while ensuring that at least one is (or preferably, two or three are) visible to every receiving antenna at all times. Also, the antennas must be self-aiming devices. This problem has been solved by phased-array antennas, which group together many smaller antennas. By comparing the slightly different signals received by each, they can track several satellites at once without having to move physically. As a new satellite rises over the horizon, the antenna array will establish a connection, and only then will it drop the link to the satellite that is setting.

LEO constellations also face the problem of jitter, or variable latency. The distance between each LEO satellite and any particular receiver is constantly changing. In addition, a satellite may spend less than a minute over that receiver, after which time the connection will have to be handed off to the next bird. If excessive jitter is generated, it could spell problems with applications such as IP telephony or streaming video. Another physical problem with LEO constellations is the sheer number of satellites they require. Not only must a vendor launch hundreds of satellites, but it must also keep them in working order. An example of a LEO system is the IRIDIUM communications network developed by Motorola, which is represented in Figure 3–31.

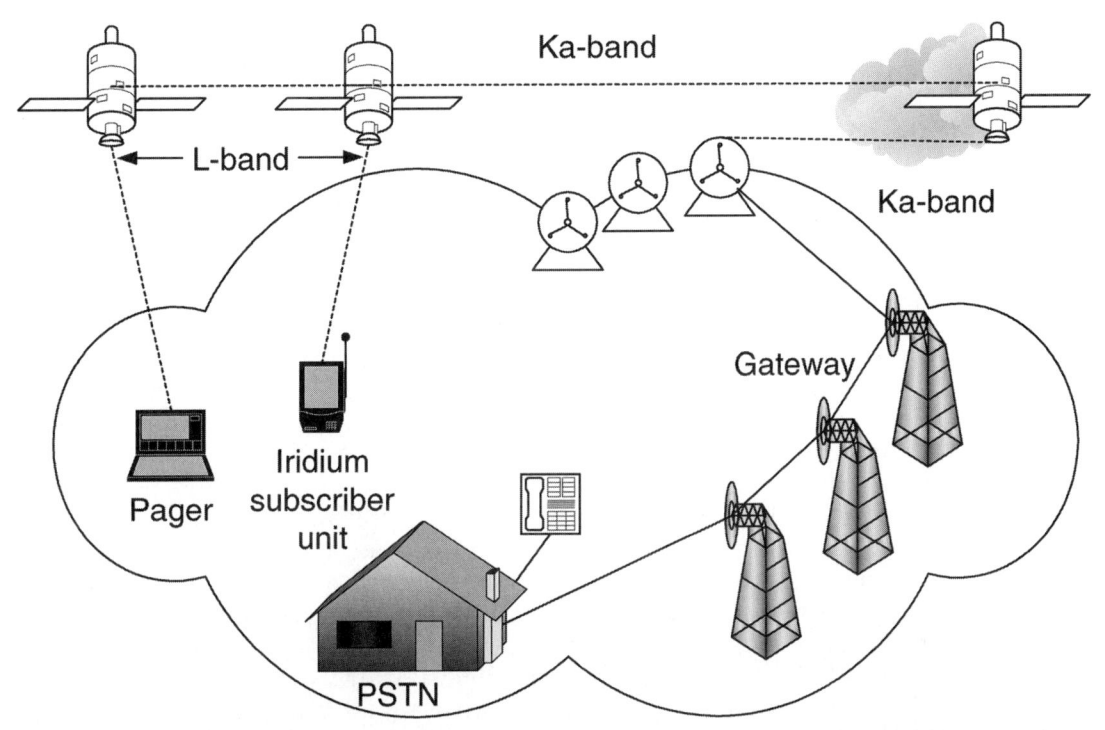

FIGURE 3–31 A Functional Diagram of the IRIDIUM Communication Network Developed by Motorola

There is a case, therefore, for launching MEO satellites, which, like LEOs, are not geostationary. Operating from an elevation between 1,800 and 6,500 miles, they take about two hours to pass over any point in their coverage area and experience a propagation delay of 0.1 s. Therefore, a constellation of some 16 satellites is sufficient to cover the globe. As satellite orbit heights from the Earth are reduced, the number of satellites needed to maintain constant communication increases.

International Wireless Communications

The scope of investment and engineering efforts required to build and maintain copper-based networks has created lofty barriers that have made high penetration rates of basic telephone service possible only in industrialized nations of the world. However, advances in technology and competitive access are driving the revolution toward wireless access infrastructure for the provision of basic telephone service, especially in developing economies.

In international wireless communication systems, analog systems could be defined as first generation and digital systems as second generation (2G). These systems, however, will not be able to provide the data rates necessary for new multimedia services. There is a growing need for next-generation systems that feature not only global services, but also a higher user-bit-rate and a transmission quality close to that of a fixed network. Thus, third-generation (3G) systems combining terrestrial and satellite systems that are expected to support global services with a high-bit-rate, high signal quality, and higher terminal mobility are already under development. Although its maximum supported bit rate has not yet been defined, it will be between 1 and 20 Mbps.

UMTS and IMT-2000

Third-generation services are designed to offer broadband access at speeds of 2 Mbps, which will allow mobile multimedia services to become possible. One of the 3G systems is the Universal Mobile Telecommunications System (UMTS) developed by the CEPT. Presently, main tasks for its standardization are being conducted by the ETSI. In 1998, ETSI decided on a single air interface solution for UMTS with the Wideband Code Division Multiple Access (W-CDMA) technology for wide-area applications and TD-CDMA technology for low-mobility indoor applications.

W-CDMA employs wider frequency bands. It is a promising technology because it is flexible, does not require frequency planning, and supports high data rates. A major weakness of W-CMDA, however, is its relative inefficiency in dealing with asymmetric traffic, which is envisioned to become more important as users move from speech (such as phone calls) to data (like Internet browsing). The other component of UMTS, mixed TD-CDMA, is, on the other hand, better equipped to deal with that issue because the ratio between uplink and downlink is not fixed in terms of resources. TD-CDMA seems to be a good complement to W-CDMA for 3G systems.

UMTS will provide wideband wireless multimedia capabilities over mobile communications networks. The technology will realize a range of new and innovative services including interactive video, Internet/Intranet, and other high-speed data communication services. UMTS is already a reality, with W-CDMA networks now operating commercially in Austria, Italy, Japan, Sweden, and the United Kingdom. The 3G licensing process is largely completed in many parts of the world, setting the stage for the worldwide deployment of UMTS systems.

An effort similar to UMTS is underway in ITU under the name of Future Public Land Mobile Telecommunication System (FPLMTS), which was recently renamed to the more catchy International Mobile Telecommunication 2000 (IMT-2000). The system will operate in the 1.885 to 2.025 GHz and 2.11 to 2.20 GHz bands. It will provide wireless access to the global telecommunication infrastructure through both satellite and terrestrial systems and serve fixed and mobile users in public and private networks. It is expected that UMTS and IMT-2000 will be compatible so as to provide global roaming. The W-CDMA format is viewed as the leading candidate to be used for the next-generation mobile communication system to meet the needs for multimedia applications.

Global System for Mobile Communications (GSM)

The 2G **Global System for Mobile Communications (GSM)** based on TDMA technology was developed in Europe. A schematic overview of the GSM system is shown in Figure 3–32. Although GSM technology has a lot of similarities to TDMA IS-136, it developed along a very different path. Unlike in the United States where the FCC moved the industry from a single analog standard to a new generation of multiple competing digital standards, in Europe the direction was reversed. Europe began with five incompatible analog air interfaces scattered around the continent. In the 1980s, momentum increased to build Europe's global influence as an economic block by integrating economically.

As part of that movement, in response to a European Commission directive, international agreements were devised to develop a single international open, non-proprietary digital cellular standard, with the most important goal being seamless roaming in all countries. New spectrum at the 900-MHz band was set aside for cellular service. In 1982, the CEPT held a meeting to begin the standardization process. During 1985, the CCITT created a list of technical recommendations

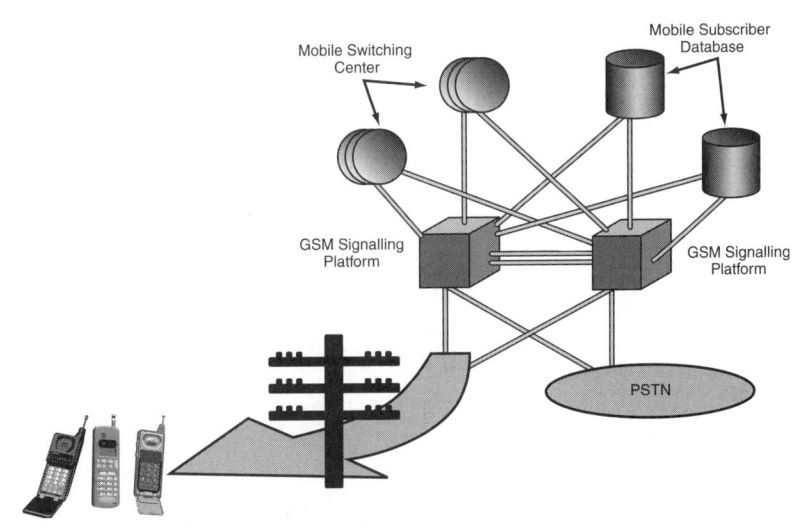

FIGURE 3–32 An Overview of the GSM System

for GSM. In 1987, all parties agreed to a compatibility specification with an air interface based on hybrid FDMA (analog) and TDMA technologies. GSM engineers decided to use wider 200-kHz channels instead of the 30-kHz channels that TDMA used; and instead of having only 3 slots like TDMA, GSM channels have 8 slots. This allows for fast bit rates and more natural-sounding voice-compression algorithms. GSM provides data services such as e-mail, fax, Internet browsing, and LAN wireless access and permits users to place a call from either North America or Europe.

In the United States, GSM specifications on the 1900-MHz band were developed starting in 1995. Commercial GSM 1900-MHz cellular systems have been operating in the United States since 1996. GSM networks offer transatlantic coverage for voice and data services such as fax, Internet access, and e-mail. Thus, GSM, which was first introduced in 1991, permits automatic roaming between North American, European, and Asian countries. As of 2003, some form of GSM digital wireless service was offered in over 193 countries and has become the de facto standard in the United States, Europe, and Asia.

CASE STUDY

University X considered all of the options for wireless networking—802.11a, 802.11b, and 802.11g—before deciding on an 802.11a solution. The 802.11a option won out because it provides greater bandwidth in two ways: more bandwidth per access point and closer grouping of access points so fewer students share a given chunk of bandwidth. 802.11a provides smaller cells with more access points in a given area, which effectively divides access, roughly by classroom, and there is less overlap. Also, 802.11a enables a greater number of channels as compared to 802.11b. According to the analysts for University X, 802.11g had one key deficit: it carries the inherent disadvantages of 802.11b with respect to the number of non-overlapping channels.

Questions

1. Why did University X select the 802.11a option?
2. Can you think of a scenario where 802.11b will be preferred over 802.11a?

Summary

- Wireless transmission includes any electromagnetic wave frequencies that lie in the range of 3 KHz to about 300 GHz; this includes the frequencies for radio and television transmission.
- Infrared transmission includes frequencies from 3 THz to about 430 THz.
- Infrared wireless does not require any form of licensing from the FCC.
- In contrast, the FCC regulates and licenses the use of the radio spectrum in the United States.
- The cellular network communicates over a series of overlapping hexagonal cells.
- Cellular technology is a type of short-wave analog or digital transmission.
- PCS cellular has expanded the horizon of wireless communication beyond the limitations of cellular system by integrating voice, data, and video.
- There are fundamentally three modes of wireless access: FDMA, TDMA, and CDMA.
- Digital systems have an advantage over analog including better coverage, more calls per channel, broadband communications, less noise and interference, and added features and functions.
- Wireless application protocol is the standard for wireless data delivery and communication over the Internet.
- Bluetooth allows short-range communication between electronic and mobile devices.
- Wireless LANs transmit and receive data over the air minimizing the need for wired connections.
- 802.11 is the preferred wireless standard for connecting wireless transceivers to a wired Ethernet of LAN.
- Communication over satellite is basically a microwave link to an earth-based station.
- The entire satellite system operates in a manner similar to a cellular telephone.
- GPS provides a worldwide navigation system formed by a constellation of 24 orbiting satellites.

Key Terms

Advanced Mobile Phone Service (AMPS) A high-capacity analog land mobile communication system that employs a cellular concept.

Base station Part of a wireless network, consisting of a transmitter, receiver, controller, and antenna system.

Bluetooth A uniting technology that allows any sort of electronic equipment from computers to cell phones to make its own connection without any wires.

Cell Used in two different contexts: a span of coverage in a wireless network, and a block of data in data transmission.

Cellular Digital Packet Data (CDPD) In a wireless network, it allows for a packet of information to be transmitted in between voice telephone calls.

Cellular network Any mobile communications network with a series of overlapping hexagonal cells in a honeycomb pattern.

Code Division Multiple Access (CDMA) A wireless digital technology that allows multiple users to share a single frequency by encrypting each signal with a different code.

Direct-Sequence Spread Spectrum (DSSS) A spread-spectrum technique that resists interference by mixing in a series of pseudo-random bits with the actual data.

Figure of merit The ratio of a satellite's receiver antenna gain to the system noise temperature.

Frequency Division Multiple Access (FDMA) The division of the frequency band into channels allocated for wireless cellular communication.

Frequency-Hopping Spread Spectrum (FHSS) A spread-spectrum technique that resists interference by moving the signal rapidly from frequency to frequency in a pseudo-random way.

Geostationary orbit (GEO) satellite A satellite that circles the equator without inclination and whose rotational period matches the Earth's own.

Global System for Mobile Communications (GSM) A second-generation global wireless system.

Improved Mobile Telephone System (IMTS) A full-duplex automatic switching system.

Mobile Switching Center (MSC) A control center for wireless services that is connected to the local exchange for wired telephones.

Satellite earth station Establishes and maintains continuous communication links with all other earth stations in the system through a satellite repeater.

Soft handoff Allows a handset to communicate with multiple base stations simultaneously in a wireless network.

Spread spectrum In a wireless network, users are separated by assigning them digital codes within a broad range of the radio frequency.

Time Division Multiple Access (TDMA) A second-generation wireless digital technology that allows users to access the assigned bandwidth on a time-sharing basis.

Wireless A communications system in which electromagnetic waves carry a signal through atmospheric space rather than along a wire.

Wireless Application Protocol (WAP) A standard for wireless data delivery and communications over the Internet.

Wireless LANs (WLANs) Transmit and receive data over the air, minimizing the need for wired connections.

Review Questions

1. Review the development of the wireless industry with a historical perspective.
2. What is the advantage of implementing a cellular network?
3. Describe the hierarchical cell structure in a PCS network.
4. Explain how analog cellular technology works.
5. Distinguish between different digital cellular technologies and their applications.
6. Discuss the specifications and applications of Bluetooth technology in wireless communications.
7. Identify the strengths, weaknesses, and applications of the following wireless LAN technologies:
    a. Microwave LANs
    b. Radio LANs
    c. Infrared LANs
    d. Broadband Wireless LANs
8. Describe the role of satellite communications in the wireless world.
9. What is the function of a satellite earth station?
10. Determine the receiver antenna gain for a satellite earth station where the system noise temperature is 30°K, to maintain a minimum figure of merit of 40.7 dB.
11. Compare and contrast GEO, GPS, and LEO satellite systems.
12. Provide an overview of the current status of international wireless communications.
13. Identify five different wireless products and describe the technology(ies) implemented in their usage.

The Technology of Communications

OBJECTIVES *After studying this chapter, you should be able to:*

- Describe the operation of a telephone circuit with respect to connection, bandwidth, and cabling.
- Explain the process of transmitting a video signal.
- Explain the purpose of modulation as related to video transmission.
- Describe the purpose and use of coaxial cable when connecting video circuits.
- Describe the delivery and distribution model of a cable television system.

OUTLINE Introduction

The Operation of Telephones

Video Transmission

CATV Networks

Introduction

We will be covering three primary communication technologies in this book: voice, video, and data. Up to this point we have covered data networking and communication technologies. Next, we will cover television and video systems, obviously one of our most important technologies.

As we begin this section of the book, we will cover just the basics of these technologies. We will later progress to cabling technologies, first covering copper wire and then fiber optics, with complete explanations of their operation and installation.

The Operation of Telephones

Modern telephones operate on essentially the same principles that were developed over 100 years ago. They use a single pair of wires that connect the phones and a power source. When phones are connected, the power source causes a current to flow in a loop, which is modulated by the voice signal from the transmitter in one handset and excites the receiver in the opposite handset (Figure 4–1).

Dialing was originally done by a rotary dial that simply switched the current on and off in a number of pulses corresponding to the number dialed. Dialing is now mostly (but not entirely) accomplished with tones.

By operating on a **current loop,** phones can be powered from a central source and extended simply by adding more wires and phones in parallel. Most phones are now electronic (that is, they use semiconductors rather than electromechanical devices), but they use the same type of wiring, frequently called current-loop wiring.

In office systems, the phones are sometimes digital, and they use twisted-pair wiring in a manner similar to computer networks but at lower speeds.

Telephone wiring is simple because the **bandwidth** of telephone signals is low, generally around 3,000 hertz (cycles per second). Bandwidth is similar to speed—a low bandwidth requires lower frequencies, and a higher bandwidth requires higher frequencies. Computer modems use sophisticated modulation techniques to send much faster digital signals over low-bandwidth phone connections.

Because of the low-bandwidth and current-loop transmission, telephone wire is easy to install and test. It can be pulled without fear, and if it is continuous, it should work.

Outside the home or office, the phone connects into a worldwide network of telephones (Figure 4–2), all connected together by a combination of copper wire and fiber-optic cables leading to switches that can interconnect any two phones for a voice conversation or data transmission. These switches create a "switched-star" network where each switch knows how to find every phone by connecting through successive switches. When a phone connection is made, the connection stays complete as long as the users require, creating a continual virtual pathway between the two phones.

Telephone cabling is simple 4-wire cables inside the home or 4-pair unshielded twisted-pair (UTP) cables inside an office. The "subscriber-loop" connections to local switches are on multipair wires, with up to 4,200 pairs per cable. Most large pair count cables are being replaced by fiber optics because those 4,200 pairs of copper wires can be replaced easily by a pair of **single-mode** optical fibers. All fibers used in telephone networks are single-mode types.

Telephone connections between local switches, the "interoffice trunks," are already predominantly optical fiber and soon will be all fiber. Long-distance has

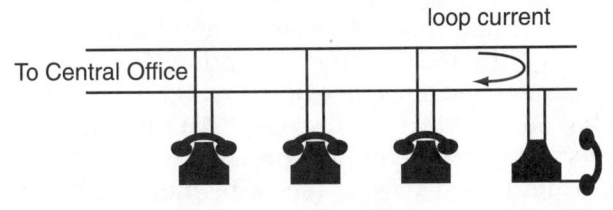

FIGURE 4–1 Telephones are connected in parallel and operate on a current loop.

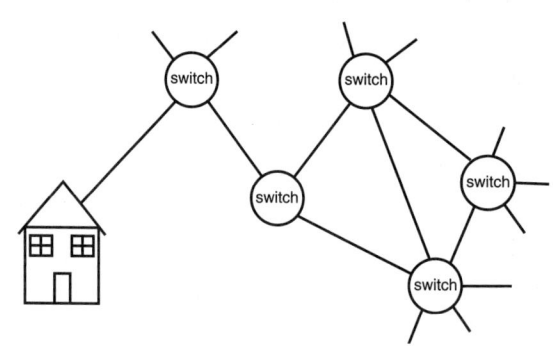

FIGURE 4–2 The telephone network is a switched-star network.

been virtually all fiber for years, except in areas of exceedingly rugged terrain or isolation where terrestrial radio or satellite links are more cost-effective. Even fiber-optic undersea links have overtaken satellites for international calls because of the higher capacity and lower cost of fiber compared to satellites.

Video Transmission

Probably the easiest way to explain basic video or TV technology is to use the fax machine as an example. Fax machines could be called slow-scan television. They send one picture frame at a time, in black and white, over telephone lines. We are all familiar with the process: the machine scans one line at a time, interprets that area as black or white space, translates the information into electronic pulses, and transfers the information to a receiver.

Black-and-white television uses the same technology as the fax machine, except at sixty frames per second and using a camera rather than a scanner. Each line of the image is scanned and translated into a complex video signal. The color process is the same, except a still more complex signal is used.

Because of the complexity of the television signal, it requires a lot of bandwidth. A standard TV channel uses 6 **MHz** (millions of cycles per second) of bandwidth. In other words, the signal requires six million cycles per second (each cycle contributing part of the signal information) to get all of the information sent fast enough. We could send video images at lower speeds, but the images would be in slow motion.

Closed-circuit TV (CCTV) used for surveillance in security systems consists of only one channel operating with a direct cable connection to a monitor. In large systems, video switches scan numerous cameras to allow the monitoring of many locations, but each remote camera is connected to the central monitor by an individual cable.

In order to use multiple channels, each channel must use separate frequencies, or their signals will all be jumbled together. For example, the FCC has assigned channel 2 the frequency range of 54 to 60 MHz, channel 3 has 60 to 66 MHz, channel 4 has 66 to 72 MHz, and so on. (UHF channels go up to 890 MHz.) Each channel transmits its programming in that 6-MHz channel.

Broadcast TV sends its signals over the air by modulating a signal on a transmitting antenna that is received by a smaller antenna on your home or TV.

Cable TV works the same way, except it sends the signal through coaxial copper cable or optical fiber, not via radio waves.

One very intimidating word used in the TV business is *modulation*. Modulation is simply the function of taking a basic 6-MHz TV signal and modifying it so that it matches one of the standard channel frequencies.

For example, the TV signal coming out of the camera at your local station operates at about 6 MHz. If we assume that your station is channel 13, it must send its programs in the 210 to 216 MHz slot. So, before sending the signal to the antenna, it must change the 6-MHz signal to a 210 to 216 MHz signal.

This is done with a special electronic device called a *modulator*. Your VCR has a modulator built into it. Because videotape cannot record the very-high frequencies used in broadcasting (in the last example, the station used 216 MHz), the VCR must record at frequencies that are too low for a television tuner to pick up. When you play a videotape, the information from the tape is modulated to channel 3 or 4, which your TV can then display.

Once we began to customize television systems in the home or office, the modulator became a very important device. It is necessary to assign the various video inputs to specific channels so that a television can use them correctly. Video distribution systems allow for custom television channels to be transmitted through the facility. This is generally done in combination with a cable TV system. Although these systems are capable of handling well over 100 stations, very few use them all. This allows the owner to transmit other signals through the unused channels. Channels are assigned with the use of a modulator, which applies the basic television signal to any channel.

Sending television signals requires much more bandwidth than that required by telephones. When the technology was first developed, the only cable capable of carrying these high-speed signals with adequately low loss was coaxial cable, also called **coax.** Coax (Figure 4–3) uses a central conductor surrounded by an insulator, then an outer conductive webbing called the shield, and finally a plastic jacket.

It is the design of the cable, with the central conductor widely and evenly separated from the outer conductor, that gives it the high bandwidth capability. In addition, the outer conductor acts to contain the signal inside the cable, reducing the emissions from the cable that cause interference in other electronics and the interference of other outside sources on the signal in the cable itself.

The signal in coax cable is a simple voltage. It can be introduced into the cable at either end, or even in the middle, where it will be carried to either end. In any application, it is important that the transmitters be selected to be appropriate to the characteristics of the coax cable, and both ends must be terminated to prevent reflections that can cause interference.

Coax cable must be installed with care. It should not be stretched or kinked. Doing so will reduce the level of bandwidth it will transmit. Connectors should be carefully installed to prevent signal leakage (the TV term analogous for signal loss), and unused ends must be properly terminated to prevent reflections.

Coaxial Cable Construction

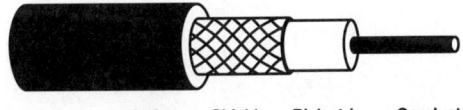

Jacket   Shield   Dielectric   Conductor

FIGURE 4–3 Video transmissions require the high bandwidth capability of coaxial cable.

CATV Networks

Television is delivered by broadcasting, or simply, by delivering the same signal to all directions and users at once. It may be delivered by terrestrial antennas to home televisions, sent to satellites that rebroadcast it to large geographic areas at once, or transmitted by coaxial cable on a CATV system.

Most CATV systems get their broadcast signals from satellites at a location called the headend, which then send it out over large, low-loss coaxial cable called RG-8 that is about an inch in diameter. These cables terminate in amplifiers that rebroadcast it over many other cables to more amplifiers that repeat the rebroadcast. Thus, the CATV system (Figure 4–4) looks like a "tree-and-branch" system, the name used to describe this network architecture.

The obvious fault of this type of system is its vulnerability to the failure of a single amplifier. If any amp fails, service is lost to all downstream customers. The solution is to separate the network into smaller segments to prevent massive failures.

Today, most CATV system operators are switching over to a hybrid fiber-coax (HFC) network (Figure 4–5), with fiber-optic cable distributing **headend** signals to local drop amplifiers for distribution to homes. By using fiber for the headend connection, the number of amplifiers connected in series is reduced to about four, minimizing the number of drops affected by any single amplifier or cable fault. This greatly enhances system reliability and minimizes maintenance costs.

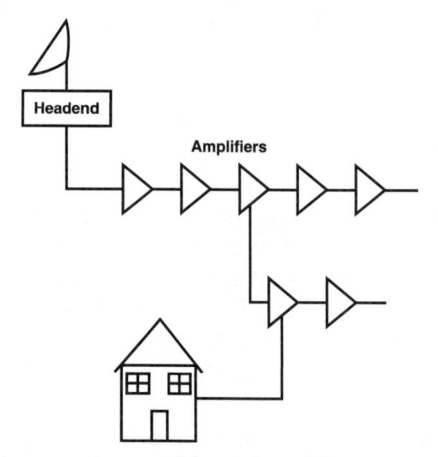

FIGURE 4–4 CATV systems are a tree-and-branch architecture.

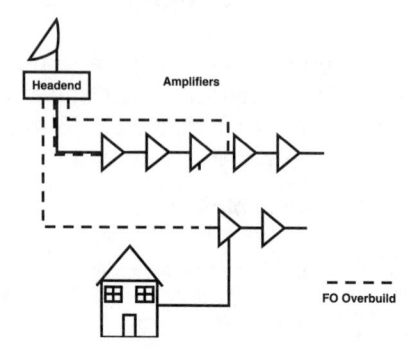

FIGURE 4–5 CATV systems use hybrid fiber-coax (HFC) networks to minimize the number of amplifiers between the headend and the subscriber.

Fiber is added to CATV systems already in place by a fiber "overbuild" where the fiber-optic cable is lashed to aerial coax or pulled alongside it in ducts. Large systems use fiber to distribute signals to several towns because fiber has the capability of transmitting long distances, thus saving on the costs of additional headends.

Once the CATV signal is split off to the house, it is carried on smaller coax cable to and throughout the home to connect all the TVs. This cable, although smaller, can still carry gigahertz bandwidth signals necessary for transmission of up to 100 or more channels of video programming.

Summary

- Telephone circuits operate on a current loop, connecting in parallel from a central source.
- The bandwidth of a standard telephone is generally around 3,000 hertz.
- While most telephone circuits are installed using unshielded twisted-pair (UTP) cable, main trunk lines from the service provider are now commonly using single-mode optical fiber.
- A telephone network is an example of a switched-star network.
- Video transmissions require a lot of bandwidth—6 MHz for a standard TV channel.
- Modulation is used to assign a video signal to a specific broadcast channel; channel frequencies are assigned by the FCC.
- Coaxial cable is the most commonly used medium for the transmission of video signals.
- Cable television networks broadcast over satellites at a location called the headend.
- Cable television signals are sent out to the customer over coaxial cables through a network of amplifiers that resembles a "tree-and-branch" structure.
- Large cable television systems are now using optical fiber to distribute transmission to local customers.

Key Terms

Bandwidth The frequency spectrum required or provided by communications networks; the range of single frequencies or bit rate within which a fiber-optic component, link, or network will operate.

Coax (coaxial) cable A cable in which a single center conductor is surrounded by a dielectric material, and a cylindrical shield that is often composed of layers of foil and metallic braid. Coax has excellent high-frequency characteristics and is most commonly used for cable television signals.

Current loop Transmission using variable current to carry information as in a simple analog telephone.

Headend The main distribution point in a CATV system.

MHz Megahertz; millions of cycles per second.

Single-mode fiber A fiber with a small core, only a few times the wavelength of light transmitted, that allows only one mode of light to propagate; commonly used with laser sources for high-speed, long-distance links.

Review Questions

1. True or False: Modern telephones operate on essentially the same principles that were developed 100 years ago.
2. What type of circuit (parallel or series) does a telephone operate under?
3. What is the bandwidth of telephone signals?

4. What machine has an operating principle similar to a black-and-white television, but slower?
5. How much bandwidth does a standard TV channel use?
6. What is the name of the type of cable capable of transmitting high-speed signals with adequately low loss?
7. List the four parts of a coaxial cable.
8. If the coaxial cable is not terminated at both ends, what type of problem might you have?
9. In a "tree-and-branch" system, what happens to everything downstream from a failed amplifier?
10. Using a coax cable capable of carrying a 1 GHz bandwidth, how many channels can you expect in your home?

Voice and Data Applications

After studying this chapter, you should be able to:

OBJECTIVES

- Understand the characteristics of an analog telephony signal.
- Describe the distribution of analog telephony signals in the home.
- Understand the functions of a private branch exchange (PBX).
- Identify the most common residential data networking applications.
- Understand how to combine wired and wireless LANs in a home data network.
- Discuss the pros and cons of Internet telephony and Voice over IP (VoIP).

Voice Applications

OUTLINE

Data Applications

Voice Over IP

Voice Applications

The telephone was the first communication device to be used in the home and is still the most common. As illustrated in Figure 5–1, a phone line enters the house via the **network interface device (NID)** and is connected to a primary protector to ensure that hazardous voltages due to lightning, for instance, do not pose a safety hazard. Traditionally, the phone line was then daisy-chained around to all the extension phones in the house.

A more modern approach is to use a **structured cabling system** to connect the telephones, as shown in Figure 5–2. In this case, the phone line is routed to the primary protector and then on to the distribution device (DD). From the DD, individual UTP cables are home-run to each telecommunications outlet (TO) in the house. The telephones may then be plugged into any available TO. Let us take a closer look at the characteristics of the telephone signal and how it is distributed throughout the house.

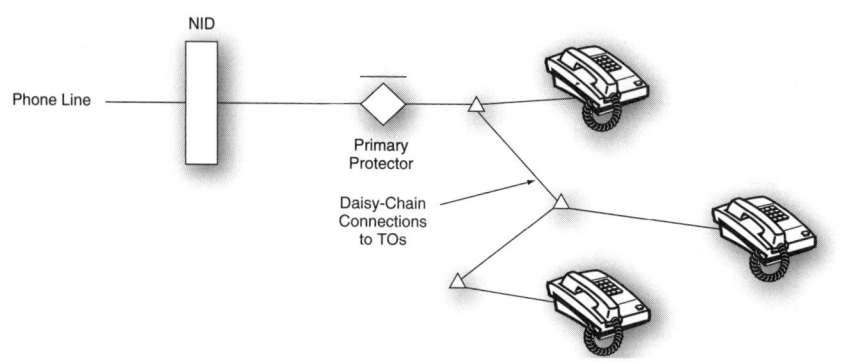

FIGURE 5–1 Traditional Voice Telephony

Analog Telephony

Traditional voice telephony uses a low-bandwidth bidirectional analog signal on a single pair of wires. The voice signals are represented by analog voltages, which are typically limited to a maximum frequency of less than 4 kHz. Voice signals propagate simultaneously in both directions on the wire. In addition, the central office (CO) places a battery voltage (–48 Vdc) across the conductors of the pair to power the telephone. When there is an incoming call, the CO applies ringing voltage to the line to operate the ringer in the phone. The ringing signal is a high-voltage (up to 90 volts or so), low-frequency (20-Hz) tone that is applied for 2 s, then removed for 4 s, and so forth.

Devices that connect to an RJ-45 modular jack must be able to tolerate the relatively high-voltage battery and ringing signals so they are not damaged if they are plugged into a traditional phone line.

The telephone signal is typically placed on Pair 1 of the RJ-45 jack (Pin 4 and Pin 5). For two-line phones, the second line is connected to Pair 2 (Pin 3 and Pin 6), as indicated in Figure 5–3. A single-line phone will simply ignore the signals on Pair 2.

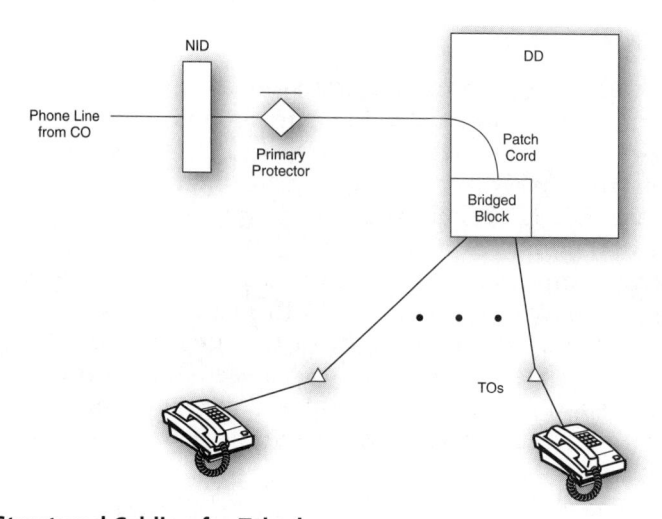

FIGURE 5–2 Structured Cabling for Telephony

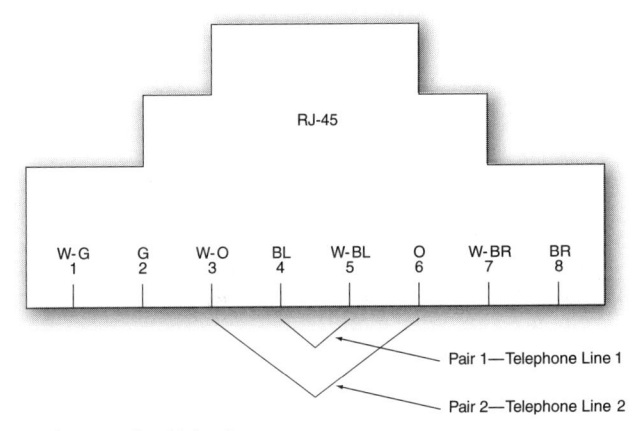

FIGURE 5–3 RJ-45 Pinouts for Telephony

Up to four analog telephone lines can be supported on a single TO by means of an **adapter** like the one illustrated in Figure 5–4. In this case, a phone line is connected to each pair. The adapter takes each of the pairs on the incoming cable and routes it to Pair 1 on one of the four output jacks. Adapters like the one in Figure 5–4 are available from a number of manufacturers.

Analog telephones are designed so that multiple extension phones can simply be bridged together, as was done with the old-fashioned daisy-chain wiring. With a structured cabling system, the bridging is done at the DD using a bridged block, as illustrated in Figure 5–5.

Figure 5–6 shows several voice modules that use a combination of 110 wiring blocks and modular jacks. Other analog telephony devices, such as voice-band modems and fax machines, are connected in the same manner as a standard telephone.

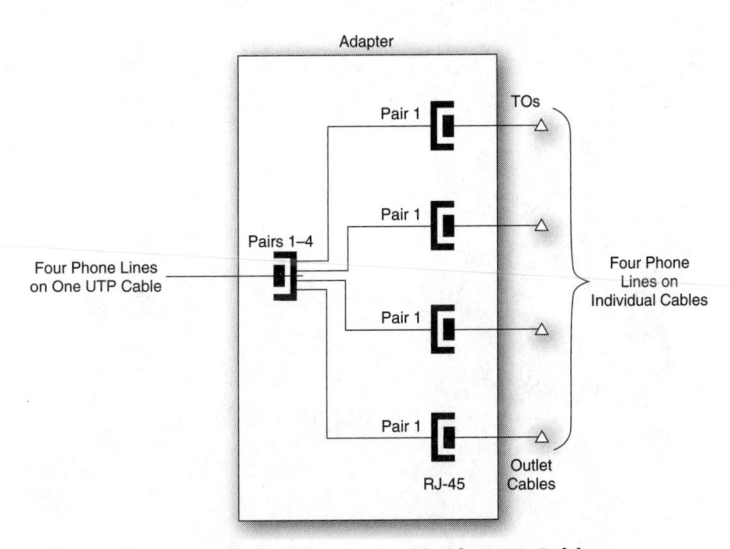

FIGURE 5–4 Adapter for Four Phone Lines on a Single UTP Cable

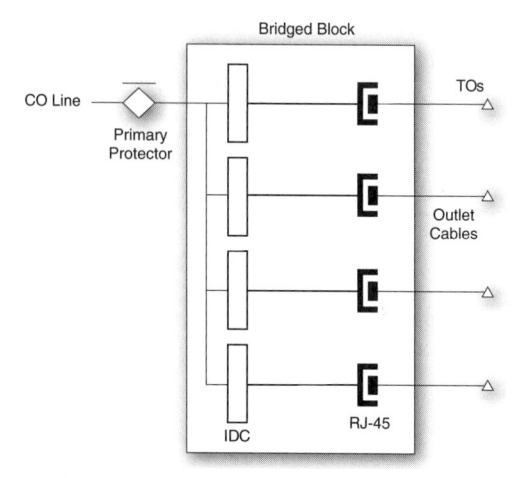

FIGURE 5–5 Bridged Connections at the DD

Key and Private Branch Exchange Systems

In larger residences with many extension phones, a **key system** or **private branch exchange (PBX)** is sometimes installed.

A key system is a small telephone switching system that allows a user to pick up (or place on hold) a call on any of several lines. The switch often consists of a set of simple electromechanical contacts that allow telephone lines to be bridged together. It usually provides an intercom line so that the local telephones can be bridged together without making an outside call. Figure 5–7 schematically illustrates the function of a key telephone system. This example shows a small key system with three telephone lines and three phones. Two of the phones have multiple-line appearances, while one of them only supports a single line. All three phones can connect to the intercom line.

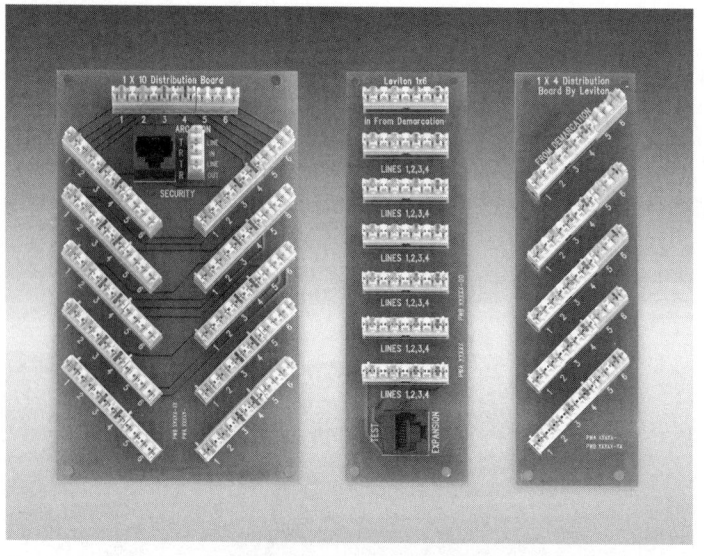

FIGURE 5–6 Examples of Voice Modules (Courtesy of Leviton Manufacturing Co., Inc.)

TABLE 5–1
Typical Key System Status Light Indications

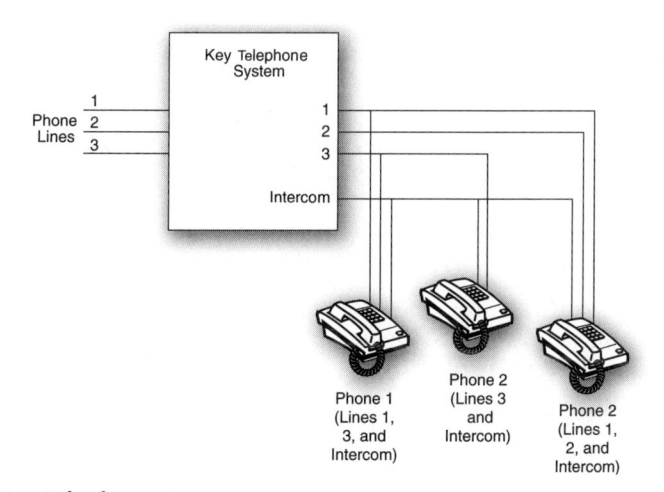

Line Status	*Indicator Light[1]*
Idle	Off
Busy	Steady on
Ringing	Flashing
On hold	Winking

FIGURE 5–7 Key Telephone System

The telephones used with key systems have a row of lighted buttons (or keys, hence, the name), which are used to select a particular line. The status of the line is indicated by the illumination of the key, as indicated in Table 5–1.

A private branch exchange (PBX) is a small switching system that can interface to telephones (often called *stations*) and telephone lines to the central office (often called *lines,* or sometimes *trunks*). The lines and stations are connected to the PBX as shown in Figure 5–8. The PBX can then connect any combination of lines or stations together. A PBX can provide a number of useful features, such as:

- *Intercom calling:* A call can be dialed from one phone in the house to another.
- *One-touch calling:* Pushing a single preprogrammed button calls a specified number.
- *Direct inward dialing (DID):* Each of the extension phones can have a unique number so that outside calls can be directed to a particular phone in the house if desired.
- *Conferencing:* Multiple stations or lines can participate in the same call.

---

[1] *Flashing* and *winking* refer to lights that are flashing 60 and 120 times per minute, respectively.

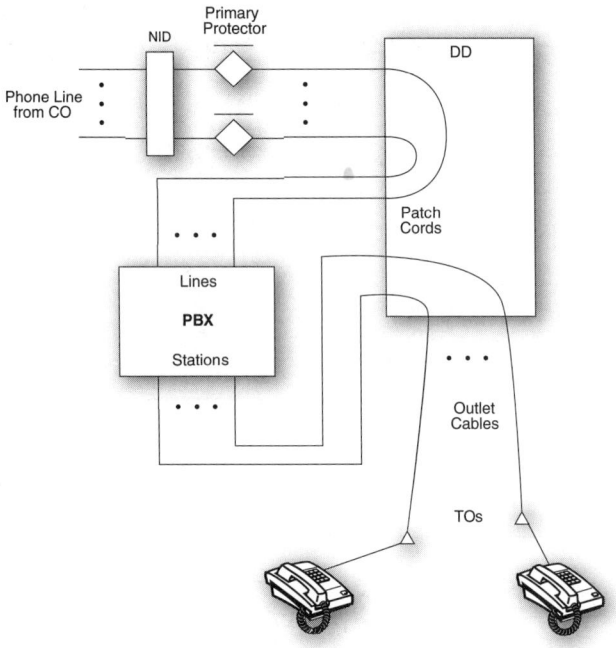

FIGURE 5–8 Connections to a PBX

- *Advanced telephone sets:* Private branch exchanges often support telephone sets with buttons, LEDs, and displays to make advanced features easier to use. These telephones often use digital signaling protocols rather than the traditional analog methods.
- Voice messaging and other features may optionally be provided by the PBX.

From an infrastructure perspective, key systems and PBXs fit in well with structured cabling and do not require any special considerations. The key system or PBX is usually located near the DD and connected as shown in Figure 5–8. The main difference from a standard telephony arrangement is that bridged, cross-connect blocks are not normally needed with a PBX or key system.

Cell Phones

Increasingly, some or all of the telephony needs of the residence are being met by cell phones. Cell phones, of course, do not place any demands on the cabling infrastructure. However, it is important to provide sufficient cabling infrastructure so that voiceband devices (such as fax machines, modems, and telephones) can be conveniently connected when desired.

Data Applications

The earliest data applications in the home used a voiceband modem for data communication. Voiceband modems are wired as indicated in the previous section on analog telephony. Note that high-speed (56-kb/s) modems normally require a direct line to the telephone office; they do not usually work behind a PBX or key system.

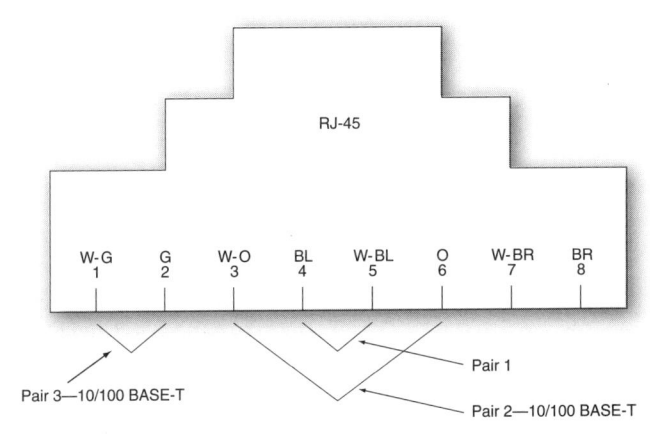

FIGURE 5–9 RJ-45 Pinouts for Ethernet

More modern data applications usually involve connecting a LAN in the home. This is most often an Ethernet or a wireless network such as Wi-Fi but may sometimes be an IEEE 1394 LAN. It is very common to have both wired and wireless LANs installed in the home.

Modern home LAN equipment is very easy to install and use. Unlike large corporate networks, home LANs generally work right out of the box and require very little administration or configuration. The main consideration is to be sure that they are wired up properly. Both 10/100-Mb/s Ethernet and the UTP version of FireWire operate on Pair 2 and Pair 3 of the RJ-45 jack, as shown in Figure 5–9.

When using 10Base-T, it is possible to share the UTP cable with an analog phone connection on Pair 1. However, this is *not* recommended, since sharing with analog signals does not work with 100Base-T and certainly not with 1000Base-T, which uses all four pairs. If the home is wired according to TIA-570B, there should be no wiring problems with any of the commonly used home networks.

Modern home LAN equipment is very easy to install and use. If the home is wired according to TIA-570B, there should be no physical-layer problems with any of the commonly used home networks.

The two main applications for home data networks (excluding entertainment applications, which are discussed in the next chapter) are Internet access and file/peripheral sharing.

Internet Access

It is increasingly common for homes to have multiple PCs. When broadband Internet access is purchased, it is natural to want to share it among all the PCs in the home. There are two ways this may be done. First, the ISP will provide you with additional IP addresses, typically for around $10 per month or so. Or you can install a router behind your Internet access equipment (cable, DSL, or satellite modem). The router uses capabilities called **network address translation (NAT)** and **network address port translation (NAPT)** to establish a private IP network within the home.[2] This private IP subnet can then provide multiple connections over a single external IP address in a manner that is transparent to

---

[2] NAT and NAPT are described in RFC 1631 and 2663, respectively.

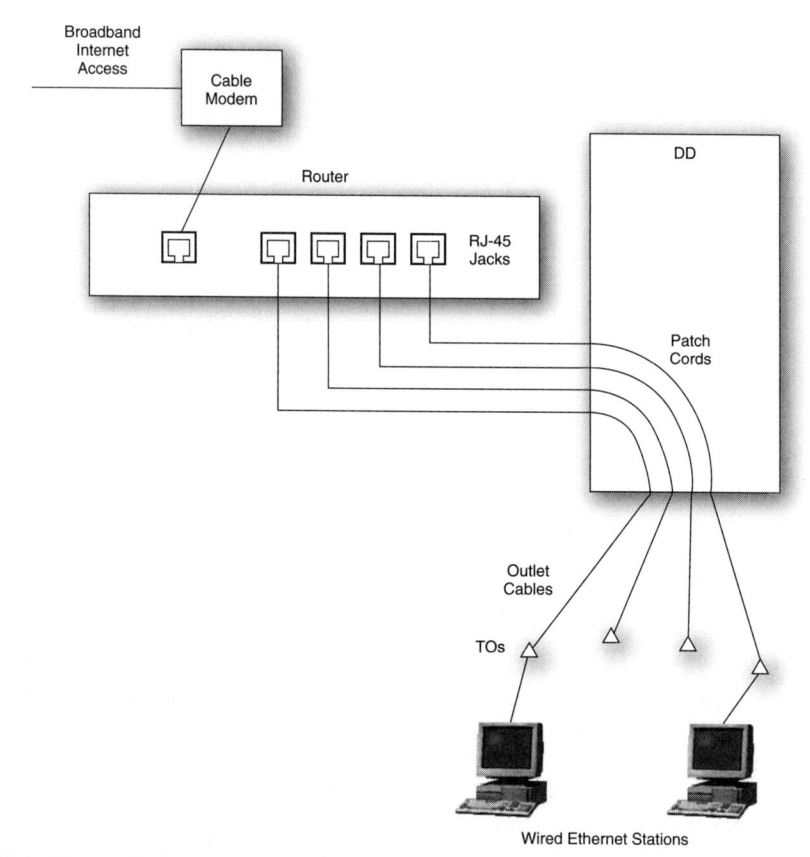

FIGURE 5–10 Typical Wired Home Router Configuration

the users. Routers for home use are usually integrated with an Ethernet switch and have become so inexpensive that this is usually the preferred method. Figure 5–10 shows a typical wired home router configuration using a cable modem for broadband access.

The cable modem and router are normally collocated with the DD, so they are directly connected together. The router's RJ-45 ports are connected to the DD, where they are cross-connected to outlet cables. The PCs may be plugged into any of the TOs that are connected to the router.

File and Peripheral Sharing

When multiple PCs are connected to a network, it is natural to want to access files and peripherals that are associated with any of the computers on the network. Sharing files and directories is one of the simplest applications. This will allow, for example, digital photos to be stored on one PC and viewed from other PCs. This capability is very easy to set up across a LAN. The exact method for doing it varies according to the operating system your PC uses. For more details, consult your computer manual or one of many reference books such as Neibauer (1999).

Another common operation is sharing a printer. Often there is a high-speed or high-quality printer attached to one PC. It is easy to set up the network so that it can be accessed from any of the PCs.

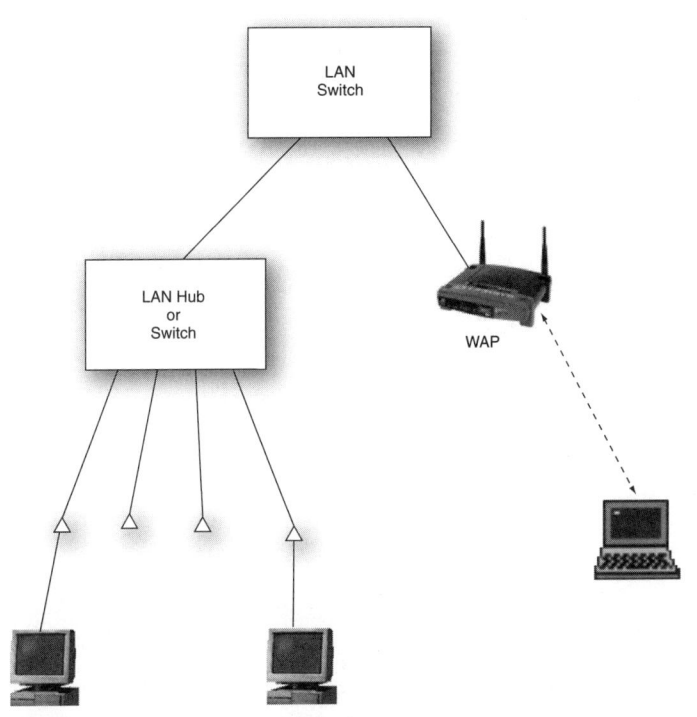

FIGURE 5-11 Connection of Wired and Wireless LANs

Sharing a storage or backup device (such as a disk drive or CD burner) is another common application. This is also easy to set up. However, depending on the speed and amount of traffic on your network, the performance of a remote storage device might be substantially less than that of a directly attached device.

Integration of Wired and Wireless Networks

Wi-Fi networks are designed to be easy to integrate with Ethernet LANs. There are two ways to do this. The first is to build separate wired and wireless networks and hook them together through a LAN switch, as shown in Figure 5-11. This technique is often used in commercial buildings but is overkill for home networks.

A more efficient configuration for home networks is to integrate the WAP and wired LAN switch (or hub) into a single unit. This is a very common configuration for home routers, which often have a WAP in addition to several wired LAN ports. Figure 5-12 shows a connection diagram for a typical system. Note that the wired part of the network is the same as shown in Figure 5-10; the only change is the addition of the WAP and wireless terminals.

Voice Over IP

Voice over Internet Protocol (VoIP), also known as **Internet telephony,** involves making phone calls over the Internet rather than over the traditional voice network. It has been billed as the most fundamental new telecommunications technology in decades. In this section, we look at the technology involved, the regulatory issues, and the pros and cons of VoIP as compared to traditional telephony.

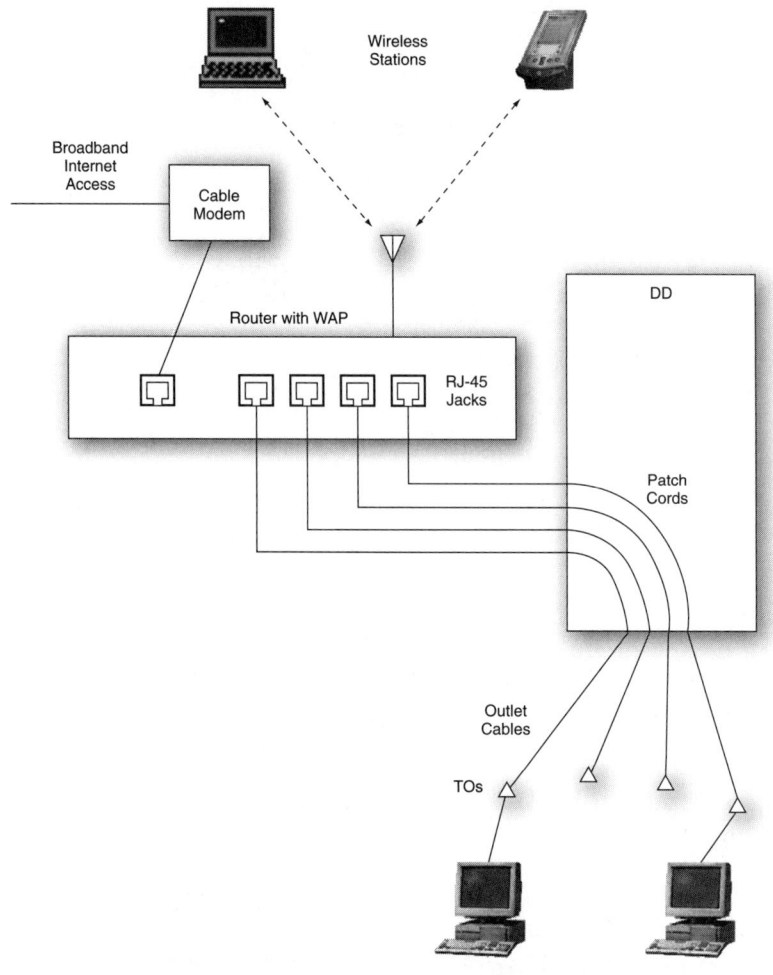

FIGURE 5–12 Connection of Integrated Wired/Wireless Router

Voice over IP (VoIP) is potentially the most revolutionary development in telecommunications since the migration from analog to digital transmission. It allows an almost unlimited variety of new services to be rolled out on the same general-purpose IP network.

Voice over IP Technology

Data networking is dominated by IP. The VoIP technology allows voice telephony to share the same networks that have been built for IP traffic. This convergence of voice and data traffic onto a single IP network is widely expected to represent the future of the communications industry.[3]

One of the main attractions for users of VoIP is that IP networks have not historically charged users based on either traffic or distance. Regardless of where a user is located or how long a session lasts, it does not cost the end user any more to

---

[3] The Chief Technical Office of AT&T has predicted that "IP will be like Pacman—it will eat everything."

access a Web site in Tokyo than one in Chicago. This is completely different from traditional telephony, where billing for long-distance calls is based on both time and distance.

Traditional telephony consists of circuit-switched connections between end points whose physical locations are known. In other words, a phone number corresponds to a particular geographical location. Signaling is done by means of loop currents, ringing voltages, touch tones, and so on.

When using VoIP, the voice signal is digitized and sent as a stream of packets. There is no longer a fixed path from source to destination; nor is there any correlation between the logical address (IP address) and the geographical locations of the end points. Because of this, it is no longer possible to use currents and tones to perform signaling.

The two primary methods of signaling in a VoIP network are **H.323** and **Session Initiation Protocol (SIP).**[4]

- H.323 is an ITU-T standard that was approved in 1996 for transmitting multimedia communication (such as audio and video) over a packet-switched network that does not guarantee quality of service (QoS). H.323 defines four major elements: terminals, gateways, gatekeepers, and multipoint control units.
- Session Initiation Protocol is an IETF specification (RFC 2543) for initiating, modifying, and terminating multimedia sessions. It is a request/response protocol with the requests being sent via any transport protocol, such as UDP or TCP. It is used for telephony, conferencing, instant messaging, and other applications.

Since VoIP represents a convergence between the voice telephony and data networking worlds, it is not surprising that one of these methods (H.323) came from the telephony industry, while the other (SIP) came from the IETF. H.323 was the first protocol to ensure interoperability among VoIP equipment from different manufacturers, and it has a significant installed base. Recently, however, SIP has been gaining momentum as it is generally considered to be simpler and more flexible. Both protocols will probably coexist in the marketplace for some time.

Voice over IP Implementation

Voice over IP has been quietly put into service in long-haul networks for a couple of years now. Many long-distance calls pass through a VoIP network without the users' knowledge. The voice quality achieved in these networks is good enough that no one notices any difference. The current fanfare about VoIP involves extending the IP connection all the way to the subscriber.

Early implementations of Internet telephony often used a headphone attached to a PC. The voice digitization and packetizing were done in the PC, and the voice stream shared the PC's broadband Internet connection. This provided telephone service that was essentially free (that is, no additional cost other than the broadband Internet connection). However, the voice quality was so low that it was often considered unacceptable, even for free. There are several reasons for this, including delay, packet loss, and echo.

Voice over IP traffic is *extremely* sensitive to delay. Even a few tens of milliseconds of delay, which would scarcely be noticed in a data connection, can wreak

---

[4] The details of SIP and H.323 are beyond the scope of this book. For a more detailed discussion, *see* Davidson et al (2000).

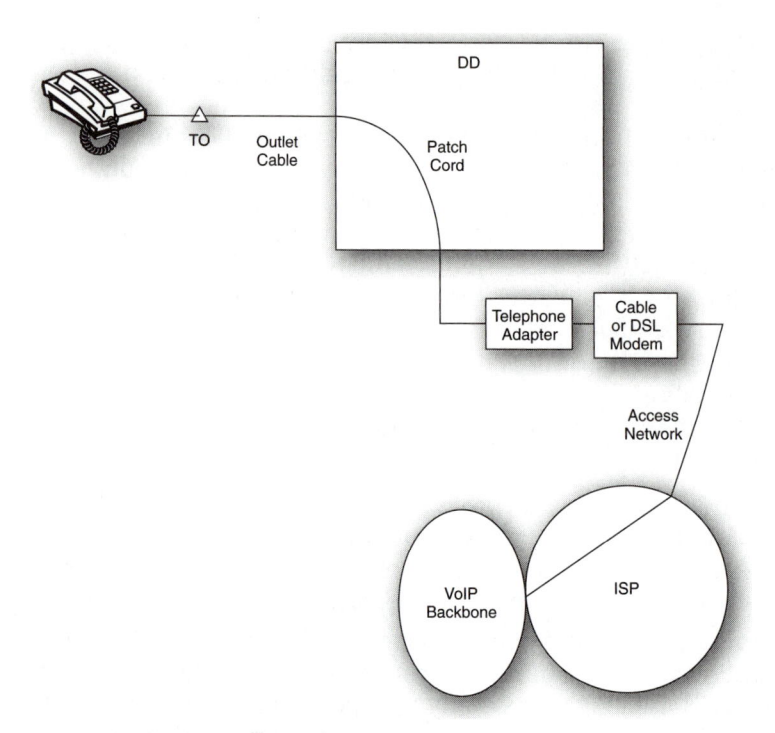

FIGURE 5–13 Residential VoIP Network Architecture

havoc with the perceived quality of a VoIP stream. Corporate VoIP networks generally use a dedicated IP network (or a VPN) that is specifically engineered for low latency.

Packet loss can also have a significant effect on VoIP quality. Studies have shown that a packet loss rate of as little as 3 to 5 percent can result in a noticeable drop in users' perception of the quality of the call.[5] Unlike many IP applications, VoIP is inherently bidirectional, so the uplink bandwidth is just as important as the downlink bandwidth. Since the asymmetrical nature of most broadband access networks means that uplink bandwidth is limited, bursts of uplink data traffic can potentially cause the loss or delay of VoIP packets.

Due to the extra delay involved in packetizing the voice samples, echo is much more of a problem with VoIP than it is with analog telephony. A small amount of echo, which would not be noticeable on an analog connection, can significantly degrade the quality of a VoIP connection.

A more modern approach to residential VoIP is being offered by a number of telecommunications carriers. This involves using a telephone adapter so that an ordinary telephone can be plugged into a VoIP network, as shown in Figure 5–13. The telephone adapter digitizes and encodes the voice and translates the signaling protocol (H.323 or SIP) into traditional ringing and tones so that the familiar user interface is preserved. The VoIP connection shares either a DSL or cable modem connection to the ISP. At the ISP's office, the VoIP connection is routed to a

---

[5] J. H. James et al., "Implementing VoIP: A Voice Transmission Performance Progress Report," *IEEE Communications Magazine,* July 2004, 36–41.

dedicated VoIP backbone to provide a low-latency path, which ensures a high-quality voice signal.

Voice over IP is extremely sensitive to delay and packet loss in the IP network. Both the network and the access line must be engineered to keep the total delay well under 100 ms.

Voice over IP Regulatory Issues

Traditional telephone services are governed by a complex web of regulations that have developed over the past century or so. However, VoIP is not, as yet, affected by these regulations. Three issues that affect the public well-being in one way or another are currently being debated.

- **E911 service:** 911 is the official national emergency number in the United States and Canada. Dialing 911 quickly connects you to a Public Safety Answering Point (PSAP) dispatcher trained to route your call to local emergency medical, fire, and law enforcement agencies. Enhanced 911 (E911) service automatically reports the telephone number and location of 911 calls. E911 service is required for all wireline phones and is rapidly being rolled out for cell phones as well.
- **Universal Service Fund (USF):** Since its inception in 1934, the policy of the FCC has been that basic telephone service should be universally available and affordable in all parts of the country. However, it is much more expensive to provide service in some areas, especially rural areas where the population is sparse and telephone lines must be run for long distances. To keep rates affordable in these areas, the USF was established. The USF is essentially a tax that is levied on long-distance carriers. The proceeds are used to subsidize telephone rates in areas where it is expensive to provide service.
- **Communications Assistance to Law Enforcement Act (CALEA):** The CALEA is a law that was passed by Congress in 1994. It requires telecommunications carriers to ensure that their digital and wireless networks can comply with authorized wiretaps and electronic surveillance. The purpose of CALEA is to assist the FBI and other law enforcement agencies in investigations of organized crime, terrorist organizations, and so on.

Under current FCC rules, VoIP is not considered to be a telecommunications service and is thus not subject to any of the preceding regulations. In early 2004 the FCC began a series of hearings to determine whether and how VoIP service should be regulated. Reasonable arguments can be made on both sides of this issue: Should VoIP be regulated like traditional telephone service, or should it be unregulated like e-mail? The FCC seems to be leaning toward not regulating VoIP, at least until the technology has had more time to develop and become widely deployed. As a larger percentage of voice calls begin using VoIP, there may be increased pressure on the FCC to make some requirements regarding E911 and CALEA.

Voice over IP Pros and Cons

At least potentially, VoIP has a number of significant advantages.

- It is currently cheaper than standard telephone service for those who do a lot of long-distance and international calling. Whether it stays cheaper or not in the long run will depend on several factors. First, if lots of VoIP traffic requires significant expansion of the VoIP backbone networks (which seems likely), the cost of this must be passed on to the subscribers. Second, if the FCC requires E911 and CALEA for VoIP, this will also raise the cost. Third, if the cost of traditional telephone service declines (which also seems likely), that will eat into VoIP's advantage.

- Voice over IP makes it much easier for service providers (including cable television companies, who can provide voice service over their networks) to develop and roll out new services without extensive network construction. This could result in a virtual cycle with many new services for subscribers and large amounts of additional revenue for service providers.
- Since it is an IP-based service, VoIP service providers can easily provide an interface to manage phone service from a PC. This could include features like a detailed call log, time-of-day service preferences (for example, forwarding or do not disturb), and so on.
- Voice over IP service can make your telephone number follow you wherever you plug in your telephone adapter (home, office, hotel, etc.).

There are also a number of potential drawbacks to VoIP service.

- You must have a broadband Internet connection to use VoIP service. It is possible that VoIP may be the killer application that causes large numbers of people to subscribe to broadband access. On the other hand, lack of broadband access may slow down the deployment of VoIP.
- Even after accounting for the cost of broadband access, the economics of VoIP are not as clear-cut as they initially appeared. As already discussed, several factors could make VoIP service more expensive. It is also likely that providers of standard telephony service will be driven to flat-rate access for long-distance service.
- To the individual subscriber, most of the preceding regulatory issues are not particularly important; USF is mainly of interest to service providers, and CALEA is most important to law enforcement agencies.[6] However VoIP's lack of E911 service may be of concern to some.
- Voice over IP is not compatible with security and fire alarm systems that automatically dial out when they detect a problem.[7] For those applications, an analog telephone is still needed. There are two problems here. First, VoIP lines do not allow the alarm system to seize the line (go "off hook") and dial a call to a central monitoring station. Second, some alarm panels use the 48-volt battery on the phone line as a source of backup power. This battery voltage is not available with VoIP.
- Phone service has traditionally continued with no interruption during power failures. This is not the case with VoIP. Since both the cable modem and the telephone adapter are run from the house alternating-current power, an **uninterruptible power supply (UPS)** must be installed to maintain service during a power failure.

---

[6] Many people probably think it is a good idea for law enforcement to be able to monitor the communications of "evil doers" but are not particularly concerned if *their* phone calls cannot be monitored.

[7] Cell phones do not work for this application either. Even if subscribers adopt cell phone and VoIP technology for their voice communications, many will still need an analog telephone line coming into the house for their alarm system.

TABLE 5–2
Pros and Cons of VoIP

Advantages of VoIP	*Disadvantages of VoIP*
Currently cheaper than traditional telephony for those who make lots of long-distance and international calls	Must have broadband Internet access to use VoIP; voice quality poor if the network has excessive delay or packet loss
Easy for service providers to roll out a huge array of new services and features	Long-term cost savings uncertain
Facilitates PC-based phone management—call log, time-of-day features, and so on	Does not provide E911 service
Telephone number that follows you wherever your adapter is plugged in	Not compatible with autodialers on security and fire alarm systems
Not constrained by the FCC regulations on traditional telephony	Phone service not continued during power outages

The pros and cons of VoIP are summarized in Table 5–2.

In the long run, it seems likely that rates for VoIP and standard telephony will tend to equalize somewhat. Prices for VoIP will probably rise as traffic increases and regulatory issues like E911 are addressed. Meanwhile, prices for standard telephony are likely to become more competitive as flat-rate pricing becomes more common. What really matters to consumers is the price and the services that are provided, not the particular technology used to make the connection. The biggest advantage of VoIP may turn out to be the almost unlimited range of services that can be provided.

Summary

This chapter reviews the basics of voice and data applications in the residential network. The major points covered in the chapter are:

- Traditional voice telephony uses a bidirectional audio frequency (<4-kHz) signal on a single pair of wires. A single-or dual-line telephone can be supported on an RJ-45 jack.
- A private branch exchange (PBX) can provide additional voice features, such as intercom calling, conferencing, or support of advanced telephone sets.
- Modern home LAN equipment is very easy to install and use. If the home is wired according to TIA-570B, there should be no problems with any of the commonly used home networks.
- The main applications of home data networks are Internet access, sharing of files and directories, and accessing printers and other peripheral equipment that is connected to another computer on the network.
- Voice over IP (VoIP) is potentially the most revolutionary development in telecommunications since the migration from analog to digital transmission. It allows telephone calls to be made over the Internet and can provide an almost unlimited variety of new services on the same general-purpose IP network.
- VoIP is extremely sensitive to delay through the network. Voice quality degrades rapidly if the delay approaches 100 ms.
- Residential VoIP is typically implemented with a standard telephone and a telephone adapter that digitizes the voice signal and handles the signaling protocols.

- Several unresolved regulatory questions regarding VoIP include E911 service, the Universal Service Fund (USF), and the Communications Assistance to Law Enforcement Act (CALEA).
- The many advantages to VoIP include the potential for lower cost and the ease of providing new features.
- Disadvantages to VoIP include the need for a broadband Internet connection, the lack of E911 service, incompatibility with auto-dialers in fire and security alarms, and service interruption during power outages.

Key Terms

Adapter A mechanical device designed to align and join two optical connectors.

Communications Assistance to Law Enforcement Act (CALEA) A law passed by Congress in 1994 that requires telecommunications carriers to ensure that their digital and wireless networks can comply with authorized wiretaps and electronic surveillance.

E911 service Enhanced 911 (E911) service automatically reports the telephone number and location of 911 calls when they are connect to a PSP.

H.323 An ITU-T standard for transmitting multimedia communications (such as audio and video) over a packet-switched network.

Internet telephony Another name for Voice over Internet Protocol (VoIP).

Key system A small (often electromechanical) telecommunications switching system that allows a user to pick up or hold a call on any several lines.

Network address port translation (NAPT) An extension to network address translation (NAT) that allows multiple port identifiers to be multiplexed into a single external IP address. NAPT is described in RFC 2663.

Network address translation (NAT) A method, often implemented in routers, that allows IP addresses to be mapped from one address realm to another. NAT is described in RFC 1631.

Network interface device (NID) The point of connection between networks. In a residence, the location of the demarcation point.

Private branch exchange (PBX) Private branch exchange.

Session Initiation Protocol (SIP) An IETF standard protocol (in the OSI applications layer) for initiating an interactive user session involving multimedia elements such as voice and video.

Structured cabling system (SCS) A cabling system that is designed to support a range of applications with standard interfaces and specified transmission performance.

Uninterruptible power supply (UPS) Alternating-current power supply with battery backup to keep critical applications running for a short time in the event of a power failure.

Universal Service Fund (USF) A fund, mandated by the FCC, that is designed to offset the higher cost of providing telephone service in some (primarily rural) areas and thus provide universal, affordable telephone service to all households.

Voice over Internet Protocol (VoIP) A technology that allows voice calls to be digitized and transmitted over the Internet using the Internet protocol.

Review Questions

1. Describe an analog telephony signal. What is present on the wires besides the voice signal?
2. What functions does a PBX provide?
3. Is it possible for a LAN connection and an analog phone line to share the same UTP cable and RJ-45 jack? Is it recommended?
4. What are the main applications of home data networks?
5. What is the function of the Network Address Translation (NAT) and Network Address Port Translation (NAPT) protocols?

6. What are two ways of sharing Internet access among multiple PCs?
7. What is the most efficient way to integrate wired and wireless LANs in a home network?
8. Give a high-level explanation of how VoIP works.
9. What are the two primary signaling protocols used with VoIP?
10. What functions does a VoIP telephone adapter provide?
11. What happens to a VoIP stream if there is excessive delay or loss of packets in the IP network? Why?
12. What are the three main regulatory constraints that apply to traditional telephony but not to VoIP?
13. If there is an alternating-current power outage, can you use a VoIP phone to call the power company?
14. What problems arise when using VoIP with the autodialers in fire and security alarms?
15. List at least three pros and three cons regarding VoIP.

Entertainment Applications

OBJECTIVES   *After studying this chapter, you should be able to:*

- Understand the architecture of home entertainment networks.
- Discuss issues related to copyright and digital rights management.
- Explain how digital audio and video are encoded and stored.
- Distinguish between the most common standards for digital audio and video.
- Discuss whole-house audio and home theater systems.
- Explain the function of an entertainment server.
- Understand the options for audio and video cabling.
- Discuss new features that can be provided on digital entertainment lnetworks.

OUTLINE   Introduction
Audio and Music
Video
Home Theater
Gaming
Entertainment Server
Future Applications

Introduction

The concept of an entertainment network is relatively new. The entertainment portion of a home network usually consists of a cable television line hooked to one or more television sets. Each television may have dedicated peripherals, such as a VCR or DVD player, plugged into it. But there is no communication among the television sets and no way to share content[1] among them.

All of this changes with the advent of high-speed digital networks in the home. With today's technology, it is possible to have digital entertainment devices (televisions, monitors, PCs, DVD players, video recorders, etc.) networked together so they can be controlled from a common point (or points) and content can be shared among them. Networked digital entertainment devices raise a number of issues in regard to control of the network, interoperability of equipment, and protection against unauthorized use of copyrighted material.

The Digital Living Network Alliance

The **Digital Living Network Alliance**[2] **(DLNA) http://www.dlna.org** is an industry consortium that was formed in 2003 to ensure the interoperability and compatibility of digital networked consumer entertainment devices. The DLNA's vision is:

> ... interoperable networked devices in the home that provide new value propositions and opportunities for consumers and product vendors. We are committed to providing a seamless interaction among Consumer Electronics (CE), mobile, and Personal Computer (PC) devices and believe that this is best accomplished through a collaborative industry effort focused on delivering an interoperability framework for networked media devices.

In other words, the DLNA works to ensure interoperability by developing guidelines for manufacturers of networked entertainment devices. The goal is to provide consumer electronic devices that are easy to install and use and that interoperate with each other and with legacy devices (such as existing television sets).

The DLNA has more than 100 members. The board of directors consists of Hewlett-Packard, Intel, Microsoft, Nokia, Panasonic, Philips, Samsung, and Sony. One of the major activities of the DLNA is to develop and validate usage scenarios so that consumer electronics from various manufacturers will work together as the user expects them to. Equipment in the entertainment network is typically one of three types.

- **Digital home server (DHS):** A device that serves as a source for content. It is for either streaming content or file transfer.
- **Digital home renderer (DHR):** A device that renders content from the DHS. To **render** refers to producing or depicting a version of the content. For example, an HDTV video of a concert would be rendered differently by an HDTV set, a color monitor, a black-and-white monitor, and a stereo amplifier.
- **Digital home controller (DHC):** A device that can control the operation of DHRs and the transfer of content between DHSs and DHRs. In some cases, the DHC might be integrated with the DHR.

A usage scenario involves considering a set of actions a user might perform and considering in detail the impact on each of the devices in the network. A simple example of a usage scenario would be something like this: A user is watching a television program. Using a universal remote control, she redirects the program

---

[1] In the context of entertainment networks, "content" refers to material to be watched or listened to, such as movies, television programs, video clips, digital photographs, recorded music, and so on.
[2] The DLNA was formerly known as the Digital Home Working Group (DHWG).

to another television set in the home, searches the **personal video recorder (PVR)** for a movie, and begins watching the movie on the first television set.

In this case, the universal remote control is the DHC, the PVR is the DHS, and both of the televisions are DHRs. While this scenario may seem simple, the flow of control commands on the network can be quite complex; it is almost impossible to guarantee interoperability of different brands of equipment without doing this type of analysis.

Providing a capability for users to freely share digital content among devices on a network leads us right into a major issue that is currently being hotly debated.

Digital Rights Management

Copyright laws give authors, musicians, and filmmakers, for example, the right to the exclusive use of their creations and any derivative works for a "limited time." That limited time was originally 14 years with an option to renew for another 14 years, but Congress has lengthened the copyright period many times. It now stands at life plus 70 years for individuals and 95 years for corporations.

In the days when content (music, video, etc.) was transmitted and stored in analog format, it was relatively difficult to make a copy and the copy was of inferior quality to the original. Copy protection was therefore not a big deal. In modern digital networks, however, making perfect copies is routine. It is just as easy to cut and paste an audio or video clip as it is a text file. In fact, just watching a digital video or listening to a digital audio recording on your PC involves making a copy in the PC's memory and this may appear (to some) to violate copyright restrictions. Methods for providing content protection in digital media are often referred to as **digital rights management (DRM).**

With the advent of file-sharing networks like Napster, Gnutella, and Kazaa, the content providers [particularly the Recording Industry Association of America (RIAA) and the Motion Picture Association of America (MPAA)] became concerned about the massive unauthorized duplication of copyrighted material. Although this concern is understandable, the entertainment industry's consistent record of opposing new technology makes its motives suspect.[3] In the past century, the entertainment industry has vigorously opposed recorded music (for violating composers' rights), radio (for violating performers' rights), cable television (for violating broadcasters' rights), and the VCR (for violating moviemakers' rights). In every case, the movie industry not only lost the battle but also proceeded to make enormous amounts of money using the technology it had opposed.

A similar scenario seems to be unfolding with digital home networks. It is important for content creators to be fairly compensated for their work; otherwise, not much content will be created. It is equally important for content consumers to be able to use the content they have purchased in a convenient manner; otherwise, they will not buy much. Since digital home networking technology is still fairly new, it will take some time for copyright law to catch up with the technology.

Copyrighted works have always been subject to **fair use** without compensation of the copyright holder. Fair use typically involves using short excerpts of the copyrighted work in criticism, reviews, news reporting, teaching, research, and parodies. The Supreme Court[4] has extended the definition of fair use to include

---

[3] The RIAA's overzealous tactics, including filing lawsuits against schoolchildren, are not helping them in the court of public opinion.

[4] Universal City Studios sued Sony claiming that the Betamax VCR was infringing the copyrights of Universal's movies. In 1984 the Supreme Court ruled in Sony's favor and established time shifting as a fair use.

making a copy for time-shifting purposes, for example, to record a television show so it can be viewed (privately and noncommercially, of course) at a later time.

Many (but not all) legal experts believe that format shifting (making a personal copy of a work in another format, such as converting an audio CD into MP3 format) and making backup copies also constitute fair use of copyrighted material.

Home networks allow digital content to be shared freely among the devices on the network. Digital rights management (DRM) technology must prevent most unauthorized uses of copyrighted material while not interfering with the fair use of material that was legally acquired.

As the DLNA has noted, consumers want to share digital content freely among the devices on their home networks. If incompatible and overly stringent copy protection schemes prevent this, then the fundamental purpose of the home network cannot be realized. Currently, there are numerous proposals to solve this problem. Some, like **Digital Transmission Content Protection (DTCP,** also known as 5C for "5 companies"), which is used with IEEE 1394 networks, are quite consumer-friendly. The DTCP proposal allows digital transmissions to be marked in one of three ways.

- *Copy Freely:* This is the default mode.
- *Copy Once:* This allows one copy to be made for backup purposes but forbids making a copy of the copy.
- *Copy Never:* This prevents the content from being recorded at all. This is usually used for premium content like pay-per-view movies.

Since IEEE 1394 uses compressed video signals, it is compatible with consumer recording equipment. In addition, DTCP has encoding rules that restrict content providers from blocking the recording of material that could be recorded during the analog VCR era.

In contrast, the **Digital Visual Interface (DVI)** uses uncompressed video that is incompatible with all consumer recording equipment. It also uses a copy protection system (**High-Definition Copy Protection (HDCP)**) that forbids all copying. It is essential that a compromise be reached on DRM issues so that home entertainment networks can achieve their full potential for consumers, equipment manufacturers, and content producers.[5]

Audio and Music

Audio systems are the most common form of entertainment devices in the home. Traditionally, audio systems have been limited to radios, portable devices, and stereos with directly connected speakers. None of these devices are usually capable of networking with other devices.

In recent years, there has been a trend toward the use of whole-house audio systems that provide speakers in a number of rooms in the house, all linked back to a central music source. This music source is often a radio receiver—either a standard AM/FM receiver or one of the relatively new satellite radio receivers (XM or Sirius) that have been designed for home use. The music source can also be a CD

---

[5] This has been a brief introduction to the very important subject of DRM. For a more thorough treatment, see *Free Culture* (Lessig 2004). For up-to-the-minute details, see the Web site of the Electronic Frontier Foundation (**http://www.eff.org**).

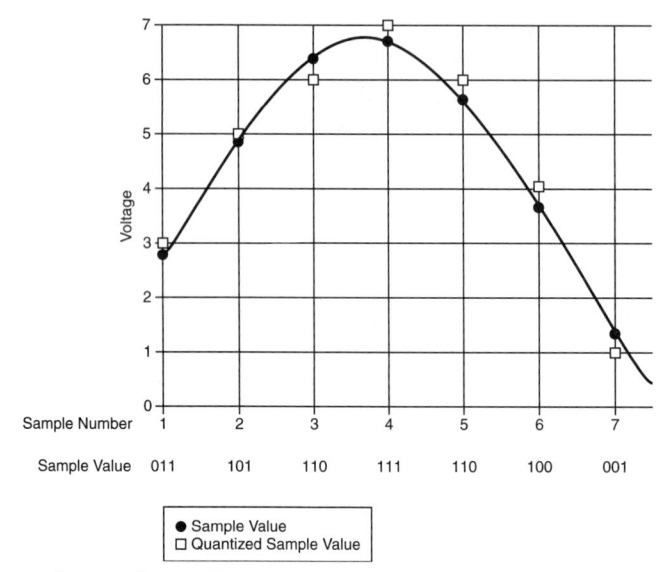

FIGURE 6–1 Sampling and Encoding an Analog Signal

player or a device that stores digital music in memory, such as a PC or an entertainment server (which we discuss later in the chapter).

Audio Encoding

Before looking at the various configurations for home audio systems, let us consider the many different ways that digital audio can be encoded. A high-fidelity audio signal consists of frequencies from near-direct-current to about 20 kHz. The

---

SAMPLING AND ENCODING

Figure 6–1 shows an analog signal. A three-step process is used to convert it to a digital bit stream.

1. First, the analog signal is *sampled* (measured) at regular time intervals. In the figure, the sampling times are numbered from 1 to 7. The number of samples per second is called the *sampling rate*. The sampling rate must be at least twice the highest frequency contained in the input signal to ensure an accurate digital representation.

2. The samples are then *quantized*, or rounded off to the nearest measurement unit. The vertical scale in the figure is in volts, so the samples are rounded off to the nearest volt. The sample values (from 1 to 7) are 3, 5, 6, 7, 6, 4, and 1.

3. Finally, the quantized samples are encoded in digital format. The simplest way is to just express the samples as a string of binary numbers. As indicated at the bottom of the figure, the binary sample values are (from 1 to 7) 011, 101, 110, 111, 110, 100, and 001.

These binary sample values are then transmitted. The receiver can store them, transmit them to someone else, or use them to reconstruct a version of the original signal. This simple encoding format is referred to as **pulse code modulation (PCM).**

simplest way to digitize the audio signal is to just sample it and digitally encode the samples. This is the technique that is used for CD audio; the signal is sampled 44,100 times per second. The voltage of each sample is then measured and recorded as a 16-bit number (that is, one of 65,536 possible levels). Each of the two stereo channels is sampled and encoded separately, so the total bit rate is: $2 \times 44,100 \times 16 = 1,411,200$ b/s (or about 1.4 Mb/s).[6] For more details on the sampling and encoding process, see the sidebar, Sampling and Encoding.

After the CD format, the most commonly used audio format is *MP3,* which is shorthand for MPEG-1, Audio Layer 3.[7] It was originally developed to supply the audio information for MPEG-1 video files. Although MPEG-1 is no longer commonly used, MP3 has become the standard way to store high-quality audio files. The data rate for MP3 signals varies according to the characteristics of the signal being encoded and the desired quality of the output. A typical MP3 data rate is 128 kb/s. Note that this is more than a factor of 10 lower than the CD audio data rate. This order of magnitude decrease in data rate is achieved by a compression algorithm that encodes the signal more efficiently than the brute-force linear encoding used by CD audio.

Digital music files that are stored in memory (either semiconductor memory or on a hard disk) are normally stored in MP3 format to save space. It is relatively simple to convert the linear coding of CD audio files into MP3 files. This process is called **ripping.**

Many other audio formats have been developed for various purposes. Of particular interest are two new formats that have been developed for DVD applications. They are known as *DVD-audio* (*DVD-A*) and *super-audio CD* (*SACD*). Although they use different encoding techniques (and are therefore incompatible with each other), they both have a typical bit rate of about 4.8 Mb/s. This is quite a bit higher than CD-audio, but there is a lot more storage space on a DVD.

Digital Music Storage Capacity

Having looked briefly at the most common digital audio formats, let us calculate the space required to store digital music files using each of the four encoding methods we have described. We consider seven different storage media.

- *3.5-inch floppy disks:* These hold about 1.5 MB. This medium is just about obsolete, as this discussion will confirm.
- *USB RAM drives:* These key-chain-style memory devices are available in a variety of capacities. For this discussion, we use a typical size of 256 MB.
- *CDs:* Audio CDs and CD-ROMs typically hold about 650 MB.
- *DVDs:* Digital versatile disks are optical disks that are read with a red laser and hold about 4.7 GB, or sometimes a little more. See the sidebar, CD and DVD Capacity, for more detail.
- *Blu-ray disks:* These are a new technology similar to DVDs, but they are read with a blue laser, which has a shorter wavelength than the red lasers used with DVDs. This allows each bit to be written in a smaller area. The capacity of a Blu-ray disk is about six or seven times that of a DVD. A typical value is 27 GB.

---

[6] This is a fairly high bit rate. For comparison purposes, a digital telephony signal is only 64 kb/s (8,000 samples per second times 8 bits per sample).

[7] MPEG stands for Motion Picture Experts Group; we discuss MPEG standards in the next section on video.

TABLE 6–1
Data Rates for Digital Audio

Format	Typical Data Rate (kb/s)	Storage Required for 1 Hour of Music (MB)
MP3	128	58
CD	1,411	635
DVD-A SACD	4,800	2,160

- *MP3 players:* They typically have a tiny hard drive that holds up to 40 GB.
- *Hard disk drives:* These are available with a capacity of 1 TB (1,000 GB) or more.

The digital audio formats, their typical data rates, and the amount of storage required to hold 1 hour of music are summarized in Table 6–1.

Using the data from Table 6–1, it is easy to calculate how much music each of the storage technologies listed can hold, as shown in Table 6–2.

A couple of points are pretty clear from Table 6–2. The first is that 3.5-inch floppy disks are essentially useless for audio storage. The second is that the other devices can all store impressive amounts of MP3 audio. The pocket-size MP3 player can store almost a month's worth of continuous music, while the terabyte hard disk can play continuous music for almost two years.

Figure 6–2 gives a graphical feel for the huge amount of MP3 data that can be easily and cheaply stored with today's technology.

Whole-House Audio

Whole-house audio is a system that distributes music to multiple rooms in the house. The two basic configurations are centralized and distributed. A centralized system has all the active equipment (amplifiers, etc.) at a central point and runs speaker wires out to the rooms being served. A distributed system has the music source(s) at a central point and has individual amplifiers in each room, usually mounted with a volume control in a single-gang electrical box. Figure 6–3 is a

TABLE 6–2
Digital Music Storage Capacities

Storage Device	Capacity	MP3	CD-audio	DVD-A SACD
Floppy disk	1.5 MB	1.6 min.	9 s	3 s
USB RAM drive	256 MB	4.4	24 min.	7 min.
CD	650 MB	11	1	18 min.
DVD	4.7 GB	82	7	2
Blu-ray disk	27 GB	469	43	13
MP3 player	40 GB	694	63	19
Hard disk	1 TB	17,361	1,575	463

> **CD AND DVD CAPACITY**
>
> CDs and DVDs are identical-looking plastic disks. They are 12 cm in diameter and 1.2 mm thick. Both are optical storage media. The difference between them is the wavelength of the laser light used to read and write data on the disk. The shorter the wavelength of the laser, the smaller the area each bit takes up and the more bits that will fit on the disk.
>
> CDs use a 780-nm infrared laser and can hold about 650 MB. (*Note:* All of the disk capacities we discuss are in decimal megabytes.)
>
> DVDs use a red laser, which operates at a wavelength of 650 nm. DVDs can have one or two layers on each side and can be single- or double-sided. A normal DVD used for video is single-layer and single-sided and has a capacity of about 4.7 GB. Single-sided, dual-layer DVDs have a capacity of about 8.5 GB. The first layer starts at the outside of the disk and spirals in, while the second layer starts in the middle and spirals out. DVD players can handle either single- or dual-layer disks. Dual-layer DVD recorders were just beginning to become widely available in 2004.
>
> Double-sided DVDs have capacities of 9.4 GB and 17 GB (for single- and dual-layer, respectively). Double-sided DVDs are often used for archival storage in libraries, for example. They are infrequently used in the consumer market for several reasons: there is very little room for a label, both sides of the disk must be protected from dirt and scratches, and the disk must be physically removed and turned over to play the other side.
>
> Higher-capacity DVDs using a 405-nm blue laser have recently been developed. There are two competing formats, which are, of course, incompatible with each other. The Blu-ray disk (supported by Sony, Dell, and H-P) holds about 23 to 27 GB on a single layer and seems to be the leading contender. The HD-DVD (High-Density DVD) was developed by the DVD Forum and holds up to 15 to 20 GB. Like CDs and standard DVDs, both Blu-ray and HDDVD disks will be available in read-only, write-once, and rewritable formats.

diagram of these two configurations. The wires in an audio system carry two different types of signals. Signals between an amplifier and a speaker are referred to as *speaker-level* signals. Speaker-level signals require special audio cable for high-quality audio transmission. Audio cable is usually stranded cable that is 16 AWG or larger. Unamplified signals between components of the audio system are called *line-level* signals. Line-level signals can be carried on standard UTP cable.

Whole-house audio systems often include a whole-house IR remote capability as well. This allows an IR remote control, like those that come with most audio and video equipment, to control the audio system from anywhere in the house.

Zones

The whole-house audio system may be divided into multiple zones, or listening areas. The music for each zone may be selected independently. Amplifiers supporting up to twelve zones are readily available. A more typical example would be a four-zone system, with separate zones for the living room, family room, kitchen, and master bedroom. Additional zones may be added for the patio or deck, home office, kids' bedrooms, and so on. Figure 6–4 shows an example of a simple, centralized, two-zone audio system.

Whether the system is single-zone or multizone is determined by the type of amplifier or receiver that is installed and the way the speaker wires are connected

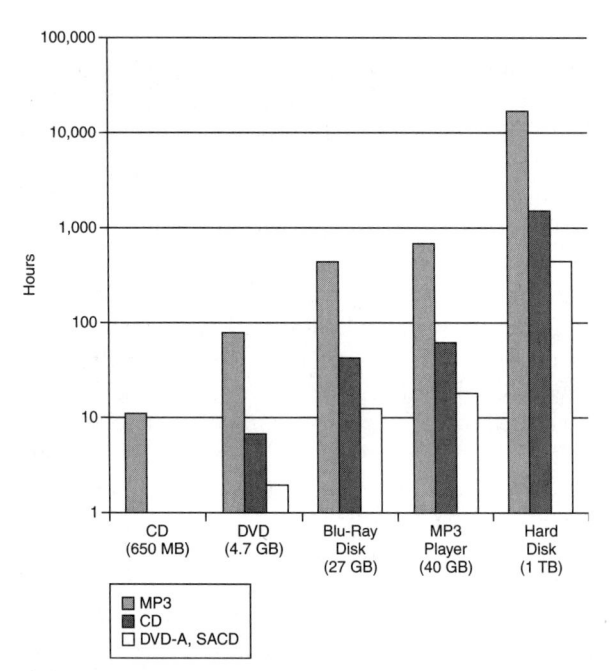

FIGURE 6–2 Digital Music Storage Capacities

to it. The speakers and their wiring are the same in either case, so the house can be prewired before the number of zones is determined.

Speakers

There are many different types of speakers to choose from. Some of the options include:

- *Freestanding speakers:* These come in a wide variety of sizes and shapes. They connect via a length of speaker wire to an audio wall outlet, which is generally provided with push-in connectors for the speaker wire, although it may also use banana plugs, spade lugs, or some other type of fastener.

- *In-wall or in-ceiling speakers:* These are permanently installed and can usually be painted to match the decor (Figure 6–5). Ceiling speakers are often preferred because they are visually more unobtrusive and do a nice job of filling the room with sound.

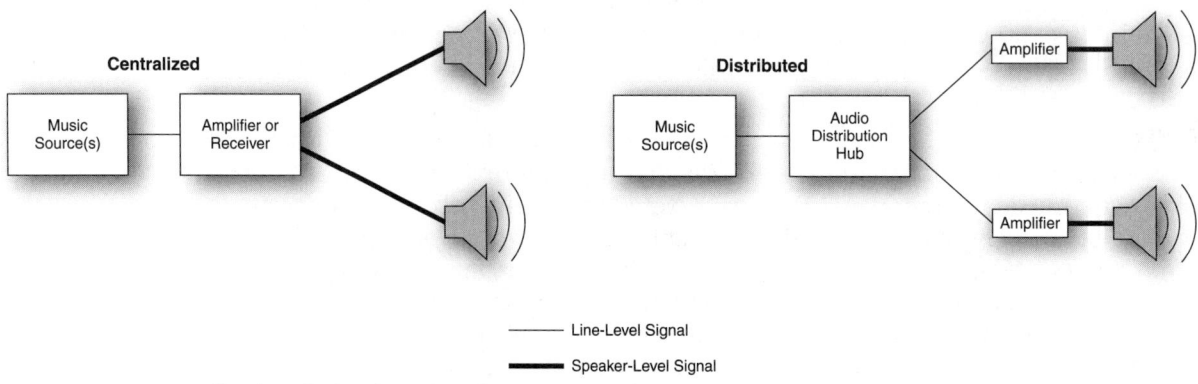

FIGURE 6–3 Centralized and Distributed Architectures for Whole-House Audio

Entertainment Applications • 121

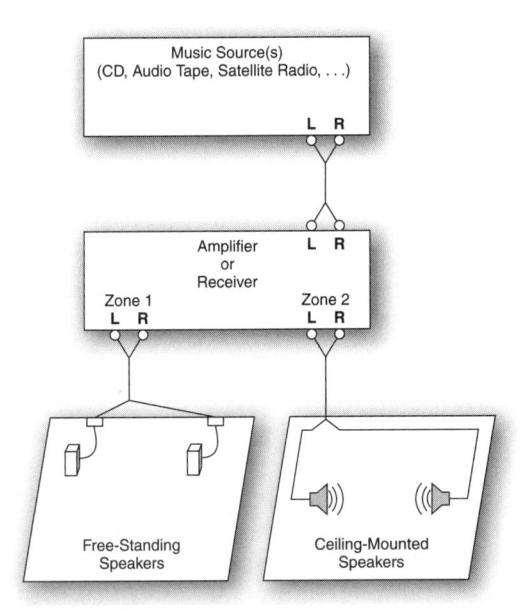

FIGURE 6–4 Centralized Two-Zone Audio System

- *Thin speakers that hang on the wall:* These connect to an audio outlet via speaker wires like freestanding speakers but do not take up any floor space. They are often decorated so that they appear to be a painting or photograph hanging on the wall. If desired, the audio outlet can be located behind the speaker so that there are no visible wires.
- *Totally in-wall speakers, such as the Induction Dynamics SolidDrive*™ (**http://www.inductiondynamics.com**): These speakers mount completely inside the wall. They literally vibrate the wallboard and turn the entire room into a speaker.
- *Wireless speakers:* These can eliminate the need for speaker wires, but the speaker must still be powered somehow. Most wireless speakers run off batteries (often quite a few batteries) with an optional alternating-current adapter.

The choice of speakers is generally based on a combination of audio performance, aesthetics, and price.

Regardless of what type of speaker is chosen (other than wireless), the impedance of the speaker is very important. For proper operation, the impedance of the

FIGURE 6–5 **In-ceiling Speakers** (Courtesy of Leviton Manufacturing Co., Inc.)

speakers must match the impedance of the amplifier. Most speakers and amplifiers have an impedance of ohms, so this is not normally a problem. However, for configurations that require more than one set of speakers for a particular signal (for example, in a single-zone system when the same signal goes to speakers in several different rooms), the combined impedance of the speakers will be too low. There are two ways to solve this problem: either use a distributed system where each set of speakers has its own amplifier, or use an impedance-matching device at the output of the centralized amplifier.

Audio Cabling

Speaker-level signals can transmit a significant amount of power to the speakers. It is important to keep noise and distortion to a minimum, since the signal is in analog form. Therefore, as already mentioned, speaker-level signals require a heftier cable than UTP. A true audiophile can spend an almost unlimited amount of money on speaker cable. For a good-quality installation of a whole-house audio system, generally, 16-AWG stranded speaker wire is fine. For long runs, the wire is generally available in 14 and 12 AWG as well.

Figure 6–6 shows an example of some audio cables with both two and four conductors. Note the size of the copper conductors as compared to the UTPs in the hybrid cable at the bottom center of the figure.

Line-level signals can be carried on standard Category 5e or Category 6 UTP cable. A common type of distributed audio system is known as A-bus and uses line-level analog signals over UTP cable to volume controls in each room, then speaker wire from the volume control to the speakers.

TIA-570B gives some guidelines on cabling for whole-house audio. TIA-570B mentions three specific areas of audio cabling, as summarized in Table 6–3. Following these recommendations will give sufficient flexibility to support either a centralized or distributed architecture, although for a specific audio system they are generally overkill.

Video

Video is the centerpiece of many home entertainment systems. Like audio, it exists in a number of different formats and can be digitally encoded in different ways. It can be rendered on many different types of display.

FIGURE 6–6 Audio Cable (Courtesy of Genesis Cable Systems)

TABLE 6–3
TIA-570B Recommendations for Audio Cabling

From	To	Cable(s)	Description
Audio-equipment	DD	6 conductors speaker wire	L/R audio channels and control voltage
		1 CAT 5e/6 UTP cable	Communication (digital radio, etc.)
DD or audio equipment	Volume control	4 conductors speaker wire	L and R audio channels
		1 CAT 5e/6 UTP cable	Digital audio or control information
Volume control	Speakers	4 conductors speaker wire	L/R audio channels
		1 CAT 5e/6 UTP cable	Future proofing

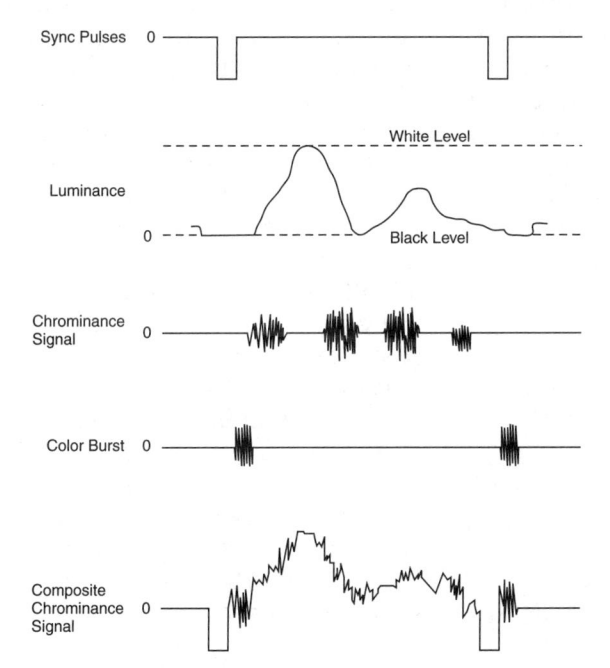

FIGURE 6–7 Baseband Analog Video Signal

Video Formats

The most basic type of video signal is analog baseband video, which consists of the luminance (brightness) and chrominance (color) signals for a single channel, multiplexed together with audio and synchronization information to form a rather complex waveform, as illustrated in Figure 6–7.

Video cameras and television screens create a picture the same way a farmer plows his fields—one line at a time.[8] The traditional method of scanning is called

---

[8] The idea of scanning a picture and sequentially transmitting it occurred to Philo T. Farnsworth as he was plowing a field on his father's potato farm. The April 2002 issue of *Wired Magazine* has an interesting article about Farnsworth's invention of television.

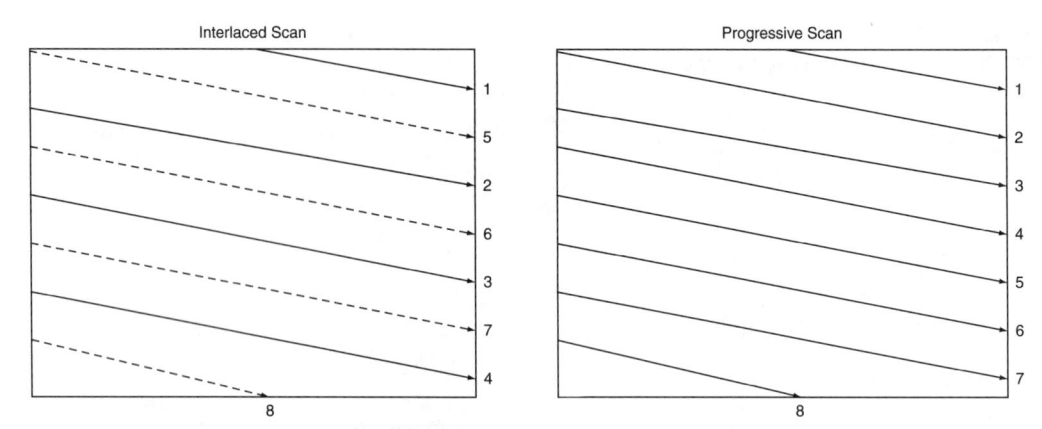

FIGURE 6–8 Interlaced and Progressive Scanning

an **interlaced scan,** as illustrated in the left half of Figure 6–8, which illustrates a very small screen with only eight scan lines. The composite waveform of Figure 6–7 represents one scan line in Figure 6–8. As shown in the figure, the solid lines are scanned first, then the dashed lines are scanned. After scanning Line 8, there is a brief delay and then scanning begins again at Line 1. Interlaced scanning makes it appear that the screen is being refreshed faster than it really is and helps to prevent visible flicker in the image. The three video specifications discussed later in this section all use interlaced scan. **Progressive scan,** as shown in the right half of Figure 6–8, simply scans sequentially down the image.

While interlaced scanning permits a slower frame rate and therefore saves bandwidth, it makes it more difficult to do some effects, such as digital cutting and pasting. This is because each frame that is displayed is really a composite of half the current frame and half the previous frame. Video equipment sometimes uses progressive scan, which means that the frame is scanned from top to bottom without interlacing. The type of scanning is often indicated by an "i" or a "p" after the frame rate. The National Television System Committee (NTSC) format, for example, uses a 60i frame rate (30 interlaced frames per second, which gives an apparent rate of 60 frames per second). Film cameras have traditionally operated at 24 frames per second. When digitized, they usually operate at a 24p frame rate.[9]

The three different specifications for baseband video signals are NTSC, PAL, and SECAM. The **National Television Standards Committee (NTSC)** format is the oldest and simplest to implement. It was developed in the United States and first broadcast on January 23, 1954. It is based on 525 lines per frame, 30 frames per second, and a 4:3 **aspect ratio** (that is, the picture is 1⅓ times as wide as it is tall). It is currently used throughout North, Central, and South America, as well as in Japan, Korea, Taiwan, and the Philippines. It employs a simultaneous amplitude and phase-modulated subcarrier to separate luminance and chrominance information and uses a bandwidth of 6 MHz.

Phase Alternate Line (PAL) was developed in Europe and first broadcast in Germany and the United Kingdom in 1967. It is currently used in most of Europe, Asia, Australia, and Africa. It uses 625 lines per frame, 25 frames per second, and a

---

[9] This has been a simplified introduction to video formats. Attempting to combine, splice, or convert video of different formats can be quite involved. For more information, see Kipp (2004). **http://www.dv.com**, or **http://www.adamwilt.com**.

TABLE 6–4
Typical U.S. Television Channel Frequencies

Band	Channels	Freq. Range (MHz)
VHF	2–13	52–210
UHF	14–83	470–890
CATV	2–78	55–547

4:3 aspect ratio. It uses a subcarrier similar to NTSC but alternates the phase of the color difference signals (hence, the name PAL) to minimize distortion between chrominance and luminance. It uses about 8 MHz of bandwidth.

Sequential Couleur a Memoire (SECAM) was developed in France and also first broadcast in 1967. The line and frame rates, aspect ratio, and bandwidth are the same as for PAL. It is not widely used.

Broadband video consists of a number of baseband video signals modulated up to carrier frequencies and multiplexed together. This is the type of television signal that is broadcast through the air or over cable television networks. Table 6–4 gives some of the channel frequency assignments in the United States.[10] Channel 2 is at the lowest frequency, from 54 to 60 MHz; remember that each NTSC channel takes 6 MHz of bandwidth. The UHF channels extend up to above 800 MHz in frequency, while the CATV channels extend up to about 550 MHz.

Digital Television

The television formats discussed are all based on analog representation of both the audio and video signals. Digital television (DTV) refers to digitally encoding both the audio and video and transmitting as a digital bit stream. This has many advantages, including higher resolution, use of (DSP-based) compression techniques, better signal-to-noise ratio, and so on. It also makes the video signal compatible with storage in digital memory (such as a disk drive) and with transmission over a digital network, including both home networks and the Internet. Finally, it takes advantage of the declining cost curve of digital electronics.

The two main types of DTV are **Standard-Definition Television (SDTV)** and **High Definition Television (HDTV).** SDTV has image quality equivalent to a DVD and the same 4:3 aspect ratio as analog television. HDTV has much better resolution and image quality—roughly equivalent to a 35-mm camera—and audio quality similar to a compact disk. HDTV also has a 16:9 aspect ratio—like a movie screen. Figure 6–9 illustrates the difference in width between SDTV and HDTV screens of the same height. It also shows how much an SDTV image must be reduced to fit on an HDTV screen.

The Advanced Television Systems Committee (ATSC) is an international, nonprofit organization that is responsible for developing digital television standards.[11]

---

[10] Note that the channel assignments may vary somewhat from one location to another and some channels may be reserved for nonbroadcast use.
[11] The ATSC was formed by a number of organizations, including the Electronic Industries Association (EIA), the Institute of Electrical and Electronic Engineers (IEEE), the National Association of Broadcasts (NAB), the National Cable Television Association (NCTA), and the Society of Motion Picture and Television Engineers (SMPTE). Their standards can be downloaded free of charge at **http://www.atsc.org**.

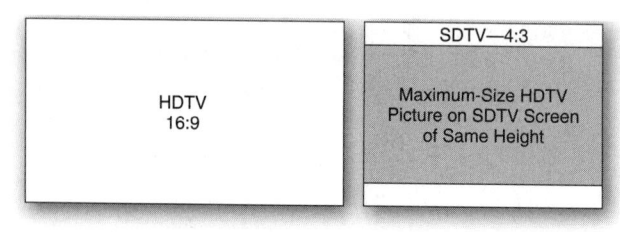

FIGURE 6–9 HDTV and SDTV Aspect Ratios

The ATSC has defined eighteen different video formats. They differ in the following parameters:

- *Frame rates:* 24p, 30p, 60p, and 60i
- *Aspect ratios:* Either 4:3 or 16:9
- *Spatial resolution:* From 640 × 480 to 1,920 × 1,080 **pixels**

The characteristics of some of the most important formats are summarized in Table 6–5. This table lists some of the most common formats; there are fourteen others as well. Note that an HDTV picture represents a lot of digital data. The 1,920 • 1,080 format, for example contains over 2 million pixels, each of which may be encoded as multiple bytes of data. So, while a picture may be worth a thousand words, it is also worth several million bytes!

Even HDTV is not the ultimate format. Higher-resolution formats are currently in the experimental stage. In late 2003, for example, the Japan Broadcasting Corporation (NHK) demonstrated a prototype of an ultra-high-definition video (UHDV) system with 16 times the resolution of HDTV. The UHDV prototype had 4,000 scan lines and 33 million pixels per screen. The images were said to be so realistic that "the eye struggles to distinguish them from reality." Eighteen minutes of UHDV footage required 3.5 terabytes of storage space.

Another system that has been proposed is a 3-D holographic television. It appears almost certain that as HDTV becomes the mainstream format, another higher-resolution video format will appear at the top end of the market.

Video Encoding

Encoding of analog video signals into a digital bit stream is similar to the encoding of audio signals but is more complex because there is more information to deal with. Video signals are typically separated into red, green, and blue components (referred to as *R*, *G*, and *B*, respectively).

TABLE 6–5
Digital Television Formats

Format	Scanning	Aspect Ratio	Pixels/Line	Lines/Frame	Pixels/Frame	Megapixels per Second
SDTV	60i	4:3	640	480	307,200	9.216
	30p	4:3	640	480	307,200	9.216
HDTV	30p	16:9	1,280	720	921,600	27.648
	60i	16:9	1,920	1,080	2,073,600	62.208

TABLE 6–6
RGB Color Combinations

Red	Green	Blue	Resulting Color
0	0	0	Black
0	0	MAX	Blue
0	MAX	0	Green
0	MAX	MAX	Cyan
MAX	0	0	Red
MAX	0	MAX	Magenta
MAX	MAX	0	Yellow
MAX	MAX	MAX	White

The R, G, and B signals can be visualized as the three coordinate axes in a color space. Any color can be made by the appropriate combination of R, G, and B, as indicated in Table 6–6.

The *RGB* signal undergoes a process known as *Gamma correction,* which results in a signal known as *R'G'B'*. The R', G', and B' signals are then combined according to the equations in Table 6–7 to form Y, Cr, and Cb. The Y signal is called the *luminance,* and the Cr (Color red) and Cb (Color blue) signals are called the *chrominance*. The luminance represents the black-and-white component of the signal, while the chrominance represents the color information.

The equations for the *YCrCb* video format, which apply to SDTV signals, are specified in ITU-R BT.601-5. For HDTV, a similar set of equations with slightly different coefficients are specified in ITU-R BT.709-5. After the *YCrCb* values are transmitted to the receiver, the R, G, and B values can be recovered using inverse equations supplied in these two standards.

At this point, it is reasonable to ask, Why bother to do all this conversion to *YCrCb* format so that you can later convert back to *RGB*? There are a couple of advantages to the *YCrCb* format.

- Separating the luminance and chrominance is the traditional way of transmitting a composite video signal, as shown in Figure 6–7. This format is backward compatible with existing monochrome monitors and televisions, which just respond to the luminance and ignore the chrominance information.

TABLE 6–7
YCrCb Video Format (SDTV)

Quantity	Equation
Luminance (Y)	$Y = 0.2989\ R' + 0.5866\ G' + 0.1145\ B'$
Chrominance (Cr)	$Cr = 128 + 0.7132\ (R' - Y)$
Chrominance (Cb)	$Cb = 128 + 0.5647\ (B' - Y)$

- The human eye responds much more strongly to luminance than it does to chrominance. Therefore, the bandwidth needed to transmit the signal can be reduced by sampling the chrominance at a lower rate than the luminance. Common sampling formats include 4:2:2 (four samples of *Y* for every two samples of *Cr* and *Cb*) and 4:1:1 (where the chrominance is sampled at 1/4 the rate of the luminance). There is also a confusingly named 4:2:0 format in which *Cr* and *Cb* are sampled at every other pixel on every other line. 4:2:0 gives the same number of samples as 4:1:1, but they are taken at different locations in the image.

Video Encoding Standards

There are a number of commonly used standards for digital video, including MPEG, ITU-R BT.601/709, DVD video, DV, IEC 61883, and DVI. Some of these are compressed video and are suitable for networking or storage in consumer equipment. Others use uncompressed video and are used in commercial (studio-grade) equipment or for local video display. In general, uncompressed video has too high a data rate to be networked or stored in consumer equipment.

Motion Picture Experts Group Series. The Motion Picture Experts Group (MPEG) is the branch of ISO/IEC that is responsible for developing international standards for motion pictures, including encoding, compression, and so on. This group has issued a series of standards, starting with MPEG-1, which was designed to store video on CD-ROMs and is no longer widely used. As previously noted, however, the audio coding format, MP3, is still widely used.

MPEG-2 is currently the most important of the series. It is widely used for both DVDs and digital television broadcasting. MPEG-2 uses five different compression techniques to reduce the bit rate by a factor of 25 to 50. It uses intra- and inter-frame compression to remove redundancy both within frames and between successive frames. The MPEG-2 algorithm is very asymmetrical; it uses a complex encoder and a relatively simple decoder. This is a good trade-off, since decoders are much more numerous. MPEG-2 provides two MP3 audio channels and also supports 5.1/7.1 audio.

MPEG-2 has many different options and can run at data rates as low as 2 Mb/s or as high as 100 Mb/s. For SDTV signals, about 4 to 8 Mb/s is more typical; while for HDTV signals, the data rates are about twice as high.

MPEG-2 is the most common video encoding format. It is used for DVDs and digital television broadcasting. MPEG-4 can provide lower bit rates and more special effects, but it is catching on slowly.

MPEG-4 is a much more advanced standard that, in addition to providing efficient video encoding for transmission and storage, addresses multimedia presentation. It uses object-based encoding, which allows computer-generated graphics, animations, synthesized voices, and so on to be efficiently integrated with the video. The processing required is more than three times as complex as MPEG-2, but the bit rate can be up to 50 percent lower. There are some applications starting to use MPEG-4, but MPEG-2 is very firmly entrenched in the market and a massive conversion has not started yet.

ITU-R BT.601/709. ITU-R BT.601 specifies parameters for uncompressed video streams that are normally used with television studio equipment. It provides a reference against which to compare other encoding techniques. It specifies a 4:2:2 sampling format with a luminance sampling rate of 13.5 MHz. For SDTV, the output can be encoded as either 8- or 10-bit samples with a resulting bit rate of 216 Mb/s or 270 Mb/s, respectively. BT.601 is so widely used that the 4:2:2 format

TABLE 6–8
DVD Region Codes

Region Code	Countries
1	United States and territories, Canada
2	Japan, Europe, South Africa, Middle East
3	Southeast Asia, East Asia
4	Australia, New Zealand, Pacific Islands, Central America, South America, Caribbean
5	Former Soviet Union, Indian Subcontinent, Africa, North Korea, Mongolia
6	China

is sometimes referred to as "601." ITU-R BT.709 is a similar specification for HDTV formats. It uses either 4:2:2 or 4:4:4 sampling at a rate of 74.25 MHz (5.5 times faster than BT.601). Data rates start at about 1.2 Gb/s and go to over 2 Gb/s.

DVD-Video. DVD-video consists of MPEG-2 coded video on a dual-layer DVD disk. To save space on the disk, DVD-video uses a couple of tricks. The first is variable bit rate encoding, where the bit rate depends on the amount of movement in the video. Scenes with little movement have more inter-frame redundancy and therefore can be compressed more. The second is coding the video at 24 frames per second and converting to 30 frames per second in the player.

DVD-video also supports region coding. A region-coded disk (as most movies on DVD are) can only be played on a DVD player that supports that particular region code. The region codes are given in Table 6–8.

DV. DV is a 1/4-inch digital tape format that is widely used in camcorders. It is specified in IEC 61834. There are a number of very similar formats, such as DVCAM, DVCPRO, and Digital 8. Consumer-grade equipment in the United States uses the NTSC format with 4:1:1 sampling. The video data rate is 25 Mb/s. The total data rate including audio and synchronization information is about 29 Mb/s.

Virtually all DV camcorders have an IEEE 1394 interface. Transmission of DV over IEEE 1394 is specified in IEC 61883-1.

IEC 61883 Series. There are several specifications in the IEC 61883 series, which deals with sending audio and video streams over an IEEE 1394 network. The title of all the 61883 specifications begin with "Consumer Audio/Video Equipment Digital Interface." There are currently six standards in the series.

1. IEC 61883-1, *General:* Defines a transmission protocol for sending audio/video (A/V) and control commands over an IEEE 1394 network
2. IEC 61883-2, *Standard Definition Digital VCR:* Uses 480-byte packets, with 0 or 1 packet per isochronous cycle with a maximum data rate of 30.72 Mb/s
3. IEC 61883-3, *High-Definition Digital VCR:* Uses 960-byte packets, with 0 or 1 packet per isochronous cycle with a maximum data rate of 61.44 Mb/s
4. IEC 61883-4, *MPEG-2 Transport Streams:* Specifies packetization, transmission timing, headers, buffer size calculations, and so on for sending MPEG-2 streams over an IEEE 1394 network

5. IEC 61883-5, *Standard Definition, High-Compression Digital VCR:* Uses 120-byte packets, with 0 or 1 packet per isochronous cycle with a maximum data rate of 15.36 Mb/s
6. IEC 61883-6, *Audio and Music Data Transmission Protocol:* Specifies isochronous transmission of audio data over an IEEE 1394 network, including time-stamping, format identification (MPEG, DVCR, etc.), synchronization of transmit and receive clocks, and so on

Digital Visual Interface. The Digital Visual Interface (DVI) is a specification for digital monitors and displays and uses uncompressed video signals with HDCP on a 24-pin connector. Note that DVI signals are not suitable for networking and cannot be recorded.

Video Storage

Let us now consider the amount of video that can be stored on common media; this is summarized in Table 6–9. A couple of points about this table should be noted. First, the data rates are typical values; they are not the best that can be achieved, nor are they the worst that may be encountered. MPEG data rates, in particular, can vary quite a bit depending on the complexity of the material to be encoded and the quality of the encoder. Second, the table clearly indicates why storing uncompressed video is not practical. BT.601, for example, requires almost 100 GB per hour for standard-definition video and more than 600 GB per hour for high-definition video.

Using the data in Table 6–9, we can calculate the video capacity of some common storage devices, as shown in Table 6–10.

Figure 6–10 gives a graphical view of the video storage capacities of DVDs, Blu-ray disks, and terabyte hard disks. Note that a terabyte hard disk can store more than 500 MPEG-4 encoded standard definition movies (each about 2 hours long).

Video Displays

From the user's perspective, there are five major types of video display: the standard cathode-ray tube (CRT), liquid crystal and plasma flat panel displays, and front and rear projection displays. There are also a few new types of displays that show some promise, such as Digital Light Processing (DLP) and Liquid Crystal on Silicon (LCOS). Our discussion is confined to the five types that are currently in common use.

TABLE 6–9
Digital Video Storage Capacities

Video Format	Typical Data Rate (Mb/s)	Storage for 1 Hour of Video (GB)
MPEG-2, SD	4	1.8
MPEG-4, SD	2	0.9
BT.601, SD	216	97
MPEG-2, HD	10	4.5
MPEG-4, HD	5	2.3
BT.601, HD	1400	630

TABLE 6–10
Video Storage Capacities

	Hours of Video			
Video Format	CD (650 MB)	DVD (4.7 GB)	Blu-ray (27 GB)	Hard Disk (1 TB)
MPEG-2, SD	0.4	2.6	15	556
MPEG-4, SD	0.7	5.2	30	1111
BT.601, SD	0.0	0.0	0.3	10
MPEG-2, HD	0.1	1.0	6	222
MPEG-4, HD	0.3	2.1	12	444
BT.601, HD	0.0	0.0	0.0	1.6

Cathode-Ray Tube Displays

The cathode-ray tube (CRT) is the ubiquitous "picture tube" that has been used on television sets since the 1950s. It was also the standard monitor for PCs until recently. A CRT creates an image by having three electron guns (red, green, and blue) shoot a beam of electrons at the phosphor coating on the inside of the screen. The electron beams scan the screen in an interlaced pattern.

Because they have been made in huge quantities for several decades, CRTs are fairly inexpensive and they produce good-quality images. They have two major drawbacks—their shape and their lack of scalability to very large sizes, both of which are due to the scanning of the electron beams. The electron guns must be located far enough behind the screen to scan it with reasonable linearity, which gives a CRT display an almost cubical shape. In other words, a CRT display is almost as deep as it is wide.

When scaling to larger-size displays, CRTs become unwieldy because they take up so much room. As the screen gets bigger, the gun moves farther away and it takes a higher voltage to control the beam. It also is difficult to maintain a sharp picture on large screens, especially around the edges. For these reasons, CRTs are

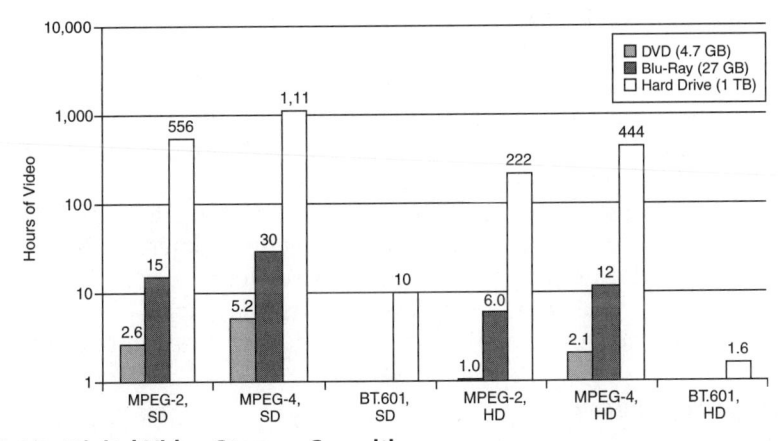

FIGURE 6–10 Digital Video Storage Capacities

rarely made larger than about 36 inches. They are also very heavy in large sizes; a typical 36-inch CRT-based television weighs well over 200 pounds.

Flat Panel Displays

Flat panel displays have become extremely popular in the last couple of years. Two main types of flat panel displays are liquid crystal displays (LCDs) and plasma displays. The main advantage of both types is that they are much thinner and use much less power than a similar-size CRT display.

Liquid crystal displays consist of two layers of glass with a liquid crystal between them. The display contains a matrix of conductors. Each intersection of conductors represents a pixel. The image is created by controlling the current to each pixel. Liquid crystal displays use much less power than other display technologies and, for that reason, are the standard display for notebook computers. Liquid crystal displays are available with either active or passive matrices; an active matrix display gives superior performance. Liquid crystal displays are also comparatively lightweight.

Drawbacks of LCDs include a relatively narrow viewing angle and limited contrast and brightness. Liquid crystal displays are typically available only in small sizes, up to about 30 inches or so. Larger sizes may be available in the future.

Plasma displays use charged gas, or plasma, to form each pixel. They are somewhat similar to an array of tiny neon signs. They are thin like LCDs but have much better brightness and contrast. A high-quality plasma panel display produces images that are absolutely stunning. Plasma displays are available in sizes up to about 60 inches. A large plasma panel display is only a few inches thick and can hang on the wall like a picture.

On the downside, plasma panel displays are quite expensive with prices (in 2004) ranging from several hundred dollars for a small 13- to 15-inch screen to several thousand dollars for a large screen. They are also rather heavy (more than 80 pounds for a 50-inch display).

Projection Displays

Projection systems can be used to make very large displays. There are two types, rear and front projection. They are often used in home theater systems but rarely anywhere else in the house.

A rear-projection system has a large chassis with an internal projector that shines an optical image on the backside of a transparent screen. They provide reasonable image quality, although not as sharp as most other display technologies. They are available in sizes up to about 65 inches or so. A typical 60-inch rear-projection television will be about five feet in both width and height and around two feet in depth, and it will weigh a couple hundred pounds. They are significantly cheaper than large plasma panels but still fairly expensive.

A front-projection system is like an old-fashioned slide projector. It is mounted in front of the viewing screen, typically on the ceiling. These systems can make huge images that fill an entire wall and have excellent image quality. They are quite expensive and usually require professional installation and alignment. Unlike other displays, they cannot be easily moved to another location once installed.

Video Display Summary

Table 6–11 summarizes the main characteristics of the five types of video display.

TABLE 6–11
Video Display Characteristics

Display	Max. Size	Image Quality	Cost	Physical Characteristics
CRT	36"	Excellent	Low	Boxy shape, almost as deep as wide
LCD	30"	Good	Moderate	Thin, low power
Plasma	60"	Outstanding	High	Thin, best-quality image, heavy
Rear projection	65"	Good	Moderate	Large boxy shape, heavy
Front Projection	Huge	Excellent	High	Permanently installed-difficult to relocate

Video Cabling

Video equipment has traditionally been connected using coaxial cable. A coax cable with the incoming television signal was hooked either directly to the television or to a set-top box (STB) that terminated the cable television signal and provided an interface to the television. Set-top boxes can also perform other functions, such as decoding premium cable television programs. Other video equipment, such as VCRs and DVD players, was typically connected to the television with short coax cords.

With the advent of digital televisions and other consumer electronics, IEEE 1394 provides an ideal medium for an entertainment network. In addition to providing from 100 to 800 Mb/s of bandwidth, IEEE 1394 supports video and audio via its isochronous transmission mode and uses consumer-friendly 5C copy-protection technology. Figure 6–11 shows an example of an IEEE 1394-based home network. This network allows digital content (both audio and video) to be shared among all the devices on the network. It can also include a wireless access point (WAP) for wireless speakers or monitors, as well as a wireless remote control. Later in this chapter, we discuss some advanced features that a network like this can provide.

Isochronous Bandwidth Allocation

On an IEEE 1394 bus like the one in Figure 6–11, bandwidth for isochronous connections is allocated according to a fairly complicated set of formulas contained in both the IEEE 1394 and ISO 61883-1 specifications. The isochronous video samples are sent in fixed-size packets. Enough bus bandwidth must be reserved for the maximum number of packets that might ever need to be sent in a frame. For a 10-Mb/s MPEG-2 HD stream, about 16.64 Mb/s would be reserved on the bus. This bandwidth reservation is automatically done by the IEEE 1394 equipment without any need for input from the user. However, it can result in using up bus bandwidth faster than the user might anticipate.

Home Theater

Home theater is an increasingly popular application that combines the latest in high-performance video and audio systems. The components of a typical home theater system include:

- A high-quality display screen, preferably high definition, which is properly sized for the room in which it is located

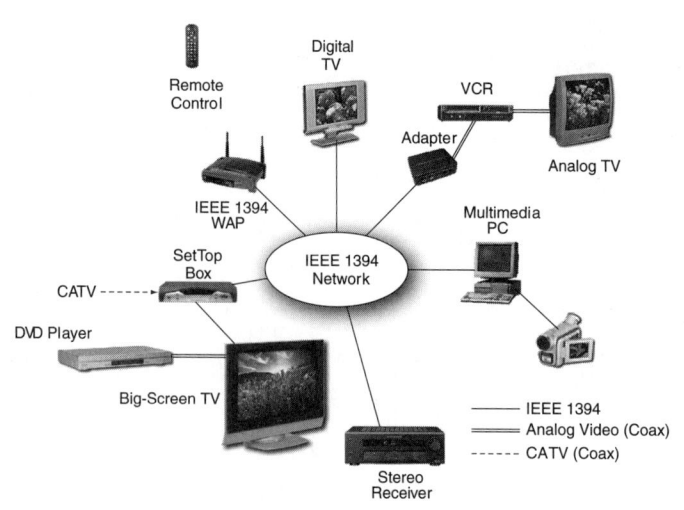

FIGURE 6–11 IEEE 1394 Home Network

- A DVD player, which usually also plays many other formats, such as CDs, MP3 CDs, JPEG[12] photo CDs, and so on
- An audio decoder system, which provides surround sound features (see the Surround Sound section that follows) and usually also includes an AM/FM tuner
- Five (or more) speakers, which must be properly situated in the room and connected to the audio system with high-quality cables that can deliver sufficient power to each speaker
- A **low frequency effects (LFE)** speaker, often called a **subwoofer,** to provide very low frequency audio
- A remote control unit to operate all the components of the system.

With the exception of the display, all the other components are usually purchased as a system. Home theater systems are available in a wide price range, from less than $100 to many thousands of dollars.

It is very important to match the home theater equipment to the room it is installed in. Sitting too far from a small display and sitting too close to a large display both give a suboptimal viewing experience. Similarly, the power required of the audio system will vary depending on the size of the room.

Home theaters are often located in the living room or family room. In upscale homes, a separate room is often dedicated to the home theater. The room can be outfitted with special lighting, movie-style seating, and so on.

Surround Sound

Audio systems originally used only a single speaker in a monaural configuration, often referred to as *mono*. In about the 1960s, stereo systems, which provided separate left and right channels, became popular. In recent years, there has been an explosion of audio configurations. Figure 6–12 shows a diagram of possible speaker configurations.

---

[12] The Joint Photographic Experts Group (JPEG) develops standards for digital photographs, just like MPEG does for movies.

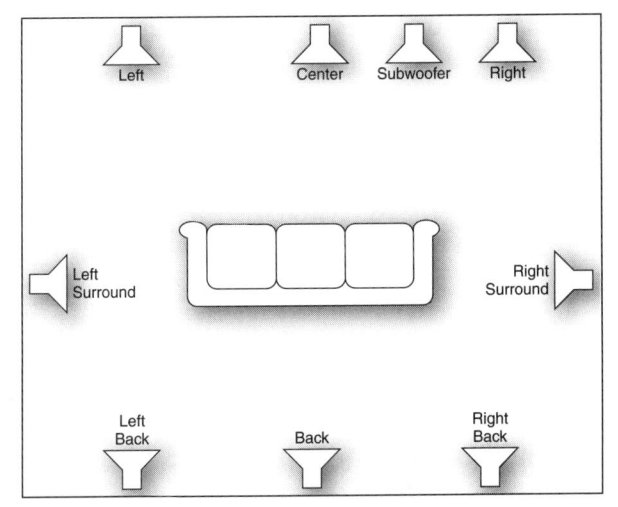

FIGURE 6–12 Surround Sound Speaker Configurations

Surround sound, which consists of two additional speakers located at the sides of the room, was introduced by Dolby Laboratories in 1982. Extended surround sound adds additional speakers at the back of the room. Wireless speakers are sometimes an attractive option for the side and back speakers because they do not require speaker cable to be run to the far side of the room. Table 6–12 lists some of the common audio configurations.

Surround sound configurations are often given a designation of the form *m.n*, as in the first column of the table. The first digit (*m*) indicates the number of standard speakers, while the second (*n*, which is almost always 1) indicates the number of subwoofers. The subwoofer is designated by ".1" because it covers a frequency range of about 3 to 120 Hz, or about 1/10 the total audio range. The subwoofer is most useful for special effects like rumbling noises and explosions.

There is not necessarily any correlation between the format in which the audio is distributed and the number of speakers used for playback. If an audio track distributed in 5.1 format is played back on a stereo system, the audio decoder

TABLE 6–12
Surround Sound Configurations

Designation	Speakers (Fig. 11–12)	Common Names
1.0	C	Mono
2.0	L, R	Stereo
5.1	L, C, R, SW, LS, RS	Surround Sound Dolby Digital Digital Theater System™ (DTS)
6.1	L, C, R, SW, LS, RS, B	Extended Surround Sound
7.1	L, C, R, SW, LS, RS, LB, RB	Dolby Digital-EX™ DTS-EX™

distributes the six audio channels between the two stereo speakers and the output sounds like stereo sound. Similarly, if a stereo track is played back on a 5.1 system, the decoder drives all six speakers from the two tracks.

5.1 surround sound is currently the most common format. It was introduced by Dolby Laboratories in 1992 and is used for the audio track on DVDs and HDTV broadcasts. There are currently no media that distribute sound in 7.1 format, which would require separate tracks for the two back speakers. In eight-speaker installations, the back channel from 6.1 audio is sent to both the LB and RB speakers.

Gaming

Video games are often thought of as kid stuff, but they are rapidly becoming a mainstream form of entertainment. The gaming industry generated about $28 billion in worldwide revenue in 2002 and, in the United States, is growing at a rate of 20 percent per year.

There are two main types of video games. Some are loaded onto a standard PC, although often a very high-powered PC with sophisticated graphics. Others run on a special gaming platform, such as the Sony PlayStation or the Microsoft X-Box. Almost everyone is familiar with the old classic video games, such as Pong, PacMan, and Frogger. Modern video games feature sophisticated surround sound audio tracks and state-of-the-art animation and special effects. They are evolving toward becoming a complete interactive experience, sort of like a movie where you participate in the action instead of just watching.

The holy grail of gaming is to have large numbers of people participating in an online game in real time. One name for this is multiple-player on-line role-playing game (MORG).[13] The following quote from the CEO of Atari describes the possibilities of MORGs.

> The golden age of movies is gone. That's it. It's a fact. What they do today to survive is they multiply the special effects to catch up with what the kids want, because they've seen it in the incredible universes of these video games. It used to be, "Well, let's make a movie and then make a video game version as a licensed product." The next step to this will be the collaboration between the stories, between the complexity of their stories and the personal expression of the video game. This product doesn't exist yet, but it will. Think about this kind of game, where you'll be in a kind of Star Wars environment, you'll have X thousand people playing together at the same time; you could just spend your day watching the screen and waiting for the stories to happen, or else you can decide to enter the game and take your own little path, all in real time. Or let's say you see a movie and your character is in the jungle, there's a snake there, you see the snake but he hasn't seen it, he's smoking a cigarette, talking to his girlfriend. You're like: "The snake! The snake!" And the character on the screen says: "A snake? Where?" But if you choose not to say anything, then he just goes on doing what he's doing. The movie people don't anticipate this revolution. They better watch their back. We're right there. Big time. (Bruno Bonnell, *The New York Times Magazine,* December 21, 2003, 38.)

Video games have definitely started to attract a lot of attention in Hollywood as the sales of video games and consoles surpassed box-office movie revenues in the United States for the first time in 2003. Video games are also attracting the attention

---

[13] It is also referred to as a massively multi-player on-line role-playing game (MMORPG) and other similar acronyms. In this book, we can us the simpler MORG (pronounced like "morgue").

TABLE 6-13
Cost of Storing a 2-Hour Movie on Disk ($1/GB)

Format	Encoding	Cost of Disk Space
SD	MPEG-2	$3.35
	MPEG-4	$1.68
HD	MPEG-2	$8.38
	MPEG-4	$4.19

of PC makers, who are producing high-end notebook and desktop computers to compete with the gaming PCs produced by companies like VoodooPC and Alienware.

The MORGs are in their infancy, but some, such as EverQuest and Ultima Online, are starting to catch on. There are even companies that allow players to buy and sell (for real, nonvirtual dollars) currency, weapons, and other virtual items used in these online games.

The most important impact on home networks is the bandwidth required to play MORGs. As higher-performance computers become cheaper, animation software becomes more realistic, and higher-speed access networks are available, MORGs will become more interesting and more popular. Their use will push both access and home networks to higher speeds.

Entertainment Server

As consumer electronics has converted to digital formats, the price of storage devices has declined (and continues to decline) substantially. Hard disks are currently available for as little as $1 per GB. As we have seen in Table 6–2 and Table 6–10, quite a bit of digital audio or video can be stored in a gigabyte of memory. With disk drives priced at $1/GB, an entire audio CD can be stored on disk for less than $1. If it is ripped into MP3 format, the storage cost drops to less than 10¢. Movies take up a little more space but can still be stored on disk for a few dollars, as shown in Table 6–13. With storage costs this low, it becomes feasible to store all the digital content in the home—including, for example, music, videos, movies, and photographs—in a single location and access them over the home network. This central storage location is referred to as an **entertainment server.**[14] An entertainment server consists of a huge disk drive with interfaces to all the networks in the home, including Ethernet, IEEE 1394, Wi-Fi, and so on.

A forerunner of the entertainment server is the personal video recorder (PVR), which allows video content to be recorded and played back later. It is much like a digital version of the VCR, although much more convenient to use because there are no tapes to fumble with, mislabel, or lose.

An entertainment server can be implemented in many ways. One example is shown in Figure 6–13. The entertainment server in this figure supports a wide variety of analog inputs and outputs for compatibility with traditional consumer electronics. It also provides digital interfaces via USB, Ethernet, IEEE 1394, Wi-Fi, and UWB. It can read (and record) CDs, DVDs, and Blu-ray disks.

Another approach to implementing an entertainment server would be to start with a PC and add functionality. There are PCs currently available with 500+-GB

---

[14] The entertainment server has also been variously referred to as a *content server, media server, media center,* and *home server.*

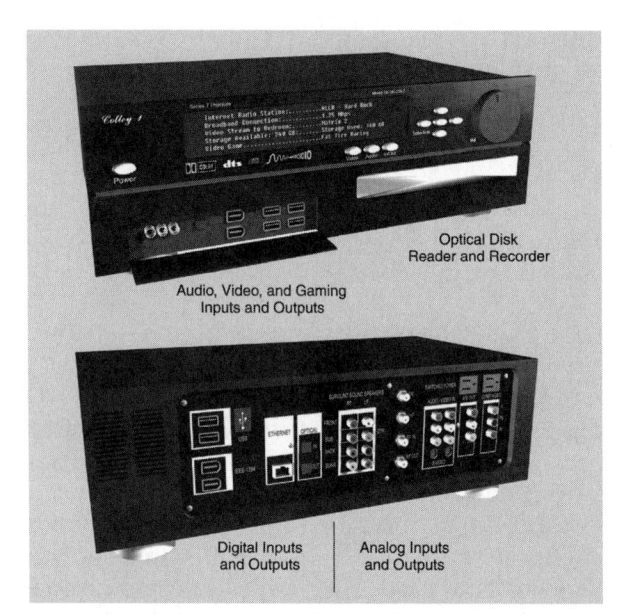

FIGURE 6–13 **Entertainment Server** (Illustration excerpted from the book Broadband Entertainment (Kipp 2004), **http://www.broadent.com**)

disk drives; built-in television tuners; interfaces to IEEE 1394, Gigabit Ethernet, USB, and Wi-Fi networks; and CATV and composite video inputs and outputs. It is also possible for residential gateway functionality to eventually be integrated with the entertainment server.

Not surprisingly, consumer electronics companies tend to favor the first approach, while PC-related companies (Microsoft, Intel, etc.) favor the second.

The entertainment server is the nerve center of the entertainment network. It provides storage for all the digital content in the home (including movies, recorded television programs, video games, home videos, photographs, ebooks, music, audio recordings, etc.). It can also provide a programmable platform for implementing advanced, customizable entertainment features.

Figure 6–14 shows how an entertainment server could be added to the home network of Figure 6–11. The entertainment server, which may have external disk drives, serves as the source of all the audio and video content for the home. Note that the CATV signal now goes only to the entertainment server, which distributes the necessary channels over the IEEE 1394 network. The DVD player and analog VCR have been removed; they are no longer necessary, although they can be left in the network if desired. In addition to these changes, a separate Ethernet for computer applications is shown. A gaming PC is shown connected to the Ethernet because it will likely have a connection to a gaming service outside the home. The entertainment server is connected to both the Ethernet and IEEE 1394 networks and can access content from any of the computers on either network.

Future Applications

Let us consider a home entertainment network like the one shown in Figure 6–14 and speculate about what types of new and exciting features could be provided if some new technology is applied in a couple of key areas. Specifically, let us discuss an enhanced user interface to a wide-screen video display and a super-duper universal wireless remote control.

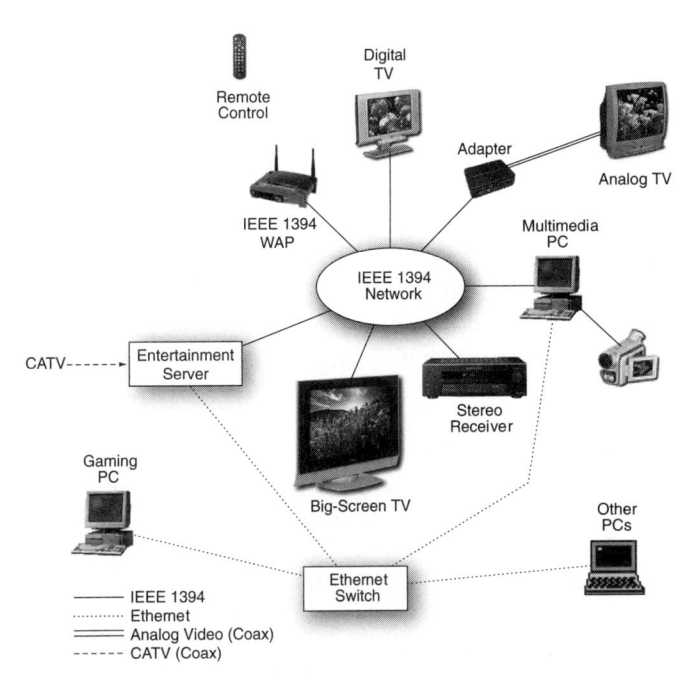

FIGURE 6–14 Home Network with an Entertainment Server

Video Display Interface

A large, digital, high-definition video display has the capability to do more than just watch a single television channel or movie. *Picture in picture* is a feature that is very popular with users but seldom implemented on television sets because it has traditionally been very expensive. On a standard television set, it requires an additional analog front end with a tuner and so on to get the second channel. With a digital television, when you are receiving digital samples over an IEEE 1394 bus all you need is a set of decoding chips and a means of multiplexing multiple channels into the video memory. This is not very hard to do, and it is all digital circuitry, so the cost will be coming down every year as semiconductor technology advances.

In addition, on a large screen you have lots of room for secondary displays while still maintaining a good-sized version of the main image. So there is no reason to limit the display to a single picture in picture; you can potentially have a bunch of different video signals displayed at the same time. For this example, let us consider a 64-inch plasma screen that can display up to sixteen independent video streams. For simplicity, we maintain the proper aspect ratio on all the images, although the system would not have to be implemented this way.

This screen could support a wide range of image configurations, as shown in Figure 6–15. The options range from a single large image to sixteen small

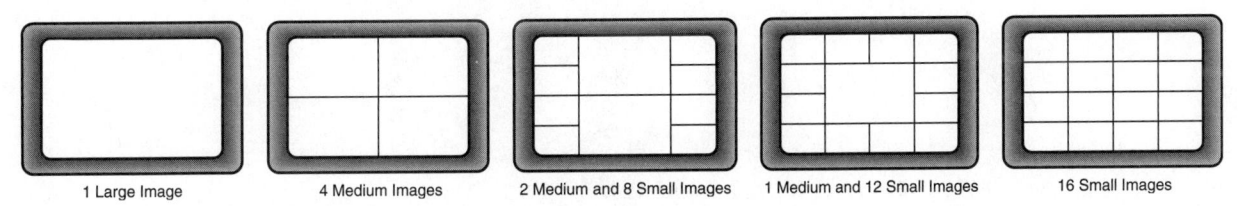

FIGURE 6–15 Potential Wide-Screen Image Configurations

images. Intermediate configurations include four medium-size images or combinations of one or two medium images with several small images. Before discussing how these configurations might be useful, let us consider remote control technology.

The Ultimate Remote Control

A typical remote control is a small, nonintelligent device with lots of tiny buttons on it. Almost every electronic device you buy comes with one, so the typical living room is littered with several of them. It is possible to buy a universal remote control to replace them all, but it generally just mimics the operation of each of the individual remote controls.

A modern home entertainment network like the one shown in Figure 6–14 consists of intelligent devices that communicate over a digital bus. The system is controlled by the entertainment server, which can be programmed with custom features. The ultimate remote control (URC) is equipped with a high-bandwidth wireless connection (such as UWB) to the main IEEE 1394 bus, so it can communicate with all the entertainment devices as well as with the entertainment server.

Figure 6–16 shows a conceptual diagram of a URC. While an actual implementation would no doubt look somewhat different, let us consider how the URC would function.

- Unlike traditional remote controls, the wide-band wireless connection is two-way so that information can be received and displayed on the URC. This also means that the URC does not have to be pointed at the device to be controlled; it must only be in communication with the WAP. The URC can even be used to control devices in another part of the house.
- The URC is equipped with a touch-screen display. The display can display not only text and graphics (such as menus, diagrams, etc.) but also provides video

FIGURE 6–16 Ultimate Remote Control (URC)

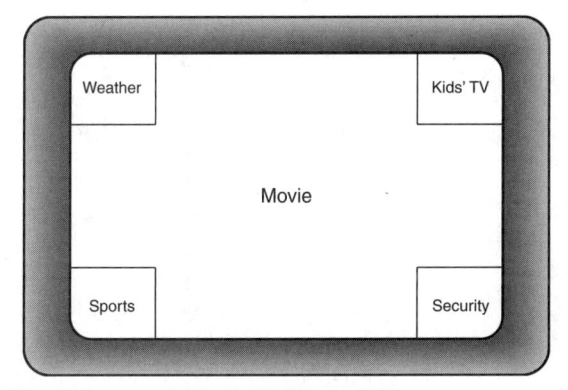

FIGURE 6–17 The Four-Corners Display Configuration

rendering capabilities. Additional relabelable buttons may be provided alongside the display.
- A keypad and fixed-function buttons are also provided for data entry, scrolling, and so on.
- Small speakers are provided for audio output, and a microphone allows for voice actuated control.

The URC provides all the features of standard remote controls plus some interesting new capabilities. Advanced features can be implemented by stepping through menus or diagrams on the touch-panel display and just pointing at the desired item. A television channel could be previewed on the URC before switching the television to the channel. Or, a second channel could be monitored on the URC while watching something else on the television display. You could even check what is being watched on a television in another room.

New Features

When you combine a large, high-definition display that can support multiple independent images with a programmable entertainment server that distributes video and audio over a digital bus and a full-featured remote control (like the URC), it becomes possible to provide lots of innovative new features.

With some custom programming in the entertainment server, any of the video configurations in Figure 6–15 could be set up with the touch of a button. Let us look at a couple of scenarios where this could be useful.

The Four-Corners Scenario

First, consider the display configuration for viewing a movie, which is shown in Figure 6–17. The four-corners configuration could, of course, be displayed with a single button push on the URC, which could turn on the home theater system as well. The movie is displayed on the large, central portion of the screen. The corners of the screen are devoted to four additional small images. Any of these small images can be turned off or temporarily displayed full screen with the touch of a button or two.

At the lower left channel is a sports broadcast, such as a football or basketball game, which can be monitored without intruding on the movie too much. If something exciting is happening, you could pause the movie and display the

FIGURE 6–18 Channel-Surfing Display Configuration

sports broadcast on the large screen for awhile.[15] The top left image can be used to monitor a weather or news channel.

The image at the top right can be used to keep an eye on what the kids are watching on their television in the playroom. At the lower right is an image from one of the home's security cameras. This window could be programmed to normally be off but to display the view from the appropriate security camera whenever a door is opened or a doorbell is rung; or it could cycle through all the security cameras that are available.

The four-corners scenario could, of course, be customized; any of the corner channels could be used to display other content or be turned off. For example, during March Madness, you could watch an NCAA basketball game on the main screen and keep track of other games on the corner images. Some may find the four-corners display more annoying than interesting, but we increasingly live in a multitasking world. People who grew up with Windows on their computer will probably be very comfortable with this type of scenario.

Channel-Surfing Scenario

Let us move on to one more scenario that most people would find very useful. Cable and satellite television systems provide dozens or hundreds of channels, but most people only watch a few of these channels with any regularity. It would be nice to be able to easily and quickly see what is on your favorite channels right now. The channel surfing scenario does just that.

In this scenario, the entertainment server is pre-programmed with the favorite channels of each family member, and the URC has a button labeled for each of them. For example, when Joe Sixpack pushes the "Joe" button, the display screen of the URC displays the image shown in Figure 6–18. The television screen is divided into sixteen images that show the actual contents of the channels. So, you can see what shows are on, whether they are in a commercial break or not, and so

---

[15] Whether this would preserve or destroy harmony in the family would depend on the personalities involved.

FIGURE 6–19 Alternate Channel-Surfing Display Configuration

on. When you decide what channel to watch, you simply touch the appropriate area on the URC screen and the television instantly displays that channel. At any time, such as during a commercial break or halftime of a sports event, pressing the "Joe" button will bring the channel-surfing screen back and a different channel can be selected.

If Mary Sixpack comes into the room, she can press the "Mary" button and her favorite channels are displayed, as in Figure 6–19. Similarly, other family members can have their own displays.

Another option, instead of (or in addition to) having a surfing configuration for each family member, would be to group the channels by content. For example, you could have a sports menu, a movie menu, a news and educational menu, and so on.

Conclusion

The four-corners and channel-surfing scenarios are just a couple of simple examples. The features that can be provided are limited only by your imagination. In addition to the large display that accepts multiple video streams, the programmable entertainment server, and the full-featured remote control, the key technical requirement is a high-bandwidth digital bus connecting the equipment. The bus must provide enough bandwidth so that all the television channels needed for the preceding scenarios can be available simultaneously on the bus. This allows the television to switch between channels on command without delays or glitches. When designing a home entertainment network, it is important to provide plenty of bandwidth so that futuristic features like those we have discussed can be supported.

With a digital entertainment network consisting of a programmable entertainment server and high-definition display screens connected by a high-bandwidth bus, it is possible to provide innovative, customizable features that were not feasible with analog equipment.

Digital home entertainment networking is undergoing tremendous innovation while the cost of digital consumer electronics is rapidly decreasing. If a consumer-friendly solution is found for the digital rights management issues, this market has almost unlimited potential for the development of new products and features.

Summary

This chapter discusses the basics of video and audio technology and how it is applied in entertainment networks. The major points covered in this chapter are:

- Home entertainment networks provide high-speed digital connectivity among entertainment devices (televisions, DVD players, video recorders, stereo equipment, PCs, etc.) so that they can share digital content and be controlled from a common point.
- Digital rights management (DRM) allows copyright owners to control the use of their creations. This must be balanced with the rights of consumers to fairly use the content (music, movies, etc.) they have purchased.
- Analog audio and video signals are converted to digital formats by a process of sampling, quantizing, and encoding. There are many formats and standards for encoding different types of content.
- Music is usually encoded in either CD or MP3 format. Newer music formats include DVD-Audio (DVD-A) and Super Audio CD (SACD). A standard CD can hold about an hour of CD-encoded music, or 11 hours of MP3 music.
- Whole-house audio is a system that distributes music to a number of rooms throughout a house. There are centralized and distributed systems. Whole-house audio systems can have multiple zones and utilize many different types of speakers.
- Audio signals can be either line level (which is compatible with UTP cable) or speaker level (which requires audio cable). TIA-570B gives recommendations for whole-house audio cabling.
- There are two main analog video specifications: NTSC (used in the Americas, Japan, and a few other parts of Asia) and PAL (used in Europe, Africa, and the rest of Asia).

- There are many different formats of digital television, most of which are categorized as either standard-definition television (SDTV) or high-definition television (HDTV). The various formats differ in terms of frame rates, aspect ratios, and spatial resolution.
- Video encoding is much more complex than audio encoding. The most commonly used video standards are MPEG-2, MPEG-4, DVD video, and DV.
- A standard DVD with MPEG-2 encoding can hold about 2½ hours of SDTV or about 1 hour of HDTV video. These capacities are approximately doubled with MPEG-4 encoding.
- There are several kinds of video display, including the CRT, LCD, plasma panel, rear projection, and front projection.
- Coaxial cable has traditionally been used for video signals. Modern digital networks support video transmission over UTP. IEEE 1394 networks are optimized for transporting digital video and audio.
- Home theaters usually include surround sound capabilities, which use speakers at the sides of and behind the viewer.
- Interactive gaming is a rapidly growing market. Multiple-player on-line role-playing games (MORGs) could increase the bandwidth needs of both home and access networks.
- An entertainment server provides a centralized place to store and access digital content, including movies, music, photographs, home videos, and so on.
- With modern digital electronic devices, including an entertainment server, connected together by a high-speed digital bus, many innovative new features can be offered that were not previously possible.

Key Terms

Aspect ratio The ratio of the width to the height of an image—typically expressed as a ratio of a whole numbers, such as 16:9.

Copyright Ownership of intellectual property within limits set by law, which typically provides the copyright holder with exclusive rights to print, distribute, copy, and make derivative works for a limited time period.

Digital Living Network Alliance (DLNA) An alliance of leading companies in the consumer electronics, mobile and personal computer industries. Its aim is to align the companies and have industry standards, which will allow products from all companies to be compatible with each other and to enable a network of electronic devices in the home.

Digital home controller (DHC) A device that can control the operation of DHRs and the transfer of content between DHSs and DHRs.

Digital home renderer (DHR) A device that renders content from the DHS.

Digital home server (DHS) A device that serves as a source for content—either streaming content or for file transfer.

Digital Rights Management (DRM) Technology to enable copy protection and secure distribution of copyrighted material.

Digital Transmission Content Protection (DTCP) A user-friendly copy protection scheme used by IEEE 1394 networks.

Digital Visual Interface (DVI) An uncompressed digital video signal standard developed for the interface between a computer and a monitor.

Entertainment server A device used to store digital content (audio, video, photos, etc.) that can be accessed by any other device on the home network.

Fair use Provisions in the copyright laws that provide some allowed uses for copyrighted works, such as quoting brief passages in a review.

High-Definition Copy Protection (HDCP) A copy prevention scheme used by DVI.

High Definition Television (HDTV) A digital television system with horizontal and vertical resolution approximately twice that of SDTV and an aspect ratio of 16:9.

Interlaced scan A video component or signal that scans even- and odd-numbered lines in alternating frames. The opposite of progressive scan.

Low frequency effects (LFE) LFE is an abbreviation that is commonly used in describing an audio track contained within a 5.1 motion picture sound mix. The signal from this track, ranging from 10 Hz to 120 Hz, is normally sent to a subwoofer.

National Television Standards Committee (NTSC) The analog television format used in North America, as well as Central/South America, Japan, Korea, Taiwan, and the Philippines.

Personal Video Recorder (PVR) A device that records television signals onto a hard disk. Similar to a VCR, except with a large disk drive instead of removable tapes.

Phase Alternate Line (PAL) The analog television format used in most of Europe, Asia, Australia, and Africa.

Pixels The smallest individually addressable units in a display or image. Short for "picture element."

Progressive scan A video component or signal that scans or displays each line of a video frame in sequence. The opposite of interlaced scan.

Pulse code modulation (PCM) Digital representation of an analog signal where the magnitude of the signal is sampled regularly at uniform intervals, then quantized to a series of symbols in a digital (usually binary) code. PCM is used in digital telephone systems and is also the standard form for digital audio in computers and various compact disc formats. It is also standard in digital video.

Render To produce a version of the original content—typically on a video screen or audio output device.

Ripping The process of uploading the content of an audio CD and converting it to MP3 format.

Sequential Couleur a Memoire (SECAM) An analog television format developed in France.

Standard-Definition Television (SDTV) A digital television system with picture quality equivalent to CDTV.

Subwoofer A type of loudspeaker dedicated to the reproduction of bass frequencies, typically from about 20 Hz to about 200 Hz.

Review Questions

1. What is the main function of an entertainment network?
2. Define the term *digital content*. Give at least five examples of digital content.
3. List two examples of fair use of digital content and two examples of copyright violations.
4. What are the three steps in making a digital representation of an analog signal?
5. Describe the difference between CDs, DVDs, and Blu-ray disks.
6. What are the two main configurations of whole-house audio systems and how do they differ?
7. What is the purpose of dividing a whole-house audio system into multiple zones?
8. Name the two types of audio signals. What types of cable are used for each?
9. Explain the difference between interlaced and progressive scanning.
10. What are at least three advantages of digital television over analog television?
11. What are the two main types of digital television and how do they differ in terms of image quality and aspect ratio?
12. List the five main video display technologies.
13. What are the six components of a typical home theater system?
14. What is the purpose of surround sound?
15. What is a MORG?
16. What are the functions of an entertainment server?

Home Automation

After studying this chapter, you should be able to:

OBJECTIVES

- List the four main home automation subsystems and discuss the purpose of each.
- Explain the connectivity requirements of each subsystem.
- Discuss the pros and cons of smart appliances.
- Explain the benefits of an integrated home automation system.
- Define home automation scenes and describe their benefits.

OUTLINE

Introduction

Climate Control

Lighting Control

Security Systems

Fire Alarm Systems

Smart Applicances

Integrated Systems

Introduction

Home automation refers to a number of systems that exist in almost every home, including heating, ventilation, and air conditioning (HVAC); lighting; fire and intrusion alarms; and sometimes others. In the past, each of these systems operated independently under the manual control of the home's occupants. There has been a trend toward adding a little intelligence to some of these systems (for example, a programmable thermostat for the HVAC system), but the various systems still operated independently of each other.

Many people assume that the purpose of a **home automation system (HAS)** is to reduce energy use or save money. While HASs may result in some savings of energy or cost, that is not their primary purpose. The main value of a HAS lies in the improved quality of life that results from the comfort, security, and convenience that results from the automated system.

The four major categories of systems we consider are:

- Climate-control systems, which include HVAC and some other auxiliary systems such as automated control of window treatments, for example
- Lighting-control systems
- Security systems, which include intrusion alarms (also known as burglar alarms), security cameras and monitoring systems, and door/access control systems
- Fire alarm systems.

These systems use four primary transmission media for communicating control information.

- Power-line carrier (PLC), such as X-10 or HomePlug
- Wireless systems, which can be infrared (IR), Wi-Fi, or UWB
- Unshielded twisted-pair cable, which can be used for many applications. (For new installations, Category 6 (or 5e) UTP should be used, although in the past, Category 3 UTP was often used for HAS applications.)
- Application-specific cable, such as 16-AWG copper wire, which is used by some devices.

The major home automation subsystems are climate control, lighting control, security, and fire alarms. The primary communication media are PLC, Wi-Fi, UTP, and application-specific cable.

Climate Control

Climate control is the most ubiquitous home automation system. Virtually every home has a heating system, and many also have air conditioning. The HVAC systems are typically controlled from a manually operated thermostat. With the increase in fuel prices over the last two or three decades, programmable thermostats (often called "set-back" thermostats) have become popular.

The benefits of intelligent HVAC control include:

- Providing a comfortable living environment in terms of temperature, humidity, and so on in all parts of the house
- Protection of the contents of the residence against temperature extremes (such as freezing pipes in the winter and overheating in the summer) when the house is unattended
- Minimizing energy consumption while accomplishing both of the preceding functions. (An HVAC system is one of the few home automation applications that can actually save a significant amount of energy.)

Heating, Ventilation, and Air-Conditioning Systems

A detailed discussion of the different types of HVAC systems is beyond the scope of this book. We consider a simple example to illustrate the general principles. Figure 7–1 illustrates a simple forced-air heating and cooling system. The system is divided into two zones serving the first and second floors. Each zone has a thermostat to set the desired temperature and a **damper** to control the flow of

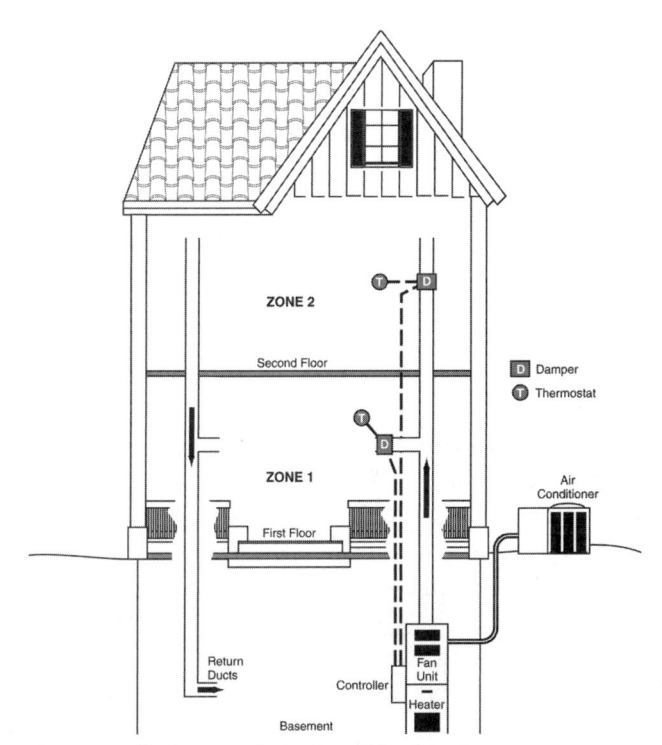

FIGURE 7–1 Heating, Ventilation, and Air-Conditioning System

air. In traditional HVAC systems, the control wiring (indicated by dashed lines in the figure) is provided by the HVAC contractor when the system is installed.

In modern systems, the thermostats and control unit may be networked to a home automation control unit so that the HVAC system can be controlled via a touchpad, PC, or other device. X-10 is probably the most common networking technology for this application, although infrared (IR), Ethernet, and Wi-Fi are also sometimes used. Relay contact closures are available for compatibility with older heating systems. In the section on Integrated Systems, we discuss the benefits of this type of networked control.

Other Climate-Control Applications

Many other types of applications can be considered as part of climate control. Some examples include:

- Automatic controls are available for all types of interior and exterior window treatments. Interior treatments include many different types of shades and blinds for windows and doors. Exterior treatments include things such as awnings and security shutters.
- For really deluxe installations, small roof-mounted weather stations are available. They can be solar powered and communicate with the home network via Wi-Fi. They enable various climate-control systems to respond to external weather conditions.
- Sprinklers and irrigation systems can be tied in to the home automation system.
- Gas fireplaces and other heating devices can also be controlled by the home automation system.

Most of these devices can be operated either on a time-of-day basis (such as automatically closing the window shades at night and opening them in the morning), under manual user control, or by some other programmed criteria. For example, the system can be programmed so that the external shutters automatically close if the weather station detects wind in excess of some predetermined velocity.

Lighting Control

Just about every house in the United States has electric lighting, but relatively few have an automatic lighting-control system. Lights are traditionally controlled by wall-mounted switches with the occasional manually operated dimmer. However, lighting-control systems are becoming much more popular in newly constructed homes. There are a number of benefits to a centralized lighting-control system, including:

- *Convenience:* A lighting-control system allows you to monitor and control all the lighting in the house from one place.
- *Aesthetics:* Lighting-control systems minimize the number of wall switches and often eliminate large banks of switches. In addition, they make extensive use of dimmers to create desired moods.
- *Safety:* Lighting-control systems can ensure that proper external and interior lighting is provided at all times.
- *Security:* Proper lighting control can also give the house a "lived-in" look when the owners are away and can provide a panic/alarm mode that instantly turns all the lights on.

While it is true that lighting-control systems can often result in some reductions in energy usage, for a residential system the savings are typically fairly small.[1] The cost of the lighting-control system is usually justified by convenience, aesthetics, safety, and security rather than by energy savings.

Another major advantage of lighting-control systems is the ability to define **scenes,** which are lighting configurations for specific purposes (such as watching a movie or eating dinner) or times of day (morning, evening, night, etc.).

Lighting-Control Technology

Several different methods are used for controlling light fixtures. The major techniques follow.

- Manual switches and dimming controls have been used for decades to control lighting fixtures. Automatic systems often include some manual switches that can be used to override the automatic settings.
- Automated switches and dimmers are controlled by messages sent from a central controller.
- Light-level sensors turn on the lights when it is dark and turn them off when it is light. They are often used for security and outdoor lighting.

---

[1] For commercial systems, however, lighting-control systems can provide significant energy savings since commercial buildings have much larger areas and are frequently not occupied during evenings and weekends.

- Clock-based switches control the lighting based on the time of day.
- Occupancy sensors turn the lights on when they detect motion in the room.

These techniques may, of course, be used in combination, such as turning the lights on when it gets dark and turning them off five hours later.

A centralized controller may be used to coordinate the action of all the individual lighting fixtures. One of the most common systems has individual X-10 controllers installed at each lighting fixture; an X-10 system controller uses PLC to communicate with them. Modern lighting-control systems make extensive use of dimming rather than just switching the lights on and off. Centralized lighting-control systems often feature a cabinet of control/dimming modules that are wired to each light fixture. This type of system makes it easy to provide scenes and other advanced features. Many manufacturers, such as Lutron, Crestron, and Leviton, make this type of system.

Communication Media for Lighting Control

Lighting systems obviously need electrical power wiring to function. Systems that have a centralized controller also need some type of path for the communication of control information to each lighting fixture. The most obvious medium to use is PLC. This can be accomplished with no additional wiring. Additional control hardware is necessary at the lighting fixture. Figure 7–2 shows how simple a PLC-based lighting-control system is from a wiring perspective. The figure also shows a hardwired keypad (or touchpad) that is often wall-mounted and is used to program and control the system. Another possibility is to have a wireless touchpad that communicates directly with the lighting controller.

Wireless communication is an attractive option for lighting control. In this case, a wireless transceiver is placed in each lighting fixture to be controlled. Communication with the lighting controller can be by IR, Wi-Fi, or UWB.

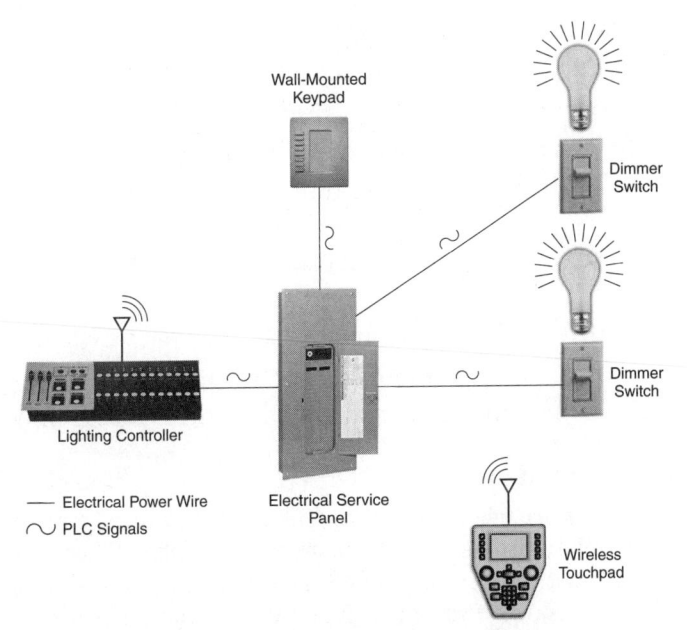

FIGURE 7–2 Lighting-Control System Using Power-Line Carrier

A wired (non-PLC) technique can also be used to communicate with each lighting fixture. Wired systems require running a separate cable, usually UTP, to each fixture to be controlled. The communication protocol can be either Ethernet or an older protocol such as RS-485.

Security Systems

Some type of security system is installed in many, if not most, homes being constructed today. The systems range from simple sensors on a few doors and windows to sophisticated systems with door/window sensors, motion detectors, automatic door locks, video surveillance cameras, and so on.

Security systems usually have a master (main) control panel, which is mounted near the DD or electric service panel. The control panel must be provided with alternating-current power and an analog telephone line for monitored systems. The control panel usually includes a battery for backup purposes and provides power to all the other components of the system. In addition, several other components may be provided, including:

- A keypad (often called the master control panel), which is located in the main part of the residence and provides the primary interface for the homeowner
- One or more audible alarms
- A "panic button," which may be used to manually indicate an alarm condition.

Although fire alarms are treated separately in the next section, they often share the preceding equipment with the security system.

Security systems consist of three main subsystems: intrusion alarms, access control, and video monitoring. Other miscellaneous types of sensors, such as carbon monoxide detectors and flood sensors for basements, may optionally be provided.

Intrusion Alarms

The purpose of an intrusion alarm is to detect the entry of unauthorized persons into the home. A number of different types of sensors can be used, such as:

- Contact-closure sensors to detect when doors or windows are opened
- Glass-breakage sensors for use on windows large enough to allow a person to enter
- Motion detectors, of which there are several kinds, including IR and laser-beam sensors
- Pressure pads to detect when someone is in a particular area, such as a hallway.

Intrusion alarms are usually turned off during the day, when the residents of the house are active, and are armed at night and when the house is unoccupied.

Access-Control Systems

Access-control systems are used to control the locks on doors, garage doors, gates, and so on, and are relatively new in the residential market. They are usually wireless devices that allow doors to be locked and unlocked remotely and are often controlled by a key-fob transmitter much like the door locks on most new cars. There are also EZPass-like modules for use in cars. One of these modules will automatically open the garage door (and, if applicable, the gate at the end of the driveway) when the car pulls into the driveway. Integration with other security systems allows features such as automatically locking all the doors when the intrusion alarm is armed, for example.

Video Monitoring

Video monitoring is the most LAN-like of the security subsystems. Although once affordable only for commercial buildings and mansions, the cost of video cameras has dropped so that video monitoring is affordable in most homes.

Traditional video cameras are connected via coaxial cable. Video **baluns** are available from a number of manufacturers that will convert the video signal into a format suitable for transmission on UTP cable. Newer models often have an RJ-45 jack to run directly on UTP using either a proprietary protocol or a standard protocol such as IEEE 1394. The camera can usually be powered and controlled (pan, tile, and zoom) over the same UTP cable. When cabling a new house, it is a good idea to run a UTP cable from the DD to each point where a security camera is likely to be installed.

Wireless security cameras (usually Wi-Fi based) are also available. They are very easy to install and relocate when desired. Unlike UTP-based cameras, they must have a separate power source (either batteries or alternating current). When wireless cameras are used, proper security precautions (such as enabling WEP) must be taken to prevent unauthorized reception of the images.

Video cameras are typically deployed at each entrance to the home. Additional cameras may be used to view the driveway, backyard, pool area, and patio or deck. Indoor cameras may be installed for purposes such as monitoring an infant's room, checking up on pets, or verifying the status of the house when it is unoccupied.

Video cameras are prime candidates for integration with other systems, such as the data network or entertainment system. If configured properly, the video cameras may be viewed remotely over a secure Web site. This allows the homeowner to remotely monitor the house while at work or away on vacation.

TIA-570B Security Cabling Recommendations

TIA-570B includes some general recommendations for security system cabling, the most important of which is to follow the applicable national and local codes. A brief summary of the TIA-570B recommendations for security cabling is given in Table 7–1. With the exception of video cameras and monitors, cabling for most security systems is application-specific and must follow both the building codes and the manufacturer's installation instructions. Note that wire larger than the 24-AWG conductors of UTP is often specified for security and other HAS devices because of the amount of current that must be supplied. It is often possible to substitute UTP for the heavier cable if the run is short, but be sure to check with the manufacturer before doing this.

In general, it is not good practice to substitute multiple smaller wires for one large wire. For example, four 24-AWG conductors have about the same cross-sectional area as one 18-AWG conductor. But if one or two of the 24-AWG wires break or become disconnected, the system controller will still think it is in communication with the device but it may not be able to source enough current over the remaining conductors to operate it. For this reason, multiple conductors should not be connected in parallel to increase the current-handling capacity.

TABLE 7–1
TIA-570B Security Cabling Recommendations

Subject	TIA-570B Recommendation
Cabling topology	Use a star-wired topology for all security devices, with the possible exception of smoke detectors, which may be daisy-chained.
Cable	Comply with *NEC® Article 725*. Security devices often use 16- to 22-AWG wire, depending on the distance and amount of current. Category 5e or 6 UTP may be used if approved by the manufacturer of the security panel.
Alerting devices	Sirens, bells, strobe lights, and so on often draw high currents (up to 3 A) and must be connected via cables that can handle this load.
Device connections	Security devices often use screw-lug terminals. If splicing to the device is necessary, insulation displacement connector (IDC) splices are preferred.
Power	The security panel must be powered from a nonswitched outlet or be directly connected to the electric service panel.

Fire Alarm Systems

Fire alarm systems are extremely important to the safety of the residents. Accordingly, they are usually mandated by local building codes. Requirements for the wiring of fire alarm systems are also specified in *Article 760* of the *NEC®*. Fire alarm system requirements are also covered in *NFPA 72—National Fire Alarm Code*. The local codes often specify exactly what types of sensors, wiring, battery backup, and so on are required and where the sensors must be located.[2] The most important principle for a fire alarm system is: Follow the local codes to the letter! Any attempt to cut corners or save a few bucks will probably result in not getting a Certificate of Occupancy until the system meets all codes.

Fire alarm systems are heavily regulated by national codes, such as *NFPA 70* (*National Electrical Code*) and *NFPA 72* (*National Fire Alarm Code*), and by local building codes.

Fire alarm systems usually require a sensor and audible alarm to be placed in each room of the house. These sensor/alarm units must be wired as per local codes, which often specify the color as well as the size and type of the wire. Wireless sensors are usually not compliant. Because of the extensive regulation of fire alarm systems, it is usually neither desirable nor possible to integrate fire alarm wiring with other systems. The fire alarm controller may sometimes be integrated with a security system or be connected to a network (such as an Ethernet) so that it can exchange information with other home automation systems.

Fire alarm systems frequently have a telephone interface so they can call in an alarm when a fire is detected. A real analog telephone line is required—VoIP will not do for this application.

---

[2] Local building codes typically reference the national codes, such as the *NEC®*, and also give specific requirements based on local conditions and construction practices. There can be big differences, for example, between a hot climate like Phoenix, a wet climate like New Orleans, and a cold climate like Minneapolis.

Smart Appliances

No discussion of home automation would be complete without mentioning **smart appliances.** As the cost of digital networking technology drops, it becomes more tempting to put it in all sorts of things that used to be just dumb, stand-alone appliances (such as washing machines, refrigerators, toasters, etc.). The question is, does this new smart appliance do anything really useful, or does it just do the same old thing in a more annoying way?[3]

Bob Lucky, former vice president of research at Bell Labs and Telcordia, has commented:

> I think of the appliances in my house, and wonder if they want to connect to the Internet. And, if so, what they would say. Perhaps my toaster would have its own Web page. Maybe all toasters come with personal Web pages. There would be a cute picture of my toaster, a little biography, pictures of recent successful toast, and maybe a summary of recent activities, along with tips that it has learned about good toasting. Best toast practices, as it were.[4]

Stuart Alsop, the well-known author and venture capitalist, has discussed the concept of a *perfect digital appliance* (PDA). The characteristics of a PDA can be expressed in pseudo-equation form:

$$PDA = (A + D + N - PC)/\$$$

Let us examine what each of the terms in this "equation" mean in the context of an appliance such as an intelligent washing machine.

- *A:* The appliance should have all the capabilities of its *analog* predecessor. If you are making an intelligent washing machine, it should be able to do everything that the old analog washing machine used to do. It should handle loads just as big, and so on.
- *D:* It should have new capabilities because it is now *digital*. A washing machine could have sensors to tell when the clothes are clean and stop washing so you do not waste energy. Maybe it would automatically dispense water and detergent based on how big the load is.
- *N:* New features should be enabled by its ability to *network*. Maybe it could talk to the water heater to make sure there is enough hot water to do the load before it starts, or to the dryer to make sure the dryer will be available when the washing cycle is completed.
- *PC:* It should not have the difficulty of use and user unfriendliness of a PC (note the minus sign in the equation). If it is as hard to use as a PC, not many people are going to buy it.
- *$:* The final criterion is cost ($). The networked digital device has to be available at a price that many people are willing to pay. Typically, for most consumer products, that is in the $500 range or below. A lot of things like VCRs and CD players did not really take off until they got down to that price range. For large

---

[3] For example, if you have an "intelligent stove" with a display screen, would you want it to emulate Amazon.com so that, every time you start cooking some eggs, the following text scrolls across the display: "People who eat eggs also enjoy sausage, bacon, and fried potatoes." Worse yet, it could use speech synthesis technology to babble the same phrase repetitively.

[4] Robert W. Lucky, Reflections column in the *IEEE Spectrum* (March 1999).

TABLE 7–2
Appliance Life Expectancies[5]

Appliance	Life Expectancy (years)
Refrigerator	14–19
Washing machine	11–14
Dryer	13
Dishwasher	11–13
Microwave	9
In-sink disposer	12
Water heater	6
Stove	11–18
Cooktop	13–21

appliances like a refrigerator, the price will be a little higher, but it cannot be outrageously more expensive than a traditional refrigerator.

Another issue is the difference in useful life expectancy between home appliances and computers. Table 7–2 gives typical life expectancies for some common home appliances.

Historically, PCs have lasted three to four years. After that, it is difficult to get new software to run on them. Clearly then, there is a large mismatch between the expected useful life of home appliances and computing equipment such as PCs. Basic networking technology, such as a 100Base-T interface, does not go out of style quickly, but sophisticated computer-like user interfaces will likely be ready for the scrap bin before the appliance hits middle age.

A classic example of an intelligent appliance is the Internet refrigerator (such as the Samsung HomePAD), also known as the *screen fridge*. It essentially consists of a standard refrigerator with a PC/television built into the door. The display screen and keypad are on the door and can be used as a family message board, shopping list, and so on, as well as providing Internet access. Some models have bar-code scanners or RFID readers built into them to identify the food that is inside. Various types of Internet fridge have been on the market for several years but have not generated significant sales.

If we consider the Internet fridge in terms of Alsop's equation, it is not hard to see why. It does not fare very well in the D, N, or $ terms. It costs more but does not really do a lot of useful things. Sure, you could log in from the office and ask it if contains a carton of milk. But it cannot tell you how much milk is in the carton or distinguish between various vegetables in the crisper drawer. While it does have some new capabilities, such as the ability to watch television

---

[5] SOURCE: Association of Home Appliance Manufacturers (**http://www.aham.org**).

or read e-mail, the refrigerator is not necessarily the most convenient place for those activities.

Another example that is possibly more promising is the smart stove (such as the Whirlpool Polara™ Refrigerated Range). The oven of the smart stove also has refrigeration capabilities. You can, for example, put a chicken and some potatoes in the oven in the evening and program it to refrigerate the food until 3:00 the next afternoon. From 3:00 to 5:00, the food is brought to room temperature, and then from 5:00 to 6:00 it is cooked. When you arrive home from work, dinner is ready. If your plans change during the day, you can communicate with the stove over the Internet and change the cooking time or cancel the cooking and hold the food in refrigerate mode. This seems to have more new useful features than the Internet fridge, but the real issue is how much more than a regular stove it costs and whether the added features are worth the difference.

Connectivity for Smart Appliances

Smart appliances typically have modest bandwidth needs and do not present any significant connectivity problems other than where they are located. An Ethernet connection (100Base-T) is adequate for almost all smart appliances. However, telecommunications outlets are not usually provided in areas where smart appliances are located. In the short term, Wi-Fi seems the most desirable means of communicating with smart appliances. In the future, if PLC becomes more widely deployed, it would also be a good option. For appliances that use a wired connection, it is usually not too difficult to run a UTP cable out to the appliance.

Integrated Systems

As we have seen in several previous scenarios, there is often an advantage to integrating some or all of the home automation subsystems under a single master controller. Figure 7–3 shows a conceptual block diagram of how various individual subsystems may be connected to a home systems controller (HSC).

The HSC can be implemented using Ethernet technology, or with IEEE 1394 as specified by VHN. It could include a wireless router (as indicated by the antenna in the figure) so that some of the components, particularly remote control devices, could operate without a wired connection. Smart appliances, if any, can also be connected. If the entertainment network is also connected, additional features can be implemented as well.

The subsystems in Figure 7–3 are potentially from different manufacturers. A more efficient solution might be to purchase an integrated HAS provided by a single manufacturer, as shown in Figure 7–4.

There are three main advantages to the type of integration shown in the previous two figures. First, it allows all the subsystems to be monitored and controlled from a common interface, such as a touchpad. Second, it allows the possibility of remote access to the HAS. Finally, it allows for much more complex (and useful) scenes to be defined and implemented.

Integrated HASs have three advantages—control from a common interface, remote access to the HAS, and provisioning of complex scenes with interaction among the various subsystems.

158 • Chapter 7

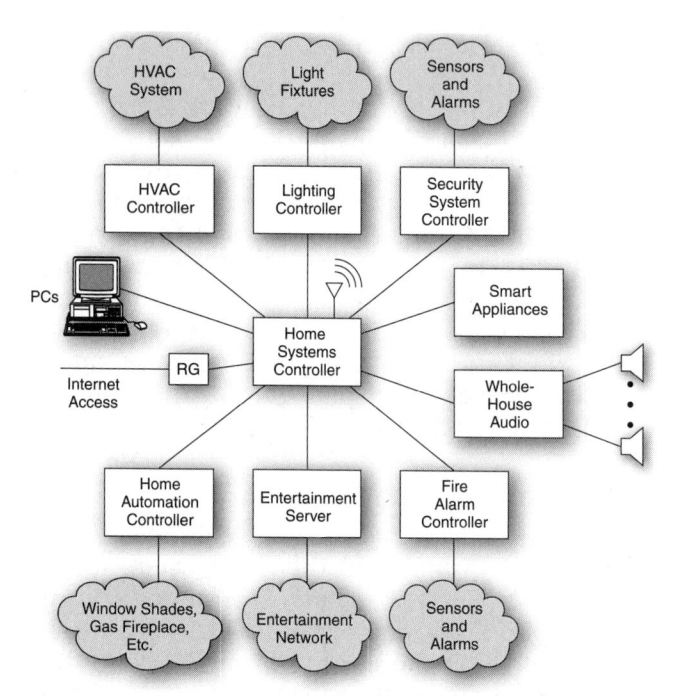

FIGURE 7–3 Integration of Home Automation Subsystems

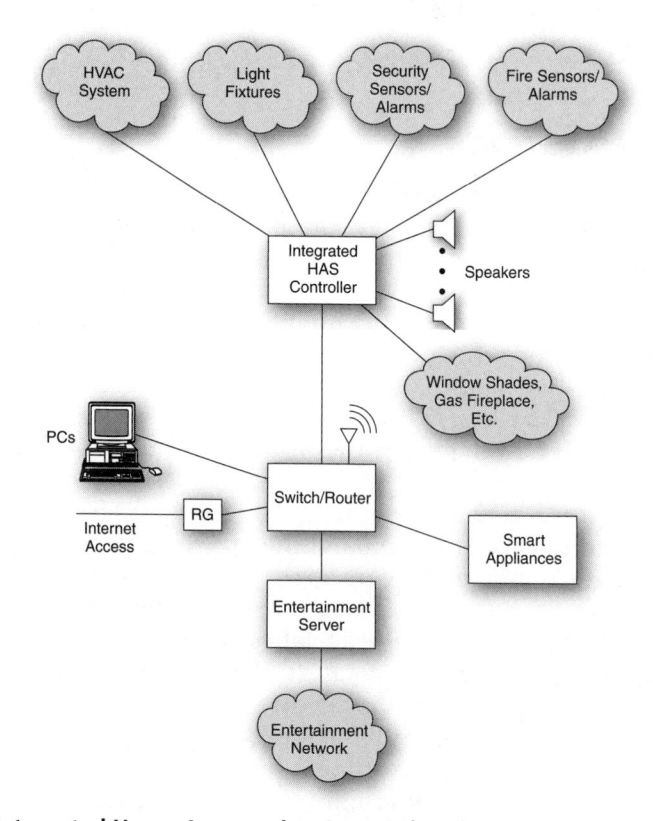

FIGURE 7–4 Integrated Home Automation System (HAS)

Remote Access to the Home Automation System

If the HAS is capable of accepting IP-based commands over a network, as shown in Figure 7–3 and Figure 7–4, then the device issuing the commands no longer has to be within the home. This opens up the possibility of being able to control the HAS from anywhere you can access the Internet.[6] Remote access to the HAS can provide a lot of useful capabilities, such as:

- The images from the security camera can be monitored from a remote site (such as your office) to verify the status of the house, check up on the babysitter, make sure that pets are safe, and so on.
- Information about any alarms that may occur (from the fire or security systems, for example) can be forwarded to you by e-mail or by text messaging to your cell phone.
- The house can be unlocked to allow access by maintenance or repair technicians, let in locked-out family members, and so on.
- Appliances in the home can be controlled, such as starting the PVR to record a television program.
- The status of smart appliances, such as an oven that is cooking dinner, can be checked.
- Lights, window shades, and the HVAC system can be adjusted either to give the house a lived-in look or to make sure they are in the proper state when you arrive home.

Home Automation Scenes

Integrated HASs allow the definition of scenes that coordinate the interaction of multiple subsystems so that commonly encountered situations can be handled with the touch of a button. The system can provide a number of predefined scenes and can be personalized to the needs and preferences of the particular home/family. Additional scenes can be custom programmed if desired. The various scenes that are available are displayed on a menu on the touchpad or remote control for the system. They are initiated merely by touching the appropriate button.

Let us consider a couple of sample scenarios with reference to the portion of a house shown in Figure 7–5. The area shown in the figure consists of a kitchen, a dining room, and a family room. The house is equipped with an integrated HAS that includes control of the lighting, HVAC, window shades, gas fireplace, whole-house audio, and home theater system.

One common activity is for the family to eat dinner in the dining room. When the touchpad button labeled "Dinner Time" is pressed, the various HAS systems respond as indicated in Table 7–3.

Another scene button is labeled "Watch Movie." When this button is touched, HAS responds as described in Table 7–4.

---

[6] For security purposes, you would probably want to implement the remote access using a virtual private network (VPN).

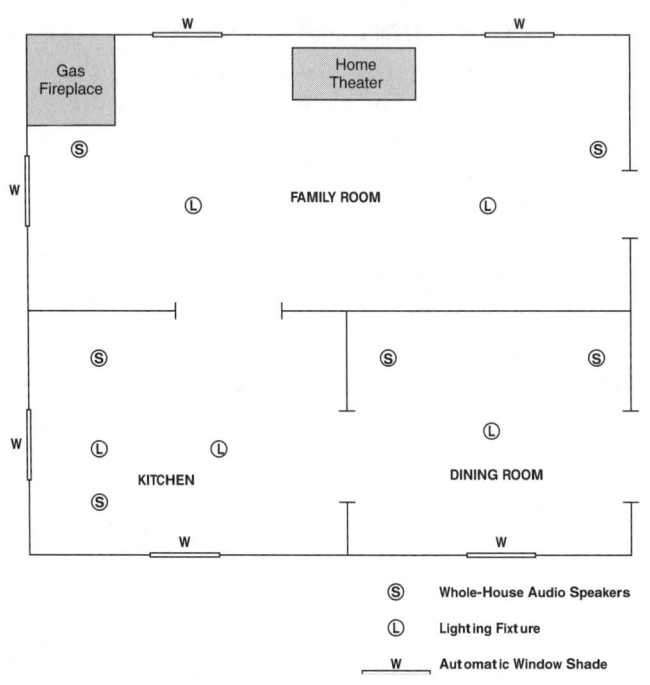

FIGURE 7–5 Sample Home Floor Plan

Many other scenarios can be envisioned. A "Night Mode" button, for example, could lock all the external doors, arm the intrusion alarm system, set all the lights in the house to the appropriate level (most of them off, maybe a few in a very dim state to serve as night lights), set the heating/cooling system for the proper temperature, and so forth.

The convenience of being able to control all the various subsystems from a single user interface (usually a touchpad or remote control) and the ability to execute complex scenarios in a simple manner are key advantages of an integrated HAS.

TABLE 7–3
"Dinner Time" Scene Description

System	Action
Lighting	Adjust to predefined level in the dining room. Dim in the kitchen and family room.
Whole-house audio	Play predefined music source (for example, favorite radio station) in dining room and kitchen. Turn off in living room.
Home theater	Turn off.
Window shades	Adjust to predefined position if desired.

TABLE 7–4
"Watch Movie" Scene Description

System	Action
Lighting	Adjust to moderately low level in the family room. Dim in the kitchen. No change in the dining room.
Whole-house audio	Turn off the zones serving the family room and kitchen. No change in the dining room.
Home theater	Turn on.
Window shades	Close the window shades in the family room. No change in the kitchen and dining room.
Other	Depending on the outside temperature, start the gas fireplace.

Summary

- Home automation systems include four major subsystems: climate control, lighting control, security, and fire alarms.
- The most common communication media for HAS devices are power lines, wireless, and UTP.
- Heating, ventilation, and air conditioning are the major components of climate-control systems; other options include automatic control of window treatments and so on.
- Lighting-control systems enhance the convenience, aesthetics, safety, and security of a home.
- The functions provided by security systems include intrusion alarms, access control, and video monitoring.
- Fire alarm systems are heavily regulated by local and national codes.
- Smart appliances have the potential to offer useful new features, but this must be traded off against the increased cost and complexity of the appliance.
- Predefined scenes allow the HAS to be customized so that several actions can be accomplished with a single command.
- Integrated HASs offer three advantages—controlling the entire system from a single touchpad, remote access, and the ability to configure complex scenes involving multiple subsystems.

Key Terms

Balun A device that connects a balanced transmission line (such as UTP) to an unbalanced transmission line (such as coaxial cable). Balun is a contraction of balanced-unbalanced.

Damper A device, such as a valve or movable plate, that controls the flow of air in a duct.

Home automation system (HAS) Equipment and infrastructure that support automatic control of home services, such as climate and lighting control, security and fire alarms, and so on.

Scenes A set of predefined HAS actions that are executed in response to a single command. Similar to a macro in computer terminology.

Smart appliance A machine built for a specific purpose (such as a refrigerator or washing machine) that contains a computer and Internet interface. Sometimes referred to as an *Internet appliance* or *network appliance*.

Review Questions

1. List the four main subsystems that comprise a HAS.
2. What are the main communication media used by HAS devices?
3. List three benefits of an intelligent HVAC controller.
4. List at least three or four other applications that may be included as part of the climate-control system.
5. List five benefits of a lighting-control system.
6. What are the five main methods of lighting control? Give an example of how two of these methods may be used in combination.
7. What is the most natural choice of a transmission medium for the control of a lighting fixture?
8. What are the three subsystems of a security system and their main functions?
9. In addition to various types of sensors, what are four components often included in a security system?
10. What transmission media are often used for security cameras?
11. According to TIA-570B, what gauge of wire is often used for connecting security devices? Why is standard 24-AWG UTP cable sometimes not adequate?
12. What codes must be met by fire alarm systems?
13. What trade-offs are involved in deciding between a standard appliance (such as a refrigerator, etc.) and a smart appliance?
14. Name three advantages of an integrated HAS over separate HAS subsystems.
15. What is the function of home automation scenes? Why do they provide more functionality in an integrated HAS?

The Construction Process

After studying this chapter, you should be able to:

OBJECTIVES

- Explain the safety rules for working on a construction site, emphasizing the importance of following proper safety procedures at all times.
- Explain why all equipment must be installed according to the manufacturer's instructions.
- Describe the construction process and the relationships among the various trade workers and others working on the job site, including officials from the authority having jurisdiction (AHJ) and those involved in the permit and inspection processes.
- Describe the general organization of the *NEC®* and locate information in the *NEC®* when given a reference number (chapter and article number).
- Use both English and SI (metric) units of measurement.
- Describe the system of listing and labeling by testing agencies.
- Use construction drawings, specifications, scope of work documents, and other construction documents as guides in wiring a dwelling.

OUTLINE

Introduction

The Construction Process

Information and Instruction

Permits and Inspections

Regional Differences

Introduction to Job Site Safety

Introduction

This chapter provides background information necessary for the electrician to understand the construction process and how a construction project is organized and controlled. Subjects such as the *National Electrical Code®* (*NEC®*), the relationship with the authority having jurisdiction (AHJ), on-the-job safety, the business and authority relationship on the job site, and technical information availability are covered within the context of a residential construction project. (*National Electrical Code®* and *NEC®* are registered trademarks of the National Fire Protection Association, Inc., Quincy, MA 02169.)

The Construction Process

Before any study of residential wiring can be undertaken, some background on the construction industry as a whole, and on the role of the electrician in particular, is necessary. The successful completion of a building involves many skilled individuals from many different trades. The workers in each trade, such as carpenters, masons, and electricians, have certain procedures, or processes, that they must follow in completing their work. They also have certain requirements of other workers that must be met before they can complete their work.

The **construction process** involves not only workers in the various construction trades but also professionals such as architects, engineers, financial managers, designers, and designated officials from government regulatory agencies. Local city or county building officials are involved in almost every aspect of the construction planning and the construction process. The agency or organization with the legal authority to approve construction projects is referred to as the **authority having jurisdiction (AHJ).** Almost all AHJs utilize building codes that set minimum standards for quality of construction. Examples of these codes are the *National Electrical Code®* (*NEC®*) and the **International Residential Code (IRC).**

These codes are published and maintained for the use of technicians and contractors as minimum standards for proper installations. The codes are usually adopted as law in the jurisdictions.

The AHJ issues **building permits** for the construction of new homes. The general contractor or builder, and many of the subcontractors also (including electrical subcontractors), must purchase these permits from the AHJ prior to beginning work. The AHJ, or its representatives, performs an **inspection** of the dwelling at critical times during the construction process to ensure that all of the code requirements are being met.

Other people also have an interest in the quality of the construction. The owner of the dwelling may inspect the job to ensure that all of the features planned and contracted for are included. In many cases, the owner has obtained a construction loan from a bank or other financial institution to provide funds for the construction of the dwelling. The bank may send inspectors to the project to ensure that the construction loan money is being used properly. Engineering firms that created the plans for the house may sometimes inspect the job to ensure that the building meets their specified engineering standards. Figure 8–1 is a diagram that shows the usual relationships among the many different people and companies involved in the construction process for a dwelling.

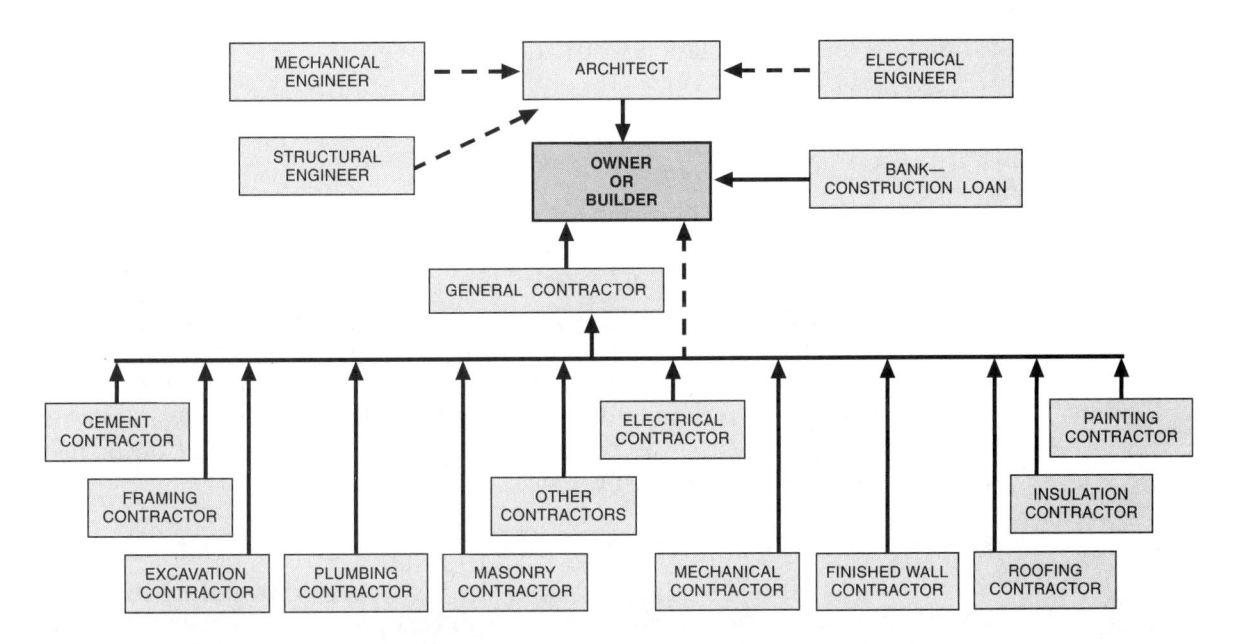

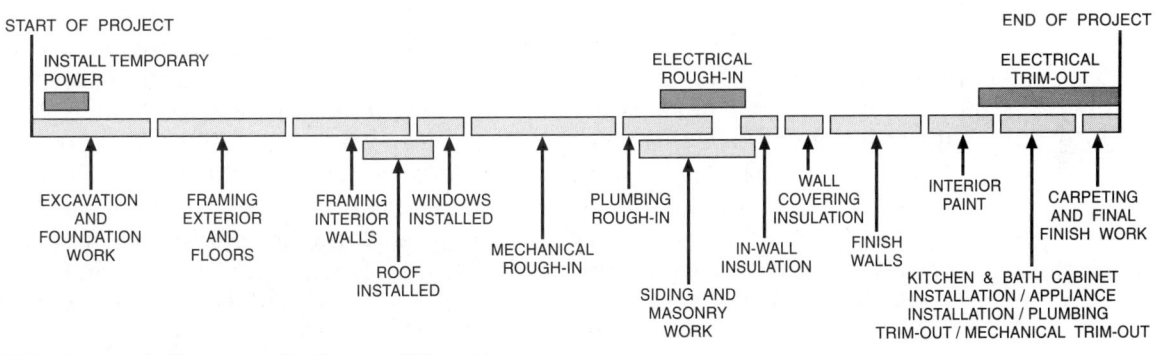

The Construction Process • 165

FIGURE 8–1 Job Site Organization and Time Line

Information and Instruction

Each of the companies involved in the construction of the dwelling must have certain information to complete the project successfully. Literally thousands of details need to be communicated from the designers and the engineers to the trade workers. Furthermore, rules related to the installation of various systems need to be closely followed. Information for the electricians on the job site usually comes from one of the following sources:

- The *NEC*®
- Construction drawings or plans

- Manufacturers' installation instructions
- The specifications
- Shop drawings, cut sheets, and submittals
- The scope of work document
- Work orders or change orders.

The *National Electrical Code*®

The *NEC*® is the document that is used in most of the United States, and in many other places in the world, as a guide for safe electrical installations. It has been in use by the electrical construction industry, governmental bodies, and electrical equipment and appliance manufacturers as a guide for over 100 years. Before any study of residential wiring can be undertaken, a firm understanding of the methods and organization used by the *Code* is essential.

Organization of the NEC®

The *NEC*® is published by the **National Fire Protection Association (NFPA).** The *NEC*® is publication number NFPA 70 and is only one of many volumes published by the Association concerning fire safety. Figure 8–2 shows how it is organized. The *NEC*®, with local amendments, is considered the electrician's law book. It provides detailed instructions for what is and is not allowed concerning electrical systems, materials, and installation requirements. The *Code* is under continuous review. An upgraded edition of the *NEC*® is published every three years. Each *Code* edition is known by the year of its adoption. The current *NEC*® is the 2005 edition.

The *NEC*® is divided into nine chapters, plus an introduction and several annexes. Each chapter covers a broad range of rules that are bound together by a common subject. The most general rules and applications are in *Chapter 1*, and the focus of the *Code* gets increasingly more specific with each subsequent chapter. The later chapters examine very specific systems and installation environments. The last chapter, *Chapter 9*, includes examples of calculations and tables with useful information. Each chapter is divided into articles, and each article is further divided into sections and subsections. All references to the *NEC*® in this book specify the article and section numbers. See Figure 8–2 for more detailed information on the organization of the *NEC*®.

The *NEC*® has two levels of control over electrical installation—mandatory and permissive. **Mandatory rules** are those rules that identify actions that are specifically prohibited or required and are characterized by the use of the terms *shall* and *shall not*. *NEC*® 210.50(B) provides a good example of mandatory rules. **Permissive rules** are defined as those that are allowed but are not required and are characterized by the use of the terms *shall be permitted* and *shall not be required*. *NEC*® 210.50(B) is a good example of permissive rules.

Measurements and the NEC®

The United States is the only industrialized country in the world that does not use the **International System of Units (SI),** also known as the metric system, as its primary measurement system. The units of measurement in this system are called SI units (for "Système International"—the system originated in France). The

CHAPTER	ARTICLE/TABLE OR ANNEX NUMBERS	CHAPTER TITLE
	Article 90	Introduction
1	Articles 100–110	General: Definitions and requirements for electrical installations
2	Articles 200–285	Wiring and Protection: Branch circuits, feeders, services, load calculations, grounding, and overcurrent protection
3	Articles 300–398	Wiring Methods and Materials: Conductors, insulation, ampacity, box fill, wiring methods, panels, services
4	Articles 400–490	Equipment for General Use: Receptacles, luminaires (lighting fixtures), appliances, motors, heating/air conditioning, switches, generators, transformers
5	Articles 500–555	Special Occupancies: Hazardous locations, health care facilities, places of assembly, theaters, recreational vehicle parks, and floating buildings
6	Articles 600–695	Special Equipment: Signs, office furnishings, elevators, welders, information technology equipment, swimming pools
7	Articles 700–780	Special Conditions: Emergency systems, low-voltage systems, Class 1, Class 2, and Class 3 circuits, instrumentation, fiber optics
8	Articles 800–830	Communication Systems: Radio and TV equipment, CATV, network-powered broadband
9	Tables 1–12(b)	Tables: Conduit fill, sizes of conductors, conductor properties, resistance
—	Annex A–G	Additional information and calculation examples

Each chapter is divided into articles—for example, *406*. The first number in the article number refers to the *NEC*® chapter number. The other two numbers identify the specific topics within that chapter.

Articles are further divided into sections by a dot (.) that separates the article number from the section number—for example, *406.8*.

A section is further divided into subsections. Three levels of subsections may be included in the Code and they are headed with:

- A capital letter inside parentheses, with (A) for the first subsection, (B) for the second subsection, and so on—for example, *406.8(B)*.
- A number inside parentheses, with (1) for the first sub-subsection, (2) for the second, and so on—for example, *406.8(B)(2)*.
- A small letter inside parentheses, with (a) for the first sub-sub-subsection, (b) for the second, and so on—for example, *406.8(B)(2)(a)*.

Lists are another organizational method used by the *NEC*®. A list is simply a series of related information. Lists are identified by numbers in parentheses, starting with (1). It is possible to distinguish an item of a list from a second-level subsection (both use a number in parentheses) because first- and second-level subsections begin with a boldface title, and lists do not have titles.

In addition to the articles, sections, subsections, and lists, there are also exceptions. Exceptions appear in the Code in *italic* type. They refer very specifically to the article, or section or subsection, that they immediately follow, and they detail situations in which the previously stated rule does not apply.

If an article is sufficiently large, it can also be divided into parts. These parts group together logically associated material within an article. The different parts of an article are labeled with Roman numerals.

Also included in the text of the *NEC*® are fine print notes (FPNs). FPNs are not enforceable as requirements of the Code. The FPNs present explanatory material that refers the reader to another place in the Code, or to another NFPA publication. FPNs can also provide the reader with guidelines for implementing the requirements of that particular section.

FIGURE 8–2 Organization of the *National Electrical Code*®

NFPA has maintained a dual measurement system for the *NEC*®. Earlier editions of the *Code* listed SI units (meters and grams) in parentheses following the English units (feet and pounds). Beginning with the 2002 edition of the *Code*, the order was reversed: The *Code* now lists the SI units first, with the English units following in parentheses.

For determining the size of a conduit or fitting, the *NEC*® uses a **trade size** system. Conduit of 1.2 in. trade size diameter is really closer to 7.8 in. in diameter. Trade size 3.4 really measures slightly over 1 in. in diameter. The dimensions referred to as 1.2 in. (0.013 m) and 3.4 in. (0.019 m) are trade size measurements. The *Code* has taken each of the English system electrical trade sizes and assigned it a **metric designator,** thus creating a kind of trade size system for SI units. The metric designator for 1.2 in. (.016 m) trade size is 16 mm. The metric designator for trade size 3.4 in. (.021 m) is 21 mm. A complete listing of the metric designator conversions can be found in *Table 300.1(C)*. In this book, the English system is used for all calculations and measurements, with the metric conversions in parentheses immediately following.

Listing and Labeling

The *NEC*® is also concerned with the systems and conductors that supply power to **utilization equipment.** Utilization equipment is anything that operates electrically. Luminaires (lighting fixtures), industrial machinery, automatic dishwashers, gasoline-dispensing pumps, and electrical signs are examples of utilization equipment. The *Code* also has rules for manufacturers of various types of utilization equipment, such as appliances and motors, but for the most part it does not dictate how the equipment is to function, or how it will be constructed.

The quality of electrical equipment is controlled by a system of testing agencies. The *NEC*® states, in *110.2*, that the various components, conduit, cable clamps, and other fittings, and equipment connected to the electrical system are permitted to be installed only if they are **approved** by the local AHJ. This approval process for each dwelling, by each jurisdiction, would be an overwhelming task. It would involve disassembling equipment in order to inspect the wiring inside of machinery, luminaires (lighting fixtures), and appliances. It would also involve the close inspection of all wire installed and various tests conducted to ensure that the conductor insulation was safe and effective. Everything would have to be inspected and tested before approval could be given.

A **fine print note (FPN)** following *110.2* directs the reader to *90.7*. The clear intent of *90.7* is to allow outside testing organizations to test the internal wiring and construction of electrical equipment. If a particular testing organization is acceptable to the local AHJ, then the inspectors can approve the installation without inspecting the internal wiring. These testing agencies are independent of the manufacturer of the equipment. If the equipment passes the tests, it is given a seal, or label, that certifies it as safe if used according to the manufacturer's instructions. The equipment is said to be **labeled.** The manufacturer can display the label of the testing agency on the equipment or appliance, on the packaging, and in its advertising. The testing agency will also put the product on a formal list to confirm that the product has indeed been tested. The product is said to be **listed.** These lists are available from the testing agency, and they are what the AHJ refers to when approving the products or equipment.

There are several testing laboratories, with Underwriters Laboratories (UL) being the largest and best known. A logo consisting of the letters "UL" in a circle is used on the equipment label to verify that the equipment has been tested and declared safe.

Construction Drawings or Plans

Construction drawings are a series of drawings that detail the design of a building. Different drawings are produced for each trade involved in the project. Included in each set of plans are drawings that apply specifically to an individual trade, such as plumbing plans or electrical plans. The construction drawings are usually letter coded according to the trade. Thus, civil engineering drawings showing the street, curb, gutters, and sidewalks are designated with the letter C. In referencing information from a particular plan, a particular plan number is used—for example, drawing C-3 would be the third drawing of the civil engineering plans.

However, construction drawings for dwellings are not usually so detailed. It is not uncommon for electricians to have only a drawing of the basic floor plan from which to design the electrical system. The floor plan shows the room arrangement and may also show the locations and types of luminaires (lighting fixtures), but true electrical plans showing receptacle and switching locations are often not included.

Drawings

A set of construction drawings for a house will typically be letter coded, designating C for civil, S for structural, A for architectural, M for mechanical, and E for electrical. The following list is an example of a complete set.

- Construction Drawing C-1 (page 175): Site Development Plan and Proximity Map
- Construction Drawing C-2 (page 176): Building Site Plans Lot 16, Block 7, Third Filing
- Construction Drawing S-1 (page 177): Foundation Plan
- Construction Drawing A-1 (page 178): Main Level Floor Plan
- Construction Drawing A-2 (page 179): Basement and Garage Dimensioned Floor Plan
- Construction Drawing A-3 (page 180): Main Floor Dimensioned Floor Plan
- Construction Drawing A-4 (page 181): East and North Elevations and Kitchen Cabinet Plans
- Construction Drawing A-5 (page 182): West and South Elevations, West Kitchen Cabinet Plan, and Bathroom Cabinet Plans
- Construction Drawing M-1 (page 183): Mechanical and Appliance Schedule
- Construction Drawing E-1 (page 184): Electrical Specifications
- Construction Drawing E-2 (page 185): Basement Electrical Plan
- Construction Drawing E-3 (page 186): Main Floor Electrical Plan
- Construction Drawing E-4 (page 187): Main Floor Telephone and Television Outlet Plan
- Construction Drawing E-5 (page 188): Luminaire (Lighting Fixture) Schedule

- Construction Drawing E-6 (page 189): Electrical Panel Schedules: Main Panel and Subpanel A
- Construction Drawing E-7 (page 190): Electrical Plan Symbols

The construction drawings will provide all of the necessary information for wiring the dwelling. It is also important to note that not all of the drawings will be using English units of measurement; while English units are predominate in most drawings, some of them may list dimensions in SI (metric) as well.

Symbols

Construction plans use **symbols** to represent actual switches and receptacles. The symbols used in plans are standard drawing symbols used throughout the country, but construction documents may include an example of each symbol with an explanation of what the symbol represents. This listing of the symbols used is called a *key*. The key to the symbols used for these drawings is presented in Construction Drawing E-7 on page 190.

Notes

In many cases, additional information, beyond what the standard symbol can represent, must be provided to the installing electricians. To notify the electrician of this information, a **flag note** is placed on the drawing adjacent to the subject equipment or device. The flag notes are numbered, and the numbers direct the electrician to a listing of explanations for each flag note located somewhere on that particular drawing. The flag notes usually apply only to the drawing that they appear on, rather than to the entire set of plans. These notes are called "flag notes" because the symbol looks like a flag or pennant.

Schedules

Construction drawings may also include other information. Sometimes a chart or table is drawn onto one or more of the drawings that lists all of the equipment being installed on the various circuits. These charts are called **schedules.** Schedules are used to organize the information and present it in one concise place, thus making it easier to find. Schedules are used for mechanical equipment, appliances, or luminaires (lighting fixtures) but can also be used for doors, wall finish, floor finish, and so on. A luminaire (lighting fixture) schedule and a mechanical and appliance schedule are presented in Construction Drawings E-5 and M-1, on pages 188 and 183, respectively.

Manufacturers' Installation Instructions

NEC® 110.3(B) states that the equipment shall be installed according to all instructions on the listing. These instructions are usually the manufacturer's **installation instructions** that accompany the product at time of sale. These instructions must be closely followed in installing that system component. The intent of the *Code* is to assure the local AHJ that an installation is safe, if the equipment is listed or labeled, and if it is installed according to the manufacturer's instructions. The installing electrician needs to understand the importance of installing the equipment according to the manufacturer's instructions. Any changes to the required installation method or any misuse of the equipment may void the testing laboratory's certification and make the equipment unlawful to install and use. Some of the actions that may void UL certification, for example, are removing parts of the equipment, drilling holes in the case or housing, and installing in unapproved locations.

Electricians must be careful to install the equipment only as directed in the manufacturer's instructions. These installation instructions should not be destroyed, but should be filed with the remainder of the project's paperwork as proof of a proper installation should any question arise. Underwriters Laboratories or any of the other testing agencies approved by the AHJ can supply additional information on listing and labeling. Underwriters Laboratories can be reached at:

Underwriters Laboratories Corporate Headquarters
333 Pfingsten Road
Northbrook, IL 60062-2096
Telephone: 847-272-8800
Fax: 847-272-8129
Internet address: **htpp://www.ul.com**

Specifications are used extensively to control the quality of the installation. The architect or the owner may want certain products or wiring methods to be used on the project. Requirements for these preferences are published in the specifications. Such requirements are also sometimes employed to exclude certain products or techniques from use. The specifications may require, for example, that conductors of 12 AWG (American Wire Gauge) minimum size must be used in wiring the house. This requirement excludes the use of 14 AWG conductors.

Specifications are sometimes listed on construction drawings. This approach is the most common in residential construction, but a separate document (sometimes a book-length document) is commonly used also. Specifications for the house are presented in Construction Drawing E-1 on page 184.

Shop Drawings/Cut Sheets, and Submittals

Much of the power consumed by a structure is used by utilization equipment other than luminaires (lighting fixtures). Appliances of all types, sizes, and electrical requirements—from exhaust fans to microwave ovens—are installed by trade workers who are not electricians. In order for the electrician to know the exact method and location of the connection and other necessary information about the size and shape of the equipment, the trades that furnish the appliances or equipment should provide the electrician with **shop drawings** or **cut sheets**. Many times the building specifications will require that each trade provide cut sheets or shop drawings to the architect for approval. These cut sheets are also provided to the general contractor. When they are provided to the general contractor or the architect, these documents are called **submittals.** They are submitted to the architect for approval prior to installation.

The Scope of Work Document

The **scope of work** is a document that lists the work to be accomplished, in a broad sense. The scope of work lists those areas that the contractor is responsible for completing and answers questions about responsibility for providing and/or installing luminaires (lighting fixtures), appliances, and other equipment. The electrical contractor uses the scope of work to help create an estimate of the installation costs. The lead person on the job uses the scope of work to determine what is included in the contract and what is not included. The general contractor uses the scope of work to ensure that the subcontractors are delivering what they contracted to install. The owner uses the scope of work to determine if there will be extra charges for some change or addition they are considering. The bank or other institution that provides construction financing uses the scope of work to determine if the dwelling will, in fact, be worth the amount loaned to the owner for construction costs.

Work Orders or Change Orders

Many times during the course of the buildings construction, changes will be made to the original design. The reasons for these changes are many and varied, but regardless of the cause, changes to the project can create havoc on the job site. Many procedures are employed by various contractors to attempt to minimize the impact of changes. The general contractor puts in place, at the start of a project, procedures to allow each contractor working on the job to know of the proposed changes, to submit an estimate of the cost impact of those changes, and to convey accurate information to the installation crew to allow for a timely and cost-effective completion of the changes. These procedures are different from one general contractor to another, but all general contractors have some sort of change order system. The changes are usually presented to the subcontractor in the form of a **work order** or a **change order.** A work order authorizes the subcontractor to perform a certain package of work.

Permits and Inspections

The contractor usually must obtain a building permit before beginning construction of a dwelling. Because the permit process varies widely throughout the country, a generalized explanation of the procedure is not possible. Usually, however, a separate permit is required for the electrical installation. Typically, a series of three inspections is required during the life of the permit. The first is usually the rough inspection, which includes the wiring within the walls, ceiling, and any other location that will be covered or otherwise not accessible later during the construction process. A second inspection is required for the dwelling when it is completed and ready for occupancy. This is sometimes called the *final inspection* or *trim inspection*. All of the electrical luminaire (lighting fixture) outlets, devices, and other systems must be installed and operating properly for the dwelling to pass this inspection. The third inspection involves the building's main electrical service. The building's electrical service is also usually required to be inspected at the time of, or before, the final inspection. The systems employed by the local AHJ to record and control the inspection process also vary widely. In many cases, the AHJ issues an inspection card to be kept on the job site and available to the inspectors at all times. As a phase of inspection is completed and approved, the inspector notes on the card what type of inspection was completed and what area of the structure was included in the inspection.

"WIREMAN'S GUIDE"
REGIONAL DIFFERENCES IN WIRING METHODS

WIRING METHODS AND CONDUCTOR TYPES ACCEPTABLE TO THE LOCAL AHJ CAN VARY WIDELY FROM ONE AREA OF THE COUNTRY TO ANOTHER. IT IS VERY IMPORTANT THAT ELECTRICIANS UNDERSTAND THAT SOME OF THE PRACTICES USED IN THEIR AREA OF THE COUNTRY MAY NOT BE ACCEPTABLE IN OTHER AREAS. THE *NATIONAL ELECTRICAL CODE®* ATTEMPTS TO ESTABLISH RULES THAT WILL BE ACCEPTABLE AND EFFECTIVE IN ALL AREAS OF THE COUNTRY, BUT LOCAL CUSTOM OR CLIMATE MAY TAKE PRECEDENCE.

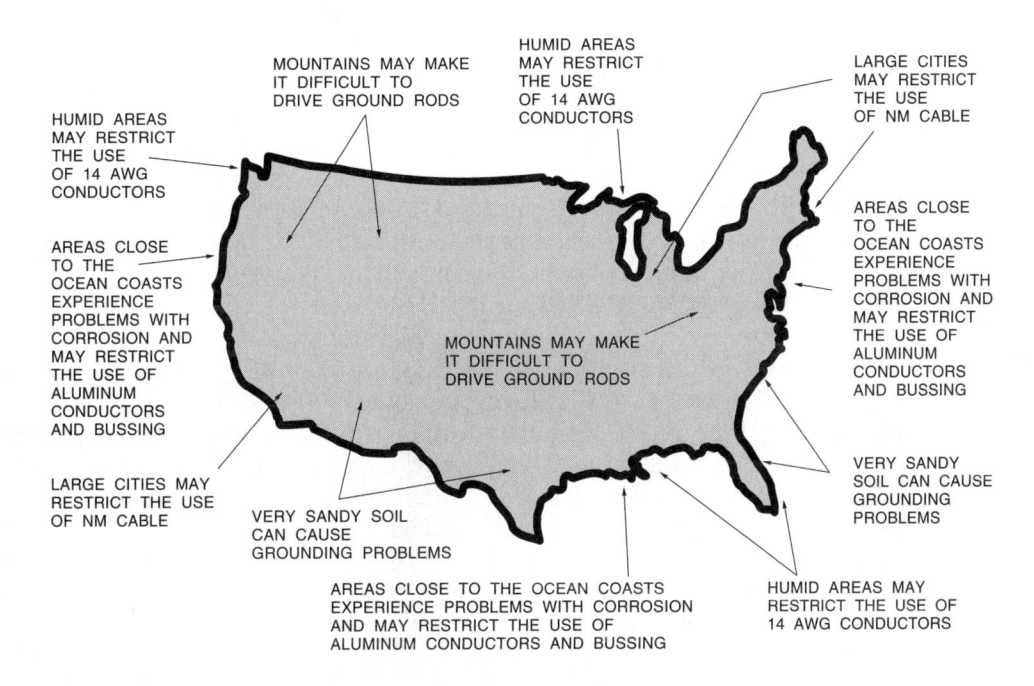

- MOUNTAINS MAY MAKE IT DIFFICULT TO DRIVE GROUND RODS
- HUMID AREAS MAY RESTRICT THE USE OF 14 AWG CONDUCTORS
- LARGE CITIES MAY RESTRICT THE USE OF NM CABLE
- HUMID AREAS MAY RESTRICT THE USE OF 14 AWG CONDUCTORS
- AREAS CLOSE TO THE OCEAN COASTS EXPERIENCE PROBLEMS WITH CORROSION AND MAY RESTRICT THE USE OF ALUMINUM CONDUCTORS AND BUSSING
- AREAS CLOSE TO THE OCEAN COASTS EXPERIENCE PROBLEMS WITH CORROSION AND MAY RESTRICT THE USE OF ALUMINUM CONDUCTORS AND BUSSING
- MOUNTAINS MAY MAKE IT DIFFICULT TO DRIVE GROUND RODS
- VERY SANDY SOIL CAN CAUSE GROUNDING PROBLEMS
- LARGE CITIES MAY RESTRICT THE USE OF NM CABLE
- VERY SANDY SOIL CAN CAUSE GROUNDING PROBLEMS
- AREAS CLOSE TO THE OCEAN COASTS EXPERIENCE PROBLEMS WITH CORROSION AND MAY RESTRICT THE USE OF ALUMINUM CONDUCTORS AND BUSSING
- HUMID AREAS MAY RESTRICT THE USE OF 14 AWG CONDUCTORS

LOCAL CLIMATE AND WEATHER CONDITIONS CAUSE THE MOST VARIATIONS IN INSTALLATION RULES.

AREAS OF THE COUNTRY THAT ARE EXTREMELY HUMID HAVE PROBLEMS WITH CORROSION. THE MOISTURE IN THE AIR CAN CAUSE BUSSING AND CONDUCTOR TERMINATIONS TO CORRODE AND FAIL. THEREFORE, THE AHJ MAY RESTRICT THE USE OF ALUMINUM CONDUCTORS, LIMIT THE MINIMUM SIZE OF CONDUCTORS TO 12 AWG, AND REQUIRE EXTRA CORROSION PROTECTION FOR CONDUITS AND RACEWAYS.

FIGURE 8–3 "Wireman's Guide" Regional Difference in Wiring Methods

Regional Differences

It is very difficult to make general statements that apply to the United States without exceptions. This is particularly true for the construction industry. Physical conditions such as weather, availability of materials, local customs, and a host of other factors allow for many variations in installation procedures and wiring methods.

The information presented in this book applies to a majority of the regions in the United States. Figure 8–3 shows some of the more likely regional differences and some of the reasons for the differences. Electricians must always be aware of the particular rules and procedures followed in their local areas, and adjustments to the information presented in this book should be made accordingly.

Introduction to Job Site Safety

Nothing is as important as job site safety. Nothing! Not money, not completion date, and not lunch, or break, or quitting time! Safety on the job site must be the first thing in each worker's and each foreperson's mind. The federal government believes in on-the-job safety so much that it established the **Occupational Safety and Health Administration (OSHA)** to oversee safety policy in the workplace. OSHA publishes safety standards and inspects places of employment, including construction sites, periodically (and when requested) to ensure compliance with these standards. Employers found to be in noncompliance with the established standards can be heavily fined, as can employees of such firms in some cases.

Employers also provide written safety policies and a company safety manual. Each employee must read, understand, and adhere to these safety policies. It is also the responsibility of each person on the site to be familiar with the safety rules established for that site by the general contractor. The wearing of **personal protective equipment (PPE),** such as hard hats and safety glasses, is very commonly required of everyone on the site. Wearing work boots, sometimes with steel toe protectors, and properly sized rugged clothing is also a common requirement. However, all of these safety rules and regulations are of no value unless they are scrupulously followed. Unless every worker on the site obeys the safety rules, it is just a matter of time before a major accident happens. The construction site is inherently a noisy, crowded, and very dangerous place, and injury or death awaits the unprepared. Construction workers must be mindful of the dangers on the site at all times, and they must take actions necessary to properly protect themselves from those dangers. Using the proper PPE is the first, the most basic, and the most effective step in job site safety.

A major hazard that exists on the job site is electricity. Almost all trade workers use electricity to run their power tools, compressors, and chargers. Furthermore, artificial lighting may be needed in basements or interior rooms with no windows to illuminate the work space. The electrical contractor usually installs temporary power at the construction site and sometimes temporary lighting. There are very strict rules concerning the temporary power that is supplied and how it is supplied. It is important to remember that workers from all trades will be using this temporary power. Many of these workers have little or no knowledge of the nature of electricity. The temporary wiring at a job site must therefore be not only safe but also relatively foolproof.

There will be times when an electrician must work on a system while it is energized. Working a circuit "hot" is something to be avoided at almost any cost. The term *hot* in the electrical trade is used to describe a circuit that is energized—that is, connected to the electrical supply. Work on an energized circuit poses needless risk of injury or death. As a rule, electricians should not work on hot circuits. Some type of disconnecting switch is required by the *Code* on all circuits for all hot conductors, except service-entrance conductors. The disconnecting switch may be a circuit breaker in a panel or may be a separate switch. These disconnecting switches must be easily accessible by the electrician and must either be able to be locked in the off position or within easy sight of the individuals working on the circuit. Just de-energizing a system or circuit is not good enough for proper safety procedures. If the system or equipment could be re-energized without the technician's being aware of it, a *lockout and tag-out* procedure must be employed. All electrical contractors have a written tag-out and lockout procedure that should be familiar to each employee.

The Construction Process • 175

SITE DEVELOPMENT PLAN AND PROXIMITY MAP

RUNNING BROOKE RANCH THIRD FILING

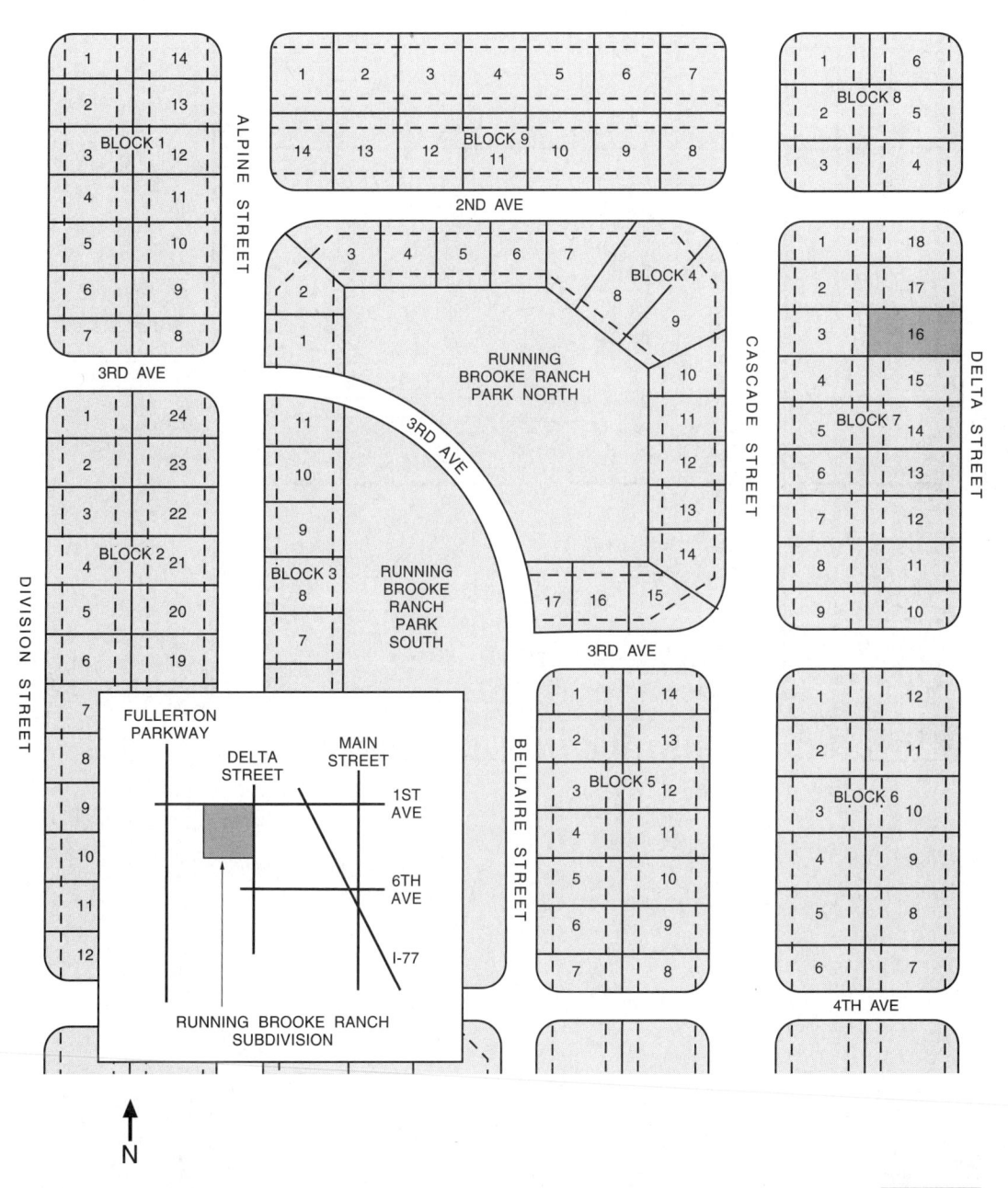

CONSTRUCTION PLAN C-1

C-1

176 • Chapter 8

BUILDING SITE PLAN
LOT 16, BLOCK 7, THIRD FILING
RUNNING BROOKE RANCH SUBDIVISION

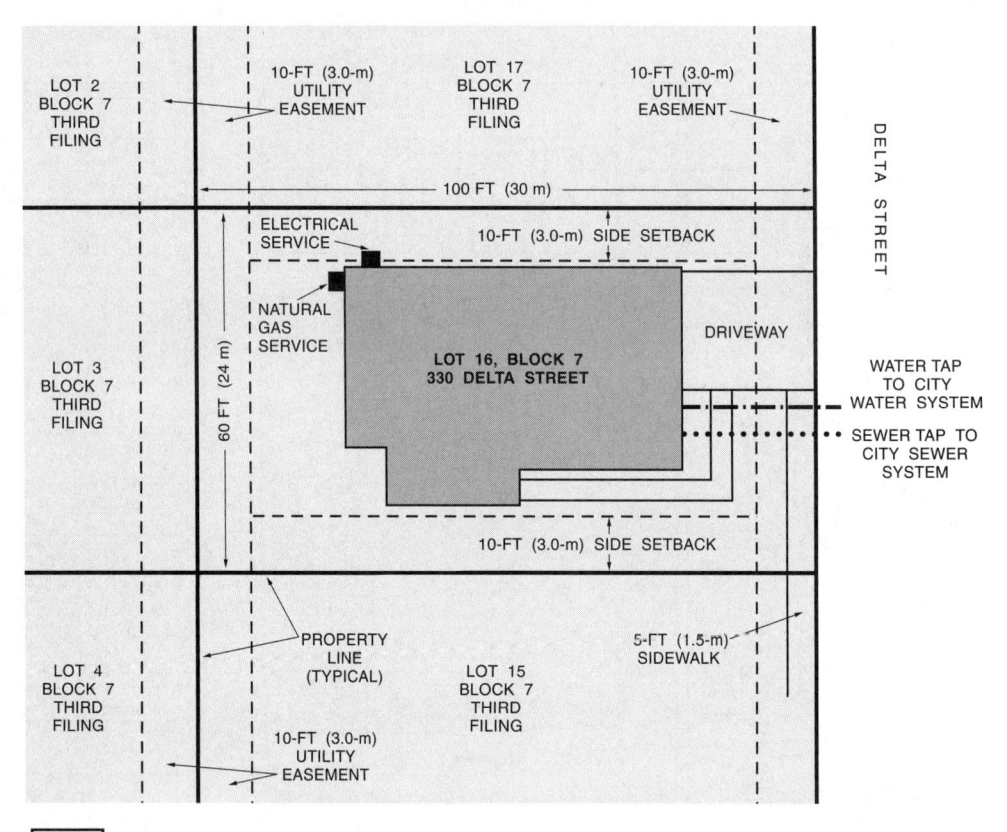

C-2 **CONSTRUCTION DRAWING C-2**

FOUNDATION PLAN

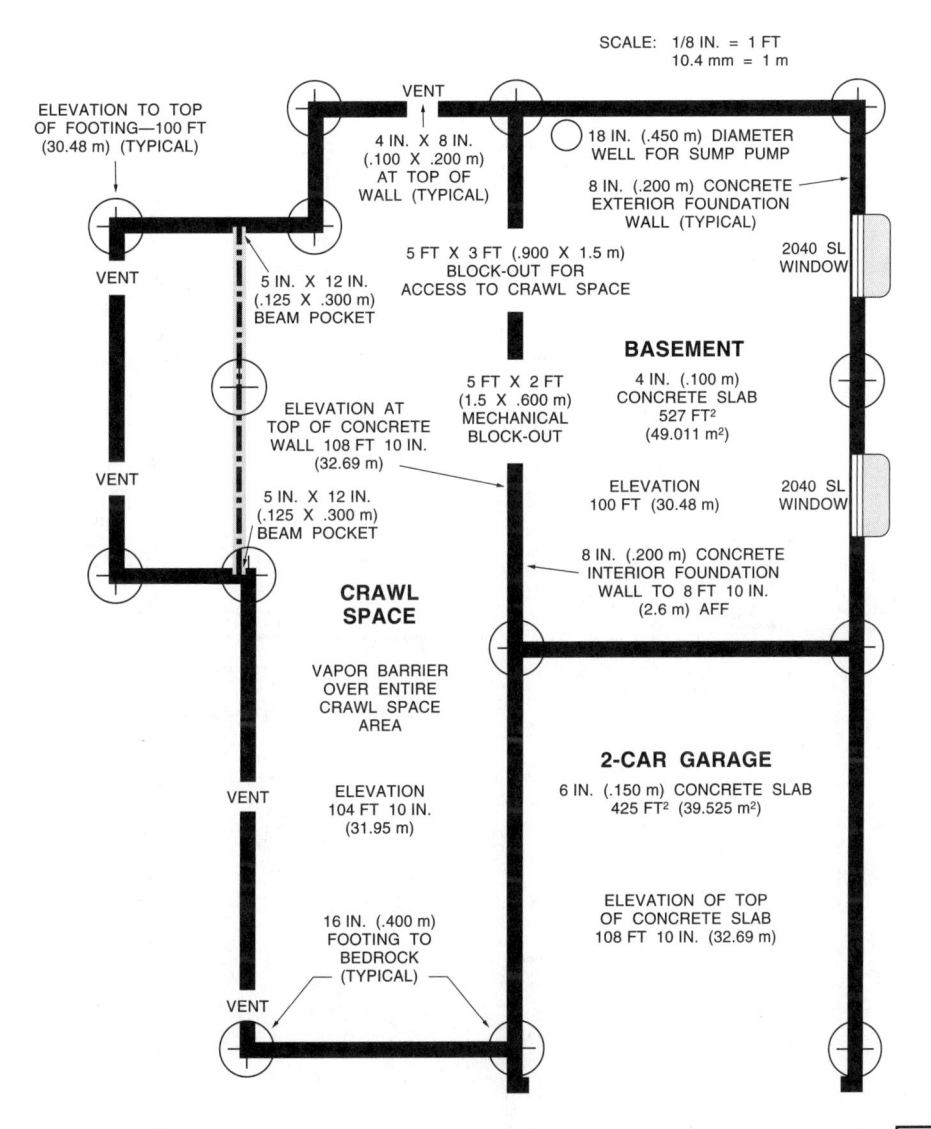

SCALE: 1/8 IN. = 1 FT
10.4 mm = 1 m

ELEVATION TO TOP OF FOOTING—100 FT (30.48 m) (TYPICAL)

VENT

4 IN. X 8 IN. (.100 X .200 m) AT TOP OF WALL (TYPICAL)

18 IN. (.450 m) DIAMETER WELL FOR SUMP PUMP

8 IN. (.200 m) CONCRETE EXTERIOR FOUNDATION WALL (TYPICAL)

2040 SL WINDOW

VENT

5 IN. X 12 IN. (.125 X .300 m) BEAM POCKET

5 FT X 3 FT (.900 X 1.5 m) BLOCK-OUT FOR ACCESS TO CRAWL SPACE

BASEMENT

5 FT X 2 FT (1.5 X .600 m) MECHANICAL BLOCK-OUT

4 IN. (.100 m) CONCRETE SLAB
527 FT² (49.011 m²)

ELEVATION AT TOP OF CONCRETE WALL 108 FT 10 IN. (32.69 m)

VENT

5 IN. X 12 IN. (.125 X .300 m) BEAM POCKET

ELEVATION 100 FT (30.48 m)

2040 SL WINDOW

8 IN. (.200 m) CONCRETE INTERIOR FOUNDATION WALL TO 8 FT 10 IN. (2.6 m) AFF

CRAWL SPACE

VAPOR BARRIER OVER ENTIRE CRAWL SPACE AREA

2-CAR GARAGE

6 IN. (.150 m) CONCRETE SLAB
425 FT² (39.525 m²)

VENT

ELEVATION 104 FT 10 IN. (31.95 m)

16 IN. (.400 m) FOOTING TO BEDROCK (TYPICAL)

ELEVATION OF TOP OF CONCRETE SLAB 108 FT 10 IN. (32.69 m)

VENT

CONSTRUCTION DRAWING S-1

S-1

178 • Chapter 8

MAIN LEVEL FLOOR PLAN

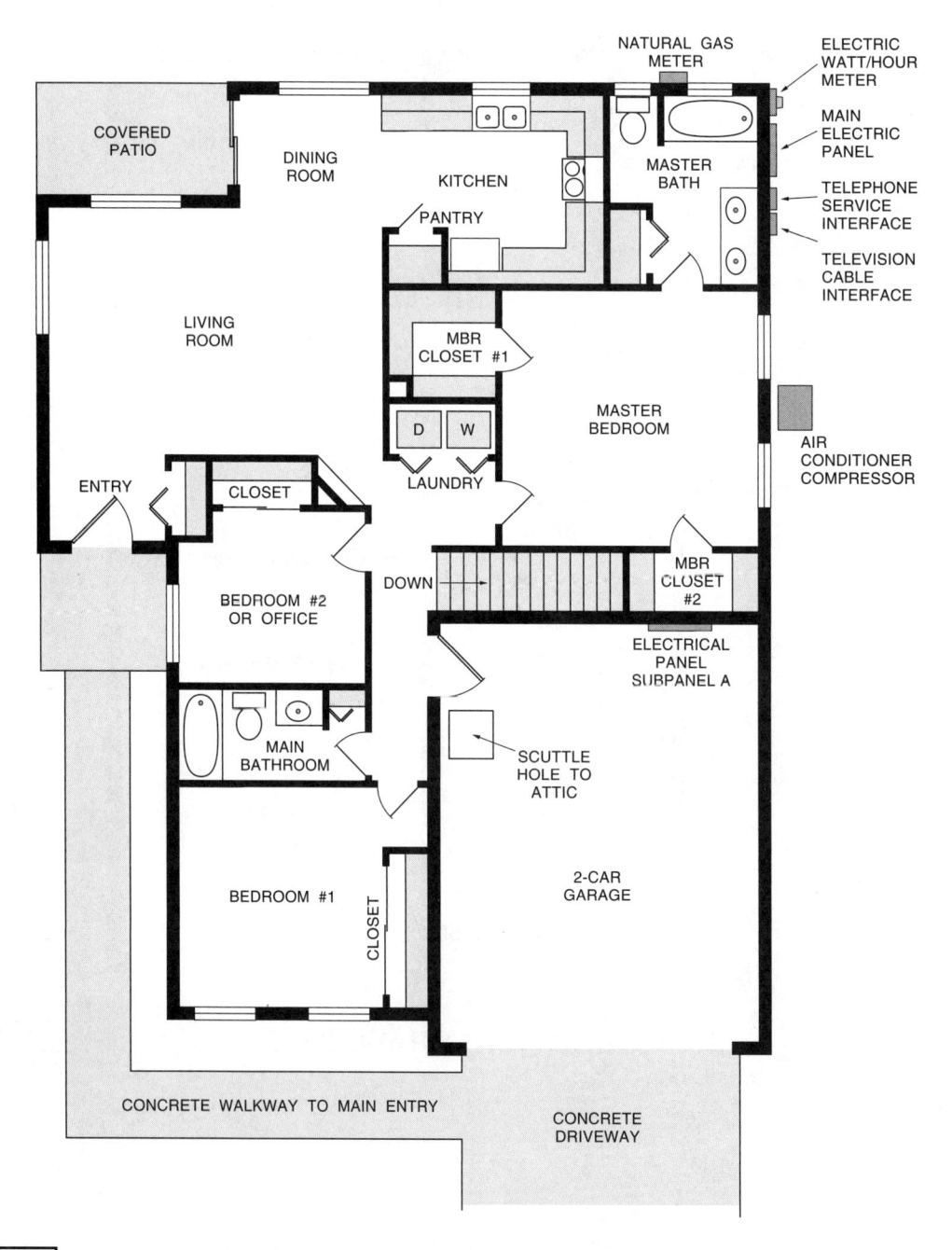

A-1

CONSTRUCTION DRAWING A-1

BASEMENT AND GARAGE DIMENSIONED FLOOR PLAN

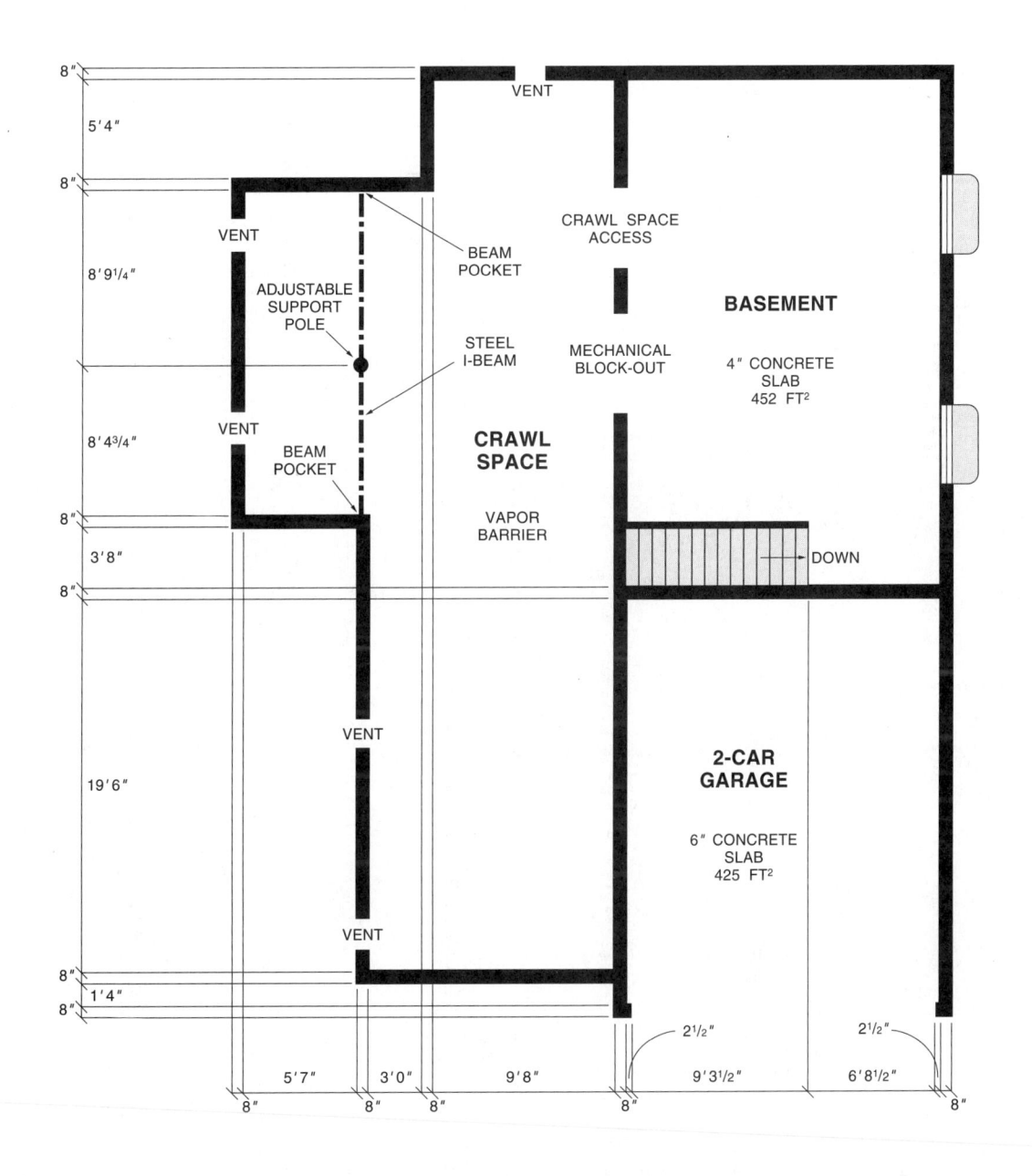

CONSTRUCTION DRAWING A-2

A-2

MAIN FLOOR DIMENSIONED FLOOR PLAN

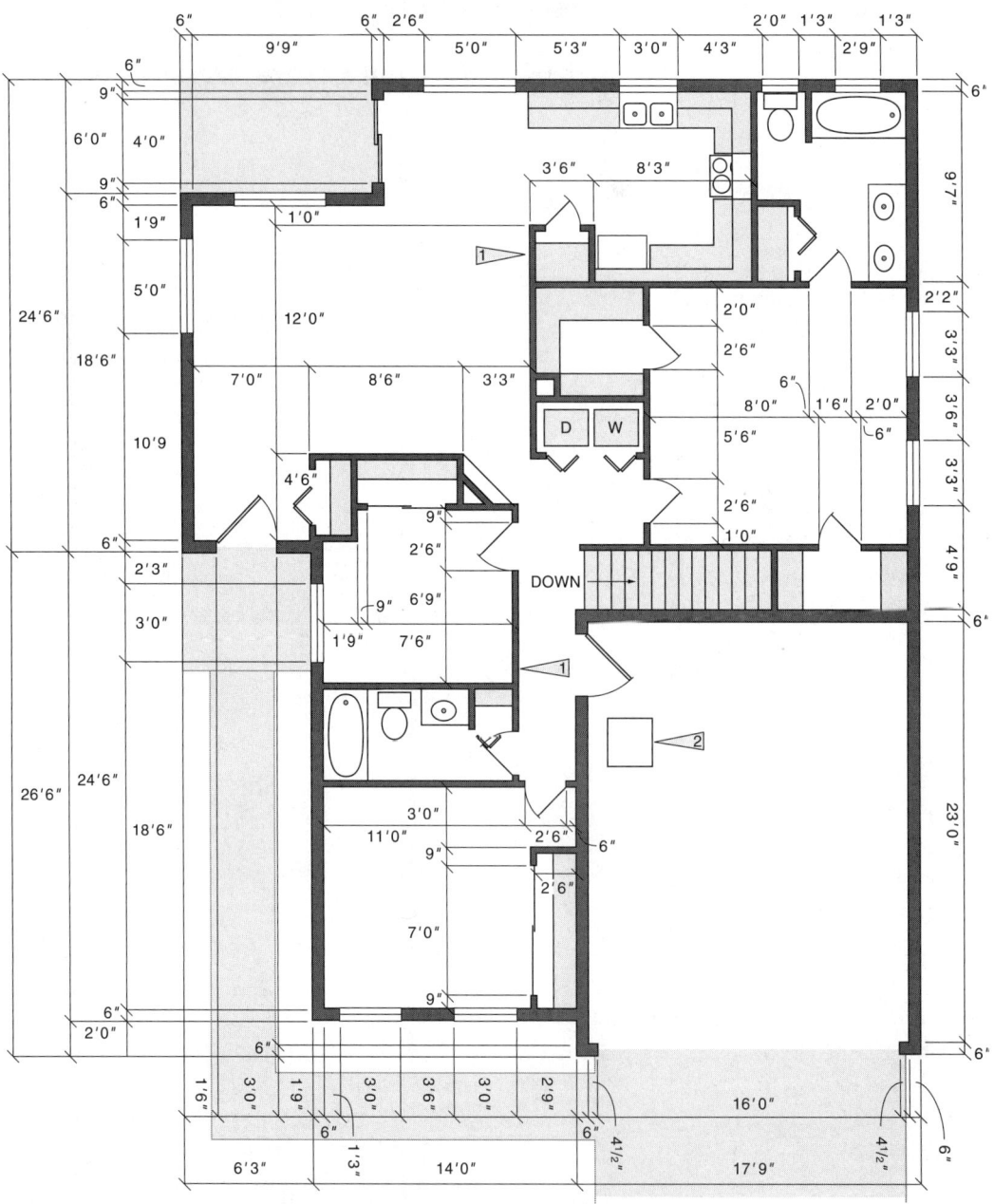

1. ALL INTERIOR WALLS ARE 5 IN. DEEP UNLESS SHOWN OTHERWISE. OUTSIDE WALLS ARE 6 IN. DEEP. ALL WALLS TO HAVE 3/4-IN. SHEETROCK.
2. SCUTTLE HOLE TO ATTIC

A-3

CONSTRUCTION DRAWING A-3

The Construction Process • 181

EAST AND NORTH ELEVATIONS
AND KITCHEN CABINET PLANS

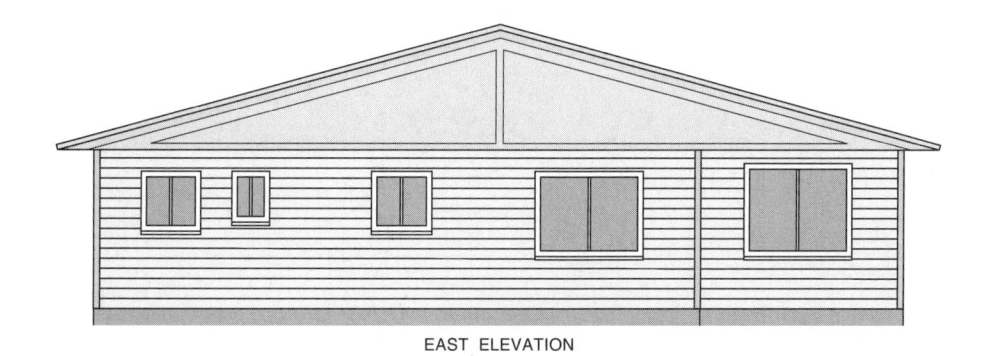

EAST ELEVATION

NORTH ELEVATION

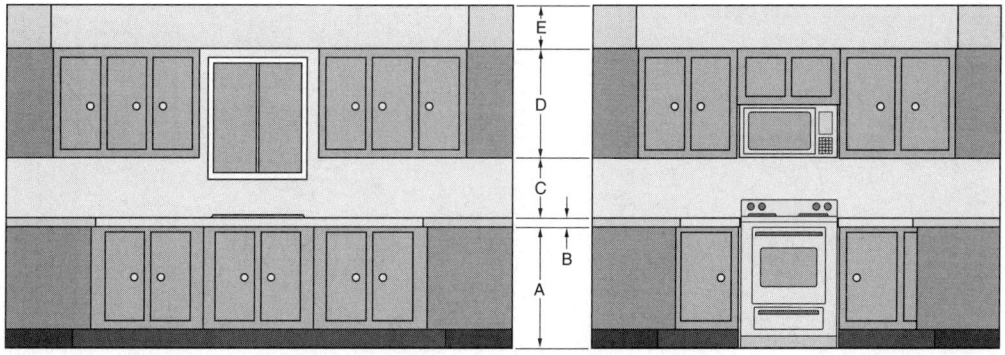

EAST KITCHEN CABINET PLAN

NORTH KITCHEN CABINET PLAN

A = 36 IN. (.900 m)
B = 2 IN. (.050 m)
C = 12 IN. (.300 m)
D = 36 IN. (.900 m)
E = 10 IN. (.250 m)

CONSTRUCTION DRAWING A-4

A-4

182 • Chapter 8

**WEST AND SOUTH ELEVATIONS,
WEST KITCHEN CABINET PLAN,
AND BATHROOM CABINET PLANS**

WEST ELEVATION

SOUTH ELEVATION

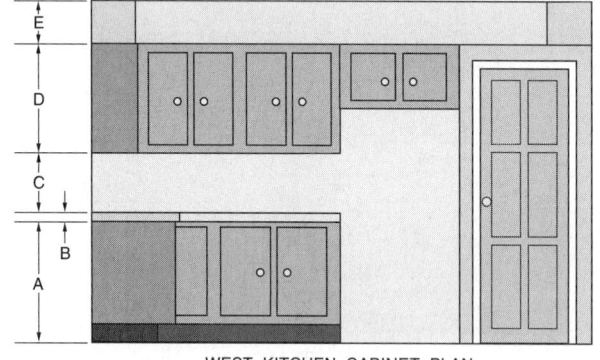

WEST KITCHEN CABINET PLAN

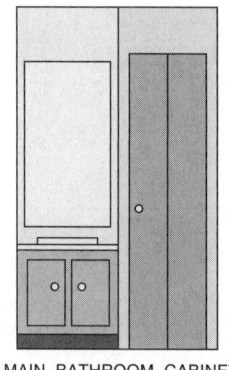

MAIN BATHROOM CABINET
AND MIRROR PLAN

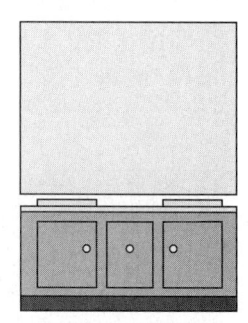

MASTER BATHROOM
CABINET AND
MIRROR PLAN

A = 36 IN. (.900 m)
B = 2 IN. (.050 m)
C = 12 IN. (.300 m)
D = 36 IN. (.900 m)
E = 10 IN. (.250 m)

A-5

CONSTRUCTION DRAWING A-5

MECHANICAL AND APPLIANCE SCHEDULE

APPLIANCE	MANUFACTURER	MODEL NUMBER	VOLTS	FLA	VA/WATTS	HP	MIN CIRCUIT AMPS
Dryer	Franklin Ind	55MFT240-2-1-F	120/240		5500		
Range	Sargent Electric	TRP11.5-7687	120/240		11500		
Microwave	Sargent Electric	NMO-5634-A	120	14.7	1764		
Disposal	Kl Klienne	998-00034-76Y	120	12.0		¾	
Dishwasher	Kl Klienne	102-00198-22P	120	13.5		¹⁄₁₀	
Garage Door Opener	Door Best	2354-2	120	12.3	1475	¾	
Sump Pump	Water Whip	99-12075	120			¾	
Baseboard Heater	Heatease of Ohio	240200T-0-0	240		2000		
Air Conditioner	Wall and Miller	9348UYTEHDG47	240				28.8 35a Fuse
Furnace	Heatease of Ohio	120G110G-2-3	120	9.2	1100	¾	

CONSTRUCTION DRAWING M-1

M-1

ELECTRICAL SPECIFICATIONS

1. ALL SERVICE EQUIPMENT, DISTRIBUTION PANELS, AND DISCONNECT SWITCHES SHALL BE MANUFACTURED BY THE SAME COMPANY. ALL EQUIPMENT SHALL BE NEW AND MANUFACTURED BY ONE OF THE FOLLOWING COMPANIES:

   GENERAL ELECTRIC
   SQUARE D
   CUTTLER-HAMMER
   SIEMENS

2. ALL WIRING DEVICES ARE TO BE SUPPLIED AND INSTALLED BY THE ELECTRICAL CONTRACTOR.

3. ALL WIRING DEVICES ARE TO BE MANUFACTURED BY LITECHTRONIX, OR EQUAL.

4. ALL SWITCHES, EXCEPT FOR SWITCHES ABOVE COUNTERTOPS, SHALL BE INSTALLED AT 44 IN. (1.118 m) ABOVE THE FINISHED FLOOR (AFF) TO THE BOTTOM OF THE BOX.

5. ALL RECEPTACLE OUTLETS, EXCEPT FOR ABOVE COUNTER RECEPTACLES IN THE KITCHEN AND THE BATHROOMS SHALL BE 13 IN. (.330 m) AFF TO THE BOTTOM OF THE BOX. RECEPTACLE OUTLETS IN THE GARAGE SHALL BE 44 IN. (1.118 m) AFF.

6. ALL RECEPTACLES INSTALLED ABOVE KITCHEN OR BATHROOM COUNTERS SHALL BE 6 IN. (150 mm) TO THE BOTTOM OF THE BOX ABOVE THE COUNTER TOP (AC).

7. ALL LUMINAIRES (LIGHTING FIXTURES) SHALL BE PROVIDED BY THE GENERAL CONTRACTOR AND INSTALLED BY THE ELECTRICAL CONTRACTOR, EXCEPT FOR RECESSED LUMINAIRE (LIGHTING FIXTURE) HOUSINGS, RECESSED LUMINAIRE (LIGHTING FIXTURE) TRIMS, AND LAMPS FOR RECESSED LUMINAIRES (LIGHTING FIXTURES), WHICH SHALL BE PROVIDED AND INSTALLED BY THE ELECTRICAL CONTRACTOR.

8. ALL KITCHEN APPLIANCES SHALL BE SUPPLIED AND INSTALLED BY THE GENERAL CONTRACTOR. WHIPS, PIGTAILS, OR CORD NEEDED TO CONNECT THE APPLIANCES TO THE ELECTRICAL SYSTEM SHALL BE PROVIDED AND INSTALLED BY THE ELECTRICAL CONTRACTOR.

9. ALL SWITCHES, RECEPTACLES, AND OTHER DEVICES SHALL BE WHITE. ALL COVER PLATES SHALL BE WHITE PLASTIC (NONMETALLIC).

10. ALL RECEPTACLES SHALL BE INSTALLED WITH THE GROUND PRONG (CONTACT) TO THE TOP OF THE DEVICE. IF THE DEVICE IS INSTALLED HORIZONTALLY, THE TOP BLADE OF THE RECEPTACLE SHALL BE THE GROUNDED CONDUCTOR BLADE.

11. ALL FIXED-IN-PLACE KITCHEN APPLIANCES SHALL BE ON THEIR OWN INDIVIDUAL CIRCUITS SIZED TO THE REQUIREMENTS OF THE PARTICULAR APPLIANCE.

12. A SUMP PUMP SHALL BE INSTALLED IN THE BASEMENT. GENERAL CONTRACTOR SHALL PROVIDE THE PUMP. THE LOCATION OF THE PUMP IS PER CONSTRUCTION DRAWING S-1. THE PUMP SHALL BE INSTALLED BY THE MECHANICAL CONTRACTOR AND WIRED BY THE ELECTRICAL CONTRACTOR PER THE REQUIREMENTS AND THE MOTOR NAMEPLATE.

13. THE INSTALLATION OF ALL WIRE, BOXES, DEVICES, LUMINAIRES (LIGHTING FIXTURES), AND OTHER ELECTRICAL EQUIPMENT SHALL BE IN STRICT ACCORDANCE WITH THE *NATIONAL ELECTRICAL CODE*®, LOCAL JURISDICTIONAL REQUIREMENTS, MANUFACTURERS INSTRUCTIONS, AND THESE SPECIFICATIONS.

14. THE ELECTRICAL CONTRACTOR SHALL PROVIDE TEMPORARY POWER, AND IF NECESSARY TEMPORARY LIGHTING, FOR THE USE OF ALL TRADES ON THE JOB SITE.

15. THE ELECTRICAL CONTRACTOR SHALL PROVIDE AND PAY FOR ANY PERMITS THAT MAY BE REQUIRED BY THE CITY BUILDING INSPECTOR S OFFICE. THE ELECTRICAL CONTRACTOR SHALL SUCCESSFULLY OBTAIN ALL REQUIRED INSPECTIONS.

E-1 **CONSTRUCTION DRAWING E-1**

BASEMENT ELECTRICAL PLAN

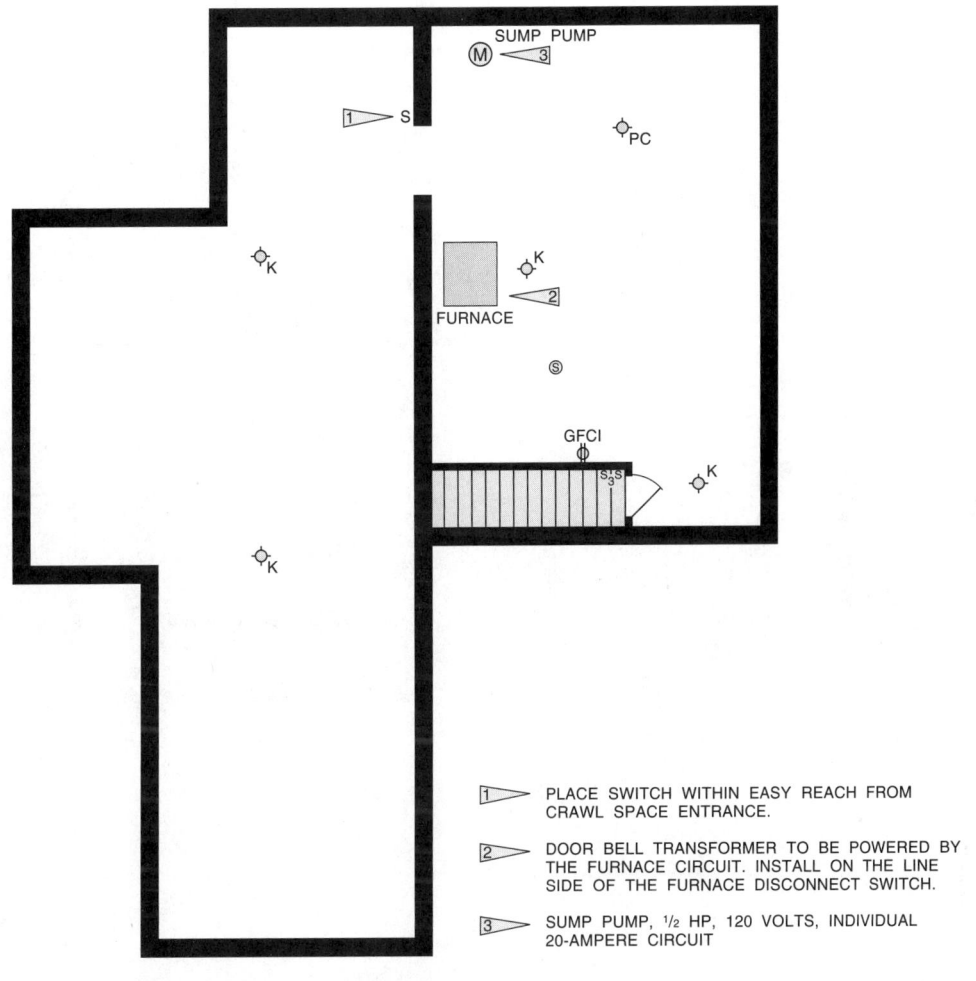

▷1 PLACE SWITCH WITHIN EASY REACH FROM CRAWL SPACE ENTRANCE.

▷2 DOOR BELL TRANSFORMER TO BE POWERED BY THE FURNACE CIRCUIT. INSTALL ON THE LINE SIDE OF THE FURNACE DISCONNECT SWITCH.

▷3 SUMP PUMP, 1/2 HP, 120 VOLTS, INDIVIDUAL 20-AMPERE CIRCUIT

CONSTRUCTION DRAWING E-2

E-2

MAIN FLOOR ELECTRICAL PLAN

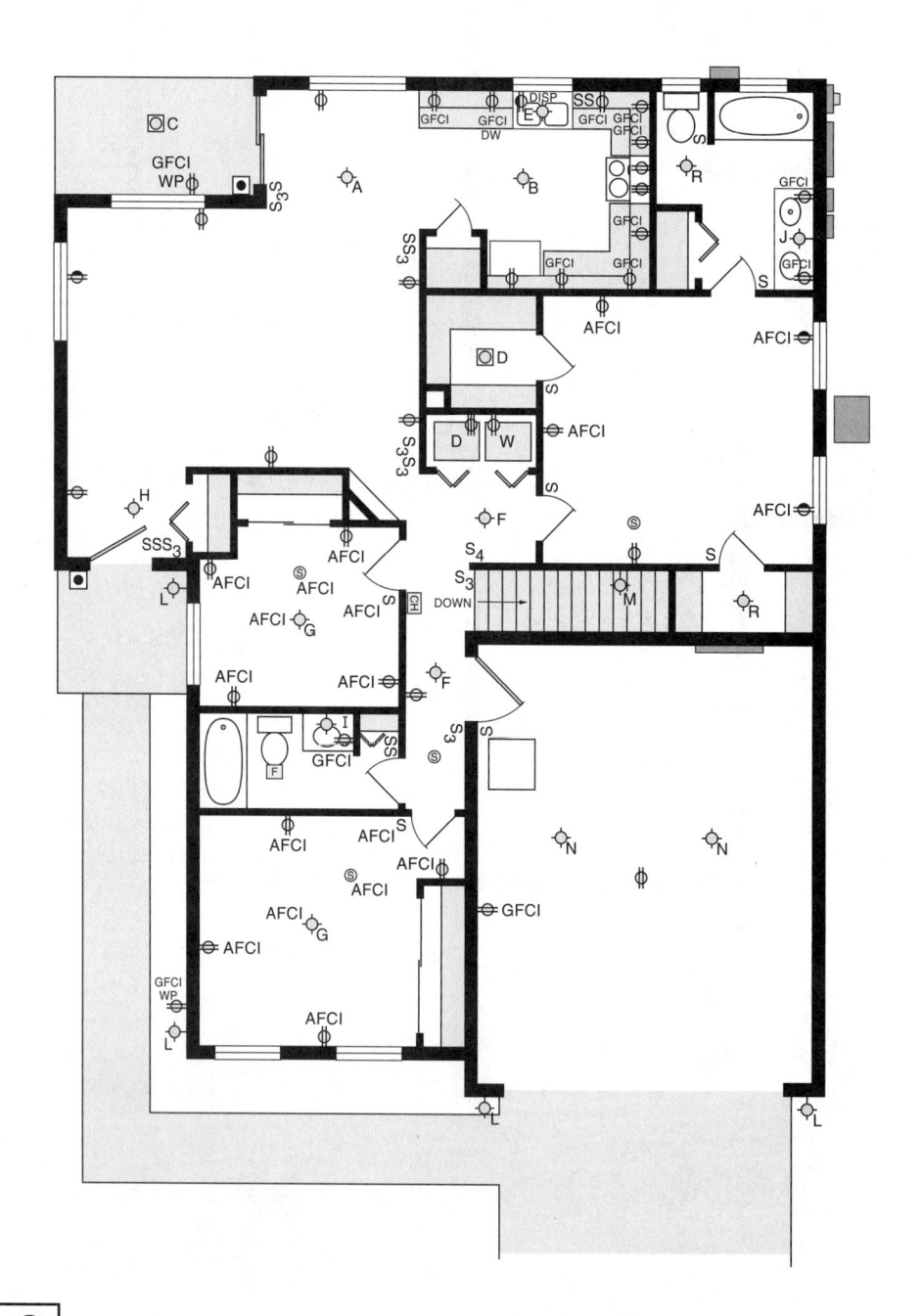

CONSTRUCTION DRAWING E-3

The Construction Process • 187

MAIN FLOOR TELEPHONE AND TELEVISION OUTLET PLAN

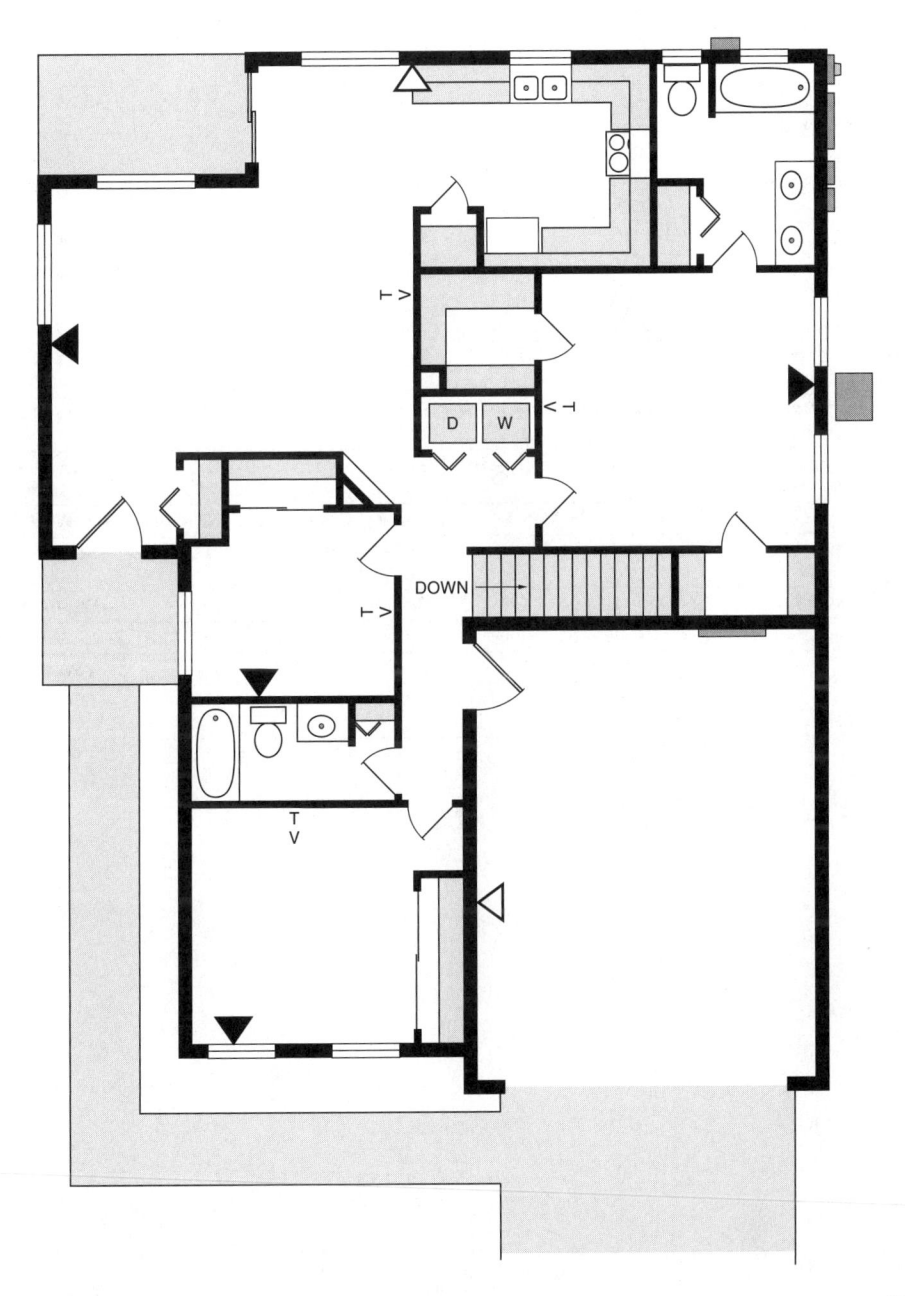

CONSTRUCTION DRAWING E-4

E-4

LUMINAIRE (LIGHTING FIXTURE) SCHEDULE

LUMINAIRE (LIGHTING FIXTURE) TYPE	MANUFACTURER	CATALOG NUMBER	MOUNTING	VOLTAGE	LAMPS
A	Hays Lighting Products and Fixtures	SCH – 231 – 0 – 9 – G – BR	Ceiling. Pendant.	120	6—40W A-17
B	Elbert Lighting	B – 84 – 3	Surface. Ceiling.	120	3—60W A-19
C	Hays Lighting Products and Fixtures	RCA – 7.5 – IC Housing T – 7.5 WP – 3 Trim	Recess. Waterproof trim.	120	1—75W PAR 38
D	Hays Lighting Products and Fixtures	RCA – 7.5 – IC Housing T – 7.5 OP – 2 Trim	Recess. Black baffle trim.	120	1—75W R-30
E	Elbert Lighting	B – 71 – 1	Surface. Ceiling.	120	1—60W A-17
F	Milwaukee Home Lighting Products	120 – 40 – BA – 0 – 8	Surface. Ceiling.	120	4—40W G-10
G	Far East Fixtures	75 – AD – R – 120 – CR	Surface. Ceiling.	120	2—60W A-19
H	Milwaukee Home Lighting Products	120 – 40 – BA – T – 8	Surface. Ceiling.	120	4—40W G-10
I	Hays Lighting Products and Fixtures	SWS – 414 – 4 – 2 – K – BR	Surface. Wall.	120	4—60W G-28
J	Hays Lighting Products and Fixtures	SWS – 414 – 6 – 2 – K – BR	Surface. Wall.	120	6—60W G-28
K		Keyless Lamp Holder	Surface. Ceiling.	120	1—75W A-19
L	Howell-Simon	DF – 02387 – P – 7	Surface. Outdoor wall mount.	120	1—75W A-19
M	Far East Fixtures	60 – AR – R – 120 – CR	Surface. Wall.	120	1—60W A-19
N	HeavyLite	120 – 2 – 40 – ST – 0 – 1	Surface. Ceiling.	120	2—32W T-8 CW
PC		Pull Chain Lamp Holder	Surface. Ceiling.	120	1—75W A-19
R	Hays Lighting Products and Fixtures	SCS – 213 – 1 – 2 – L – BR	Surface. Ceiling.	120	1—60W A-19

[E-5] **CONSTRUCTION DRAWING E-5**

ELECTRIC PANEL SCHEDULES

MAIN PANEL

Main Circuit Breaker: 150 AMPERES, 120/240 VOLTS
Panel Bussing: 200 AMPERES

CIRCUIT NUMBER	DESCRIPTION	CIR BKR		CIR BKR	DESCRIPTION	CIRCUIT NUMBER
1	AIR CONDITIONER	40			SPACE	2
3	AIR CONDITIONER	40			SPACE	4
5	SUBPANEL A	125			SPACE	6
7	SUBPANEL A	125			SPACE	8

SUBPANEL A

Main Lugs
Panel Bussing: 200 AMPERES

CIRCUIT NUMBER	DESCRIPTION	CIR BKR		CIR BKR	DESCRIPTION	CIRCUIT NUMBER
1	REFRIGERATOR	15		20	SMALL APPLIANCE	2
3	KIT, DR, LR, LIGHTING / LR, OUTSIDE RECEPT.	15		20	SMALL APPLIANCE	4
5	BR LIGHTING & RECEPT.	15		20	MICROWAVE / HOOD	6
7	MBR, MBR BATH LIGHTING & RECEPT.	15		20	DISHWASHER	8
9	RANGE	40		20	DISPOSAL	10
11	RANGE	40		20	BATHROOM RECEPTACLES	12
13	DRYER	30		15	GARAGE / OUTSIDE FRONT RECEPT.	14
15	DRYER	30		20	CLOTHES WASHER	16
17	FURNACE	20		15	BSMT LIGHTING & RECEPT.	18
19	SMOKE DETECTOR AND ALARMS	15		15	BASEBOARD HEATER	20
21	GARAGE DOOR OPENER	15		15	BASEBOARD HEATER	22
23	SPACE			20	SUMP PUMP	24

CONSTRUCTION DRAWING E-6

ELECTRICAL PLAN SYMBOLS

Symbol	Description	Symbol	Description
⏀	INDIVIDUAL RECEPTACLE, 125 VOLTS, 15 OR 20 AMPERES	S SAFCI	SINGLE-POLE SWITCH SINGLE-POLE SWITCH, AFCI-PROTECTED
⏀	DUPLEX RECEPTACLE, 125 VOLTS, 15 OR 20 AMPERES	S_3	3-WAY SWITCH
⏀	SPLIT-WIRED DUPLEX RECEPTACLE, 125 VOLTS, 15 OR 20 AMPERES	S_4	4-WAY SWITCH
⏀ GFCI	GFCI-PROTECTED DUPLEX RECEPTACLE, 125 VOLTS, 15 OR 20 AMPERES	⊕	CEILING-MOUNTED LUMINAIRE (LIGHTING FIXTURE)
⏀ WP	WATERPROOF DUPLEX RECEPTACLE, 125 VOLTS, 15 OR 20 AMPERES	⊕	WALL-MOUNTED LUMINAIRE (LIGHTING FIXTURE)
⏀ GFCI WP	WATERPROOF AND GFCI-PROTECTED DUPLEX RECEPTACLE, 125 VOLTS, 15 OR 20 AMPERES	▢	RECESSED LUMINAIRE (LIGHTING FIXTURE)
⏀ AFCI	AFCI-PROTECTED DUPLEX RECEPTACLE, 125 VOLTS, 15 OR 20 AMPERES	Ⓢ AFCI	SMOKE DETECTOR, 120 VOLTS W/BATTERY BACKUP AFCI-PROTECTED
⏀ AFCI	SPLIT-WIRED AFCI-PROTECTED DUPLEX RECEPTACLE, 125 VOLTS, 15 OR 20 AMPERES	F	BATH FAN
⏀	INDIVIDUAL RECEPTACLE, 240 VOLTS, 30 OR 50 AMPERES	CH	DOORBELL CHIME, 12 VOLTS
		▣	DOORBELL PUSHBUTTON, 12 VOLTS
		T V	CATV (TELEVISION) OUTLET
		▼	TELEPHONE OUTLET—DESK
		▽	TELEPHONE OUTLET—WALL

E-7   **CONSTRUCTION DRAWING E-7**

However, sometimes under the right set of conditions, it is unavoidable that an electrician must work on a system that is energized. In these cases, only qualified persons can work on the circuit or equipment. *NEC® 100* defines a qualified person as *one familiar with the construction and operation of the equipment and the hazards involved.* A qualified person is one who has the training, experience, and knowledge to determine the hazards involved. Asking an inexperienced or untrained person to work on energized circuits is a recipe for disaster. The simple admonition "Be careful" or "Don't touch that" certainly does not constitute enough training to make an individual qualified. Electrical contractors also have a written policy with the proper procedures to follow when working on a hot system. Electricians must know and follow these safety procedures explicitly.

Summary

- The various documents that provide information about a job include specifications, schedules, construction drawings, scope of work documents, and cut sheets.
- Local government is involved with a construction project from permit to inspection, through the listing and labeling process, and it can make changes to the rules established by the *NEC®* or the UBC.
- Safety is the first, the middle, and the last consideration on a construction site. Safety rules must be followed at all times.

Key Terms

Approved (See *NEC® Article 100*): Referring to equipment, devices, raceways, and other electrical materials that are acceptable to the AHJ for installation and use.

Authority having jurisdiction (AHJ). (See *NEC® Article 100*): The local governmental authority that is charged with the regulation of construction projects. The AHJ may be a city, county, or sometimes a state organization, that is usually answerable to a legislative body that assigns the authority.

Change order Authorization to proceed with a change—an addition or a deletion—to scheduled work. In many cases a change order takes the place of a work order.

Construction drawings A series of drawings, sometimes called *blueprints*, that show the intended design of a building or other structure. There are usually different construction drawings for each of the different trades involved in the construction process.

Construction process Collectively, the procedures followed by the various trades for the successful completion of the construction project.

Cut sheets (*See* shop drawings.)

Fine print note (FPN) A type of entry in the *NEC®* that provides explanatory information but is not formally enforceable as part of the *NEC®*.

Flag note An icon used on construction drawings that alerts the construction team to the existence of additional information or requirements listed elsewhere on the drawings.

Inspection The process whereby the AHJ enforces established installation and construction standards. The AHJ periodically reviews the progress of construction and either approves or rejects the quality of the construction and ensures compliance with minimum standards. These reviews are accomplished by physically inspecting the building, and the person who performs the inspections is usually referred to as an *inspector*.

Installation instructions The directions provided by the manufacturer concerning the procedures to be followed in preparing electrical equipment, devices, or other materials for use. These instructions must be closely followed in installing electrical materials;

otherwise, unsatisfactory operation may result in fire or other safety hazards.

International Residential Code (IRC) A set of minimum rules for the installation of building systems and the construction of new dwellings. It is the most widely applied standard for construction in the United States; however, it is not recognized in all areas of the country.

International System of Units (SI) The measurement system, sometimes called the metric system in the United States, defined by the use of meters and kilograms, that uses base 10 unit division (SI units). The SI system is widely employed in almost all areas of the world except in the United States. The *NEC®* uses the SI system as the primary measurement system, with the English system, defined by the use of fee and pounds, used as a secondary system.

Labeled (See *NEC® Article 100*): Referring to materials and equipment that have been found to meet certain requirements for safety and function by a testing agency recognized by the AHJ. An identification label, marking, or decal from the testing agency is attached to the materials to identify them as having been tested.

Listed (See *NEC® Article 100*): Referring to materials and equipment that have been found to meet certain requirements for safety and function by a testing agency recognized by the AHJ. The materials are placed on a list to identify them as having been tested.

Mandatory rules Describing the rules and procedures listed in the *NEC®* with the terms *shall* and *shall not*. These rules must be followed to comply with the requirements of the *NEC®*.

Metric designator A dimension corresponding to trade size, as employed for equipment and raceways, measured using the SI system.

National Electrical Code® **(*NEC®*)** A set of minimum rules for the design and installation of electrical systems and devices published by the National Fire Protection Association as document NFPA 70. It is the most widely applied standard for electrical installations in the United States; however, it is not recognized in all areas of the country.

National Fire Protection Association (NFPA) The organization devoted to fire safety and prevention that publishes, along with many other documents, the *NEC®*.

Occupational Safety and Health Administration (OSHA) A branch of the federal government that is charged with improving workplace safety and encouraging the establishment of safe workplace practices.

Permissive rules Referring to the rules and procedures listed in the *NEC®* that are allowed but that are not strictly required. These rules are identified in the *NEC®* with the terms *shall be permitted* and *shall not be required*.

Personal protective equipment (PPE) Safety equipment, such as hardhats, gloves, safety glasses, and work boots, that is provided by individual construction workers for their own use.

Schedule A layout, usually in the form of a table, that is part of the construction drawings and provides detailed information concerning materials, devices, or equipment to be installed. For example, detailed information about the various luminaires (lighting fixtures) to be installed is often conveyed using a schedule.

Scope of work A construction document that describes the work that is to be accomplished and the companies or trades that are to complete the work. This document provides a framework for the other construction documents.

Shop drawings/cut sheets Drawings, usually provided by the manufacturer of equipment or materials to be installed in the building or structure, that show details of fabrication, such as dimensions, color, and materials used.

Specifications A construction document that usually controls the quality of the construction installations. It may list the brand names of the materials to be used during construction, certain procedures to be employed, and directions about communications between the various construction trades and the management team.

Submittals Cut sheets or shop drawings submitted to an architect or engineer for approval prior to the equipment's installation in the building or structure. Submittals are usually required by the building specifications.

Symbols Icons used on construction drawings to represent various design features such as switches, receptacles, and luminaires (lighting fixtures).

Trade size A system employed in the *NEC®* to define certain standard sizes of electrical equipment and raceways. For example, 1.2-in. (16-mm) internal diameter trade size conduit can actually measure between .526 in. (13.36 mm) and .660 in. (16.76 mm).

Utilization equipment (See *NEC® Article 100*): Equipment that requires electricity to function. Utilization equipment includes appliances and luminaires (lighting fixtures) but does not include raceways, boxes, or devices.

Work order An authorization to proceed with scheduled work. Work orders usually involve work that has been priced and contracted or that has been agreed to be completed on a "time and materials" basis.

Review Questions

1. Who is the authority having jurisdiction (AHJ)?
2. What organization publishes the *NEC*®?
3. What construction drawings contain the dimensions for walls and rooms?
4. What is more important on a job site, getting the job done on time and under budget or safety?
5. What is the purpose of specifications?
6. What is the purpose of the panel schedule?
7. What governmental agency is concerned with improving job site safety?

Boxes

OBJECTIVES *After studying this chapter, you should be able to:*

- State the reason for installing boxes in electrical systems.
- Identify the types of boxes used in dwellings.
- Properly size a box from a description of the conductors, fittings, and devices installed in the box.
- Select the proper box for installation.
- Explain the difference between lighting outlet boxes and device boxes.
- Describe conduit bodies and state where they are used.

OUTLINE
Introduction
Functions of Boxes
Types of Boxes
Sizing Boxes
Installing Boxes
Conduit Bodies

Introduction

This chapter deals with boxes and box installation. Boxes are required at virtually all junction points and splice points, at all outlets, and at all switch points. The various types of boxes, as well as their uses and installation, are discussed in this chapter.

Functions of Boxes

The electrical system must be easily accessible to electricians to install devices, receptacles, and switches, and to effect repairs or additions. Access to the electrical system is provided by the use of **boxes.** All splices, taps, and terminations (with very few exceptions) of the wiring system must be accomplished using an approved electrical box or equipment wiring enclosure. Boxes also serve an important function in the event of a fire by containing the fire inside the housing for some time. In many cases, the fire is discovered before it can spread to other parts of the structure.

Conductor insulation must be protected against excessive heating due to the level of current flow and ambient temperature influences. Boxes are another potential source of heating problems for the electrical system. Current-carrying conductors and ambient temperature influences contribute heat to the inside of the box. Wiring for a device such as a dimmer switch not only takes up space in the box, thus allowing less available air to cool the conductors, but also contributes to the heating of the conductors. Obviously, the number of conductors allowed in the box must be limited in order to control the conductor operating temperature within the box.

Types of Boxes

The *NEC®* divides boxes into two distinct groups according to the size of the largest conductor that enters the box. This two-group system is used because the *NEC®* is concerned with the heating of the conductors inside the device and outlet boxes. Air cannot circulate easily when the box is crowded with conductors and devices, and an overloaded box can create very high temperatures. When conductors become larger, the concern changes to the amount of bending space available within the box. Larger conductors take up a considerable amount of room. Boxes that contain no conductor larger than 6 AWG are called device, outlet, and junction boxes, and boxes with conductors 4 AWG and larger are called junction and pull boxes.

Boxes for Conductors 6 AWG and Smaller

Electrical outlet and device boxes come in many different sizes, configurations, and materials. Metal boxes are very commonly used, but boxes constructed from plastic and fiber compounds (nonmetallic) are most prevalent in residential construction today. The boxes designed for use with receptacle outlets and switching devices can be divided into four major groupings: (1) **device boxes,** (2) **lighting outlet boxes,** (3) **square boxes,** and (4) **waterproof boxes.** Table 9–1 presents a brief look at the various boxes for small conductors and some of their uses and limitations.

Device Boxes

This group of boxes contains a variety of boxes, some for surface mounting and others for flush mounting in walls or ceilings, and they can be metal or nonmetallic. Metal boxes can be single-gang or multigang, or they can be gangable (allowing connection of two or more boxes together to make a multigang box). Nonmetallic boxes can be single-gang or multigang boxes. Some of these device boxes are designed for remodeling work, in which the boxes will be cut into existing walls, and others are designed strictly for installation onto structural framing members during construction. Figure 9–1 shows a selection of typical metal device boxes, and Figure 9–2 shows a selection of nonmetallic device boxes. Figure 9–3 shows a typical box rough-in in a dwelling.

One feature that all device boxes have in common is the shape of the opening in the front of the box. The opening is rectangular, nominally 3 in. (75 mm) by 2 in. (50 mm) for each gang. A three-gang device box (or three single-gang

TABLE 9-1
Selected Boxes: Their Uses and Limitations

Feature	Metal Octagon Box	Metal Square Box	Metal Device Box	Metal Masonry Box	Metal Handy Box	Waterproof Device Boxes	Waterproof Round Box	Nonmetal Square Box	Nonmetal Round Box	Nonmetal Device Box
Surface mounted										
Junction or pull box	X	X			X	X	X	X		
Support luminaire (lighting fixture)	X	X				X	X	X(1)		
Support device	X	X(1)			X	X		X(1)		
		X(1)								
Mounted flush inside of wall or ceiling										
Junction or pull box	X	X(2)	X	X				X(2)	X	X
Support luminaire (lighting fixture)	X	X(2)	X	X				X(2)	X	X
Support device	X	X(2)	X	X	X			X(2)	X	
		X(2)						X(2)		X
Available with										
Built-in AC, MC, or NM cable clamps	X	X	X		X			X	X	X(3)
Without built-in AC, MC, or NM cable clamps	X	X	X	X	X			X		X(4)
Additional exterior cable clamp	X	X	X		X					
Side mounting bracket	X	X	X		X			X	X	X
Face mounting bracket	X	X	X		X			X	X	X
Extension box	X	X			X	X		X		
Gangable or multigang		X(5)(6)	X			X		X(5)		X
Threaded hole size for mounting to box	8-32	8-32(7)	6-32	6-32	6-32	6-32	8-32	8-32(7)	8-32	6-32

(1) With proper industrial cover.
(2) With proper plaster ring installed.
(3) Most single-gang boxes do not have clamps.
(4) Single gang only.
(5) One-gang or two-gang plastic rings available.
(6) Multigang box available with multigang plaster ring.
(7) The box itself has 8-32 holes. Plastic ring can have 6-32 or 8-32 threaded holes.

Boxes • 197

"WIREMAN'S GUIDE"
SELECTED METAL DEVICE BOXES

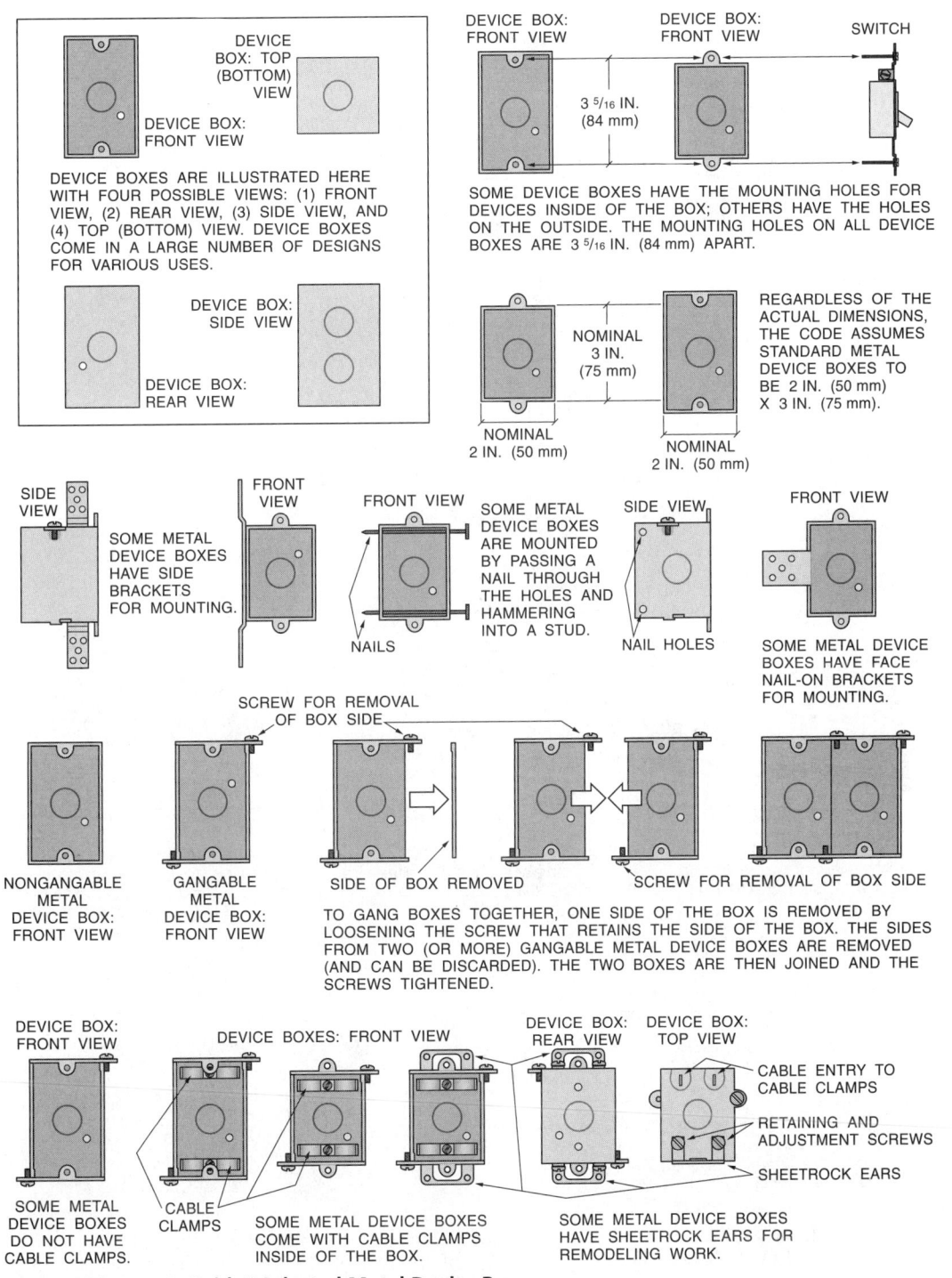

DEVICE BOX: TOP (BOTTOM) VIEW
DEVICE BOX: FRONT VIEW

DEVICE BOXES ARE ILLUSTRATED HERE WITH FOUR POSSIBLE VIEWS: (1) FRONT VIEW, (2) REAR VIEW, (3) SIDE VIEW, AND (4) TOP (BOTTOM) VIEW. DEVICE BOXES COME IN A LARGE NUMBER OF DESIGNS FOR VARIOUS USES.

DEVICE BOX: SIDE VIEW
DEVICE BOX: REAR VIEW

DEVICE BOX: FRONT VIEW
DEVICE BOX: FRONT VIEW
SWITCH
3 5/16 IN. (84 mm)

SOME DEVICE BOXES HAVE THE MOUNTING HOLES FOR DEVICES INSIDE OF THE BOX; OTHERS HAVE THE HOLES ON THE OUTSIDE. THE MOUNTING HOLES ON ALL DEVICE BOXES ARE 3 5/16 IN. (84 mm) APART.

NOMINAL 3 IN. (75 mm)
NOMINAL 2 IN. (50 mm)
NOMINAL 2 IN. (50 mm)

REGARDLESS OF THE ACTUAL DIMENSIONS, THE CODE ASSUMES STANDARD METAL DEVICE BOXES TO BE 2 IN. (50 mm) X 3 IN. (75 mm).

SIDE VIEW
FRONT VIEW
SOME METAL DEVICE BOXES HAVE SIDE BRACKETS FOR MOUNTING.

FRONT VIEW
SOME METAL DEVICE BOXES ARE MOUNTED BY PASSING A NAIL THROUGH THE HOLES AND HAMMERING INTO A STUD.
NAILS

SIDE VIEW
NAIL HOLES

FRONT VIEW
SOME METAL DEVICE BOXES HAVE FACE NAIL-ON BRACKETS FOR MOUNTING.

SCREW FOR REMOVAL OF BOX SIDE

NONGANGABLE METAL DEVICE BOX: FRONT VIEW
GANGABLE METAL DEVICE BOX: FRONT VIEW
SIDE OF BOX REMOVED
SCREW FOR REMOVAL OF BOX SIDE

TO GANG BOXES TOGETHER, ONE SIDE OF THE BOX IS REMOVED BY LOOSENING THE SCREW THAT RETAINS THE SIDE OF THE BOX. THE SIDES FROM TWO (OR MORE) GANGABLE METAL DEVICE BOXES ARE REMOVED (AND CAN BE DISCARDED). THE TWO BOXES ARE THEN JOINED AND THE SCREWS TIGHTENED.

DEVICE BOX: FRONT VIEW
DEVICE BOXES: FRONT VIEW
DEVICE BOX: REAR VIEW
DEVICE BOX: TOP VIEW

CABLE ENTRY TO CABLE CLAMPS
RETAINING AND ADJUSTMENT SCREWS
SHEETROCK EARS

SOME METAL DEVICE BOXES DO NOT HAVE CABLE CLAMPS.
CABLE CLAMPS
SOME METAL DEVICE BOXES COME WITH CABLE CLAMPS INSIDE OF THE BOX.
SOME METAL DEVICE BOXES HAVE SHEETROCK EARS FOR REMODELING WORK.

FIGURE 9–1 "Wireman's Guide" Selected Metal Device Boxes

198 • Chapter 9

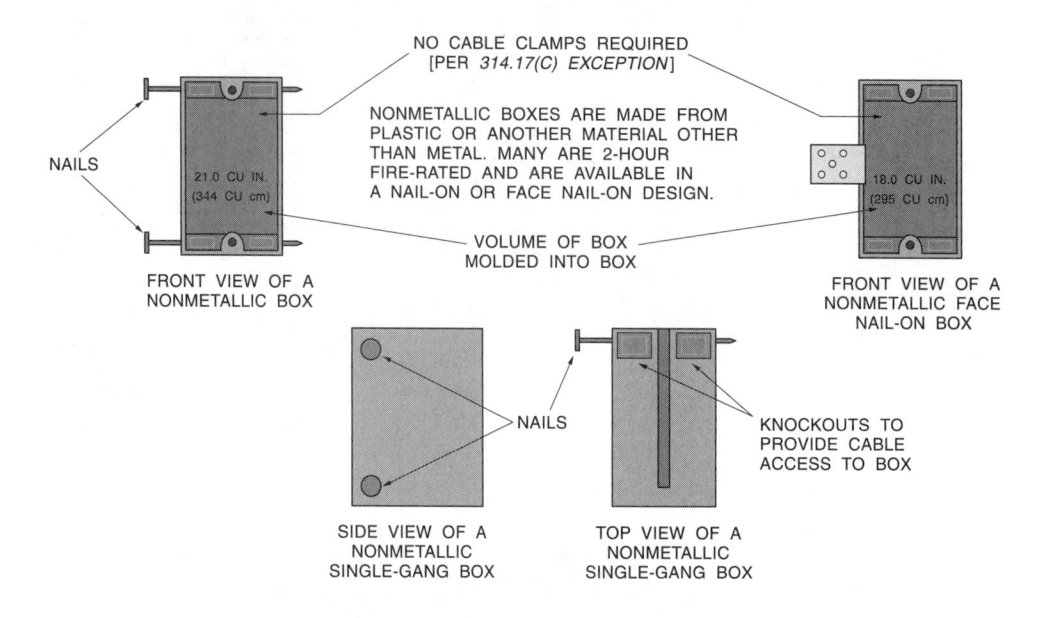

"WIREMAN'S GUIDE"
SELECTED NONMETALLIC BOXES

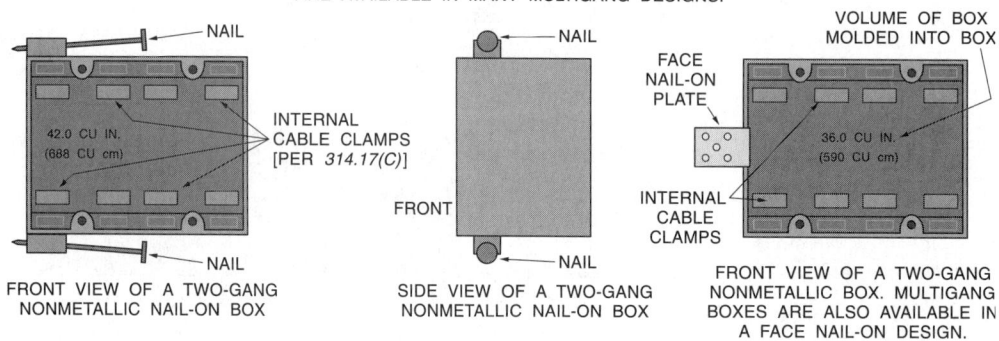

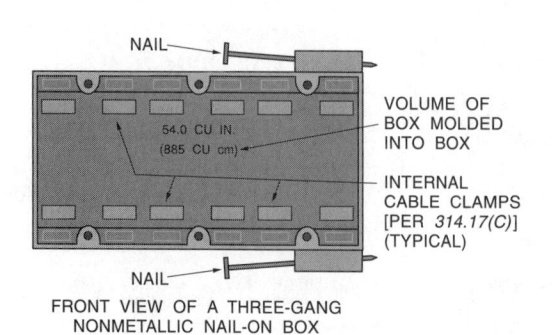

FIGURE 9–2 "Wireman's Guide" Selected Nonmetallic Boxes

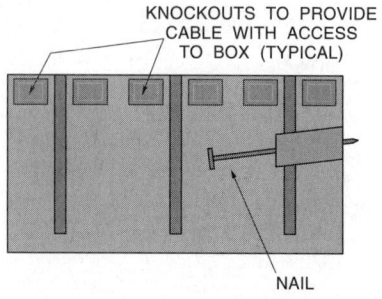

FIGURE 9–3 A Typical Nail-On Box Rough Installation

metal device boxes ganged together) has a nominal opening of 3 in. (75 mm) by 6 in. (150 mm). Receptacle devices and switching devices are designed to fit into a space of that size at trim.

Another feature that all device boxes have in common is that they provide threaded size 6-32 holes to facilitate the installation of receptacles and switches (devices). The 6-32 screw is used to mount these devices, and the devices usually fit inside the box for a flush finish.

Device boxes are not intended to support luminaires (lighting fixtures). Lighting outlet boxes are provided with size 8-32 threaded holes. The larger-sized screw is needed to support the weight of the luminaire (lighting fixture). Size 6-32 screws have not been listed as a supporting means for luminaires (lighting fixtures), and the *Code* does not allow it, with one exception, as shown in Figure 9–4.

Lighting Outlet Boxes

Like device boxes, boxes for lighting outlets are available in metal and nonmetal designs, and they are available with special brackets for remodeling installations or in a nail-on design intended for new construction. Figure 9–5 shows examples of lighting outlet boxes. Figure 9–6 shows a typical nail-on lighting outlet box, and Figure 9–7 shows a standard lighting outlet box supported by a hanger bar for exact positioning. Lighting outlet boxes differ from device boxes in several significant ways:

- The openings in the boxes, and the boxes themselves, are round or octagonal instead of rectangular.
- The round and octagonal boxes are available in a 3-in. (.075-m)-diameter and 4-in. (.100-m)-diameter sizes.
- Lighting outlet boxes have threaded holes for the mounting of the luminaire (lighting fixture) with 8-32 screws.
- Lighting outlet boxes are not designed to enclose the luminaire lighting (fixture), in contrast to device boxes as used with receptacles and switches.
- There are no gangs with lighting outlet boxes. Each box is intended for only one luminaire (lighting fixture).

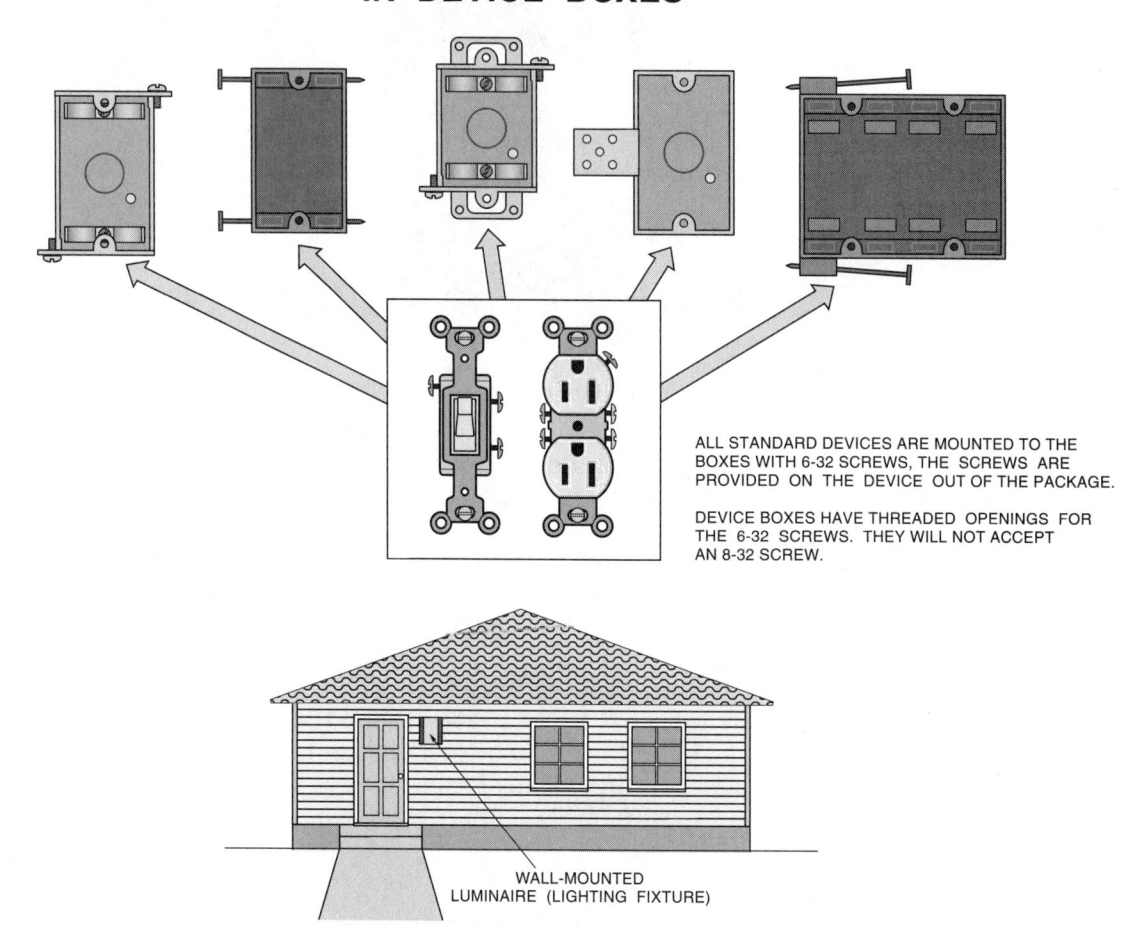

"WIREMAN'S GUIDE"
SWITCHES AND RECEPTACLES INSTALLED IN DEVICE BOXES

ALL STANDARD DEVICES ARE MOUNTED TO THE BOXES WITH 6-32 SCREWS. THE SCREWS ARE PROVIDED ON THE DEVICE OUT OF THE PACKAGE.

DEVICE BOXES HAVE THREADED OPENINGS FOR THE 6-32 SCREWS. THEY WILL NOT ACCEPT AN 8-32 SCREW.

WALL-MOUNTED LUMINAIRE (LIGHTING FIXTURE)

ACCORDING TO *314.27,* LUMINAIRES (LIGHTING FIXTURES) MUST BE MOUNTED TO BOXES DESIGNED FOR THIS USE—THAT IS, LIGHTING OUTLET BOXES WITH 8-32 THREADED HOLES. HOWEVER, THE EXCEPTION TO *314.27* ALSO ALLOWS MOUNTING OF A WALL-MOUNTED LUMINAIRE (LIGHTING FIXTURE), WEIGHING LESS THAN 6 LB (3 KG) TO A DEVICE BOX OR DEVICE PLASTER RING AS LONG AS THE LUMINAIRE (LIGHTING FIXTURE) IS MOUNTED USING AT LEAST TWO 6-32 SCREWS. IN THE DRAWING, THE INSTALLATION IS OUTDOORS. THE *CODE* DOES NOT REQUIRE THAT THE LUMINAIRE (LIGHTING FIXTURE) BE OUTDOORS FOR THIS EXCEPTION. TO APPLY, INTERIOR INSTALLATIONS ARE ALSO ACCEPTABLE.

FIGURE 9–4 "Wireman's Guide" Switches and Receptacles Installed in Device Boxes

Metal lighting outlet boxes are available with face nail-on and side nail-on bracket designs. Lighting outlet boxes can sometimes be installed on the bottom of a ceiling joist or truss. The luminaire (lighting fixture) mounts over the box using a canopy or domed cover. A luminaire (lighting fixture) with a domed cover can be used with a 1.2-in.-deep (.013-m-deep) lighting outlet box under certain conditions, as shown in Figures 9–8 and 9–9.

There are many different ways in which luminaires (lighting fixtures) are connected to a box. Sometimes the luminaire (lighting fixture) screws directly into the 8-32 threaded holes in the box. In many cases, room must be allowed in the box for a luminaire (lighting fixture) stud or nipple that attaches to the box with a special bracket. The luminaire (lighting fixture) attaches to the special bracket, and the stud or nipple provides a pathway for the conductors.

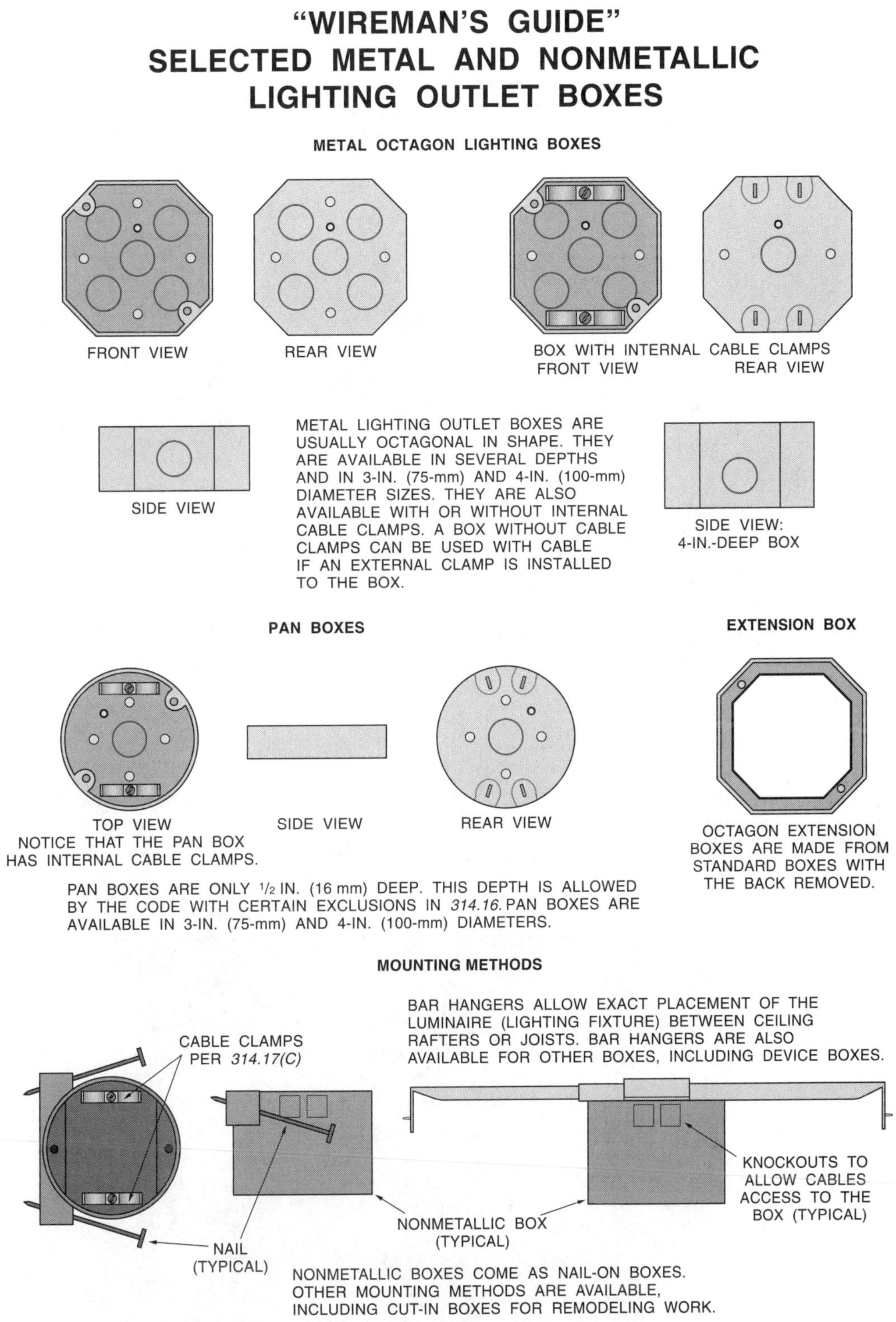

Boxes • 201

"WIREMAN'S GUIDE"
SELECTED METAL AND NONMETALLIC LIGHTING OUTLET BOXES

METAL OCTAGON LIGHTING BOXES

FRONT VIEW

REAR VIEW

BOX WITH INTERNAL CABLE CLAMPS
FRONT VIEW   REAR VIEW

SIDE VIEW

METAL LIGHTING OUTLET BOXES ARE USUALLY OCTAGONAL IN SHAPE. THEY ARE AVAILABLE IN SEVERAL DEPTHS AND IN 3-IN. (75-mm) AND 4-IN. (100-mm) DIAMETER SIZES. THEY ARE ALSO AVAILABLE WITH OR WITHOUT INTERNAL CABLE CLAMPS. A BOX WITHOUT CABLE CLAMPS CAN BE USED WITH CABLE IF AN EXTERNAL CLAMP IS INSTALLED TO THE BOX.

SIDE VIEW: 4-IN.-DEEP BOX

PAN BOXES

TOP VIEW
NOTICE THAT THE PAN BOX HAS INTERNAL CABLE CLAMPS.

SIDE VIEW

REAR VIEW

PAN BOXES ARE ONLY 1/2 IN. (16 mm) DEEP. THIS DEPTH IS ALLOWED BY THE CODE WITH CERTAIN EXCLUSIONS IN *314.16*. PAN BOXES ARE AVAILABLE IN 3-IN. (75-mm) AND 4-IN. (100-mm) DIAMETERS.

EXTENSION BOX

OCTAGON EXTENSION BOXES ARE MADE FROM STANDARD BOXES WITH THE BACK REMOVED.

MOUNTING METHODS

CABLE CLAMPS PER *314.17(C)*

NAIL (TYPICAL)

NONMETALLIC BOX (TYPICAL)

BAR HANGERS ALLOW EXACT PLACEMENT OF THE LUMINAIRE (LIGHTING FIXTURE) BETWEEN CEILING RAFTERS OR JOISTS. BAR HANGERS ARE ALSO AVAILABLE FOR OTHER BOXES, INCLUDING DEVICE BOXES.

KNOCKOUTS TO ALLOW CABLES ACCESS TO THE BOX (TYPICAL)

NONMETALLIC BOXES COME AS NAIL-ON BOXES. OTHER MOUNTING METHODS ARE AVAILABLE, INCLUDING CUT-IN BOXES FOR REMODELING WORK.

FIGURE 9–5 "Wireman's Guide" Selected Metal and Nonmetallic Lighting Outlet Boxes

FIGURE 9–6 A Standard Nail-On Lighting Box Installation

Square Boxes

A square box is a versatile type of box that can be used for both surface and flush installations, can be used for devices and for luminaires (lighting fixtures), and can be a one-gang or a two-gang device box. Square boxes are available in both metal and nonmetal designs and can be readily obtained with side-mounting brackets or face-mounting brackets or with no bracket at all.

Square boxes owe their versatility to the use of special rings, called **plaster rings** or mud rings, and special covers, called industrial covers, in addition to a simple blank cover. These special covers install to the front of the square box and are used to mount devices in surface installations. The plaster rings are used for flush installations and can be single-gang or two-gang openings with 6-32 threaded holes for mounting the devices. Other plaster rings have a round opening with 8-32 threaded holes for use with luminaires (lighting fixtures). Several selected square boxes are shown in Figure 9–10, a square box is shown installed on

FIGURE 9–7 A bar hanger lighting box installation is used when the exact location of the luminaire (lighting fixture) is critical.

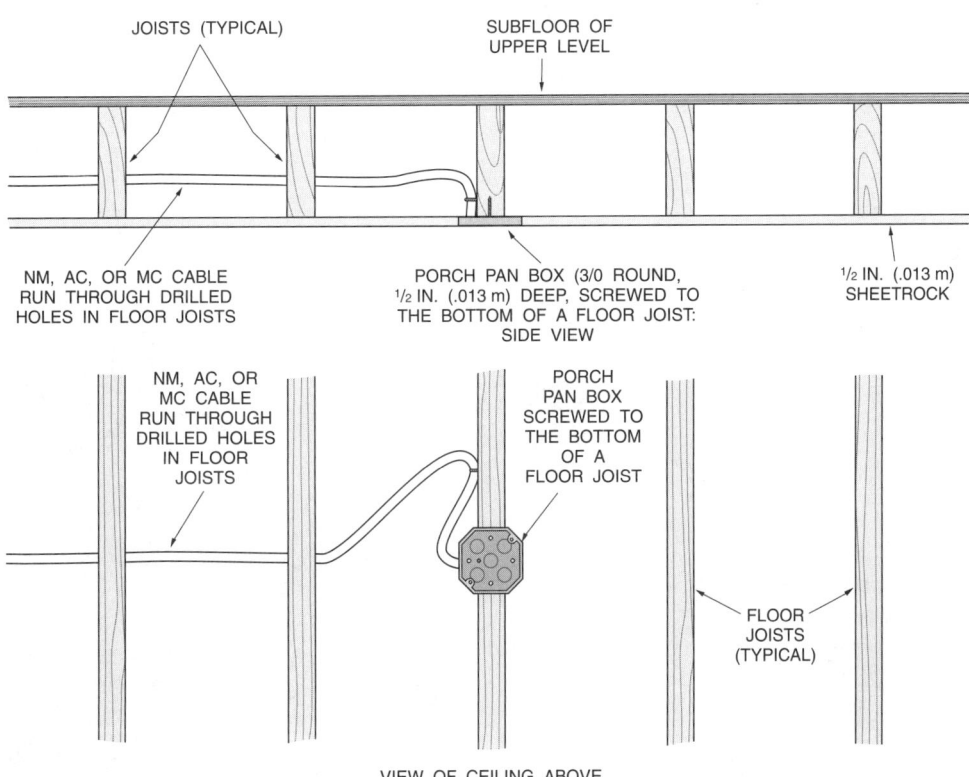

"WIREMAN'S GUIDE"
ATTACHING A PAN BOX TO THE BOTTOM OF A FLOOR JOIST

ACCORDING TO *314.24* BOXES CANNOT BE LESS THAN ½ IN. (13 mm) DEEP. WITH PAN BOXES, HOWEVER, THERE IS NOT ENOUGH ROOM FOR THE CONDUCTORS IN THE BOX IF ALL CONDUCTORS ARE COUNTED ACCORDING TO *314.16*. THE EXCEPTION TO *314.16(B)(1)* ALLOWS THE ELIMINATION OF SOME CONDUCTORS FROM THE CONDUCTOR COUNT FOR INSTALLING A LUMINAIRE (LIGHTING FIXTURE) WITH A CANOPY OR DOMED COVER. USE OF A PAN BOX IS OFTEN ACCEPTABLE TO THE AHJ BECAUSE OF THIS EXCEPTION. WHEN INSTALLED IN THIS MANNER, THE PAN BOX DOES NOT EXTEND BEYOND THE LOWER EDGE OF THE SHEETROCK, AND THE BOTTOM OF THE JOIST PROVIDES A VERY SECURE MOUNTING.

FIGURE 9–8 "Wireman's Guide" Attaching a Pan Box to the Bottom of a Floor Joist

FIGURE 9–9 A Pan Box Attached to the Bottom of a Floor Joist

203

204 • Chapter 9

"WIREMAN'S GUIDE"
4-SQUARE AND 4¹¹/₁₆-SQUARE BOXES

A LARGE SELECTION OF SQUARE BOXES IS AVAILABLE. SQUARE BOXES ARE VERY VERSATILE; THEY CAN BE USED AS SURFACE- OR FLUSH-MOUNTED JUNCTION BOXES, SURFACE- OR FLUSH-MOUNTED DEVICE BOXES, OR SURFACE- OR FLUSH-MOUNTED LIGHTING OUTLET BOXES, DEPENDING ON THE PLASTER RING, INDUSTRIAL COVER, OR BLANK COVER USED. SQUARE BOXES ARE ALSO AVAILABLE WITH MOUNTING DESIGNS SUCH AS SIDE AND FACE BRACKETS. THEY ARE AVAILABLE WITH OR WITHOUT CABLE CLAMPS, IN THREE DIFFERENT DEPTHS—1¹/₄ IN. (32 mm), 1¹/₂ IN. (38 mm), AND 2¹/₈ IN. (54 mm)—AND IN SEVERAL DIFFERENT KNOCKOUT PATTERNS AND SIZES.

4-SQUARE BOX

A 4-SQUARE BOX WITH ¹/₂-IN. TRADE SIZE (16 mm) KNOCKOUTS. THIS SIZE SQUARE BOX IS CALLED A 4-SQUARE BOX BECAUSE IT MEASURES 4 IN. (100 mm) ON EACH SIDE. THIS BOX IS A 1¹/₄-IN. (35-mm)-DEEP BOX. NOTICE THE 8-32 THREADED HOLES IN OPPOSITE CORNERS FOR SECURING THE PLASTER RING OR COVER.

FRONT VIEW
8-32 THREADED HOLE
10-32 THREADED HOLE FOR GROUNDING SCREW
4 IN. (100 mm)
¹/₂-IN. (16-mm) KNOCKOUT (TYPICAL)
8-32 THREADED HOLE
4 IN. (100 mm)

REAR VIEW

SIDE VIEW

4-SQUARE COMBINATION BOX

FRONT VIEW

A 4-SQUARE BOX WITH ¹/₂-IN. AND ³/₄-IN. TRADE SIZE (16-mm AND 21-mm) KNOCKOUTS. IT IS ALSO CALLED A 4-SQUARE COMBINATION (COMBO) BOX.

SIDE VIEW   SIDE VIEW

4-SQUARE COMBINATION BOXES ARE AVAILABLE IN 1¹/₂ IN. (38 mm) AND 2¹/₈ IN. (54 mm) DEPTHS.

A 2¹/₈-IN. (54-mm)-DEEP BOX ALLOWS UP TO 1¹/₄-IN. TRADE SIZE (35-mm) KNOCKOUTS.

A 4-SQUARE BOX WITH CABLE CLAMPS. THERE ARE STANDARD KNOCKOUTS ON THE BOX AS WELL AS CABLE CLAMPS.

FRONT VIEW

SIDE VIEW

A FRONT VIEW OF A 4-SQUARE BOX WITH A FACE NAIL-ON BRACKET AND A SIDE VIEW OF A 4-SQUARE BOX WITH A SIDE NAIL-ON MOUNTING BRACKET.

SIDE VIEW

FRONT VIEW

4¹¹/₁₆ SQUARE BOX

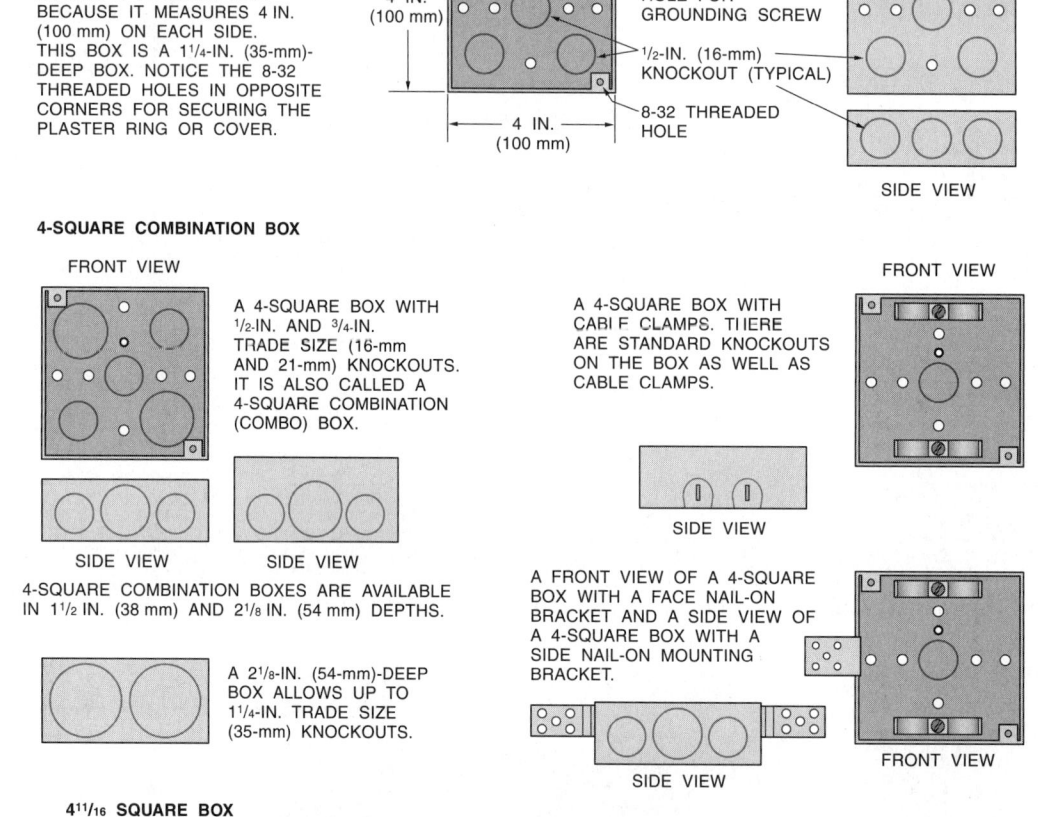

FIGURE 9–10 "Wireman's Guide" 4-Square and 4-11/16 Square Boxes

the surface in Figure 9–11, and a selected assortment of plaster rings and industrial covers are shown in Figure 9–12. An example of a metal square box with a plaster ring installed using a bar hanger for exact location is shown in Figure 9–13, and a nonmetallic square box with a single-gang plaster ring is shown in Figure 9–14.

Waterproof Boxes

Weatherproof boxes are designed to be installed on the exterior surface of dwellings and other structures. If a receptacle outlet is to be installed flush with

FIGURE 9–11 A Typical Surface-Mounted 4-Square Box

the building surface, a normal single-gang device box is installed to house the device. A special waterproof cover is used to keep out moisture. If the receptacle is to mount on the exterior surface of the building, a waterproof box is needed to house the device and to keep out moisture. Weatherproof boxes are available as device boxes and lighting outlet boxes. They come with a variety of knockout configurations with both 1.2-in. (.016-m) and 3.4-in. (.021-m) threaded knockouts on the ends and also, if desired, on the sides of the boxes. Weatherproof device boxes are readily available in one-gang and two-gang sizes with waterproof covers for receptacles and switches. Weatherproof boxes are usually made from a cast aluminum alloy that prevents rusting.

Boxes for Conductors 4 AWG and Larger

Branch circuits are not typically found in a dwelling that requires 4 AWG or larger conductors. They are sometimes used in service conductor installations or feeders, but boxes for these larger conductor sizes are also large. These larger boxes are available in certain common sizes, such as 8 in. × 8 in. × 6 in. (.200 m × .200 m × .150 m), or 12 in. × 12 in. × 8 in. (.300 m × .300 m × .200 m), but they can easily be manufactured in any size, thickness, configuration, or shape desired. These boxes can be weatherproof (National Electrical Manufacturers Association [NEMA] 3R) or dry location (NEMA 1) boxes, or of any other type desired. Larger boxes are intended for conductors only, and no devices should be installed in this type of box.

Sizing Boxes

For sizing boxes, the *NEC*® uses the same two-group system. The rules for sizing smaller boxes are intended to limit heat build-up inside the box. The rules for sizing larger boxes are concerned with providing adequate bending space within the box.

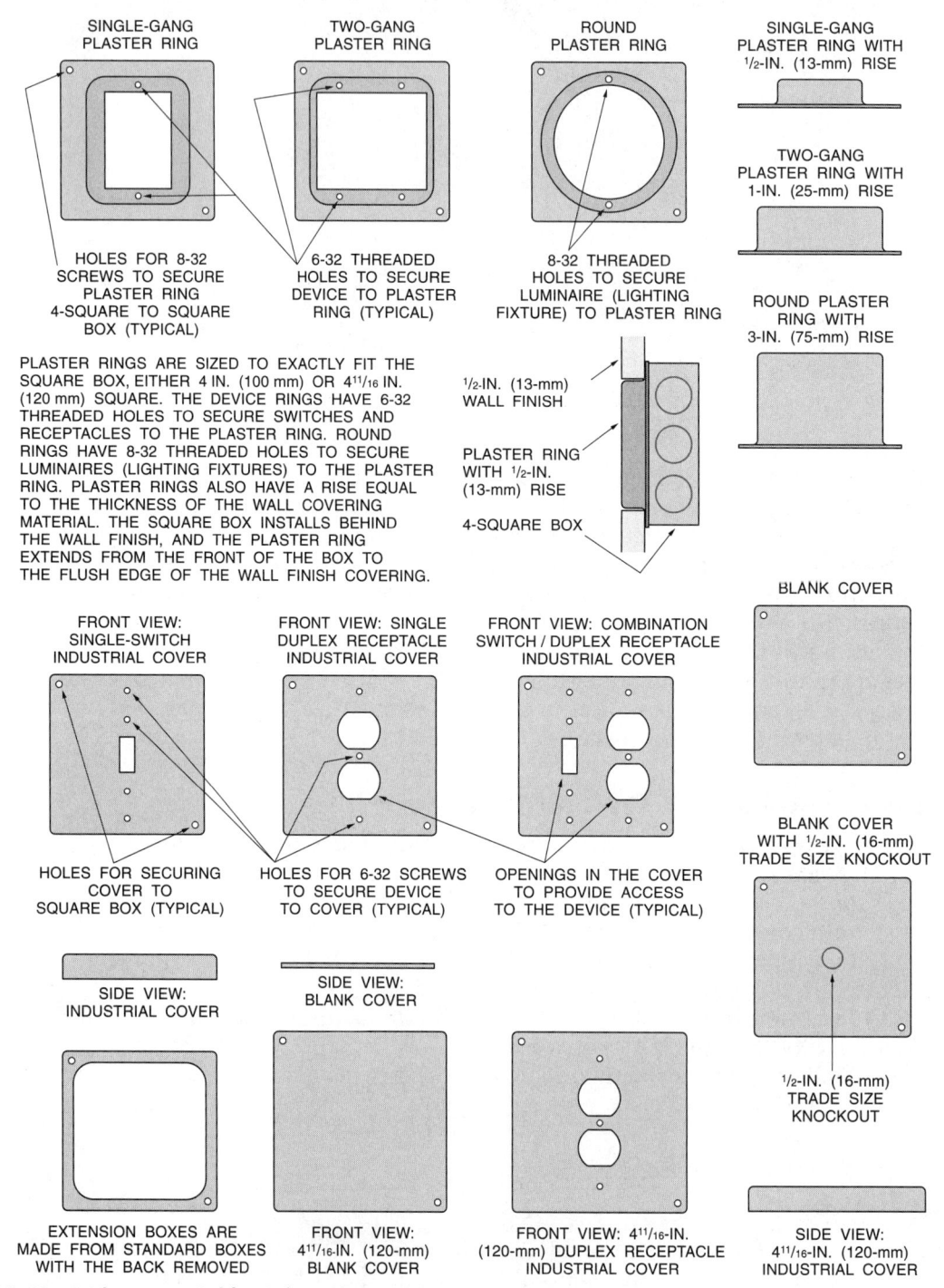

FIGURE 9–12 "Wireman's Guide" Selected Square Box Plaster Rings and Covers

FIGURE 9–13 A Metallic 4-Square Box with a Single-Gang Plaster Ring

Sizing Boxes for Conductors 6 AWG and Smaller

The rules for sizing boxes for 6 AWG and smaller conductors can be found in *314.16* of the *Code*. Determining the number of conductors allowed in a given box requires the following information:

- The volume of the box—that is, the space, measured in cubic inches or cubic centimeters, enclosed within the box. This volume can be obtained in either of two places. For boxes that the *Code* considers to be standard boxes, the volume can be obtained from *Table 314.16(A)* in the "Minimum Volume" column. For boxes that are not standard boxes, the volume is displayed somewhere on the box itself. It may be stamped into the metal or molded into the box when it is manufactured, or it may be printed or stamped on the box.
- Box fill—that is, the number and the size of the conductors that will be installed in the box. This information will come from the routing of the cables and the details of the circuiting.

FIGURE 9–14 A Nonmetallic 4-Square Box with a Single-Gang Plaster Ring

- Conductor volume requirements—the volume that each conductor size is required to receive according to *Table 314.16(B)*. Obviously, a 6 AWG conductor will occupy more volume than a 14 AWG conductor of the same length.

The process for calculating the number of wires allowed in the box involves three steps:

1. Multiply the number of each of the conductor sizes by the required volume allowance from *Table 314.16(B)*.
2. Add the products. This number should represent the minimum volume that a box must have to house the conductors.
3. Compare the allowable volume of the various boxes with the calculated volume needed for all of the conductors. Any box that has a volume larger than the calculated necessary volume can be used.

Care must be taken with this calculation, however. The *NEC*® considers other items installed in the box to be conductors for the purposes of box fill calculations. In general,

- A device counts as two conductors. Because the box can house different sizes of conductors, the device is counted as the same wire size as that of conductors connected to the device. If the device is on a 15-ampere lighting circuit, the switch counts as two 14 AWG conductors.
- All of the equipment-grounding conductors together count as a single conductor. If different sizes of conductors are present in the box, the equipment-grounding conductor counts as the largest equipment-grounding conductor in the box.
- Any cable clamps that are installed in the box count as one of the largest conductors in the box. If the clamp is outside the box, it is not included in counting conductors.
- Any luminaire (lighting fixture) supports, luminaire (lighting fixture) studs, or nipples installed in the box will count as one of the largest conductors in the box. The domed cover of a luminaire (lighting fixture) may count as additional volume for the box. Figure 9–15 shows a sample box fill calculation.

Sizing Boxes for Conductors 4 AWG and Larger

In general, *314.28* stipulates the requirements for sizing junction and pull boxes containing conductors 4 AWG and larger. The *Code* is concerned with the distance from the conductor's point of entry into the box to the opposite wall of that box. As the cable size increases, the distance to the opposite wall of the box becomes greater.

Three general configurations are detailed by the *Code*:

1. *Straight pulls*, in which the conductors enter the box and exit the box on the opposite wall of the box, as shown in Figure 9–16. The rules are covered in detail in *314.28(A)(1)*.
2. *Angle pulls*, including U-pulls, in which the conductors enter a box on one wall and exit the box on an adjacent wall (a wall other than the opposite wall). This configuration is shown in Figure 9–17. Also considered in this figure are rows of cable or conduit entries in the same box. If a box is used with two or more rows of openings, the box size is controlled by the row that requires the largest box. The rules for angle pulls are covered in *314.28(A)(2)*.

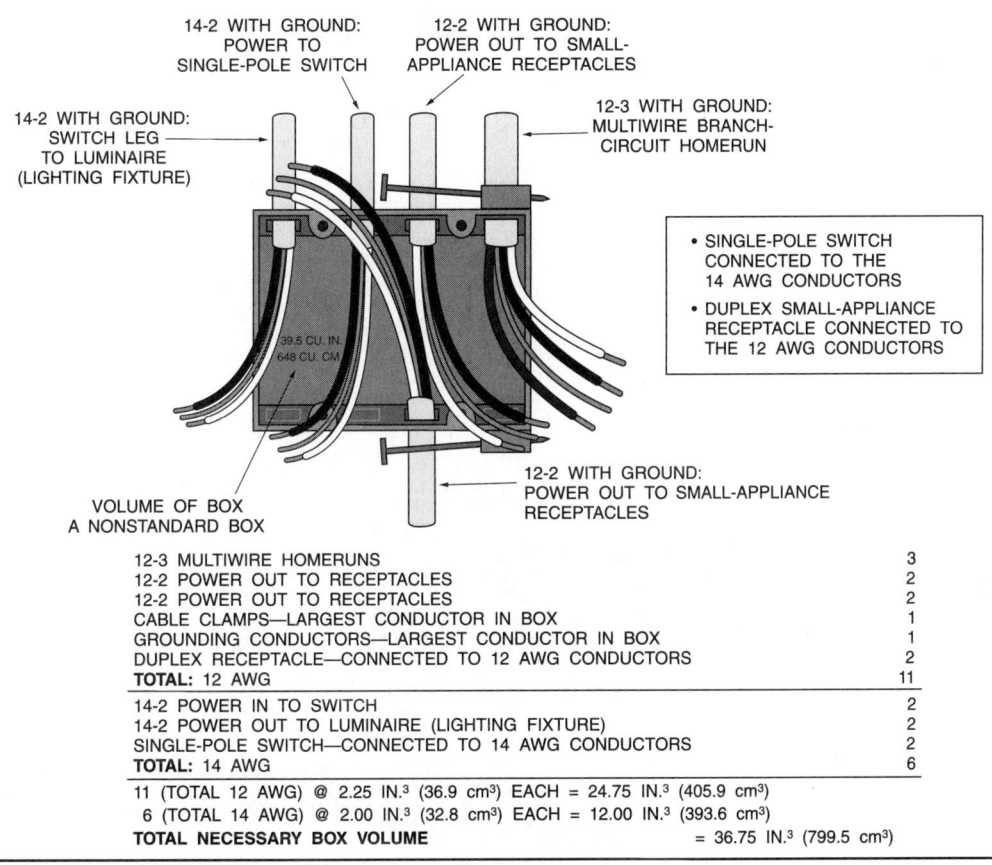

"WIREMAN'S GUIDE" SAMPLE DEVICE FILL CALCULATION

14-2 WITH GROUND: POWER TO SINGLE-POLE SWITCH

12-2 WITH GROUND: POWER OUT TO SMALL-APPLIANCE RECEPTACLES

14-2 WITH GROUND: SWITCH LEG TO LUMINAIRE (LIGHTING FIXTURE)

12-3 WITH GROUND: MULTIWIRE BRANCH-CIRCUIT HOMERUN

- SINGLE-POLE SWITCH CONNECTED TO THE 14 AWG CONDUCTORS
- DUPLEX SMALL-APPLIANCE RECEPTACLE CONNECTED TO THE 12 AWG CONDUCTORS

VOLUME OF BOX A NONSTANDARD BOX (39.5 CU. IN. 648 CU. CM)

12-2 WITH GROUND: POWER OUT TO SMALL-APPLIANCE RECEPTACLES

12-3 MULTIWIRE HOMERUNS	3
12-2 POWER OUT TO RECEPTACLES	2
12-2 POWER OUT TO RECEPTACLES	2
CABLE CLAMPS—LARGEST CONDUCTOR IN BOX	1
GROUNDING CONDUCTORS—LARGEST CONDUCTOR IN BOX	1
DUPLEX RECEPTACLE—CONNECTED TO 12 AWG CONDUCTORS	2
TOTAL: 12 AWG	**11**
14-2 POWER IN TO SWITCH	2
14-2 POWER OUT TO LUMINAIRE (LIGHTING FIXTURE)	2
SINGLE-POLE SWITCH—CONNECTED TO 14 AWG CONDUCTORS	2
TOTAL: 14 AWG	**6**

11 (TOTAL 12 AWG) @ 2.25 IN.^{3} (36.9 cm^3) EACH = 24.75 IN.^{3} (405.9 cm^3)
 6 (TOTAL 14 AWG) @ 2.00 IN.^{3} (32.8 cm^3) EACH = 12.00 IN.^{3} (393.6 cm^3)
TOTAL NECESSARY BOX VOLUME = 36.75 IN.^{3} (799.5 cm^3)

TO OBTAIN THE VOLUME OF THE BOX:
1. FOR STANDARD BOXES [THOSE BOXES LISTED IN TABLE 314.16(A)], THE VOLUME OF THE BOX CAN BE TAKEN FROM THE TABLE UNDER THE COLUMN HEADED "MINIMUM VOLUME."
2. FOR NONSTANDARD BOXES, THE VOLUME WILL BE FOUND STAMPED, ETCHED, PRINTED, OR MOLDED INTO THE BOX SURFACE, USUALLY INSIDE OF THE BOX.

THE TOTAL VOLUME OF THE BOX INCLUDES THAT VOLUME ADDED BY EXTENSION BOXES, RAISED COVERS (SUCH AS INDUSTRIAL COVERS), AND RAISED PLASTER RINGS. THE ADDED VOLUME CAN BE INCLUDED ONLY IF THE ACTUAL VOLUME IS MARKED ON THE EXTENSION, COVER, OR RING OR IF THE EXTENSION, COVER, OR RING IS MADE FROM A STANDARD BOX. SQUARE AND OCTAGONAL EXTENSION BOXES ARE USUALLY MADE FROM STANDARD BOXES. INDUSTRIAL COVERS AND PLASTER RINGS ARE NOT MADE FROM STANDARD BOXES, AND THEREFORE THEIR VOLUME MUST BE MARKED ON THE COVER OR RING.

FIGURE 9–15 "Wireman's Guide" Sample Device Fill Calculation

3. *Front-access angle pulls*, in which the conductors enter the back of a box that has a removable cover. The minimum depth of the box is the value given in *Table 312.6(A)*, in the "1 Wire Per Terminal" column. This special type of angle pull is detailed in Figure 9–18.

Installing Boxes

Boxes can be installed two ways: (1) surface mounted or (2) flush mounted. Some device boxes are intended for use with only one mounting method, such as nail-on nonmetallic device boxes. Others, such as square boxes, can be installed either surface mounted with the use of an industrial cover or flush mounted with the use

210 • Chapter 9

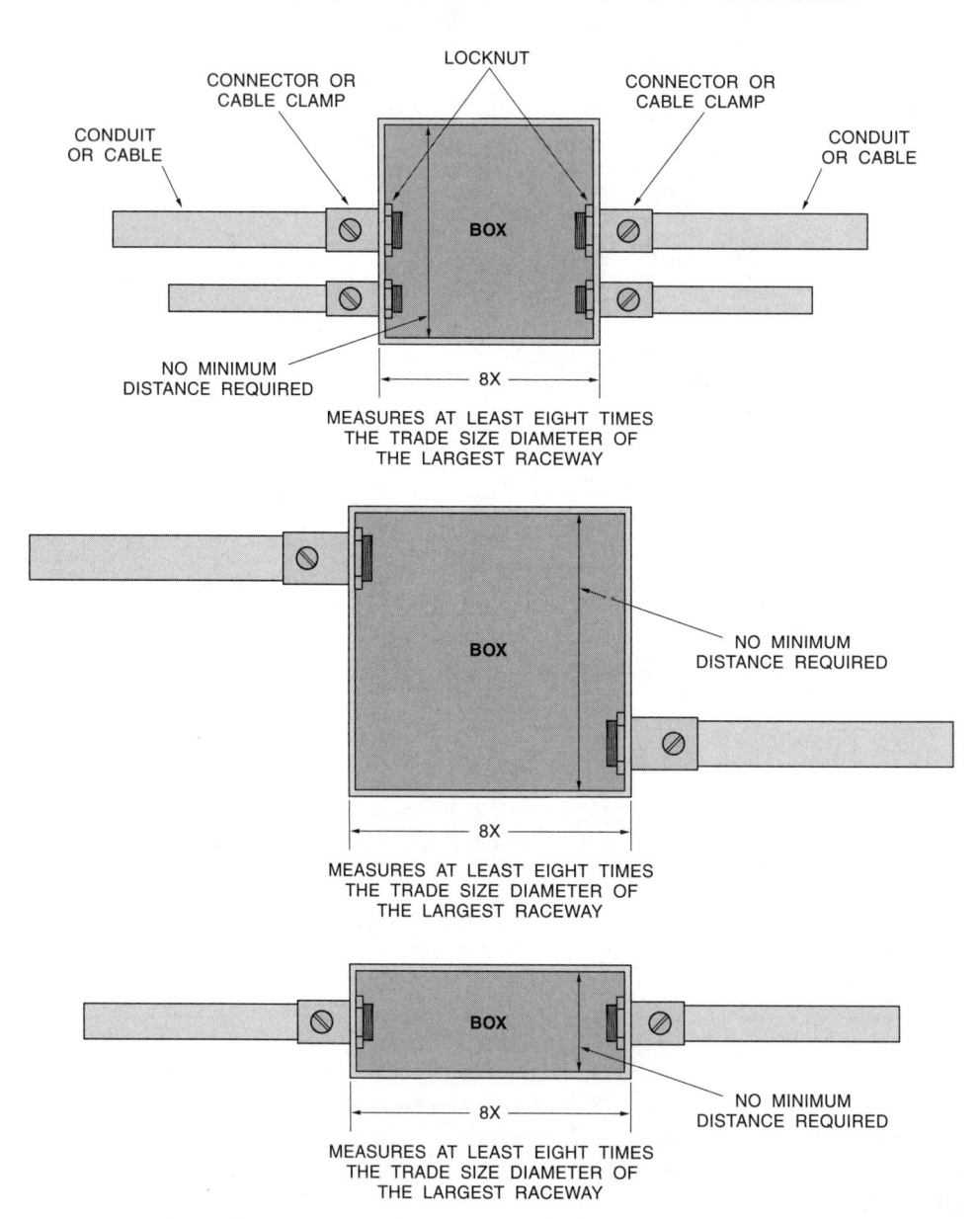

"WIREMAN'S GUIDE"
BOX SIZING FOR STRAIGHT PULLS FOR BOXES WITH CONDUCTORS 4 AWG OR LARGER

A STRAIGHT PULL IS A CONFIGURATION IN WHICH THE CABLE OR CONDUIT EXITS (LOAD SIDE) ON THE OPPOSITE WALL OF THE BOX FROM THAT OF THE ENTRY (LINE SIDE) CONDUIT OR CABLE. THE BOX IS SIZED TO THE LARGEST CABLE OR CONDUIT ENTERING THE BOX. ALL OF THE EXAMPLES SHOWN ARE CONSIDERED STRAIGHT PULLS. NOTICE THAT THERE IS NO MINIMUM DISTANCE REQUIRED BETWEEN THE WALLS ADJACENT TO THE ENTRY AND EXIT WALLS OF THE BOX.

FIGURE 9–16 "Wireman's Guide" Box Sizing for Straight Pulls for Boxes with Conductors 4 AWG or Larger

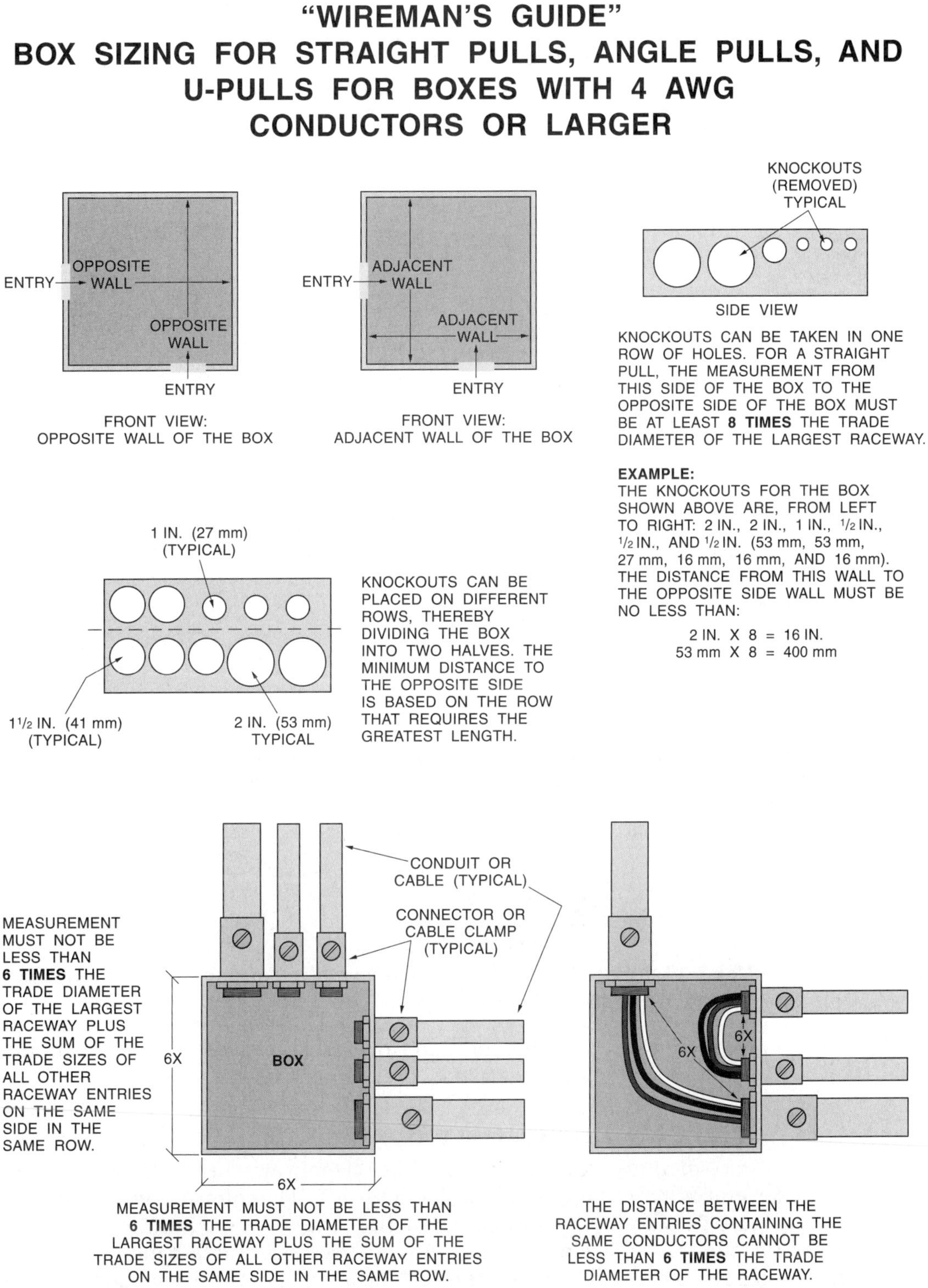

"WIREMAN'S GUIDE"
BOX SIZING FOR STRAIGHT PULLS, ANGLE PULLS, AND U-PULLS FOR BOXES WITH 4 AWG CONDUCTORS OR LARGER

FRONT VIEW: OPPOSITE WALL OF THE BOX

FRONT VIEW: ADJACENT WALL OF THE BOX

SIDE VIEW

KNOCKOUTS CAN BE TAKEN IN ONE ROW OF HOLES. FOR A STRAIGHT PULL, THE MEASUREMENT FROM THIS SIDE OF THE BOX TO THE OPPOSITE SIDE OF THE BOX MUST BE AT LEAST **8 TIMES** THE TRADE DIAMETER OF THE LARGEST RACEWAY.

EXAMPLE:
THE KNOCKOUTS FOR THE BOX SHOWN ABOVE ARE, FROM LEFT TO RIGHT: 2 IN., 2 IN., 1 IN., 1/2 IN., 1/2 IN., AND 1/2 IN. (53 mm, 53 mm, 27 mm, 16 mm, 16 mm, AND 16 mm). THE DISTANCE FROM THIS WALL TO THE OPPOSITE SIDE WALL MUST BE NO LESS THAN:

2 IN. X 8 = 16 IN.
53 mm X 8 = 400 mm

KNOCKOUTS CAN BE PLACED ON DIFFERENT ROWS, THEREBY DIVIDING THE BOX INTO TWO HALVES. THE MINIMUM DISTANCE TO THE OPPOSITE SIDE IS BASED ON THE ROW THAT REQUIRES THE GREATEST LENGTH.

MEASUREMENT MUST NOT BE LESS THAN **6 TIMES** THE TRADE DIAMETER OF THE LARGEST RACEWAY PLUS THE SUM OF THE TRADE SIZES OF ALL OTHER RACEWAY ENTRIES ON THE SAME SIDE IN THE SAME ROW.

MEASUREMENT MUST NOT BE LESS THAN **6 TIMES** THE TRADE DIAMETER OF THE LARGEST RACEWAY PLUS THE SUM OF THE TRADE SIZES OF ALL OTHER RACEWAY ENTRIES ON THE SAME SIDE IN THE SAME ROW.

THE DISTANCE BETWEEN THE RACEWAY ENTRIES CONTAINING THE SAME CONDUCTORS CANNOT BE LESS THAN **6 TIMES** THE TRADE DIAMETER OF THE RACEWAY.

FIGURE 9–17 "Wireman's Guide" Box Sizing for Straight Pulls, Angle Pulls, and U-Pulls for Boxes with 4 AWG Conductors or Larger

"WIREMAN'S GUIDE"
BOX SIZING WITH ENTRY OPPOSITE
A REMOVABLE COVER FOR
BOXES WITH CONDUCTORS 4 AWG OR LARGER

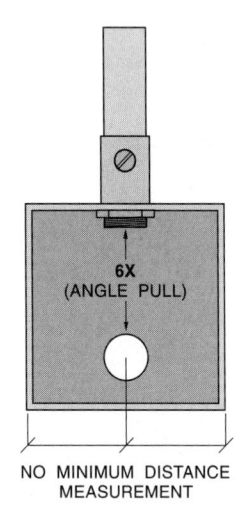

BOXES THAT HAVE A CONDUIT OR CABLE THAT ENTERS A WALL OPPOSITE A REMOVABLE COVER ARE SIZED BY THE "ONE WIRE PER TERMINAL" COLUMN IN *TABLE 312.6(A)* HOWEVER, *TABLE 312.6(A)* IS BASED ON THE SIZE OF THE CONDUCTORS ENTERING ON THE WALL OPPOSITE THE REMOVABLE COVER. CONDUCTOR SIZE MUST BE CONVERTED TO CONDUIT SIZE. THE CONDUIT SIZE IS DETERMINED FROM THE SMALLEST CONDUIT OR THE SMALLEST CABLE CONNECTOR SIZE REQUIRED TO HOUSE THAT NUMBER AND SIZE OF THE CONDUCTORS INSTALLED IN THE CONDUIT OR CABLE.

6X (ANGLE PULL)

NO MINIMUM DISTANCE MEASUREMENT

THE DISTANCE FROM THE ENTRY WALL OF THE BOX TO THE REMOVABLE COVER IS DETERMINED FROM *TABLE 312.6(A)*.

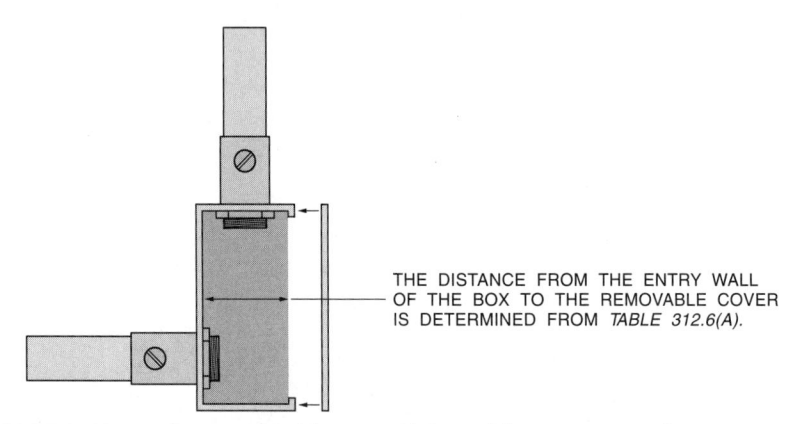

FIGURE 9–18 "Wireman's Guide" Box Sizing with Entry Opposite a Removable Cover for Boxes with Conductors 4 AWG or Larger

of a plaster ring. Large junction and pull boxes can also be surface mounted or flush mounted, depending on the covers employed.

Surface-Mounted Device Boxes

Surface-mounted device boxes must be securely attached to the building surface. Several fastening methods can be employed in mounting boxes, as shown in Figure 9–19 for hollow wall construction and in Figure 9–20 for solid wall construction. A surface-mounted device box can be either a handy box or a square box. Handy boxes allow for the direct installation of the device, and the covers are designed for surface mounting. Surface-mounted square boxes do not allow for the direct installation of a device but must employ industrial covers to secure the device in place.

Boxes • 213

"WIREMAN'S GUIDE"
THREE COMMON TYPES OF
HOLLOW WALL FASTENERS

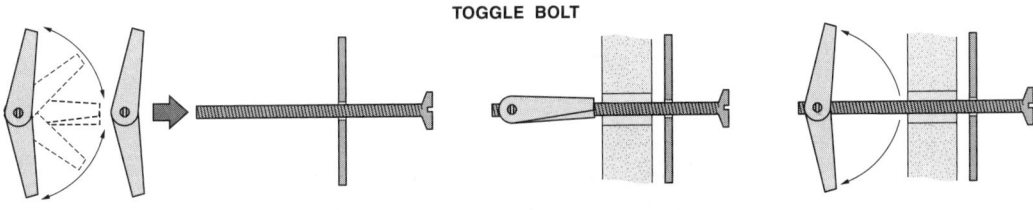

TOGGLE BOLT

A TOGGLE BOLT CONSISTS OF A BOLT AND A SPRING-LOADED, HINGED WING NUT. A HOLE IS DRILLED IN THE HOLLOW WALL LARGE ENOUGH TO ACCOMMODATE THE WING NUT. THE BOLT IS INSERTED THROUGH THE MOUNTING HOLE IN THE ITEM TO BE ATTACHED (THE BOX STRAP), AND THE WING NUT IS THREADED ONTO THE BOLT. THE BOLT ASSEMBLY IS THEN INSERTED THROUGH THE HOLE IN THE HOLLOW WALL. INSIDE THE WALL, THE SPRING LOAD ON THE WING NUT CAUSES IT TO OPEN. THE BOLT IS THEN TIGHTENED, DRAWING THE WING NUT FAST AGAINST THE INSIDE OF THE HOLLOW WALL COVERING.

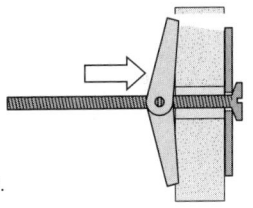

AUGER ANCHOR

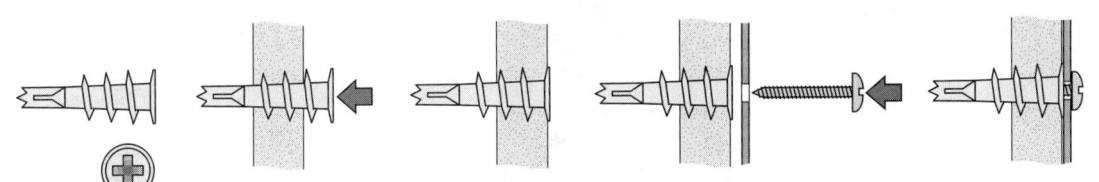

AN AUGER ANCHOR IS A ONE-PIECE METALLIC OR NONMETALLIC SLEEVE WITH A HOLE THROUGH THE MIDDLE FOR THREADING A TAPPING SCREW. A HOLE IS DRILLED IN THE HOLLOW WALL LARGE ENOUGH TO ACCOMMODATE THE SHANK OF THE ANCHOR, AND THE ANCHOR IS SCREWED INTO THE HOLE. THE THREADS ON THE ANCHOR DIG INTO THE WALLBOARD, THEREBY HOLDING THE ANCHOR FAST. THE ITEM TO BE ATTACHED IS THEN PLACED OVER THE ANCHOR, AND A SHEET METAL OR TAPPING SCREW IS SCREWED INTO THE ANCHOR. THIS TYPE OF ANCHOR CAN BE USED ONLY WHEN THE WALL IS CONSTRUCTED OF A SOFT MATERIAL, SUCH AS SHEETROCK, SO THAT THE AUGER THREADS CAN EFFECTIVELY DIG INTO THE WALL. THIS TYPE OF ANCHOR IS INTENDED FOR LIGHT-DUTY APPLICATIONS ONLY.

EXPANSION ANCHOR

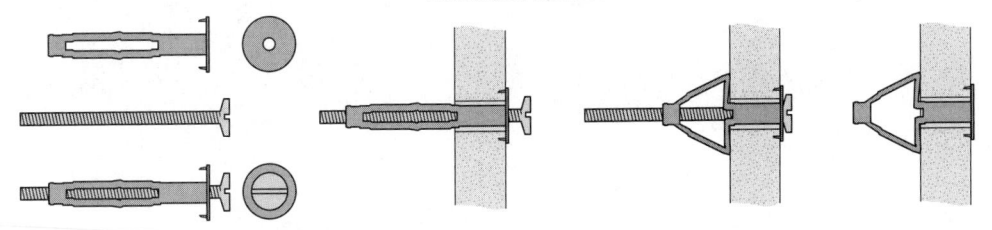

AN EXPANSION ANCHOR CONSISTS OF A BOLT AND A NUT ATTACHED TO A COLLAPSIBLE METAL FRAME, OR CAGE, WITH A RETAINING FLANGE ON THE OUTSIDE END. A HOLE IS DRILLED IN THE HOLLOW WALL LARGE ENOUGH TO ACCOMMODATE THE ANCHOR. THE ANCHOR IS THEN INSERTED INTO THE HOLE AND TIGHTENED. AS THE BOLT IS TIGHTENED, THE FRAME COLLAPSES, CONSTRICTING AGAINST THE INSIDE OF THE WALLBOARD. THE BOLT IS THEN REMOVED AND INSERTED INTO THE HOLE OF THE ITEM TO BE ATTACHED. THE BOLT IS THEN RE-INSERTED INTO THE HOLE, THREADED INTO THE NUT, AND TIGHTENED.

FIGURE 9–19 "Wireman's Guide" Three Common Types of Hollow Wall Fasteners

"WIREMAN'S GUIDE"
THREE COMMON TYPES OF SOLID WALL FASTENERS

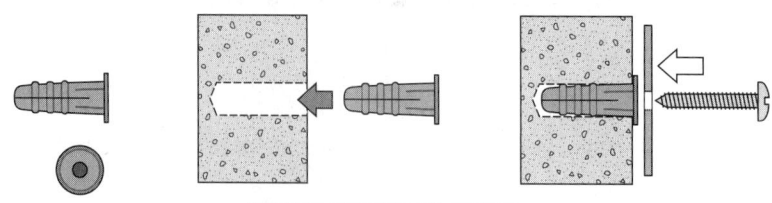

PLASTIC ANCHOR OR SHIELD

A HOLE OF THE PROPER SIZE FOR THE ANCHOR TO BE INSTALLED IS DRILLED INTO THE SOLID WALL. THE ANCHOR IS INSERTED INTO THE HOLE AND TAPPED INTO PLACE WITH A HAMMER. A TAPPING SCREW OF THE PROPER SIZE IS THEN INSERTED THROUGH THE HOLE IN THE ITEM TO BE FASTENED (BOX OR STRAP) AND SCREWED INTO THE ANCHOR. THIS TYPE OF ANCHOR IS INTENDED FOR LIGHT-DUTY APPLICATIONS ONLY, AND PLASTIC SHIELDS SHOULD NEVER BE USED IN SHEETROCK OR PLASTER WALLS. THESE ANCHORS COME IN VARIOUS SIZES, AND THE SCREW SHOULD ALWAYS BE SIZED ACCORDING TO THE INSTALLATION INSTRUCTIONS ACCOMPANYING THE ANCHORS. NO SPECIAL TOOLS ARE REQUIRED FOR THIS TYPE OF ANCHOR.

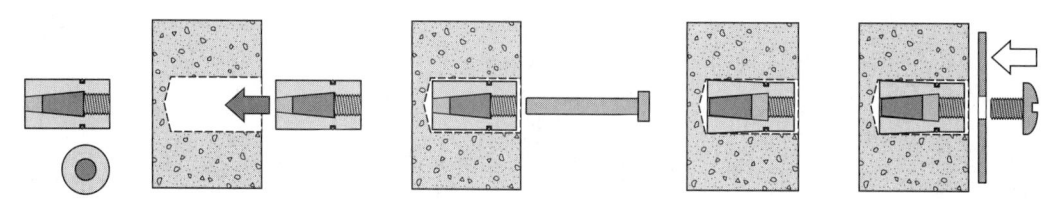

DROP-IN ANCHORS

A DROP-IN ANCHOR CONSISTS OF A METAL SLEEVE WITH A THREADED HOLE THROUGH THE CENTER. INSIDE OF THE THREADED HOLE IS A METAL WEDGE. A PROPERLY SIZED HOLE IS DRILLED INTO THE SOLID WALL AND THE DROP-IN ANCHOR IS INSERTED INTO THE HOLE. A SPECIAL SETTING TOOL IS THEN INSERTED INTO THE HOLE, AND A HAMMER IS USED TO STRIKE THE TOOL, DRIVING THE WEDGE FARTHER INTO THE HOLE. THE METAL SLEEVE IS CAUSED TO EXPAND BY THE WEDGE, THEREBY CONSTRICTING THE SLEEVE AGAINST THE SIDES OF THE SOLID WALL. THE ITEM TO BE FASTENED IS THEN PLACED OVER THE HOLE, AND A PROPERLY SIZED MACHINE SCREW IS INSERTED. THE SCREW IS THEN TIGHTENED. DROP-IN ANCHORS ARE AVAILABLE IN VARIOUS SIZES AND ARE DESIGNED FOR MEDIUM- TO HEAVY-DUTY APPLICATIONS. IT IS ALWAYS NECESSARY TO USE THE PROPER-SIZE ANCHOR FOR THE JOB AND THE PROPER-SIZE MACHINE SCREW FOR THE ANCHOR. A SETTING TOOL IS CONTAINED IN EACH BOX OF ANCHORS. THE PROPER SETTING TOOL MUST ALWAYS BE USED TO ENSURE SATISFACTORY RESULTS.

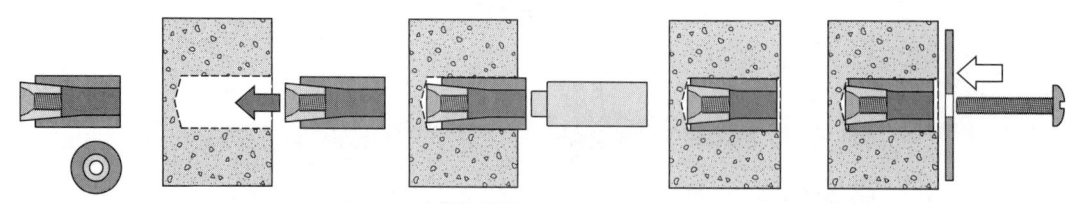

LEAD SHIELD ANCHORS

LEAD SHIELD ANCHORS CONSIST OF A LEAD OUTER SLEEVE SURROUNDING A STEEL CORE. THE STEEL CORE IS FLARED AT ONE END WITH A THREADED HOLE THROUGH THE CENTER. A PROPERLY SIZED HOLE IS DRILLED IN THE SOLID WALL, AND THE LEAD ANCHOR IS INSERTED INTO THE HOLE. IT IS VERY IMPORTANT THAT THE CORRECT END BE INSERTED, OR THE ANCHOR WILL NOT FASTEN PROPERLY. THE END OF THE ANCHOR THAT IS FLARED IS INSERTED INTO THE HOLE. A SPECIAL SETTING TOOL IS THEN USED WITH A HAMMER TO DRIVE THE LEAD INTO THE HOLE. THE LEAD CONSTRICTS AROUND THE FLARED END OF THE STEEL CORE AND THE WALL OF THE HOLE TO HOLD THE ANCHOR FAST. A PROPERLY SIZED MACHINE SCREW IS THEN USED TO FASTEN THE BOX OR STRAP TO THE WALL. LEAD ANCHORS ARE AVAILABLE IN SEVERAL SIZES AND ARE RATED FOR MEDIUM- TO HEAVY-DUTY FASTENING. WHEN INSTALLED PROPERLY, THEY WILL SUPPORT A CONSIDERABLE AMOUNT OF WEIGHT. A SETTING TOOL IS INCLUDED IN EACH BOX OF ANCHORS. BECAUSE OF THE RELATIVELY HIGH COST OF LEAD, THESE ANCHORS TEND TO BE MORE EXPENSIVE THAN OTHER TYPES.

FIGURE 9–20 "Wireman's Guide" Three Common Types of Solid Wall Fasteners

When a box is surface mounted, the cable or conduit must enter from the rear of the box in order to be hidden from view. If the wiring method is surface mounted—for example, as with electrical metallic tubing (EMT)—then the box must also be surface mounted.

Flush-Mounted Device Boxes

Most types of device boxes are designed to be installed flush with the outer surface of a wall or ceiling. *NEC® 314.20* requires that the outer edge of the box be no more than 1.4 in. (6.4 mm) from the finished wall surface if the wall is constructed of noncombustible materials, and that it be exactly flush with the outer wall surface if the surface of the wall is constructed of wood or other flammable materials. Obviously, accuracy in installing boxes is very important, as shown in Figure 9–21. The electrician must be aware of the type of wall finish—Sheetrock, plaster, or wood paneling—and the thickness of the wall covering in order to properly install the device box. If the box is recessed too far into the wall, it will be in violation of the *NEC®*. If the box is installed too far from the edge of the stud or framing member, it will protrude from the wall surface, causing problems when the device is installed. The box must also be mounted square to the stud or joist and plumb with the wall.

Many flush-mounted device boxes are available with the means for mounting, usually nails, included in the box from the manufacturer. Nails are acceptable for mounting the boxes on wooden studs or joists, but if nonmetallic framing methods are used, screws or listed cups must be employed to mount the boxes. The box must be secure and rigidly attached. Any movement in the box will cause problems when the receptacle, switch, or other device is installed at the time of trim and will cause the box to eventually become separated from the mounting surface with use over time.

If there is any gap or opening between the wall surface and the edge of the box greater than 1.8 in. (3 mm), the gap must be filled in or repaired to eliminate the opening. The covers for flush device installations are constructed so that the cover plate is larger than the device box. This design allows the cover plate to hide the edge between the box and the wall surface, ensuring a clean appearance.

Surface-Mounted Junction or Pull Boxes

Surface-mounted junction and pull boxes must also be rigidly attached to the building surface. The cover for the box must be of the same size as the box so that there is no overhang to the cover, as shown in Figure 9–22. As with surface-mounted device boxes, the cables or conduits must enter through the back of the box in order to be hidden from view. If the wiring method is surface mounted, the junction or pull box must also be surface mounted.

Flush-Mounted Junction or Pull Boxes

Many large junction and pull boxes can be flush mounted. Several considerations arise with such installations: (1) Can the box be rigidly attached to a stud or other framing members within the wall or ceiling? The box must be attached on opposite sides in order to be rigidly attached. (2) Is the wall deep enough to accommodate the depth of the box? (3) Is the box cover larger than the box itself? If the cover is of the same size as the box, as with surface-mounted boxes, the edge of the box and the edge of the wall covering will present an unfinished appearance that will be unacceptable in most installations. Figure 9–22 includes more information about flush-mounted junction and pull boxes.

216 • Chapter 9

"WIREMAN'S GUIDE"
INSTALLING DEVICE BOXES

BOXES MUST BE MOUNTED SQUARE TO THE STUD OR JOIST AND PLUMB WITH THE WALL.

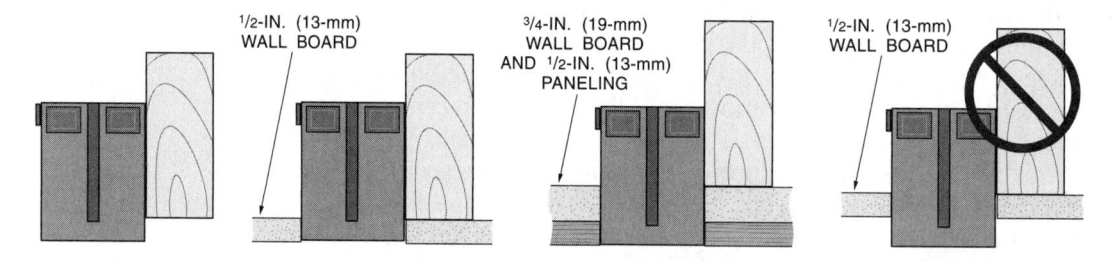

THE BOX SHOULD BE INSTALLED SO THAT THE FRONT EDGE OF THE BOX WILL BE FLUSH WITH THE FINISHED WALL SURFACE. THE ELECTRICIAN MUST BE AWARE OF WHAT TYPE OF WALL FINISHING IS TO BE INSTALLED AND THE THICKNESS OF THE FINISH. IF THE BOX IS INSTALLED SO THAT IT EXTENDS PAST THE FINISHED WALL, TRIM PROBLEMS WILL RESULT.

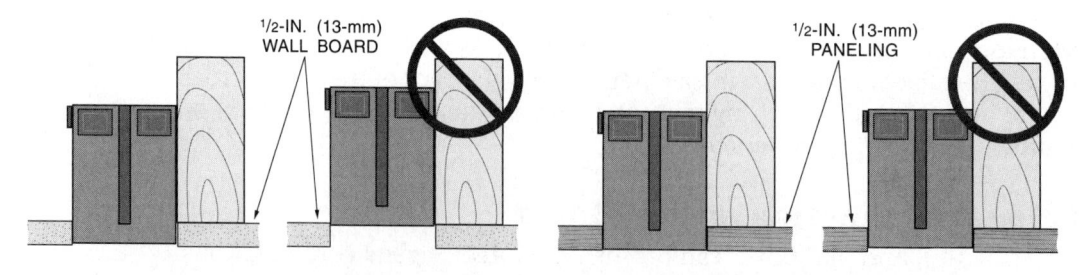

IN NONFLAMMABLE WALLS SUCH AS SHEETROCK AND PLASTER, THE FRONT EDGE OF THE BOX CAN BE NO MORE THAN 1/4 IN. (6.4 mm) FROM THE FINISHED WALL SURFACE. IN FLAMMABLE WALLS SUCH AS PANELING OR WOODEN SIDING, THE FRONT EDGE OF THE BOX MUST BE FLUSH WITH THE FINISHED WALL SURFACE. THE BOX IS DESIGNED TO RETARD THE SPREAD OF FIRE SHOULD A PROBLEM OCCUR WITHIN THE BOX. IF ANY FLAMMABLE WALL SURFACE IS EXPOSED TO THE INSIDE OF THE BOX, FIRE COULD SPREAD MORE EASILY.

FIGURE 9–21 "Wireman's Guide" Installing Device Boxes

Boxes • 217

"WIREMAN'S GUIDE"
COVERS AND LARGE JUNCTION OR
PULL BOX INSTALLATION

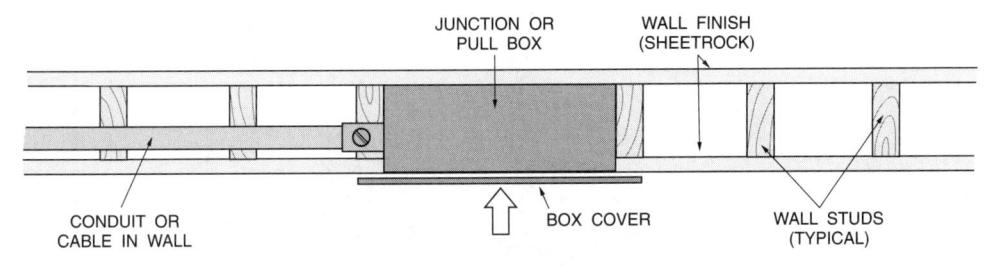

FOR FLUSH MOUNTING OF LARGE JUNCTION OR PULL BOXES, THE BOX COVER MUST BE LARGER THAN THE BOX TO PROVIDE A NEAT AND PROFESSIONAL APPEARANCE.

IF THE BOX COVER IS THE SAME SIZE AS THE BOX, THE UNFINISHED EDGE OF THE WALL COVERING WILL SHOW WHERE IT WAS CUT TO ACCEPT THE BOX. THIS TYPE OF COVER IS DESIGNED FOR SURFACE MOUNTING OF THE BOX.

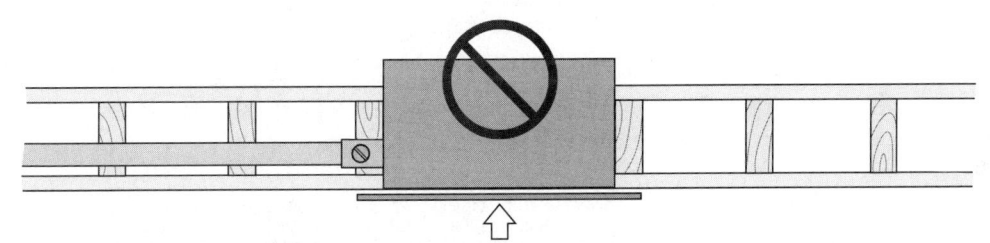

FOR FLUSH MOUNTING OF LARGE JUNCTION OR PULL BOXES, CARE MUST BE TAKEN TO ENSURE THAT THE BOX IS NOT TOO DEEP FOR THE WALL WHERE IT IS TO BE INSTALLED.

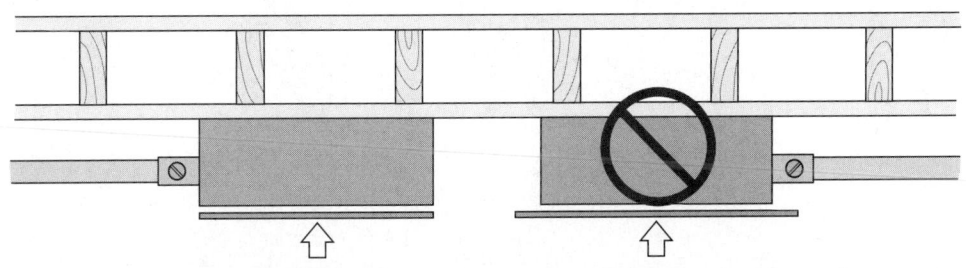

FOR SURFACE-MOUNTED LARGE JUNCTION OR PULL BOXES, THE COVER MUST BE OF THE SAME SIZE AS THAT OF THE BOX, OR IT WILL OVERHANG THE BOX, PRESENTING A POOR APPEARANCE AND A HAZARD FOR SCRATCHING OR SNAGGING.

FIGURE 9–22 "Wireman's Guide" Covers and Large Junction or Pull Box Installation

218 • Chapter 9

"WIREMAN'S GUIDE"
CONDUIT BODIES

CONDUIT BODIES ARE INSTALLED IN CONDUIT RUNS TO TURN SHARP CORNERS, TO PROVIDE WIRE PULLING POINTS, AND FOR MAKING SPLICES. CONDUIT BODIES CAN BE MANUFACTURED FROM CAST ALUMINUM, CAST IRON, STEEL, OR PVC AND ARE AVAILABLE IN ALL SIZES OF CONDUIT, FROM 1/2 IN. (16 mm) TO 6 IN. (155 mm).

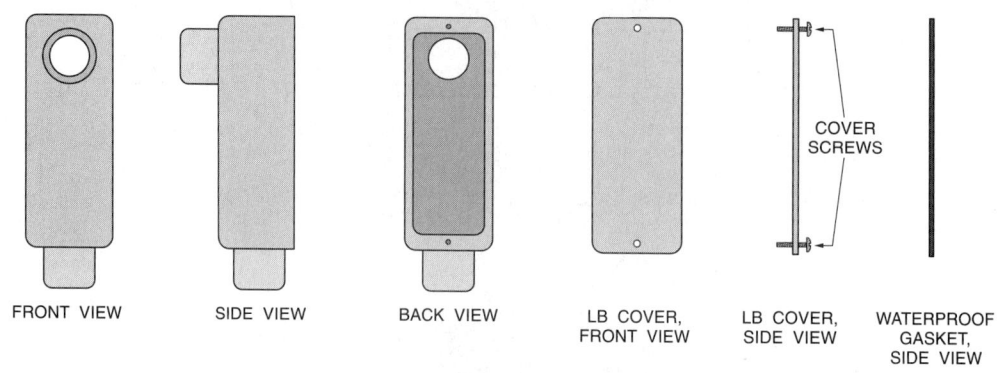

FRONT VIEW   SIDE VIEW   BACK VIEW   LB COVER, FRONT VIEW   LB COVER, SIDE VIEW   COVER SCREWS   WATERPROOF GASKET, SIDE VIEW

TYPE OF LB CONDUIT BODY

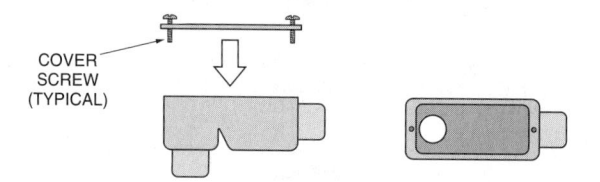

COVER SCREW (TYPICAL)

SIDE VIEW AND BACK VIEW OF A SHORT LB (SLB) CONDUIT BODY. THESE CONDUIT BODIES ARE USED FOR CONDUCTORS UP TO 6 AWG. CONDUCTORS LARGER THAN 6 AWG REQUIRE A FULL-SIZED LB CONDUIT BODY.

SOME OTHER AVAILABLE CONDUIT BODIES

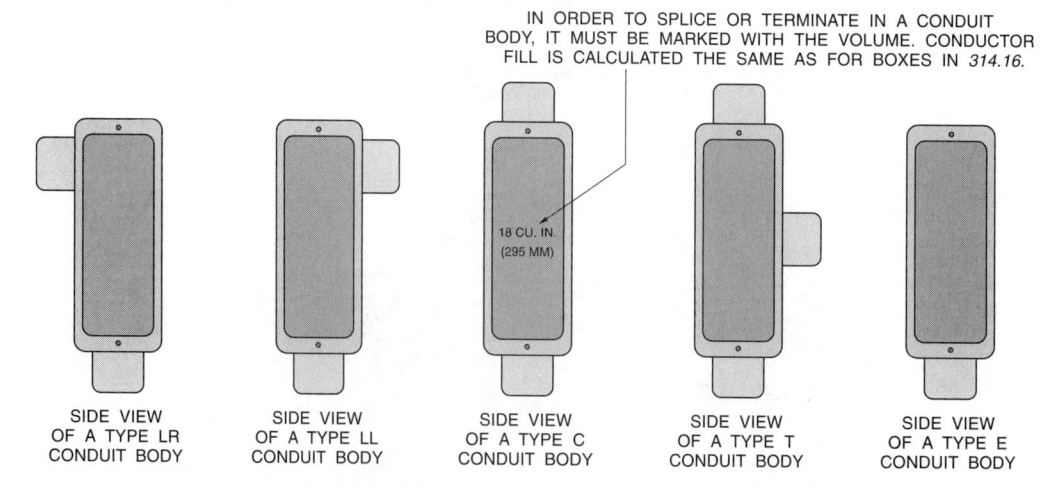

IN ORDER TO SPLICE OR TERMINATE IN A CONDUIT BODY, IT MUST BE MARKED WITH THE VOLUME. CONDUCTOR FILL IS CALCULATED THE SAME AS FOR BOXES IN *314.16*.

18 CU. IN. (295 MM)

SIDE VIEW OF A TYPE LR CONDUIT BODY   SIDE VIEW OF A TYPE LL CONDUIT BODY   SIDE VIEW OF A TYPE C CONDUIT BODY   SIDE VIEW OF A TYPE T CONDUIT BODY   SIDE VIEW OF A TYPE E CONDUIT BODY

SOME OTHER AVAILABLE CONDUIT BODIES

FIGURE 9–23 "Wireman's Guide" Conduit Bodies

Conduit Bodies

Conduit bodies are fittings that aid in running conduit around tight corners or for tapping off from a conduit run. Residential electricians occasionally need to install conduit in remodeling work or outside circuit installations. Conduit bodies are covered in *314.16(C)*.

Conduit bodies are available as **short conduit bodies** as well as regular conduit bodies. Short conduit bodies are intended for the installation of conductors 6 AWG or smaller. No taps or splices are to be accomplished in a short conduit body. Regular conduit bodies can include splices and taps if the conduit body is durably and legibly marked with the volume. The number of conductors allowed in a conduit body is the same as the number of conductors allowed in the conduit to which it is attached, although this may not be the case with some larger (conductor size 300 kcmil and up) conduit bodies. Care should be used in sizing the conduit bodies for large conduit runs. Figure 9–23 shows some examples of conduit bodies.

Summary

- Boxes must be used for all devices, outlets, and switch points.
- Boxes for conductors 6 AWG and smaller are termed device boxes and are sized by the number of conductors, devices, and fittings to be housed in the box.
- Device boxes have provisions for mounting devices using 6-32 screws.
- Device boxes can be metallic or nonmetallic in nature.
- Device boxes are available as single-gang boxes, or as multigang boxes to house several devices in one box.
- Lighting outlet boxes have provisions for mounting luminaries with 8-32 screws.
- Lighting outlet boxes can be metallic or nonmetallic in nature.
- Square boxes are a special kind of box that is versatile and can be used for surface and concealed installations and as device boxes or lighting outlet boxes.
- Boxes containing conductors larger than 6 AWG are termed pull boxes and are sized for the type of pull, straight pull versus angle pull, accomplished by the conductors in the box.
- Conduit bodies can be used instead of boxes in making sharp bends or for tapping a conduit.

Key Terms

Boxes Housings in the electrical circuit that contain splices and terminations. Boxes can be metallic or nonmetallic and may house devices or simply contain conductors, but they provide a barrier between the electrical system of the structure and the living or working space of that structure.

Conduit bodies (See *NEC® Article 100*): A type of raceway fitting that allows for a rapid change in direction of wiring, or for the tapping of a raceway, without the use of a box. Conduit bodies have removable covers that allow access to the interior of the fitting to facilitate the installation of the conductors.

Device boxes Electrical boxes intended to house and make available devices such as switches and receptacles.

Lighting outlet boxes Boxes that are designed to supply outlets for luminaires (lighting fixtures). These boxes are usually round in shape, can be metallic or nonmetallic, and have 8-32 threaded holes for the connection of the luminaire (lighting fixture) support hardware.

Plaster ring An accessory that is used with square boxes to allow them to be used in flush installations.

Short conduit bodies A type of conduit body designed for use with smaller sizes of raceways but with limited bending space of the conductors. The length of short conduit bodies is considerably less than the length of a regular conduit body of the same raceway trade size.

Square boxes Electrical boxes that are square in shape and can be metallic or nonmetallic. Square boxes can be used for flush and surface installations by employing plaster rings or industrial covers and are the most common type of box used in commercial electrical work.

Waterproof boxes Electrical boxes designed to be used in wet and damp areas, such as outdoors, where moisture can enter the raceway system or the boxes themselves, thereby causing faulting problems.

Review Questions

1. How is the volume of a device box or outlet box determined?
2. Device or outlet boxes contain conductors smaller than _____ AWG.
3. Metallic device boxes are limited to _____ gangs.
4. Metallic square boxes are limited to _____ gangs.
5. For concealed work, square boxes need to have _____ used with the box.
6. What is the maximum number of 14 AWG conductors allowed in a 3-in. × 2-in. × 31.2-in. (75-mm × 50-mm × 90-mm) device box?
7. A device installed in a device box is counted as _____ conductor(s).
8. For U-pulls or angled pulls in boxes with 250-kcmil conductors installed, the distance between raceway entries containing the same conductors must be _____ times the trade diameter of the raceway or cable.
9. How are cable clamps counted when installed inside of device boxes?
10. Are internal clamps available on metallic device boxes?

Residential Electrical Cabling Installation

OBJECTIVES

After studying this chapter, you should be able to:

- Apply the *NEC®* and IRC requirements for notching and drilling studs, joists, and rafters.
- State the requirements for installing conductors in environmental airspaces as applied to residential wiring.
- Apply the rules for bundling cables through holes or notches.
- Apply the rules for running cables parallel to framing members.
- Describe the proper way to splice and pigtail conductors with wire nuts or splice caps.
- Identify conductors for makeup using a uniform system of markings.
- Explain the importance of conductor organization in junction or device boxes.

OUTLINE

Introduction

NEC® Requirements for Drilling or Notching Studs, Rafters, and Joists

Requirements for Drilling or Cutting Studs, Joists, and Rafters

Environmental Airspaces

Bundling Cables

Cables Run Parallel to Framing Members

Connecting Conductors with Wire Nuts or Splice Caps

Pigtail Connections

Identification of Conductors

Organizing the Box

Introduction

This chapter covers the requirements for drilling or notching of wall studs and ceiling joists or rafters for the installation of electrical wiring in dwelling units. The *NEC*® has rules for keeping conductors away from the edges of studs, joists, and rafters to help protect them from physical damage. This code is not the only regulation that places limits on the notching or drilling of studs, joists, and rafters, however. The International Residential Code (IRC) also has requirements for drilling and notching framing members. Other codes, such as the International Residential Code and NFPA building codes, are also being introduced in the United States.

This chapter also examines the connection of conductors using **splices** and **pigtails** and the proper procedures to use when such connections are made. In addition, the organization of the conductors in the various boxes is covered. **Box makeup** is one of the most important functions of the electrician during rough-in, and proper organization and marking of the various conductors are central to the makeup process.

NEC® Requirements for Drilling or Notching Studs, Rafters, and Joists

The *NEC*® requirements for **drilled holes** or **cut notches** in framing members for dwellings are intended to protect electric cable from physical damage. Damage may occur during the construction process but may also occur during the life of the structure as the result of penetration by nails or screws during remodeling, the installation of wall-mounted shelving, the hanging of pictures, or any number of other events. According to *300.4(A)*, any cable must be kept at least 11.4 in. (32 mm) from the edge of a **stud, rafter,** or **joist.** This restriction provides an area in the center of the framing member that satisfies the requirements of the *NEC*®. This area is called the *drilling zone* and is shown in Figure 10–1. Figure 10–2 shows an NM cable running through a drilled hole. If the hole must be drilled closer than 11.4 in. (32 mm) to the edge of a stud, rafter, or joist, or if a notch must be cut, the cable must be protected by a steel plate at least 1.16 in. (1.6 mm) thick. Such a plate is called a **notch plate** and is illustrated in Figures 10–3 and 10–4.

Requirements for Drilling or Cutting Studs, Joists, and Rafters

The IRC has an entirely different interest in the notching and drilling of framing members. The IRC is concerned with a building's structural integrity—whether the building can withstand the stresses likely to be encountered in use of the structure. Any notch or hole in a stud, rafter, or joist potentially weakens the structure; therefore, the IRC requirements for cutting notches and drilling holes are not intended to protect wiring but are intended to protect the building. The residential electrician must be aware of these IRC requirements and must follow them when drilling in preparation for cable installation.

Residential Electrical Cabling Installation • 223

"WIREMAN'S GUIDE"
NEC® DRILLING REQUIREMENTS

1¼ IN. (32 mm) MINIMUM

DRILLING ZONE

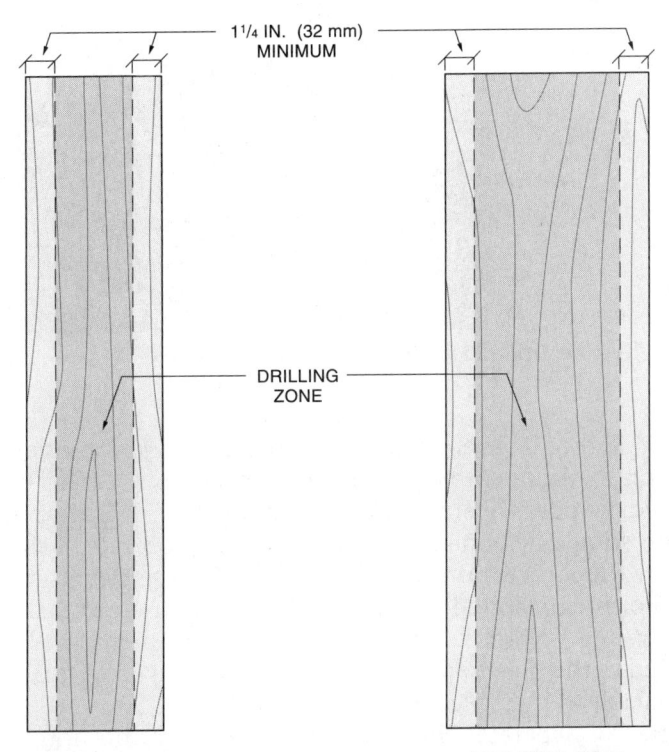

2 X 4 WALL STUD:
THE ACTUAL MEASUREMENTS OF A 2 X 4 STUD ARE APPROXIMATELY 1½ IN. (38 mm) X 3½ IN. (90 mm)

2 X 6 WALL STUD:
THE ACTUAL MEASUREMENTS OF A 2 X 6 STUD ARE APPROXIMATELY 1½ IN. (38 mm) X 5½ IN. (140 mm)

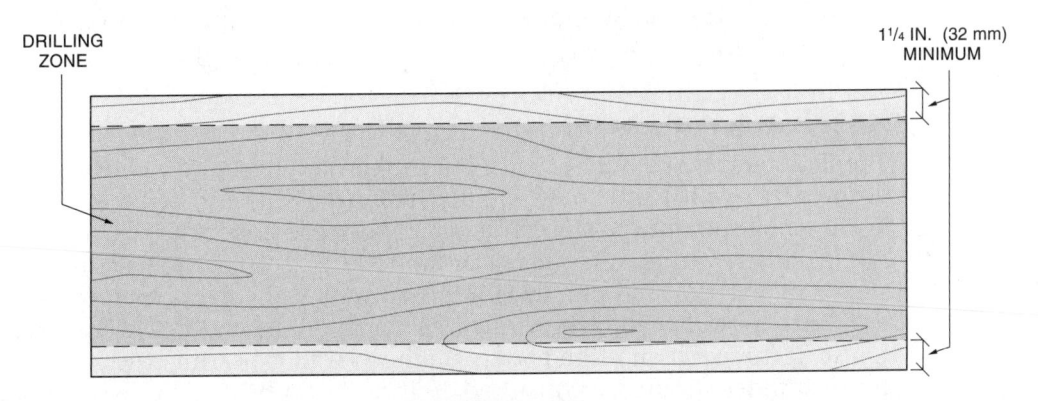

DRILLING ZONE

1¼ IN. (32 mm) MINIMUM

2 X 8 FLOOR JOIST:
THE ACTUAL MEASUREMENTS OF A 2 X 8 JOIST ARE APPROXIMATELY 1½ IN. (38 mm) X 7½ IN. (191 mm)

FIGURE 10–1 "Wireman's Guide" *NEC*® Drilling Requirements

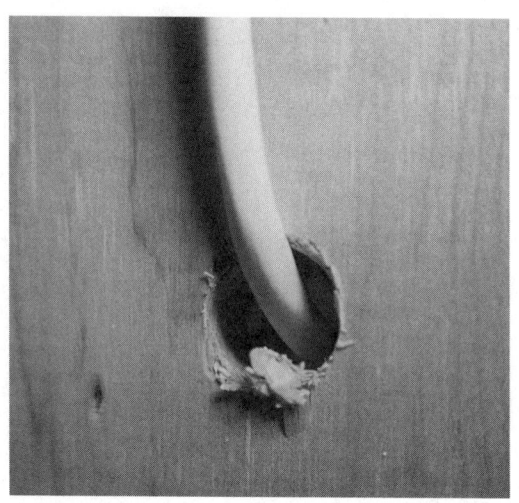

FIGURE 10–2 A Drilled Hole Containing an NM Cable

Not all dwellings are constructed using wooden studs and rafters. Several construction methods employ manufactured systems such as floor trusses and manufactured wooden beams, among others. The drilling requirements for these systems should be obtained from the general contractor, the IRC, or the manufacturer of the structural members. Some of these structural members cannot be drilled or notched at all. Some dwellings may use metal studs instead of wooden studs. When NM cable is installed in holes in metal studs, bushings need to be installed in the hole to protect the cable. Bushings for standard-size holes are readily available at most electrical supply houses.

The IRC separates the functions of drilling and notching. Drilling involves making round holes, with wood on all sides, whereas notching involves removing some of the framing member's edge. Generally, notches weaken the structure of the dwelling more than holes do; therefore, they are more restricted in size and location. Structural members of a dwelling are divided by the IRC into two distinct groups: (1) wall supports (studs) and (2) floor supports or ceiling supports (joist or rafter).

Notching Wall Studs

The IRC states that a notch can be cut into any wall stud to a depth of 25 percent of the stud's width and anywhere up the length of the stud, except back to back in the same spot on the stud. The cutting is to be done with a saw. If the stud is not in a load-bearing wall, the notch can take up to 40 percent of the depth of the stud. Any exterior wall of a dwelling is automatically considered to be a bearing wall. Interior walls can be bearing or nonbearing, depending on the design of the house. If there is a question about whether a wall is bearing, the general contractor or builder should be consulted. Figure 10–5 presents more information on notching studs.

Notching Joists and Rafters

The requirements for notching floors and ceilings are discussed in *IRC R502.8.1* and *R802.7.1*. The size, or depth, of the notch varies with the distance that the

Residential Electrical Cabling Installation • 225

"WIREMAN'S GUIDE" NOTCH PLATES

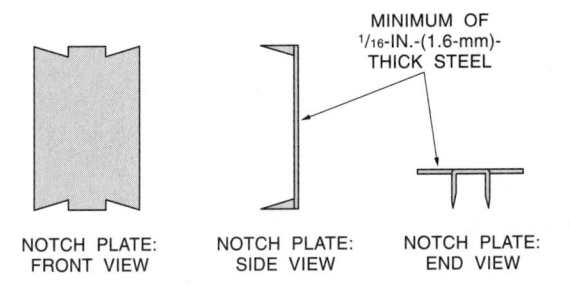

MINIMUM OF 1/16-IN.-(1.6-mm)-THICK STEEL

NOTCH PLATE: FRONT VIEW

NOTCH PLATE: SIDE VIEW

NOTCH PLATE: END VIEW

NOTCH PLATES ARE INSTALLED OVER A NOTCH OR A HOLE THAT ALLOWS THE CABLE TO BE WITHIN 1 1/4 IN. (32 mm) OF THE EDGE OF THE STUD. THE PLATE MUST BE AT LEAST 1/16 IN. (1.6 mm) THICK AND COVER THE ENTIRE CABLE WHILE IT PASSES THROUGH THE FRAMING MEMBER. A NAIL OR SCREW WILL NOT BE ABLE TO PENETRATE THE PLATE, THEREBY PROTECTING THE CABLE FROM PHYSICAL DAMAGE.

NONMETALLIC-SHEATHED CABLE 14-2 600 VOLTS

CABLE 14-2 600 VOLTS NONMETALLIC

NOTCH PLATE

CABLE

NONMETALLIC-SHEATHED CABLE 14-2 600 VOLTS

FRONT VIEW
2 X 4 WALL STUD

1 1/4 IN. (32 mm)

CUT NOTCH WITH CABLE INSTALLED

NOTCH PLATE

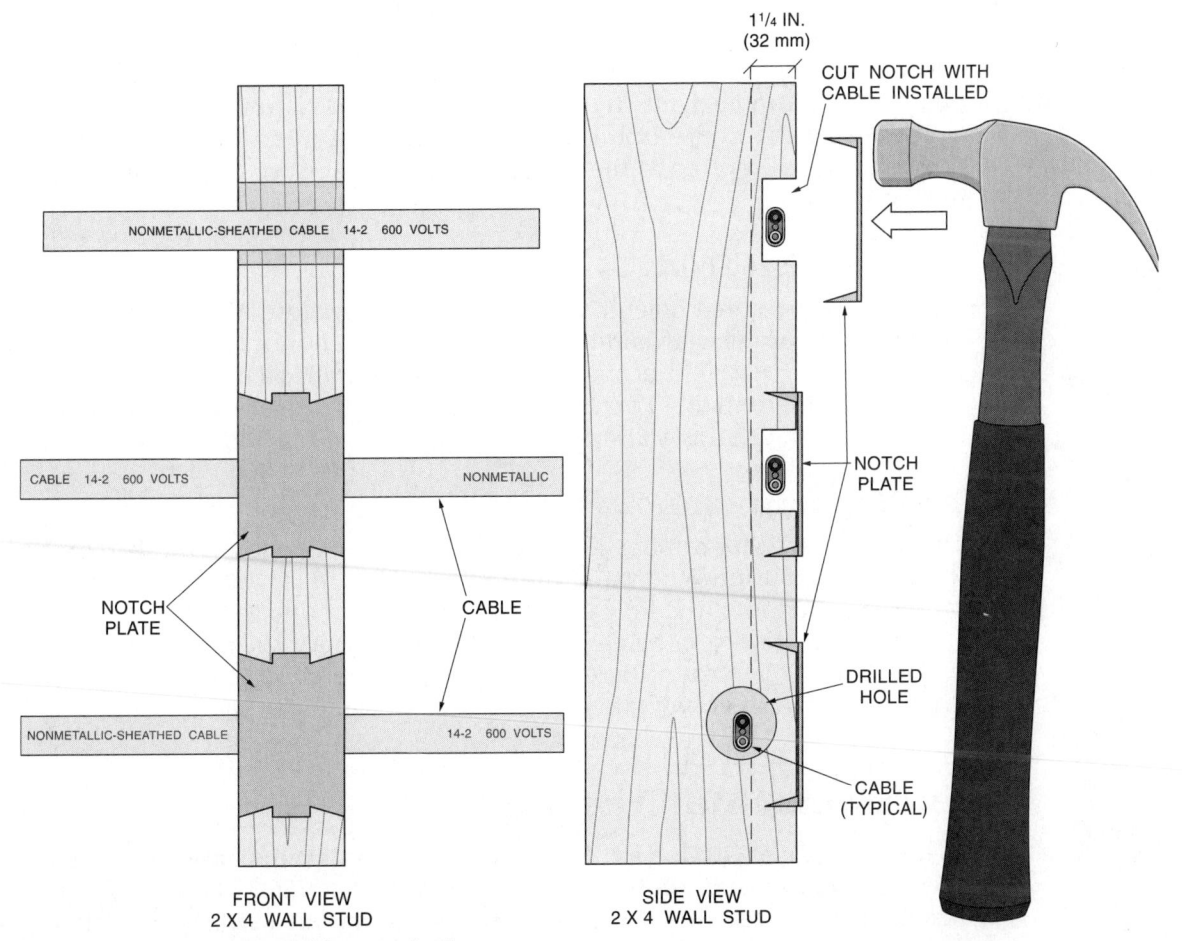

DRILLED HOLE

CABLE (TYPICAL)

SIDE VIEW
2 X 4 WALL STUD

FIGURE 10–3 "Wireman's Guide" Notch Plates

FIGURE 10–4 If the cable is within 1-1/4 in. (32 mm), a notch plate must be installed to protect the cable from nails.

joist or rafter has to span and with the load that the joists or rafters have to support. Notches can be made in the edges of joists and rafters as long as the notch is not deeper than one-sixth the depth of the joist or rafter. Furthermore, the notch cannot be located in the middle third of the joist or rafter span, as shown in Figure 10–6.

Drilling Wall Studs

The edge of any bored hole in a stud wall cannot be closer than 5.8 in. (16 mm) from the edge of a stud. In addition, a bored hole in a nonbearing wall cannot take up more than 60 percent of the stud depth. If the wall is a bearing wall, the hole cannot take more than 40 percent of the stud depth. Figure 10–7 presents more information on drilling wall studs.

Drilling Joists and Rafters

The requirements for drilling floors and ceilings are discussed in *IRC R502.8.1* and *R802.7.1*. Holes cannot be made within 2 in. (50 mm) of the edges of joists and rafters. Furthermore, the hole cannot be larger than one-third the depth of the joist or rafter. However, there is no restriction on where the holes can be located in the span of the joist or rafter as there is with notches. More information on drilling holes in joists and rafters is presented in Figure 10–8.

Environmental Airspaces

In dwelling units with forced air heat, a number of **environmental airspaces** present special considerations in installing cable. *NEC® 300.22* outlines the installation of electrical conductors into spaces that are used for the delivery or return of heated or cooled air. The *Code* is very careful not to allow nonmetallic wiring methods in airspaces where the air will circulate through the building. Not only can this circulating air increase the spread of a fire, but the burning of

Residential Electrical Cabling Installation • 227

"WIREMAN'S GUIDE" IRC NOTCHING REQUIREMENTS FOR STUD WALLS

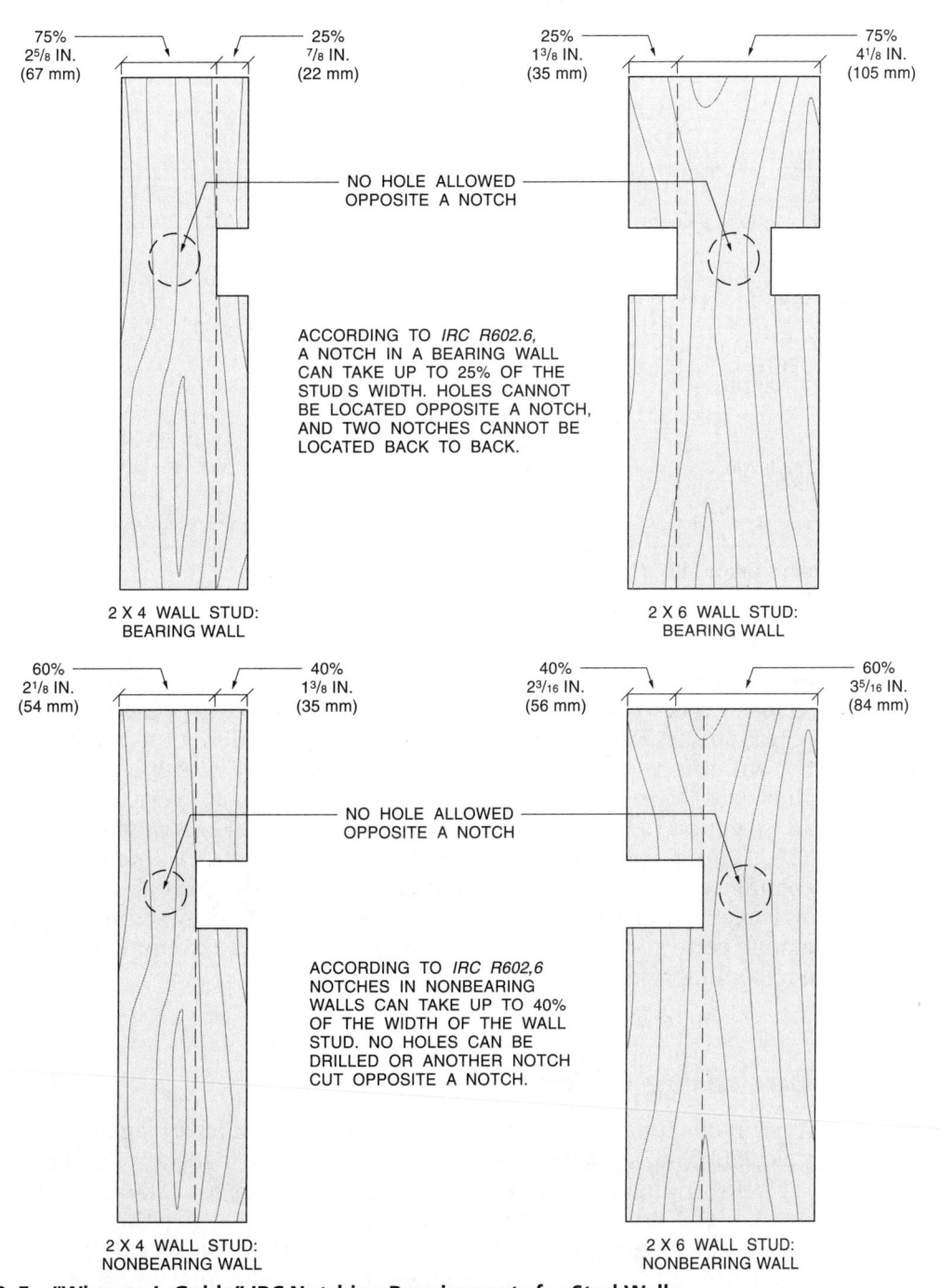

75% — 2 5/8 IN. (67 mm)
25% — 7/8 IN. (22 mm)
25% — 1 3/8 IN. (35 mm)
75% — 4 1/8 IN. (105 mm)

NO HOLE ALLOWED OPPOSITE A NOTCH

ACCORDING TO *IRC R602.6*, A NOTCH IN A BEARING WALL CAN TAKE UP TO 25% OF THE STUD'S WIDTH. HOLES CANNOT BE LOCATED OPPOSITE A NOTCH, AND TWO NOTCHES CANNOT BE LOCATED BACK TO BACK.

2 X 4 WALL STUD: BEARING WALL

2 X 6 WALL STUD: BEARING WALL

60% — 2 1/8 IN. (54 mm)
40% — 1 3/8 IN. (35 mm)
40% — 2 3/16 IN. (56 mm)
60% — 3 5/16 IN. (84 mm)

NO HOLE ALLOWED OPPOSITE A NOTCH

ACCORDING TO *IRC R602.6* NOTCHES IN NONBEARING WALLS CAN TAKE UP TO 40% OF THE WIDTH OF THE WALL STUD. NO HOLES CAN BE DRILLED OR ANOTHER NOTCH CUT OPPOSITE A NOTCH.

2 X 4 WALL STUD: NONBEARING WALL

2 X 6 WALL STUD: NONBEARING WALL

FIGURE 10–5 "Wireman's Guide" IRC Notching Requirements for Stud Walls

"WIREMAN'S GUIDE" IRC NOTCH REQUIREMENTS FOR JOISTS AND RAFTERS

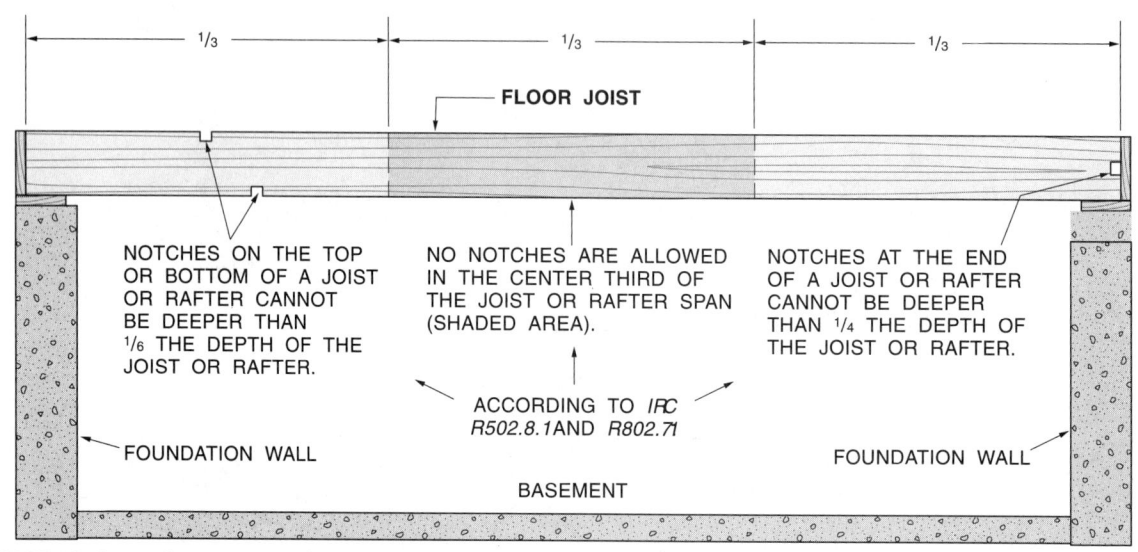

FIGURE 10–6 "Wireman's Guide" IRC Notch Requirements for Joists and Rafters

nonmetallic conduit or cabling systems contributes to the distribution of poisonous gases. When chloride-based chemicals such as PVC burn, they give off a poisonous gas. The environmental air system can distribute this gas throughout the building very quickly. Therefore, usually only metallic wiring systems are allowed to be installed in environmental airspaces. However, dwelling units are an exception to the general rule. The *Exception to 300.22(C)* says that it is acceptable to install nonmetallic cables in return air ducts for forced-air heating and cooling systems in houses as long as the cables run through the environmental airspaces using the shortest route. Figure 10–9 shows the method usually employed to convert a joist space into a return airspace for a forced-air heating and cooling system.

Bundling Cables

A primary aim of the *NEC*® is to protect conductors against overheating. Calculations of ampacity and insulation ratings are made to keep conductor heating below some maximum level. The same applies to installing cables through notches and holes. When cables are tightly bundled together, air circulation is restricted and conductors can overheat. Therefore, it is usually advisable for the installing electrician to drill several sets of relatively smaller holes rather than one set of large holes to accommodate the cables using the same routing. Figure 10–10 gives more detailed information on bundling cables, and Figure 10–11 shows cables in a dwelling that are closely bundled.

Residential Electrical Cabling Installation • 229

"WIREMAN'S GUIDE"
IRC DRILLING REQUIREMENTS FOR STUDS

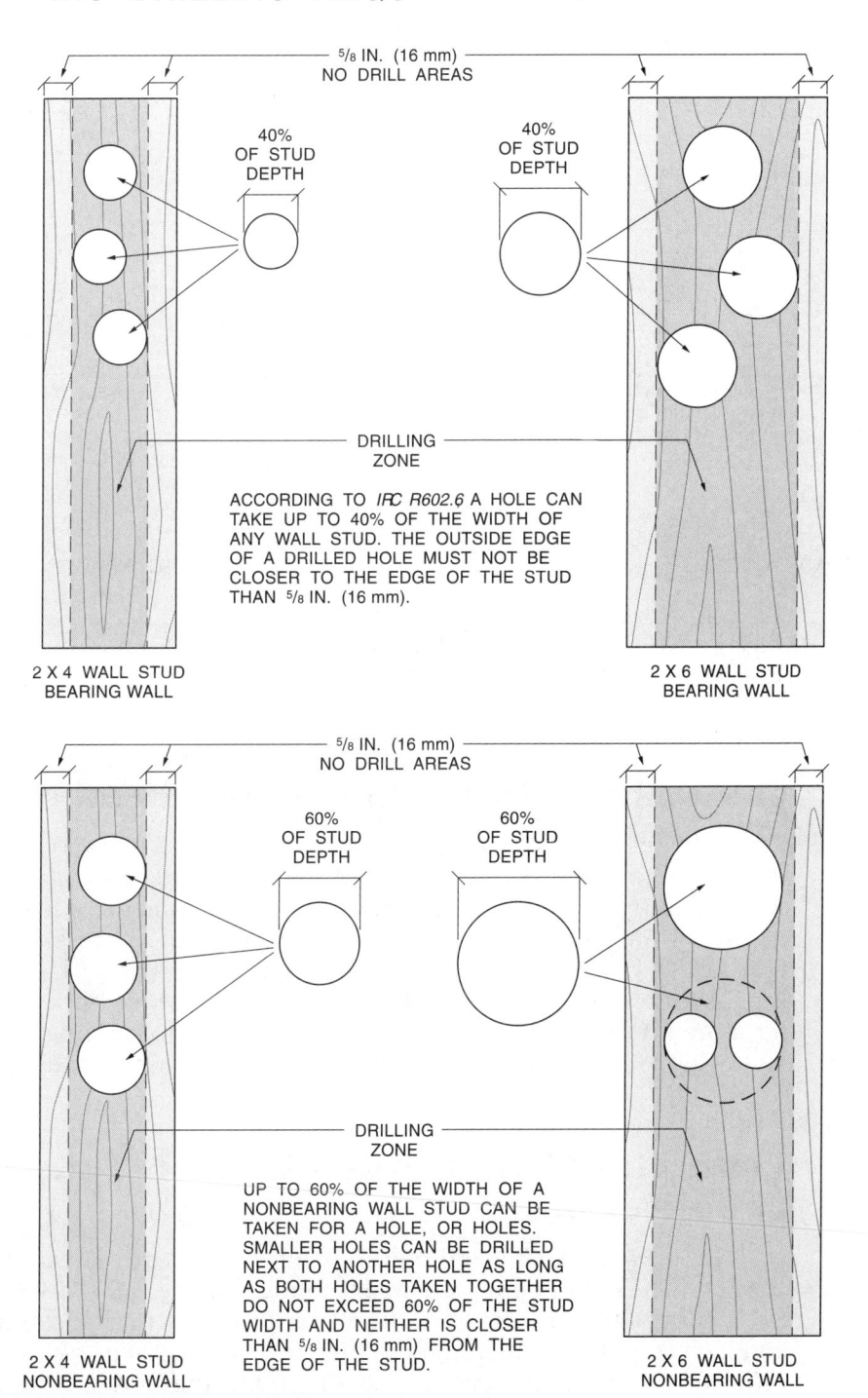

⁵⁄₈ IN. (16 mm) NO DRILL AREAS

40% OF STUD DEPTH

40% OF STUD DEPTH

DRILLING ZONE

ACCORDING TO *IRC R602.6* A HOLE CAN TAKE UP TO 40% OF THE WIDTH OF ANY WALL STUD. THE OUTSIDE EDGE OF A DRILLED HOLE MUST NOT BE CLOSER TO THE EDGE OF THE STUD THAN ⁵⁄₈ IN. (16 mm).

2 X 4 WALL STUD
BEARING WALL

2 X 6 WALL STUD
BEARING WALL

⁵⁄₈ IN. (16 mm) NO DRILL AREAS

60% OF STUD DEPTH

60% OF STUD DEPTH

DRILLING ZONE

UP TO 60% OF THE WIDTH OF A NONBEARING WALL STUD CAN BE TAKEN FOR A HOLE, OR HOLES. SMALLER HOLES CAN BE DRILLED NEXT TO ANOTHER HOLE AS LONG AS BOTH HOLES TAKEN TOGETHER DO NOT EXCEED 60% OF THE STUD WIDTH AND NEITHER IS CLOSER THAN ⁵⁄₈ IN. (16 mm) FROM THE EDGE OF THE STUD.

2 X 4 WALL STUD
NONBEARING WALL

2 X 6 WALL STUD
NONBEARING WALL

FIGURE 10–7 "Wireman's Guide" IRC Drilling Requirements for Studs

"WIREMAN'S GUIDE" IRC DRILLING REQUIREMENTS FOR JOISTS AND RAFTERS

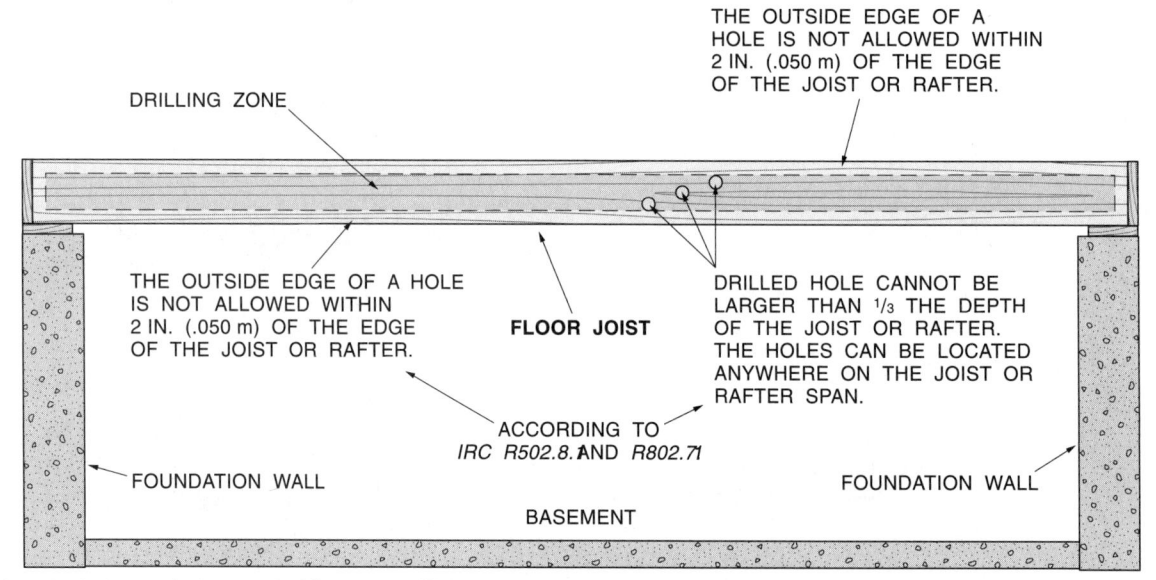

FIGURE 10–8 "Wireman's Guide" IRC Drilling Requirements for Joists and Rafters

Cables Run Parallel to Framing Members

Cables are subject to the same damage when they are installed parallel to a framing member as when they are installed through bored holes or placed in a notch. Screws and nails can penetrate the cables during construction as well as during the life of the structure. Cables installed parallel to framing members—cables that run up or down wall studs or horizontally along a joist or stud—must be kept at least 1-1/4 in. (32 mm) from the edge of the member. See Figure 10–12 for more information on installing cables parallel to framing members. Figure 10–13 shows cables installed parallel to framing members in a typical dwelling.

Connecting Conductors with Wire Nuts or Splice Caps

NEC® 110.14(B) lists several methods for splicing or connecting conductors together. The most popular method is to use a device called a **wire connector.** Such fittings have a coil of wire embedded inside a plastic cap to dig into the conductors when they are joined, thereby holding them tightly together. The different sizes of splice caps are listed for a maximum number and size of conductors that must not be exceeded if the cap is to make a good connection. Many electrical contractors require that the conductors be mechanically connected by twisting the conductors together before the splice cap is installed for

Residential Electrical Cabling Installation • 231

"WIREMAN'S GUIDE"
JOIST SPACE USED AS ENVIRONMENT AIRSPACE

SIDE VIEW OF COLD AIR RETURN METAL DUCT

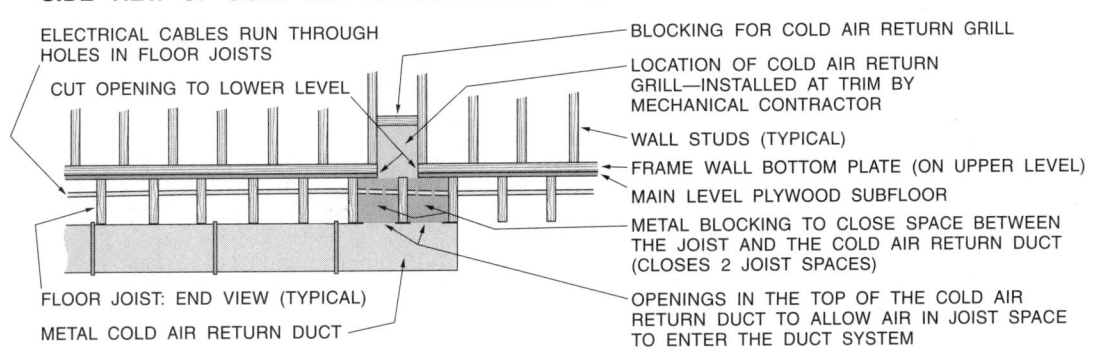

- ELECTRICAL CABLES RUN THROUGH HOLES IN FLOOR JOISTS
- CUT OPENING TO LOWER LEVEL
- BLOCKING FOR COLD AIR RETURN GRILL
- LOCATION OF COLD AIR RETURN GRILL—INSTALLED AT TRIM BY MECHANICAL CONTRACTOR
- WALL STUDS (TYPICAL)
- FRAME WALL BOTTOM PLATE (ON UPPER LEVEL)
- MAIN LEVEL PLYWOOD SUBFLOOR
- METAL BLOCKING TO CLOSE SPACE BETWEEN THE JOIST AND THE COLD AIR RETURN DUCT (CLOSES 2 JOIST SPACES)
- FLOOR JOIST: END VIEW (TYPICAL)
- METAL COLD AIR RETURN DUCT
- OPENINGS IN THE TOP OF THE COLD AIR RETURN DUCT TO ALLOW AIR IN JOIST SPACE TO ENTER THE DUCT SYSTEM

LOOKING DOWN FROM THE MAIN FLOOR WITH SUBFLOORING REMOVED

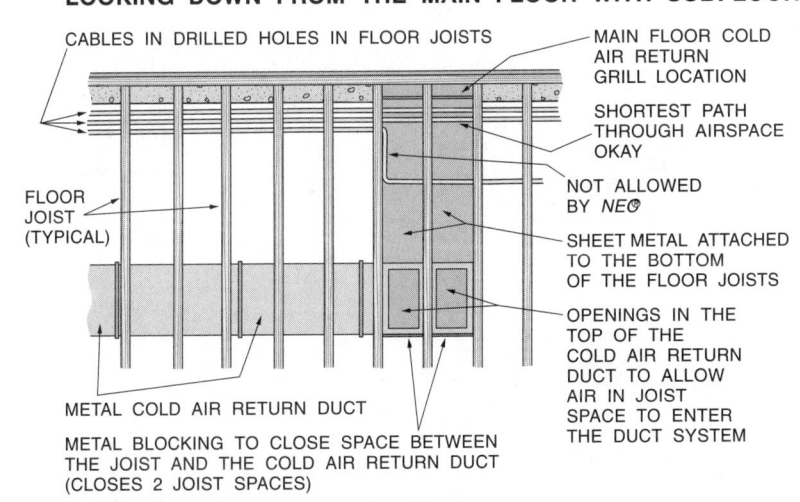

- CABLES IN DRILLED HOLES IN FLOOR JOISTS
- MAIN FLOOR COLD AIR RETURN GRILL LOCATION
- SHORTEST PATH THROUGH AIRSPACE OKAY
- NOT ALLOWED BY NEC
- SHEET METAL ATTACHED TO THE BOTTOM OF THE FLOOR JOISTS
- FLOOR JOIST (TYPICAL)
- OPENINGS IN THE TOP OF THE COLD AIR RETURN DUCT TO ALLOW AIR IN JOIST SPACE TO ENTER THE DUCT SYSTEM
- METAL COLD AIR RETURN DUCT
- METAL BLOCKING TO CLOSE SPACE BETWEEN THE JOIST AND THE COLD AIR RETURN DUCT (CLOSES 2 JOIST SPACES)

MECHANICAL OR HEATING AND COOLING CONTRACTORS SOMETIMES USE THE SPACE BETWEEN FLOOR JOISTS AS PART OF THE COLD AIR RETURN SYSTEM WHEN HEATING AND COOLING WITH FORCED AIR. THESE JOIST SPACES BECOME WHAT IS TERMED ENVIRONMENT AIRSPACES, AND ANY NONMETALLIC WIRING SYSTEM IN THOSE SPACES MUST TAKE THE SHORTEST ROUTE THROUGH THE SPACES, ACCORDING TO 300.22(C), EXCEPTION.

AN OPENING IS CUT INTO THE SUBFLOOR TO ALLOW ACCESS TO THE COLD AIR RETURN BETWEEN THE JOISTS IN THE LOWER LEVEL. THIS OPENING IS USUALLY INSIDE AN UPPER-LEVEL WALL STUD SPACE. THE WALL STUD SPACE IS BLOCKED OFF ABOVE THE GRILL LOCATION TO CLOSE THE COLD AIR RETURN SYSTEM.

THE LOCAL AHJ SHOULD BE CONSULTED TO ENSURE THAT THERE ARE NO LOCAL RESTRICTIONS REGARDING INSTALLING CABLES IN ENVIRONMENT AIRSPACES.

END VIEW OF COLD AIR DUCT AND JOIST SPACE

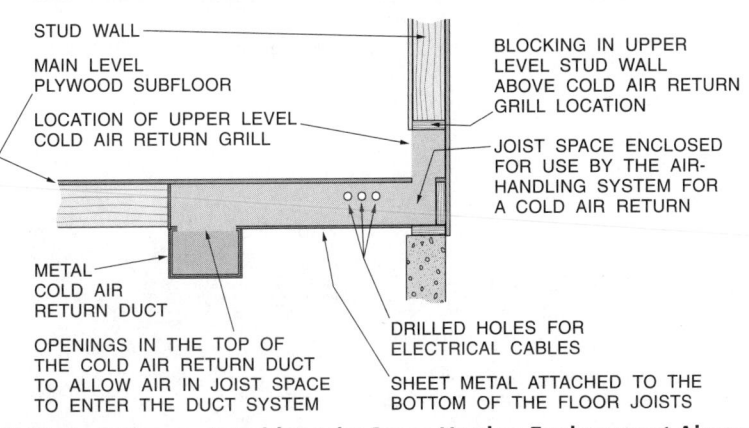

- STUD WALL
- MAIN LEVEL PLYWOOD SUBFLOOR
- LOCATION OF UPPER LEVEL COLD AIR RETURN GRILL
- BLOCKING IN UPPER LEVEL STUD WALL ABOVE COLD AIR RETURN GRILL LOCATION
- JOIST SPACE ENCLOSED FOR USE BY THE AIR-HANDLING SYSTEM FOR A COLD AIR RETURN
- METAL COLD AIR RETURN DUCT
- OPENINGS IN THE TOP OF THE COLD AIR RETURN DUCT TO ALLOW AIR IN JOIST SPACE TO ENTER THE DUCT SYSTEM
- DRILLED HOLES FOR ELECTRICAL CABLES
- SHEET METAL ATTACHED TO THE BOTTOM OF THE FLOOR JOISTS

FIGURE 10–9 "Wireman's Guide" Joist Space Used as Environment Airspace

232 • Chapter 10

"WIREMAN'S GUIDE" BUNDLING CABLES

NEC 334.80 STATES THAT TYPE NM CABLE SHALL BE CONSTRUCTED FROM CONDUCTORS WITH 90C INSULATION BUT THAT THE AMPACITY MUST COME FROM THE 60C COLUMN OF THE AMPACITY TABLES. ON THE 90C COLUMN OF *TABLE 310.16*, THE MAXIMUM ALLOWABLE AMPACITY OF 14 AWG 90C COPPER CONDUCTORS IS 25 AMPERES, AND FOR 12 AWG IT IS 30 AMPERES. *TABLE 310.15(B)(2)(a)* SPECIFIES THAT UP TO NINE CONDUCTORS CAN OCCUPY THE SAME RACEWAY WITH A 70% REDUCTION IN ALLOWABLE AMPACITY. THAT WOULD MEAN THAT 14 AWG CONDUCTORS ARE ALLOWED TO CARRY 17.5 AMPERES (25 A X .70 = 17.5 A) AND 12 AWG CONDUCTORS CAN CARRY UP TO 21.0 AMPERES (30 A X .70 = 21.0). SINCE THERE IS A 15-AMPERE LIMIT ON 14 AWG CIRCUIT AMPACITY AND A 20-AMPERE LIMIT ON 12 AWG CIRCUIT AMPACITY REQUIRED BY *240.4(D)*, IT WOULD SEEM THAT CABLES WITH A TOTAL OF NINE CURRENT-CARRYING CONDUCTORS WOULD BE ALLOWED TO OCCUPY THE SAME SET OF HOLES. THIS COULD BE FOUR 12-2 NM CABLES, THREE 14-2 CABLES AND ONE 14-3 CABLE, OR ANY OTHER POSSIBLE COMBINATION THAT TOTALS NINE OR FEWER CURRENT-CARRYING CONDUCTORS. THIS ANALYSIS MAY NOT SATISFY THE LOCAL INSPECTOR. IN MANY LOCALES, THE AHJ MAY CONSIDER THE SHEATHING ON THE CABLES A HINDRANCE TO COOLING THE CONDUCTORS AND WILL NOT ALLOW EVEN THIS MANY CABLES BUNDLED TOGETHER. *NEC® 310.15(B)(2) EXCEPTION 5* ALLOWS UP TO 20 CURRENT-CARRYING CONDUCTORS IN AC AND MC CABLES TO BE BUNDLED TOGETHER WITHOUT HAVING TO ADJUST THE ALLOWABLE AMPACITY. IF THERE IS A QUESTION CONCERNING HOW THE LOCAL ELECTRICAL INSPECTOR WILL ENFORCE THIS PROVISION OF THE CODE, THE AHJ SHOULD BE CONSULTED FOR MORE DETAILS.

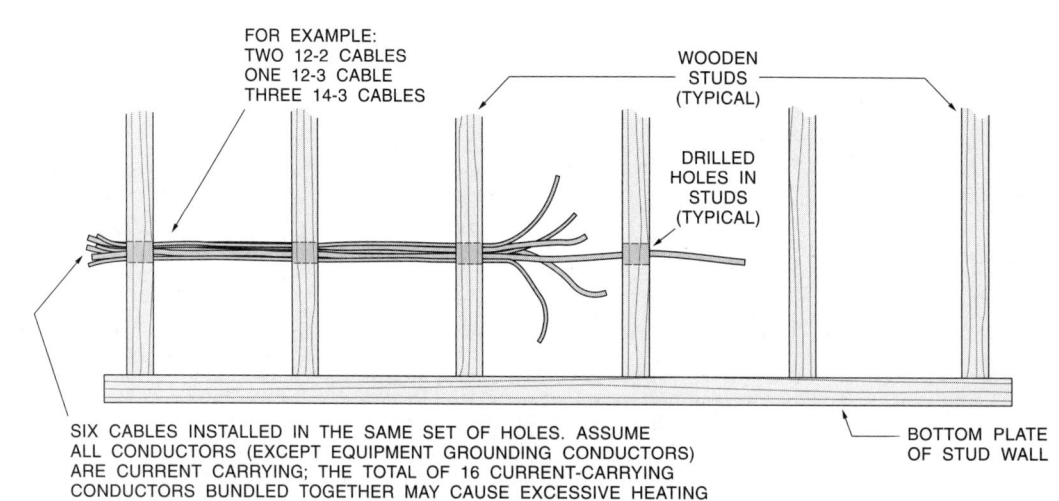

FOR EXAMPLE:
TWO 12-2 CABLES
ONE 12-3 CABLE
THREE 14-3 CABLES

WOODEN STUDS (TYPICAL)

DRILLED HOLES IN STUDS (TYPICAL)

BOTTOM PLATE OF STUD WALL

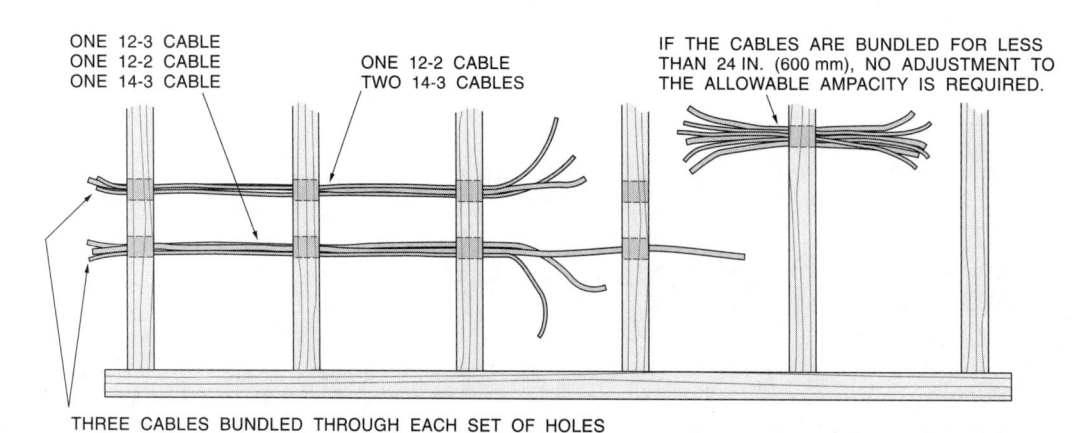

SIX CABLES INSTALLED IN THE SAME SET OF HOLES. ASSUME ALL CONDUCTORS (EXCEPT EQUIPMENT GROUNDING CONDUCTORS) ARE CURRENT CARRYING; THE TOTAL OF 16 CURRENT-CARRYING CONDUCTORS BUNDLED TOGETHER MAY CAUSE EXCESSIVE HEATING OF THE CONDUCTORS.

ONE 12-3 CABLE
ONE 12-2 CABLE
ONE 14-3 CABLE

ONE 12-2 CABLE
TWO 14-3 CABLES

IF THE CABLES ARE BUNDLED FOR LESS THAN 24 IN. (600 mm), NO ADJUSTMENT TO THE ALLOWABLE AMPACITY IS REQUIRED.

THREE CABLES BUNDLED THROUGH EACH SET OF HOLES

BY SPLITTING THE SIX CABLES INTO BUNDLES OF THREE CABLES EACH, THE POSSIBILITY OF OVERHEATING THE CONDUCTORS IS GREATLY REDUCED.

FIGURE 10–10 "Wireman's Guide" Bundling Cables

FIGURE 10–11 Bundled Cables Installed in TGI-Type Floor Joists

added protection against a loose connection. Figure 10–14 presents more information on use of these wire-splicing devices. Figure 10–15 is a close-up view of splices in conductors.

Pigtail Connections

Pigtail is the name given to a conductor that originates in a box, that does not leave the box, and that is included with other conductors in a splice. Pigtails provide an easy and effective means for tapping into a conductor system (such as the power conductors to connect a device or luminaire [lighting fixture]). For example, in a box containing a cable that provides line-side power (phase conductor) and neutral (grounded circuit conductor), another cable containing a load-side power and neutral, and a cable to supply a luminaire (lighting fixture) with switched power, there needs to be a way to tap into the power available in the power cables and to use that tap to operate the switch. A pigtail is an effective means of accomplishing the tap.

More than one pigtail can be taken from a splice bundle. In multigang switch boxes, several switches may all be powered by the same circuit. With such an arrangement, multiple pigtails are necessary in order for the power cable to provide the power to each of the switches. With other arrangements, a pigtail may be needed from each of the circuit conductors, power conductors, neutral conductors, equipment-grounding conductors, and switch leg conductors. Figure 10–16

234 • Chapter 10

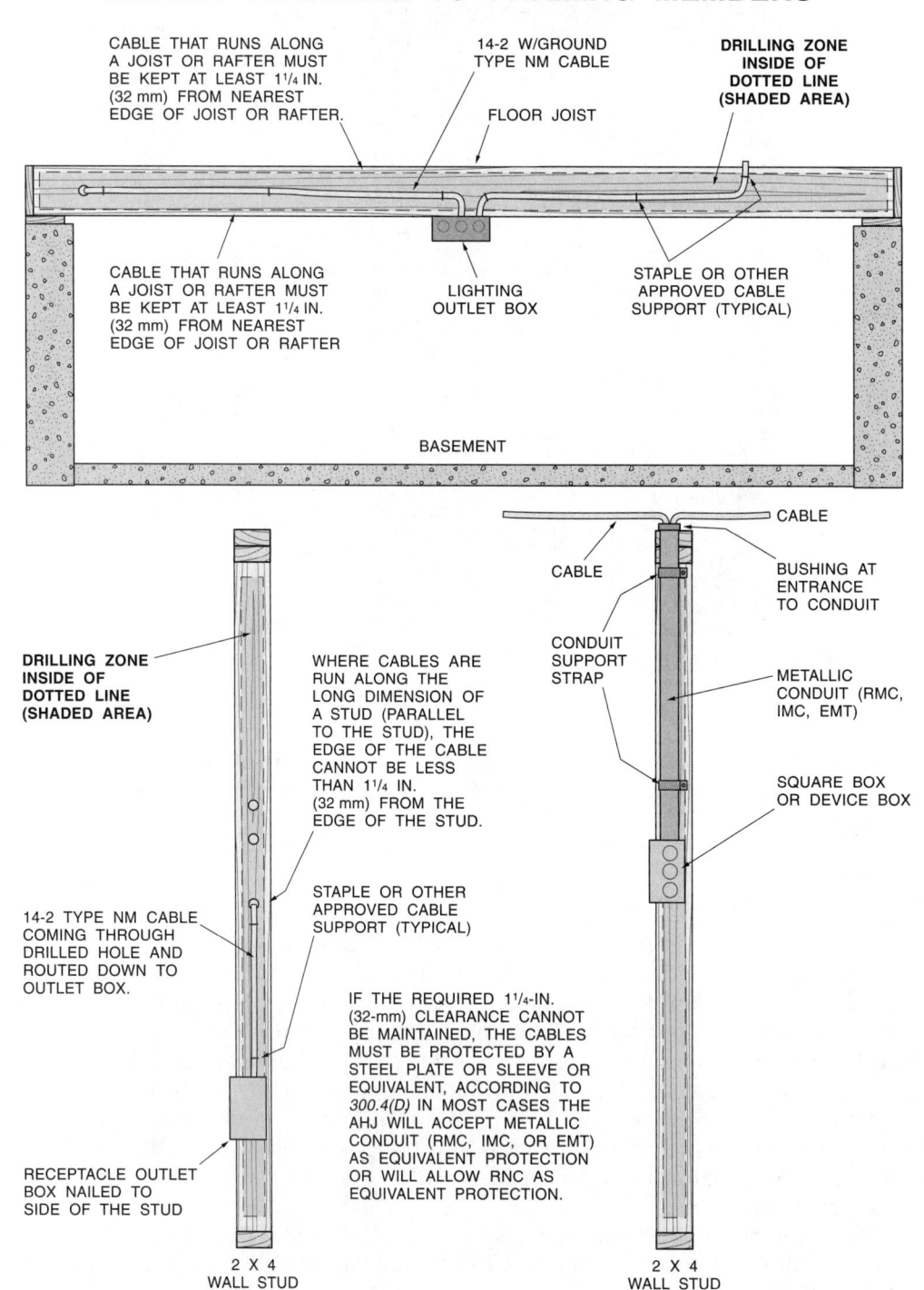

FIGURE 10–12 "Wireman's Guide" Cables Parallel to Framing Members

FIGURE 10–13 Cables Run Parallel with Framing Members

details how to make a pigtail. The equipment-grounding conductor is used as an example of the pigtail assembly process, but the same techniques are also used in creating pigtails of other types of conductors.

Identification of Conductors

There are several different uses for cable in any device or switch point box. The cable may contain a phase conductor (power) and a grounded circuit conductor (neutral), or a switch leg and a neutral cable, or a switch leg and power for the switch, or a cable that contains travelers with a neutral, a switch leg, or a power feed to the switch. Each of these conductors must be positively identified at each box during cable installation in order to ensure proper box makeup.

Box makeup is the act of preparing the box to receive the switch or receptacle to be installed at trim. The electrician who installs the devices (several weeks or months after rough-in) must know exactly what each conductor is being used for, and what conductors to connect to which terminals on the switch or receptacle. A white wire may be a traveler, or it may be a power (phase) conductor in some circuits. A red conductor may be a switch leg, a power, or a traveler. A black conductor may be a switch leg or a power conductor. A system to positively identify each conductor in a box is needed in order to eliminate confusion and error-plagued installations. The particular system used is not of major importance, but a systematic approach to conductor identification is essential. It is also important that everyone working on the project use the same coding system, at both rough-in and trim-out. The system presented in this chapter is simple and is the standard in many parts of the country and with many different contractors, although it is not the only system employed. This system can be used with wiring methods using Type NM cable as well as AC and MC cable, or other wiring methods. The system specifies five rules to be followed in making up a box at rough-in. The terms used in these five rules are the same terms that are employed in the field to

236 • Chapter 10

"WIREMAN'S GUIDE"
SPLICING WITH WIRE NUTS OR SPLICE CAPS

WIRE CONNECTIONS MAY BE MADE EITHER BY SOLDERING OR BY THE USE OF SOLDERLESS CONNECTORS. THE *CODE* ALLOWS SOLDERED CONNECTIONS BUT REQUIRES THAT THE CONNECTION BE MECHANICALLY SOUND BEFORE THE SOLDER IS APPLIED—THE SPLICE CANNOT DEPEND SOLELY ON THE SOLDER FOR THE CONNECTION. SOLDERLESS CONNECTIONS MAY BE MADE BY EXOTHERMIC WELDING OR WITH THE USE OF CONNECTORS SUCH AS LUGS OR CRIMP-ON SLEEVES. THE MOST POPULAR CONNECTORS ARE SCREW-ON CONNECTORS, COMMONLY REFERRED TO AS WIRE NUTS OR SPLICE CAPS.

WIRE NUTS OR SPLICE CAPS COME IN A NUMBER OF SIZES THAT CAN ACCOMMODATE ALMOST ANY SPLICING REQUIREMENT FOR THE CONDUCTORS COMMONLY FOUND IN DWELLING UNITS, EXCEPT FOR THE SERVICE AND DISTRIBUTION SYSTEM. YELLOW IS THE SMALLEST PRACTICAL WIRE NUT FOR EVERYDAY WIRING. THERE IS A SMALLER SERIES OF WIRE NUTS INCLUDING A BLUE WIRE NUT FOR SPLICING CONDUCTORS SMALLER THAN 14 AWG. THERE IS ALSO A BLUE WIRE NUT THAT IS LARGER THAN THE GRAY SIZE. YELLOW AND RED ARE THE MOST COMMONLY USED SIZES FOR 12 AWG AND 14 AWG CONDUCTORS. THERE IS EVEN A GREEN ONE WITH A HOLE IN THE SMALL END TO ASSIST IN THE MAKEUP OF THE EQUIPMENT GROUNDING CONDUCTORS.

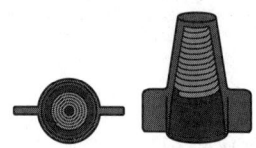

A CUT-AWAY SIDE VIEW OF A TYPICAL WIRE NUT IS SHOWN AT LEFT. IT CONSISTS OF A METAL COIL, OR CONE, THAT IS TIGHTLY WRAPPED WITH SHARP EDGES EXPOSED TO THE INTERIOR OF THE CONE. PLASTIC OR OTHER INSULATING MATERIAL IS MOLDED OVER THE METAL CONE SO THAT THE CONE IS HELD FAST BY THE PLASTIC. THE LARGE END OF THE CONE IS LEFT EXPOSED TO RECEIVE THE WIRES.

WIRE NUTS ARE SAFE, EASY TO USE, AND RELATIVELY INEXPENSIVE. THE WIRES TO BE SPLICED ARE FORMED INTO A NEAT BUNDLE WITH THE INSULATION REMOVED FROM ALL OF THE CONDUCTORS FOR APPROXIMATELY 1/2 IN. (.013 m) TO 1 IN. (.0254 m) AND THE TOPS OF ALL THE CONDUCTORS AT THE SAME LEVEL. THE PROPERLY SIZED WIRE NUT IS THEN TWISTED ONTO THE EXPOSED CONDUCTORS WITH A CLOCKWISE TURN AND DOWNWARD PRESSURE APPLIED UNTIL THE WIRE NUT IS TIGHT. MAKE SURE TO FOLLOW ANY AND ALL INSTALLATION INSTRUCTIONS THAT MAY ACCOMPANY THE WIRE NUTS AT PURCHASE. ALSO, IT IS VERY IMPORTANT THAT THE WIRE NUT BE PROPERLY SIZED.

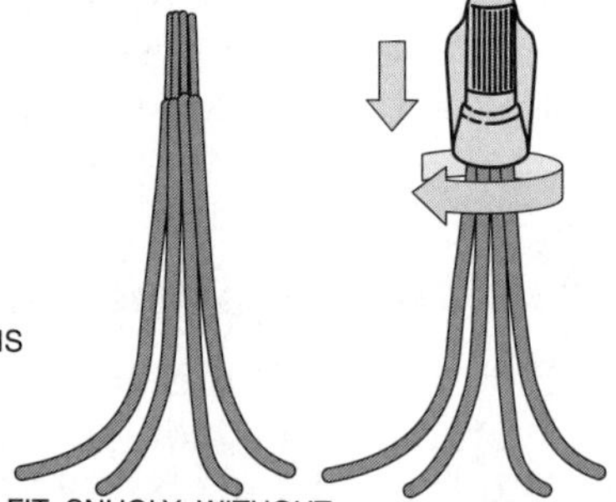

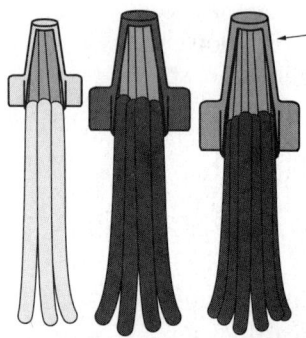

PROPERLY SIZED WIRE NUTS FIT SNUGLY WITHOUT BEING FORCED ONTO THE WIRES. THE LARGER THE CONDUCTOR SIZE, OR THE MORE CONDUCTORS IN THE SPLICE, THE LARGER THE WIRE NUT HAS TO BE. PROPERLY SIZED AND INSTALLED WIRE NUTS WILL PROVIDE A SAFE AND SECURE SPLICE.

IN MANY LOCALES, THE AHJ MAY REQUIRE THAT THE CONDUCTORS BE TWISTED TOGETHER BEFORE THE WIRE NUT IS INSTALLED. THE SPLICE MUST BE MECHANICALLY SECURE BEFORE THE WIRE NUT IS INSTALLED. THIS HAS THE EFFECT OF MAKING THE WIRE NUT NOTHING MORE THAN AN INSULATING FITTING AND A BACKUP CONNECTION METHOD TO THE TWISTED SPLICE.

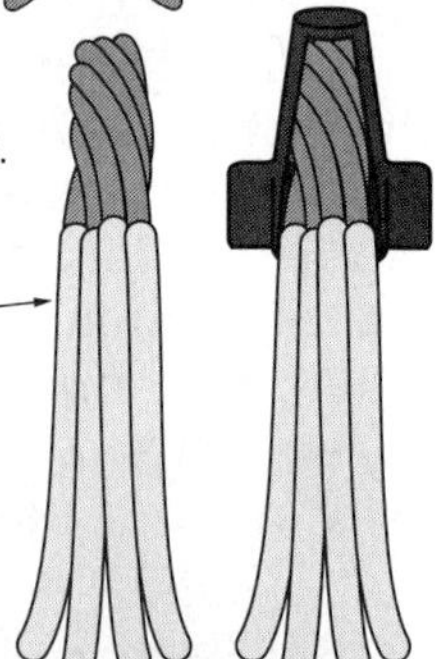

SOME COMMON PROBLEMS EXPERIENCED WITH WIRE NUT INSTALLATION:
- OVERSIZING THE WIRE NUT. TOO MUCH SPACE INSIDE THE WIRE NUT CAN YIELD A LOOSE CONNECTION.
- OVERSTRIPPING THE CONDUCTORS. THE BARE CONDUCTOR IS THEN TOO LONG TO BE COVERED BY THE WIRE NUT, THUS LEAVING BARE CONDUCTORS EXPOSED.
- UNDERSIZING THE WIRE NUT. USE OF TOO MANY CONDUCTORS OR CONDUCTORS THAT ARE TOO LARGE CAN YIELD A LOOSE CONNECTION.

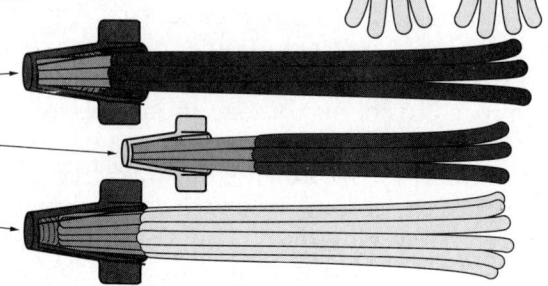

FIGURE 10–14 "Wireman's Guide" Splicing with Wire Nuts or Splice Caps

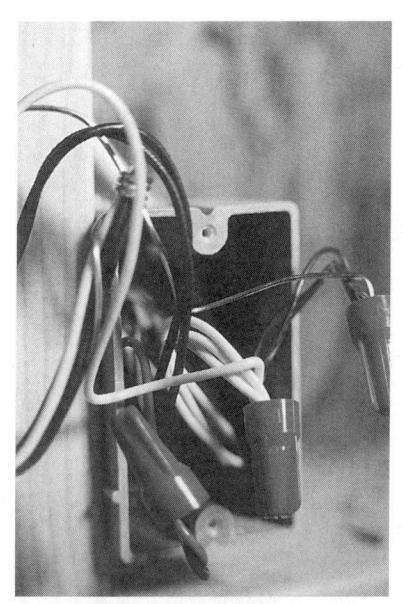

FIGURE 10–15 Conductor Makeup in a Box. These conductors are spliced using wire connectors.

identify each conductor. Brief definitions of these terms as used in this identification system follow:

- *Power:* The phase conductor—*not* the phase conductor plus the neutral. Neutrals are considered separately.
- *Neutral:* The grounded circuit conductor.
- *Switch leg:* The conductor that carries the power to the equipment or appliance using the power (the load) if the load is controlled by a switch of any kind. Again, the neutral required by the load is considered separately.
- *Traveler:* Two of the three conductors needed to complete a three-way or four-way switching system. A traveler is not to be connected to the common terminal of a three-way switch.
- *In:* The line side of the box—the conductors entering the box. For example, power in means the line-side phase conductor.
- *Out:* The load side of the box. For example, the switch leg out is the conductor that powers the load when the switch is closed. In most circumstances, a power in is accompanied by a neutral in and a power out is accompanied by a neutral out, but not always.

These six terms describe all of the possible combinations of conductors that can exist in a box. The following five rules apply to roughing-in cables, either non-metallic-sheathed or metallic-sheathed:

- **Rule 1:** If the conductor is a power in, a power out, a neutral in, a neutral out, or an equipment-grounding conductor, no further identification is necessary. The neutrals are identified by their white or gray coloring, the power conductors are identified by their black or red coloring, and the equipment-grounding conductor is identified by the green coloring or by the absence of insulation (bare conductor).

238 • Chapter 10

"WIREMAN'S GUIDE"
PIGTAILS AND GROUND SCREW CONNECTIONS

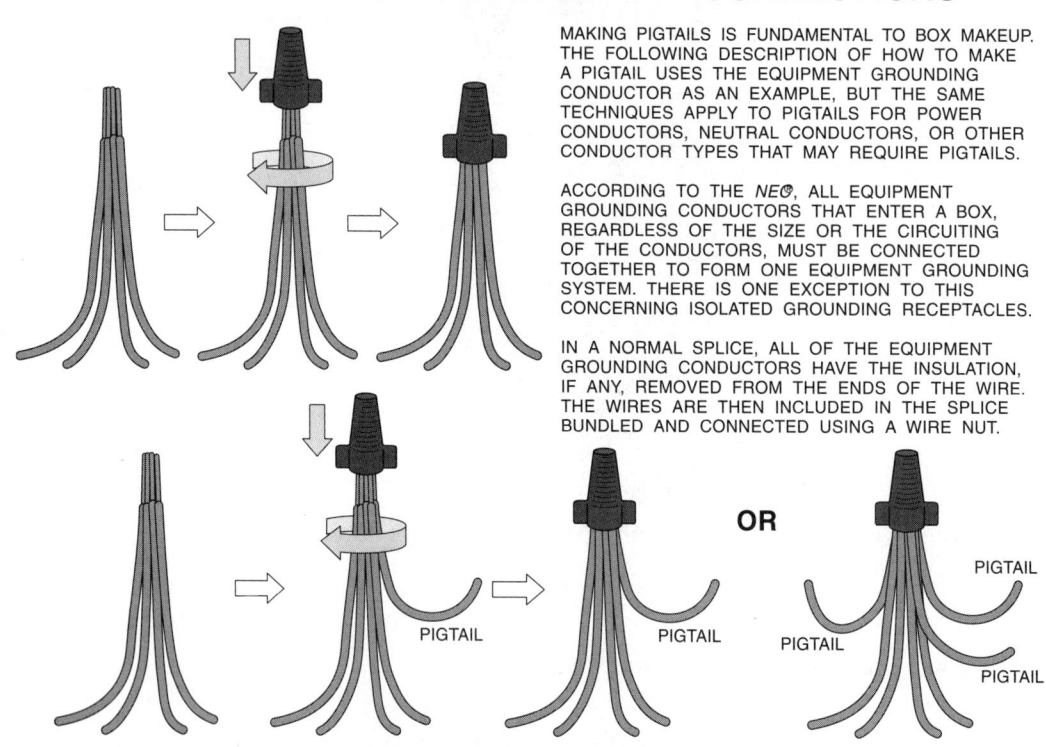

MAKING PIGTAILS IS FUNDAMENTAL TO BOX MAKEUP. THE FOLLOWING DESCRIPTION OF HOW TO MAKE A PIGTAIL USES THE EQUIPMENT GROUNDING CONDUCTOR AS AN EXAMPLE, BUT THE SAME TECHNIQUES APPLY TO PIGTAILS FOR POWER CONDUCTORS, NEUTRAL CONDUCTORS, OR OTHER CONDUCTOR TYPES THAT MAY REQUIRE PIGTAILS.

ACCORDING TO THE *NEC*, ALL EQUIPMENT GROUNDING CONDUCTORS THAT ENTER A BOX, REGARDLESS OF THE SIZE OR THE CIRCUITING OF THE CONDUCTORS, MUST BE CONNECTED TOGETHER TO FORM ONE EQUIPMENT GROUNDING SYSTEM. THERE IS ONE EXCEPTION TO THIS CONCERNING ISOLATED GROUNDING RECEPTACLES.

IN A NORMAL SPLICE, ALL OF THE EQUIPMENT GROUNDING CONDUCTORS HAVE THE INSULATION, IF ANY, REMOVED FROM THE ENDS OF THE WIRE. THE WIRES ARE THEN INCLUDED IN THE SPLICE BUNDLED AND CONNECTED USING A WIRE NUT.

ONE OR MANY PIGTAILS CAN BE CONNECTED TO THE SPLICE BUNDLE. THIS WIRE IS THEN CONNECTED TO THE EQUIPMENT GROUNDING SYSTEM, AND THE END OF THE PIGTAIL CAN BE CONNECTED TO A BOX OR TO A DEVICE. IF MORE THAN ONE PIGTAIL IS INCLUDED IN THE BUNDLE, THERE COULD BE A SEPARATE PIGTAIL FOR EACH OF THE ITEMS THAT NEED CONNECTION TO THE GROUNDING SYSTEM.

THE *NEC* ALSO REQUIRES THAT ALL METALLIC BOXES BE CONNECTED TO THE EQUIPMENT GROUNDING CONDUCTOR SYSTEM. THIS IS ACCOMPLISHED BY INSTALLING AN EQUIPMENT GROUNDING SCREW INTO A THREADED HOLE IN THE BACK OF THE METAL BOX. THERE ARE TWO POSSIBILITIES OR A COMBINATION OF POSSIBILITIES MAKING THESE CONNECTIONS:
1. A SEPARATE WIRE FOR THE PIGTAIL IS INCLUDED IN THE SPLICE WITH THE OTHER CONDUCTORS. THE PIGTAIL WILL THEN BE CONNECTED TO THE BOX USING THE GROUNDING SCREW.
2. STRIP ABOUT 1 IN. (.0254 m) OF INSULATION OFF THE CENTER OF ONE OF THE EQUIPMENT GROUNDING CONDUCTORS, AND INSTEAD OF GOING DIRECTLY TO THE SPLICING BUNDLE, LOOP THE CONDUCTOR AROUND THE GROUNDING SCREW FIRST. WHEN THE SCREW IS TIGHTENED IT WILL SECURE THE BOX TO THE EQUIPMENT GROUNDING SYSTEM.

THE *CODE* ALSO REQUIRES THAT AN EQUIPMENT GROUNDING SYSTEM CONNECTION BE MADE TO ALL DEVICES INSTALLED IN THE BOX. THIS IS BEST ACCOMPLISHED USING A PIGTAIL. IF THE DEVICE IS REMOVED FOR MAINTENANCE OR INSPECTION THE EQUIPMENT GROUNDING CONNECTION WILL REMAIN INTACT.

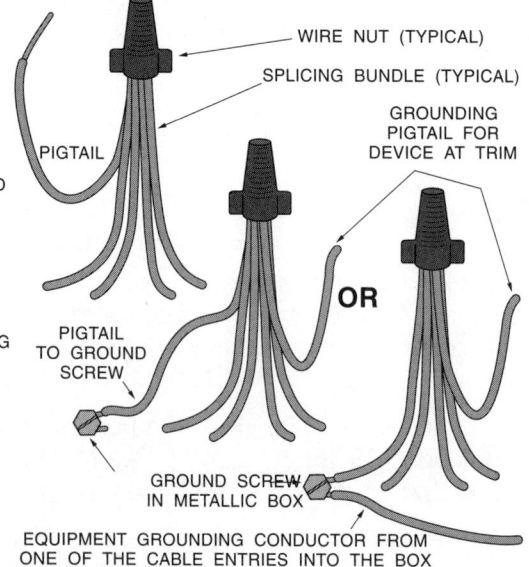

FIGURE 10–16 "Wireman's Guide" Pigtails and Ground Screw Connections

- **Rule 2:** If the conductor is a switch leg or if the conductor is common in a three-way or four-way switching system, strip about 1.2 in. (13 mm) of insulation from the end of the conductor.
- **Rule 3:** If the conductor is one of a pair of travelers, locate the other traveler of the pair and lightly twist (using approximately one twist for each inch [26 mm] of conductor) the two travelers together.
- **Rule 4:** If the conductor is an out (load side) of a ground-fault circuit-interrupter device or the out (load side) of an arc-fault circuit-interrupter device, strip about 1.2 in. (13 mm) of the insulation off both the load-side power and the loadside neutral conductors. Associate the power in and the neutral out by lightly twisting (using one twist for each inch [26 mm]) the two neutral and the two power conductors together.
- **Rule 5:** If the conductor is white or gray in color but is being used as something other than a neutral in a single-pole or two-pole, three-way or four-way switching system, permanently re-identify the conductor with tape, paint, or other effective means at each box where the conductor is visible, and at each termination, to indicate its use.

It is obvious from the rules that the electrician doing the rough-in must be aware of the use assigned to each conductor as the cables are being pulled. The rough-in electricians must have a wiring plan, and that plan must be consistently followed or errors will undoubtedly occur. For example, if a switch leg for a luminaire (lighting fixture) is not properly identified when installed into the box, it is possible for the switch leg to be mistaken for a power conductor; then the luminaire (lighting fixture) that it supplies will not be switched and will burn all the time. More information about the conductor identification system is presented in Figure 10–17.

Organizing the Box

The system for box organization presented in this book (see Figure 10–18) is certainly not the only system. However, it is reasonably efficient and provides satisfactory results. Regardless of the system employed, good box organization is critical to device installation and an efficient trim. Conductors occupy a considerable amount of space within the box, and many devices such as GFCI receptacles and dimmer switches take up a lot of space. The conductors should not have to be excessively compressed to make room in the box for the device, and the device should not have to be forced into the box because of the location of the conductors. Additionally, the trim electrician should be able to pull out of the box just those conductors that are needed to terminate to the device, and the other conductors should remain in the box where they were placed at makeup during rough-in. When the device is connected, it and the conductors it is connected to should install into the box with minimal effort.

An efficient way for the conductors attached to the device to be stored in the box is in accordion fashion. This arrangement allows the conductors some flexibility so that the device can easily be pulled from the box and easily returned to the box. Use of this method allows the device to be easily positioned both horizontally and vertically for proper alignment with the mounting screw holes of the box. This method also allows the trim electrician to remove only those conductors necessary for connection to the device, and does not require expenditure of time on unrolling or reshaping the conductors.

"WIREMAN'S GUIDE"
CONDUCTOR IDENTIFICATION AND MAKEUP

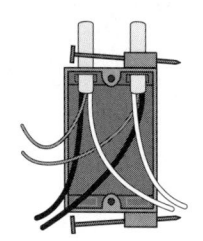

RULE 1: IF THE CONDUCTOR IS A LINE-SIDE POWER, A LOAD-SIDE POWER, A LINE-SIDE NEUTRAL, A LOAD-SIDE NEUTRAL, OR AN EQUIPMENT GROUNDING CONDUCTOR, NO FURTHER IDENTIFICATION IS NECESSARY.

THE BOX AT LEFT CONTAINS TWO 14-2 CABLES. IT CAN BE IDENTIFIED AS A RECEPTACLE OUTLET BOX BECAUSE ALL OF THE POWERS AND NEUTRALS ARE UNMARKED.

TO COMPLETE THE MAKEUP OF THE BOX: CONNECT AND PIGTAIL THE EQUIPMENT GROUNDING CONDUCTORS. THE TWO POWER CONDUCTORS, THE TWO NEUTRAL CONDUCTORS, AND THE EQUIPMENT GROUNDING CONDUCTOR PIGTAIL WILL TERMINATE ON THE RECEPTACLE.

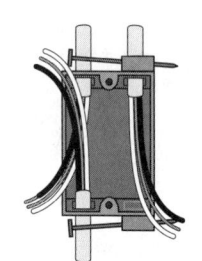

RULE 2: IF THE CONDUCTOR IS A SWITCH LEG, OR IF THE CONDUCTOR IS A COMMON IN A THREE-WAY OR FOUR-WAY SWITCHING SYSTEM, REMOVE APPROXIMATELY 1/2 IN. (13 mm) OF INSULATION FROM THE END OF THE CONDUCTOR.

THE BOX AT LEFT CONTAINS THREE 14-2 CABLES. IT CAN BE IDENTIFIED AS A SINGLE-POLE SWITCH POINT BOX BECAUSE ONE OF THE BLACK CONDUCTORS HAS HAD THE INSULATION REMOVED TO IDENTIFY IT AS A SWITCH LEG.

TO COMPLETE THE MAKEUP OF THE BOX: CONNECT AND PIGTAIL ALL OF THE EQUIPMENT GROUNDING CONDUCTORS. CONNECT ALL OF THE NEUTRAL CONDUCTORS. CONNECT AND PIGTAIL THE TWO UNIDENTIFIED POWER CONDUCTORS. THE POWER PIGTAIL, THE SWITCH LEG, AND THE EQUIPMENT GROUNDING CONDUCTOR PIGTAIL ALL TERMINATE ON THE SINGLE-POLE SWITCH.

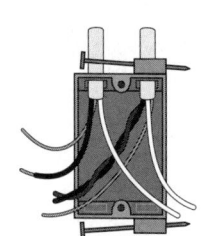

RULE 3: IF THE CONDUCTOR IS ONE OF A PAIR OF TRAVELERS, LOCATE THE OTHER TRAVELER AND LIGHTLY TWIST THE TWO TRAVELERS TOGETHER.

THE BOX AT LEFT CONTAINS ONE 14-2 CABLE AND ONE 14-3 CABLE. THE TRAVELERS ARE TWISTED TOGETHER TO IDENTIFY THEM AS A TRAVELER PAIR. NOTICE THAT THE POWER CONDUCTOR FROM THE 2-WIRE CABLE HAS THE INSULATION REMOVED FROM THE END, IDENTIFYING IT AS THE COMMON CONDUCTOR TO THE TRAVELERS.

TO COMPLETE THE MAKEUP OF THE BOX: CONNECT AND PIGTAIL THE EQUIPMENT GROUNDING CONDUCTORS. CONNECT THE TWO NEUTRAL CONDUCTORS. LIGHTLY TWIST THE COMMON CONDUCTOR AROUND THE TWO TRAVELERS TO ASSOCIATE THE COMMON WITH THE TRAVELERS (IN MULTIGANG BOXES THERE MAY BE MORE THAN ONE TRAVELER PAIR AND MORE THAN ONE COMMON). THE TWO TRAVELERS, THE COMMON, AND THE EQUIPMENT GROUNDING PIGTAIL ALL TERMINATE ON THE THREE-WAY SWITCH.

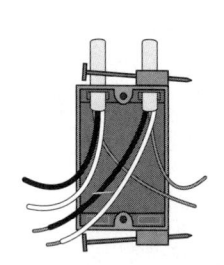

RULE 4: IF THE CONDUCTOR IS A LOAD-SIDE CONDUCTOR OF A GFCI, REMOVE ABOUT 1/2 IN. (16 mm) OF THE INSULATION FROM THE END OF BOTH THE POWER CONDUCTOR AND THE NEUTRAL CONDUCTOR. ASSOCIATE THE LOAD-SIDE POWER AND THE LOAD-SIDE NEUTRAL CONDUCTORS BY LIGHTLY TWISTING THEM TOGETHER. IF THE CONDUCTOR IS A LINE SIDE FOR A GFCI DEVICE THAT HAS NO LOAD CONDUCTORS (ONLY ONE CABLE IN THE BOX), REMOVE APPROXIMATELY 1/2 IN. (16 mm) OF THE INSULATION ON BOTH THE POWER AND THE NEUTRAL AS NOTIFICATION THAT THIS IS A SPECIAL RECEPTACLE.

THE BOX AT LEFT CONTAINS TWO 14-2 CABLES. THE POWER AND THE NEUTRAL FROM ONE OF THE CABLES HAVE HAD THE INSULATION REMOVED FROM THE END, IDENTIFYING THEM AS LOAD-SIDE CONDUCTORS FROM A GFCI RECEPTACLE. THE DETERMINATION OF WHICH DEVICE TO INSTALL CAN BE MADE BY THE LOCATION OF THE BOX.

TO COMPLETE THE MAKEUP OF THE BOX: CONNECT AND PIGTAIL THE EQUIPMENT GROUNDING CONDUCTORS. THE TWO POWER CONDUCTORS, THE TWO NEUTRAL CONDUCTORS, AND THE EQUIPMENT GROUNDING PIGTAIL WILL ALL TERMINATE ON THE DEVICE.

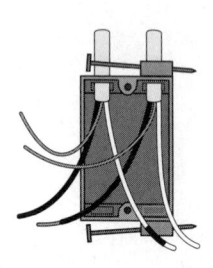

RULE 5: IF THE CONDUCTOR IS WHITE OR GRAY IN COLOR BUT IS BEING USED AS SOMETHING OTHER THAN A GROUNDED CIRCUIT CONDUCTOR, PERMANENTLY RE-IDENTIFY THE CONDUCTOR WITH TAPE, PAINT, OR OTHER EFFECTIVE MEANS.

THE BOX TO THE LEFT CONTAINS TWO 14-2 CABLES. THE NEUTRAL CONDUCTOR FROM ONE OF THE CABLES IS MARKED WITH BLACK TAPE AND THE POWER CONDUCTOR FROM THE SAME CABLE IS MARKED AS A SWITCH LEG. THE WHITE CONDUCTOR IS BEING USED AS A POWER TO SUPPLY A SINGLE-POLE SWITCH (IN ANOTHER BOX) AND THE RETURN BLACK CABLE IS BEING USED AS THE SWITCH LEG. THIS INDICATES THAT THE RECEPTACLE TO BE INSTALLED IN THIS DEVICE BOX IS SWITCH CONTROLLED.

TO COMPLETE THE MAKEUP OF THE BOX: CONNECT AND PIGTAIL THE EQUIPMENT GROUNDING CONDUCTORS AND CONNECT THE RE-IDENTIFIED WHITE CONDUCTOR WITH THE LINE-SIDE POWER CONDUCTOR FROM THE OTHER CABLE. THE REMAINING NEUTRAL CONDUCTOR, THE SWITCH LEG, AND THE EQUIPMENT GROUNDING PIGTAIL ARE TO BE TERMINATED ON THE DEVICE.

FIGURE 10–17 "Wireman's Guide" Conductor Identification and Makeup

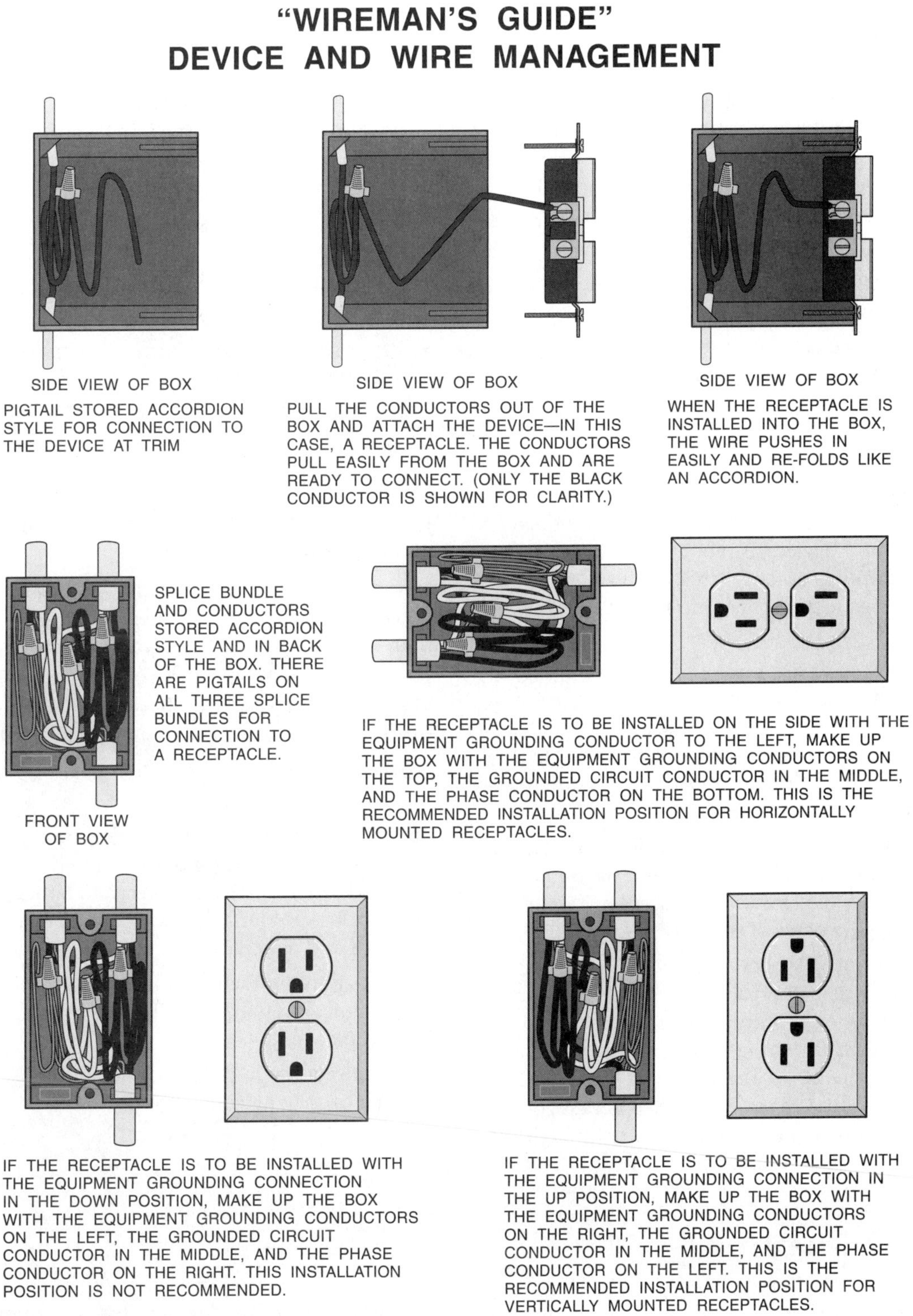

Residential Electrical Cabling Installation • 241

"WIREMAN'S GUIDE" DEVICE AND WIRE MANAGEMENT

SIDE VIEW OF BOX
PIGTAIL STORED ACCORDION STYLE FOR CONNECTION TO THE DEVICE AT TRIM

SIDE VIEW OF BOX
PULL THE CONDUCTORS OUT OF THE BOX AND ATTACH THE DEVICE—IN THIS CASE, A RECEPTACLE. THE CONDUCTORS PULL EASILY FROM THE BOX AND ARE READY TO CONNECT. (ONLY THE BLACK CONDUCTOR IS SHOWN FOR CLARITY.)

SIDE VIEW OF BOX
WHEN THE RECEPTACLE IS INSTALLED INTO THE BOX, THE WIRE PUSHES IN EASILY AND RE-FOLDS LIKE AN ACCORDION.

FRONT VIEW OF BOX
SPLICE BUNDLE AND CONDUCTORS STORED ACCORDION STYLE AND IN BACK OF THE BOX. THERE ARE PIGTAILS ON ALL THREE SPLICE BUNDLES FOR CONNECTION TO A RECEPTACLE.

IF THE RECEPTACLE IS TO BE INSTALLED ON THE SIDE WITH THE EQUIPMENT GROUNDING CONDUCTOR TO THE LEFT, MAKE UP THE BOX WITH THE EQUIPMENT GROUNDING CONDUCTORS ON THE TOP, THE GROUNDED CIRCUIT CONDUCTOR IN THE MIDDLE, AND THE PHASE CONDUCTOR ON THE BOTTOM. THIS IS THE RECOMMENDED INSTALLATION POSITION FOR HORIZONTALLY MOUNTED RECEPTACLES.

IF THE RECEPTACLE IS TO BE INSTALLED WITH THE EQUIPMENT GROUNDING CONNECTION IN THE DOWN POSITION, MAKE UP THE BOX WITH THE EQUIPMENT GROUNDING CONDUCTORS ON THE LEFT, THE GROUNDED CIRCUIT CONDUCTOR IN THE MIDDLE, AND THE PHASE CONDUCTOR ON THE RIGHT. THIS INSTALLATION POSITION IS NOT RECOMMENDED.

IF THE RECEPTACLE IS TO BE INSTALLED WITH THE EQUIPMENT GROUNDING CONNECTION IN THE UP POSITION, MAKE UP THE BOX WITH THE EQUIPMENT GROUNDING CONDUCTORS ON THE RIGHT, THE GROUNDED CIRCUIT CONDUCTOR IN THE MIDDLE, AND THE PHASE CONDUCTOR ON THE LEFT. THIS IS THE RECOMMENDED INSTALLATION POSITION FOR VERTICALLY MOUNTED RECEPTACLES.

FIGURE 10–18 "Wireman's Guide" Device and Wire Management

Summary

- The requirements for the drilling and notching of framing members are necessary to protect the cables from physical damage and to ensure the structural integrity of a building.
- Different rules apply for drilled holes and cut notches.
- Cables should not be bundled when installed in holes or notches.
- A system needs to be used for identifying conductors for box makeup and trim.
- The system must positively identify each conductor, power, switch leg, neutral, and equipment-grounding conductor at every outlet or switch point box.
- Cables are usually spliced using wire connectors.
- Circuit taps can be accomplished by using pigtails from splices.
- Proper box makeup and wire management will ensure a trouble-free and efficient trim.

Key Terms

Box makeup   The act of preparing the conductors contained in a device box for the installation of a device. Makeup is accomplished with the intent of making subsequent installation of the device as easy as possible.

Cut notches   Sections along the edge of a joist or stud for the installation of cables or raceways. Notches must be cut with a saw, and usually the opening must be covered with a plate to protect the cable or raceway. Notches must meet the requirements of the *NEC®* and the UBC.

Drilled holes   Holes that are drilled in framing members for the installation of electrical conductors or cables. Holes that are drilled in studs or joists must meet the requirements of the *NEC®* and the UBC.

Environmental airspaces   Airspaces within a structure that are intended as part of the structure's heating and cooling systems. These areas are used to circulate the air through the furnace or air conditioner and then return the air to the rooms of the building. Wiring within these areas is restricted because of the possibility of the rapid spreading of fire through the air-handling system should a fire occur.

Joist   A framing member that makes up part of a flooring system.

Notch plate   A metal plate that is installed over a notch or over a hole that is closer than 11.4 in. (32 mm) from the edge of a framing member in order to protect the cables or raceway.

Pigtail   (as referring to splices) An extra conductor that is added to a splice for connection to a device.

Rafter   A framing member that makes up part of a roof support system.

Splice   The act of connecting two individual conductors together to form one continuous conductor, or the location of that connection. Splices usually occur in boxes and are accomplished using proper methods and materials.

Stud   A framing member that makes up a part of a wall. Framed walls also usually have a top plate and a bottom plate.

Wire connector   A fitting allowing for a solderless method of splicing conductors.

Review Questions

1. According to the *NEC®*, a notch plate must be installed to protect the cable when the edge of a drilled hole is within _____ inches to the edge of a wall stud.

2. According to the UBC, a notch can remove up to _____ percent of the depth of a wall stud if it is a bearing wall.

3. According to the UBC, a notch cannot be located within the middle _____ of a joist or rafter.

4. Holes cannot be made within _____ inches of the edge of a joist or rafter.

5. According to the *NEC*®, a cable that is run parallel to a stud must be located at least _____ inches from the edge of the stud.

6. Explain the requirements for running nonmetallic cables within an environmental air space.

7. What is the purpose of a wire connector?

8. The method of tapping a splice is called a _____.

9. According to the conductor identification system used in this book, how are power (phase) conductors identified?

10. True or False: A white conductor is only used as a grounded circuit conductor, and cannot be re-identified for other uses.

11. True or False: According to the *NEC*®, no hole can be drilled less than 5/8 in. (16 mm) from the edge of a wall stud.

12. True or False: Conductors must be twisted together before splicing with a wire connector.

13. True or False: No hole larger than 2 in. (50 mm) can be drilled in a floor joist or ceiling rafter.

14. True or False: According to the UBC, no hole can be drilled so that its edge is closer than 2 in. (50 mm) from the edge of a rafter or joist.

15. True or False: The length of the rafter determines the maximum size of a drilled hole.

Cabling Standards

OBJECTIVES   *After studying this chapter, you should be able to:*

- Describe the standards that apply to residential cabling systems.
- Understand the architecture of residential cabling systems.
- Compare the architectures of residential and commercial building cabling systems.
- Distinguish between the various categories of UTP cable.
- Distinguish between the various types of coaxial cable.
- Explain the difference between a permanent link and a channel.
- Understand the grades of residential cabling.
- List the NFPA fire ratings for copper and fiber cables.
- Understand the grounding requirements for residential cabling.

OUTLINE   Introduction

TIA-568 Series

TIA-570-B

Environmental and Safety Issues

International Cabling Standards

Introduction

The Telecommunications Industry Association (TIA) has issued a number of standards that specify how cabling technology should be used to provide the telecommunications infrastructure in various types of buildings. For residential buildings, the most important of these standards is TIA-570-B, Residential Telecommunications Infrastructure Standard (April, 2004). TIA-570-B makes use of technology and architecture that are defined in the Commercial Building

Telecommunications Cabling Standard (TIA/EIA-568-B, May 2001). This chapter discusses both of these standards and some other closely related material.

TIA-568 Series

EIA/TIA-568 was originally issued in July 1991 and was the first standard to address structured cabling systems. In 1995 the document was reissued as TIA/EIA-568-A. In the latest revision, the document was split into a series of specifications, as follows:

- TIA/EIA-568-B.1, General Requirements (May 2001)
- TIA/EIA-568-B.2, Balanced Twisted-Pair Cabling Components (May 2001)
- TIA/EIA-568-B.2-1, Addendum 1: Transmission Performance Specifications for Four-Pair 100-Ω Category 6 Cabling (February 2003)
- TIA/EIA-568-B.3, Optical Fiber Cabling Components Standard (April 2000).

Partitioning the document in this way allows each specification to be independently updated and reissued.

Terminology and Architecture

The original TIA-568[1] specification was a landmark document. To a large extent, this original document defined the architecture, technology, and terminology that have been used for structured cabling systems since that time. The acceptance of this specification has been so profound and universal that it is difficult to remember that, prior to TIA-568, there were many different cabling architectures and media in use (such as Ethernet coaxial cable, IBM token ring, DECnet, etc.).

As the name implies, the TIA-568 standards are aimed at commercial buildings, such as office buildings, and so on. However, many of the features of the commercial standard are carried over into residential cabling. Some of the major features of the TIA-568 series of standards that affect residential cabling are discussed in the following paragraphs.

Hierarchical Star Topology

TIA-568 defined the hierarchical star topology that is illustrated in Figure 11–1. This figure is a simplified version of the commercial cabling architecture—which can include other functionality such as additional equipment rooms, intermediate cross-connects, zone cabling, and so on—but it is adequate to illustrate the concepts needed for residential cabling.

In this architecture, the cabling spans from the telecommunications outlet (that is, the modular jack) in the work area to the main cross-connect in the equipment room. The cabling is divided into several subsystems. Two of the most important are the horizontal and backbone subsystems. The horizontal subsystem extends from the cross-connect in the **telecommunications room (TR)**—formerly called the **telecommunications closet**—to the **telecommunications outlet (TO)**. The horizontal subsystem is limited to 100 m in length—90 m for permanently installed cable and 10 m total for cords at both ends.

---

[1] For notational convenience, the cabling standards are usually referred to as TIA-xxx rather than the more complete TIA/EIA-xxx, or the even more complete ANSI/TIA/EIA-xxx.

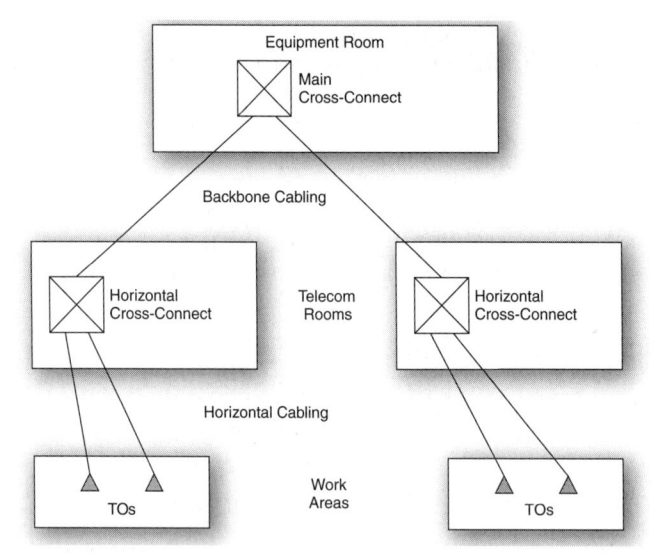

FIGURE 11–1 Hierarchical Star Topology

The backbone subsystem consists of the cables that connect the various telecommunications rooms and equipment rooms together. Backbone cabling distances depend on the media, and application and can be quite a bit longer than horizontal distances.

In commercial buildings, the horizontal subsystem is usually composed of four-pair UTP cable. At the horizontal cross-connect, the cables can be connected to a LAN hub or other data networking equipment in the telecommunications room or they can be connected to backbone cables that run to the main equipment room. The backbone subsystem is usually composed of a mix of four-pair UTP cable, large pair-count UTP cable, and fiber-optic cable.

The cabling system has a multistage tree architecture. Other network topologies, as illustrated in Figure 11–2, can easily be formed by making the appropriate cross-connections.

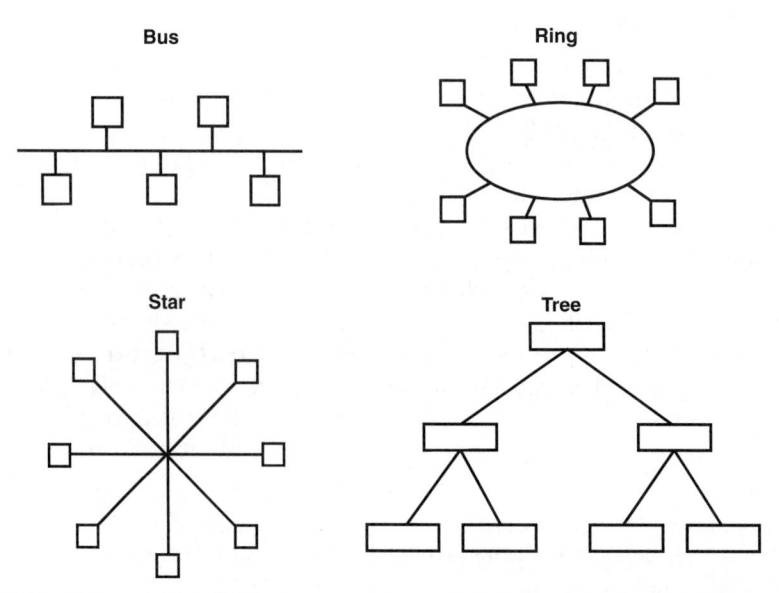

FIGURE 11–2 Network Topologies

Cabling Standards • 247

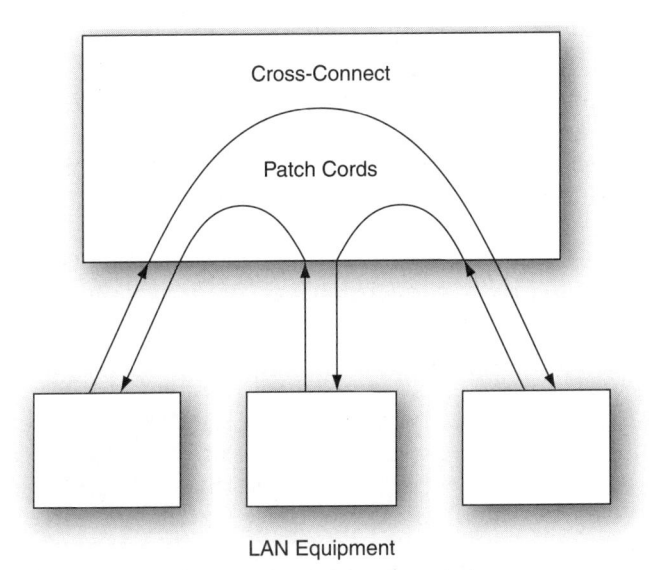

FIGURE 11–3 Making a Ring Topology Via Cross-Connection

For example, Figure 11–3 shows how a ring network can be made by cross-connection of star-wired end points. This example shows three pieces of equipment each having a transmit and a receive pair. By connecting each transmitter to the receiver of the next unit, a logical ring is constructed.

Permanent Link and Channel Definitions

It is very important to be able to test the performance of cable after it is installed and to unambiguously specify what configuration was measured. For this purpose, TIA-568 defines two configurations known as the **permanent link** and the **channel,** which are illustrated in Figure 11–4.

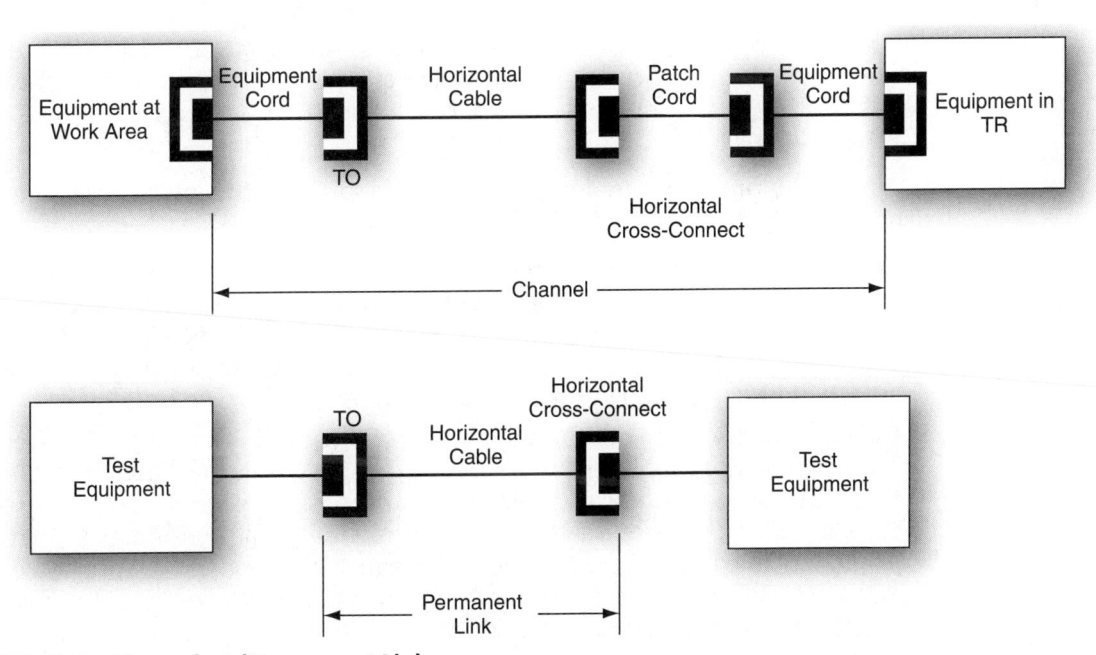

FIGURE 11–4 Channel and Permanent Link

The channel configuration includes all the cable, connecting hardware, and cords that are present when the cabling system is being used. The channel has a maximum length of 100 m. Channel measurements give the best indication of the overall performance of the cabling system.

The permanent link configuration includes the telecommunications outlet, the horizontal cable, and the cross-connect block on which the cable terminates, but does not include any equipment cords or patch cords. The permanent link therefore has a maximum length of 90 m. The purpose of the permanent link is to allow the contractor to perform a test when the installation of the cabling system is completed. Patch cords and equipment cords are usually not added to the system until much later when the telecommunications and data networking equipment is installed. The permanent link test allows the cabling to be verified before the equipment is installed. It is usually followed by a channel test after the equipment installation. Since the channel includes more connectors and cordage than the permanent link, specifications for the channel will always be worse (that is, higher attenuation, lower NEXT loss, etc.) than equivalent specifications for a permanent link.

Unshielded Twisted-Pair Categories

The original TIA-568 standard contained only a single specification for four-pair UTP cable. As network speeds increased and better cables were designed, it became obvious that additional specifications for higher-performance UTP were needed. The TIA-568-A standard provided three different performance specifications for UTP cables. These were called categories of UTP. For historical reasons, Category 1 and Category 2 were not used. The three categories defined in TIA-568-A are:

- **Category 3 UTP** is the same as the original UTP specification from TIA-568. Its performance is specified up to 16 MHz. The highest bit-rate application for this cable was 10-Mb/s Ethernet (10Base-T).

- **Category 4 UTP** extended the performance specification to 20 MHz. The primary application for this cable was a 16-Mb/s token ring. This cable was never widely used and has been deprecated in the standard.

- **Category 5 UTP** extended the performance specification to 100 MHz. It featured a major improvement in cross-talk performance. The primary application at that time was 100-Mb/s Ethernet (100Base-T). Category 5 UTP was very widely deployed.

All of the preceding cable categories have since been deprecated in the residential cabling standard. Since 1995, two new types of cable have been introduced.

- **Category 5e UTP** was introduced to facilitate the use of LANs that transmitted on multiple pairs. It is essentially the same as Category 5, except that it specifies the same values for PSNEXT as Category 5 specifies for NEXT. This is equivalent to about a 3-dB improvement in NEXT performance. Category 5 has been deprecated from TIA-568 and replaced by Category 5e UTP.

- **Category 6 UTP** was introduced in 2002 with the publication of TIA-568-B.2-1. This specification had been in the works for a long time, and there had already been Category 6 products available for a couple of years when it finally was issued. Category 6 UTP represents a major improvement in cable performance. There are several notable improvements. The performance of Category 6 is specified out to 250 MHz, giving it 2.5 times the useful bandwidth of Category 5e. The cross-talk specifications are several decibels better than those in Category 5e. For the first time, the Category 6 standard includes balance specifications (specifically, LCL).

TABLE 11–1
Insertion Loss

Frequency (MHz)	Category 3 PL (dB)	Category 3 Ch (dB)	Category 5e PL (dB)	Category 5e Ch (dB)	Category 6 PL (dB)	Category 6 Ch (dB)
1	3.5	4.2	2.1	2.2	1.9	2.1
16	13.0	14.9	7.9	9.1	7.0	8.0
100			21.0	24.0	18.6	21.3
200					27.4	31.5
250					31.1	35.9

> **UNSHIELDED TWISTED-PAIR CATEGORIES**
>
> Five categories of UTP cable have been defined. Category 3, Category 4, and Category 5 are obsolete and should not be used for new residential installations. Category 5e and Category 6 UTP are compliant with TIA-570-B. Category 6 is recommended for all new installations. Work has begun to define Augmented Category 6 UTP.

For the purposes of this book, we are more concerned with the end-to-end performance of the cabling system. For this, the permanent link and channel specifications are more relevant. TIA-568-B.1 gives channel and permanent link specifications for Category 3 and 5e UTP and, in an informative annex, for Category 5. TIA-568-B.2-1 gives similar specifications for Category 6. The following parameters are specified for permanent link and channel testing:[2] insertion loss (also known as attenuation), NEXT, PSNEXT, ELFEXT, PSELFEXT, return loss, propagation delay, and delay skew.

The standards referenced thus far give an equation describing each of the preceding parameters for each category of cabling, followed by a table with up to a dozen or so data points at sample frequencies. The sample frequencies used are 1, 4, 8, 10, 16, 20, 25, 31.25, 62.5, 100, 200, and 250 MHz. Most of the low-frequency values are there for historical reasons (that is, they used to be considered high frequencies). To get a feel for how the different categories of cables perform, and also to see the difference between channel and permanent link specifications, a few data points for each category are given in the following tables.

Table 11–1 gives the insertion-loss specification for permanent links (PL) and channels (Ch) at a few selected frequencies for Category 3, Category 5e, and Category 6 UTP. Remember that lower values of insertion loss represent better cable performance.

Table 11–2 and Table 11–3 give specifications for NEXT and PSNEXT loss, respectively. For these parameters, higher values represent better cable performance.

---

[2] Only insertion loss and NEXT loss are specified for Category 3.

TABLE 11–2
Near-End Cross-talk Loss (NEXT)

Frequency (MHz)	Category 3 PL (dB)	Category 3 Ch (dB)	Category 5e PL (dB)	Category 5e Ch (dB)	Category 6 PL (dB)	Category 6 Ch (dB)
1	40.1	39.1	60.0	60.0	65.0	65.0
16	21.0	19.3	45.2	43.6	54.6	53.2
100			32.3	30.1	41.8	39.9
200					36.9	34.8
250					35.3	33.1

Notice that, as with insertion loss, the permanent link values are always a little better than the channel values. This difference increases with frequency, as the additional connectors and cords affect the performance more at higher frequencies. Also note the huge difference in cross-talk performance when going from Category 3 to Category 5e, and from Category 5e to Category 6. For example, at 100 MHz, a Category 6 channel or permanent link has 10 dB better PSNEXT performance than an equivalent Category 5e configuration.

Table 11–4 and Table 11–5 give specifications for ELFEXT and PSELFEXT loss, respectively. For these parameters, like NEXT/PSNEXT, higher values represent better cable performance.

Return-loss values are given in Table 11–6. Higher values represent better cable performance.

Propagation delay and delay skew specifications are given in Table 11–7. All propagation delay measurements are made at a frequency of 10 MHz. The specifications for Category 5e and Category 6 are identical. Lower values are, of course, better for these parameters. Due to the tighter twisting in Category 6 cable, the helical path through the cable is longer, so the NVP of the cable must be higher to meet the same propagation delay specification.

TABLE 11–3
Power-Sum Near-End Cross-talk Loss (PSNEXT)

Frequency (MHz)	Category 3 PL (dB)	Category 3 Ch (dB)	Category 5e PL (dB)	Category 5e Ch (dB)	Category 6 PL (dB)	Category 6 Ch (dB)
1	N/A	N/A	57.0	57.0	62.0	62.0
16			42.2	40.6	52.2	50.6
100			29.3	27.1	39.3	37.1
200					34.3	31.9
250					32.7	30.2

TABLE 11–4
Equal-Level Far-End Cross-talk Loss (ELFEXT)

Frequency (MHz)	Category 3 PL (dB)	Category 3 Ch (dB)	Category 5e PL (dB)	Category 5e Ch (dB)	Category 6 PL (dB)	Category 6 Ch (dB)
1	N/A	N/A	58.6	57.4	64.2	63.3
16			34.5	33.3	40.1	39.2
100			18.6	17.4	24.2	23.3
200					18.2	17.2
250					16.2	15.3

TABLE 11–5
Power-Sum Equal-Level Far-End Cross-talk Loss (PSELFEXT)

Frequency (MHz)	Category 3 PL (dB)	Category 3 Ch (dB)	Category 5e PL (dB)	Category 5e Ch (dB)	Category 6 PL (dB)	Category 6 Ch (dB)
1	N/A	N/A	55.6	54.4	61.2	60.3
16			31.5	30.3	37.1	36.2
100			15.6	14.4	21.2	20.3
200					15.2	14.2
250					13.2	12.3

TABLE 11–6
Return Loss

Frequency (MHz)	Category 3 PL (dB)	Category 3 Ch (dB)	Category 5e PL (dB)	Category 5e Ch (dB)	Category 6 PL (dB)	Category 6 Ch (dB)
1	N/A	N/A	19.0	17.0	19.1	19.0
16			19.0	17.0	20.0	18.0
100			12.0	10.0	14.0	12.0
200					11.0	9.0
250					10.0	8.0

TABLE 11-7
Maximum Propagation Delay and Delay Skew (at 10 MHz)

Parameter	Category 3 PL (ns)	Category 3 Ch (ns)	Category 5e PL (ns)	Category 5e Ch (ns)	Category 6 PL (ns)	Category 6 Ch (ns)
Maximum propagation delay	N/A	N/A	498	555	498	555
Maximum delay skew	N/A	N/A	44	50	44	50

From the data in Table 11-1 and Table 11-2, we can calculate the ACR, as shown in Table 11-8.

The channel ACR data are plotted in Figure 11-5. A few points are worth noting relative to the ACR figures.

- Note the extremely poor performance of Category 3 cabling. Even at the relatively low frequency of 16 MHz, a Category 3 channel only has 4.4 dB of ACR. This is why Category 3 cable has virtually disappeared from the marketplace.
- Category 6 cabling is a tremendous improvement over that at Category 5e. At a frequency of 100 MHz, a Category 6 channel has 12.5 dB more ACR than Category 5e.
- Category 6 has extremely good high-frequency performance. At 250 MHz, the channel ACR is slightly negative; but, with cross-talk cancellation circuitry in the interface, it could support a signaling rate of 500 MHz.

Augmented Category 6

To support the demanding performance requirements of 10-Gb/s networks, the TIA began defining Augmented Category 6 cable in 2004. For convenience, we refer to Augmented Category 6 cable as Category 6a. While at the time of this writing the specifications are just beginning to be written, a few things can be said. First, the transmission parameters for Category 6a will be specified to 500 MHz, which is double the frequency range of Category 6. The Category 6 specifications will, for the most part, probably be extrapolated out to the new maximum frequency. Insertion loss and return loss will be key parameters of the new specification. For the first time, alien cross-talk requirements will be specified—definitely for ANEXT

TABLE 11-8
Attenuation to Cross-talk Ratio (ACR)

Frequency (MHz)	Category 3 PL (dB)	Category 3 Ch (dB)	Category 5e PL (dB)	Category 5e Ch (dB)	Category 6 PL (dB)	Category 6 Ch (dB)
1	36.6	34.9	57.9	57.8	63.1	62.9
16	8.0	4.4	37.3	34.5	47.6	45.2
100			11.3	6.1	23.2	18.6
200					9.5	3.3
250					4.2	−2.8

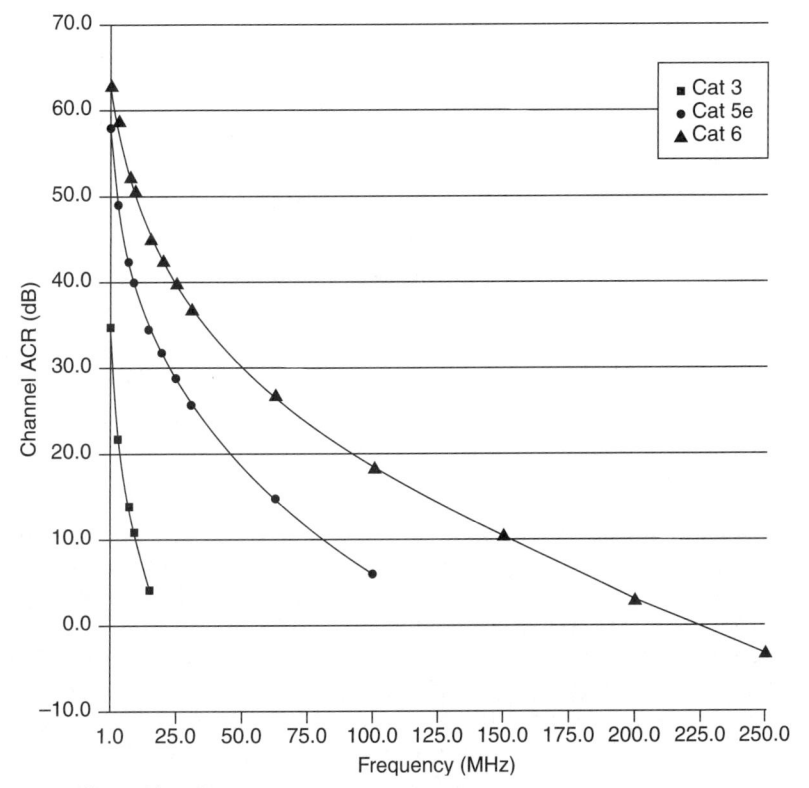

FIGURE 11–5 Channel ACR

and possibly for AFEXT as well. A lot of work is yet to be done on measurement techniques for alien cross-talk.

Fiber Specifications

TIA-568-B.3 gives component specifications for fiber cables, connectors, patch panels, cords, and splices. As with UTP, our main focus here is on the performance of installed fiber links, which is covered in TIA-568-B.1. Since fiber links can be connected together for much longer runs than copper cables can, the test configuration for installed fiber is a little different than the permanent link and channel that were defined for UTP. For fiber installations, testing is done on a **passive link,** as shown in Figure 11–6.

The optical fiber passive link consists of a length of duplex fibers with connectors on each end. For the horizontal subsystem, one end of the fiber terminates at the telecommunications outlet and the other at the horizontal cross-connect. Test equipment is attached to the connectors at each end, and the attenuation is measured at either 850 nm or 1,300 nm. For a horizontal fiber segment, the attenuation of the passive link must be less than 2.0 dB.

TIA-570-B

The Residential Telecommunications Cabling Standard (TIA-570-B) is the primary standard specifying requirements for cabling in a residential building. The first edition (EIA/TIA-570) was issued in May 1991. At that time, residential cabling was not a particularly hot topic and it did not generate much interest. In September 1999, a revision (TIA/EIA-570-A) was issued to update the standard and

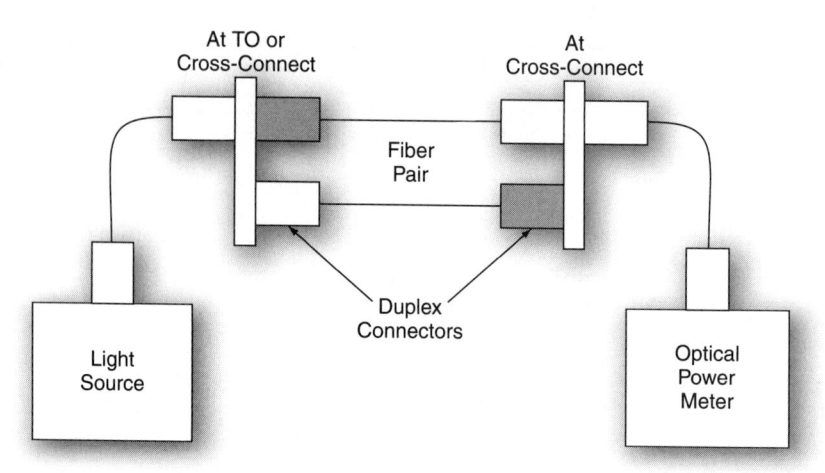

FIGURE 11–6 Optical Fiber Passive Link

incorporated a number of new requirements and technologies. By this time, residential cabling was attracting more attention. TIA-570-B (which was approved by the TIA in early 2004) incorporates some new material on residential applications and adds Category 6 UTP as a recognized medium.

TIA-570-B covers both single- and multi-dwelling units. In this chapter, we only address single-dwelling units. Topics covered in TIA-570-B include residential infrastructure, cable and connecting hardware specifications, installation requirements, and requirements for some common residential applications (such as whole-house audio). Each of these is discussed in subsequent sections.

Residential Infrastructure

Figure 11–7 (which was adapted from TIA-570-B) illustrates the architecture of a typical residential structured cabling system. Let us walk through this figure from left to right and identify each of the elements of the system.

- On the left side of the figure are **access lines** from various service providers into the house. These are typically lines from telephone companies, cable television operators, and so on. These are usually copper cables and may be either UTP or coax.

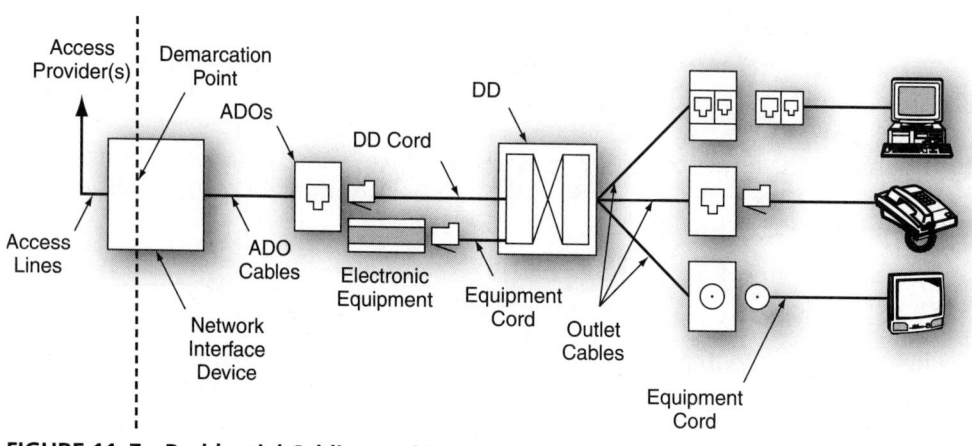

FIGURE 11–7 Residential Cabling Architecture

- The access lines terminate in the **network interface device (NID).** The NID contains a **demarcation point** between the exterior and the interior of the residence. Problems that occur outside the demarcation point are the service provider's responsibility. Problems that occur inside the demarcation point are the responsibility of the residence owner. The NID also usually contains electrical protection devices for the incoming cables. In single-unit residences, the NID is normally located on the outside of an exterior wall so that it can be accessed without entering the home.
- The NID is connected by one or more cables to the auxiliary disconnect outlet (ADO). There is typically one cable and ADO for each incoming service, for example, phone line, cable television, and so on. The ADO allows incoming services to be easily disconnected for troubleshooting or other reasons. The ADO is usually collocated with and may physically be a part of the DD.
- The ADO is connected by a cord to the distribution device (DD). The DD is the nerve center of the cabling system. It provides the capability to terminate and cross-connect (or interconnect; see the sidebar discussion, Cross-connects, Interconnects, and Bridging) all the various cables within the residence. It may also contain other equipment such as video amplifiers, LAN hubs, and so forth. The DD is sometimes called the service center or some other vendor-specific name. Both the ADO and the DD must be located in an accessible place inside the residence. When possible, the DD should be centrally located to minimize the length of the outlet cables. The DD is normally enclosed by a cabinet or cover. The size of the DD should be appropriate for the number of cables that will terminate in it, including an allowance for future expansion. An electric power outlet must be available within 5 feet of the DD.
- From the DD, outlet cables (OCs) run to the telecommunications outlets (TOs), which are installed in various locations throughout the residence. The outlet cables are most commonly UTP but may also be coax or fiber. Note that all the cables are directly connected to the DD using the home-run topology.
- The telecommunications outlet (TO) is the interface to users of the cabling system. Different types of outlets are provided for UTP, coax, and fiber cable runs. Users connect their equipment (telephones, televisions, PCs, etc.) to the outlet via an equipment cord. For UTP, the required TO is an eight-pin modular jack as specified in IEC 60603-7 with T568A pinouts as specified in TIA-568-B.2. For coax, the F connector is specified as the TO. For fiber, no particular connector is specified; any connector that meets the requirements of TIA-568-B.3 may be used.

In many respects, the residential cabling architecture looks like a scaled-down version of the commercial premises architecture specified in TIA-568-B where there is only a single cross-connect. Table 11–9 provides a comparison and contrast between commercial and residential cabling.

Recognized Cables

The residential cabling standard recognizes three types of communications cable:

- Four-pair, 100-ohm UTP (Category 5e or Category 6)[3]

---

[3] Since July 2000 the FCC has mandated that all telephone cable installed in residences must be Category 3 or better. Prior to that time, there were no performance requirements on residential cabling, and homes were often wired with quad cable.

TABLE 11–9
Comparison of Commercial and Residential Cabling

Function	*Commercial Premises Cabling*	*Residential Cabling*
Standard	TIA-568-B	TIA-570-B
User Interface	Telecommunications Outlet (TO)	Telecommunications Outlet (TO)
Cable to TO	Horizontal Cable	Outlet Cable (same 90-m distance limitation)
Cross-Connects	Main cross-connect (equipment room) and horizontal cross-connects (telecommunications rooms)	Distribution Device (DD)
Connection between Cross-Connect Panels	Backbone Cabling	N/A In single-unit dwellings
Service Disconnect	N/A	Auxiliary Disconnect Outlet (ADO)

- Series 6 coaxial cable (Series 59 coax may be used for baseband CCTV applications only.)
- Multimode optical fiber (either 50/125 µm or 62.5/125 µm). Single-mode fiber may be used for "special cases future applications."

Thus, TIA-570-B allows a fair amount of leeway in terms of what cables to install. In practice, the preceding recommendations can be simplified. The author's recommendations are:

- For UTP cable, use Category 6 exclusively. The added performance more than makes up for the small increment in cost.
- For coax, use Series 6 exclusively, since Series 59 only supports baseband signals.
- If you plan to install multimode fiber, use 50/125-µm fiber due to its higher bandwidth.
- At this time there does not, in general, appear to be any reason to install single-mode fiber in a residence.

Grades of Cabling

TIA-570-B specifies two grades of cabling (cleverly called Grade 1 and Grade 2) to "assist in the selection of cabling." The description of each grade follows.[4]

- Grade 1 "provides a generic cabling system that meets the minimum requirements for basic telecommunications services. As an example, this grade provides for telephone, satellite, community antenna television (CATV) and data services."
- Grade 2 "provides a generic cabling system that meets the requirements for basic and advanced telecommunications services, such as high-speed Internet and in-home generated video. This grade provides for both current and developing telecommunications services."

---
[4] From Clause 3.3.2 of TIA-570-B.

CROSS-CONNECTS, INTERCONNECTS, AND BRIDGING

Cross-connects and interconnects are two different means to provide for the termination of cables with the flexibility to rearrange the connections. Both techniques are illustrated in Figure 11–8. The top of the figure shows a cross-connect configuration. Both the service cables and the outlet cables are permanently punched down to the cross-connect block. Connections between the two are made by means of patch cords that are plugged into modular jacks on the block. This example shows Service Cable 1 connected to Outlet Cable B and Service Cable 2 connected to Outlet Cable A.

The bottom of the figure shows an interconnect configuration. In this case, the service cables are permanently punched down on the block. The outlet cables are field-terminated with modular connectors, which are then plugged into the jacks on the block. This allows connections to be made without an additional cord. The interconnect example shows the same connections as the preceding cross-connect example. Cross-connecting and interconnecting are done with the same block; the only difference is how the service and outlet cables (and patch cords) are connected to the block.

The advantages of a cross-connect are:

- It allows the outlet cables to be permanently terminated on the back of the block where they are out of the way and not subject to damage.
- It uses factory-terminated patch cords and thus does not require modular plugs to be installed in the field.

The main advantages of an interconnect are:

- It is physically smaller and uses roughly half as many blocks.
- It does not require patch cords.

For large installations with frequent rearrangement, such as the equipment room in commercial buildings, interconnects become unwieldy and cross-connects are clearly preferable. For small installations like a typical home, the trade-offs between cross-connects and interconnects are not as clear. The choice depends on whether you would like to save a little space and cost or have cleaner and simpler administration.

A related concept is bridging. This refers to hooking several cables to the same point. For example, in voice telephone, it is very common to have the same phone line appear at several locations in the house. This can be accomplished by using a bridged block, as shown in Figure 11–9. In this case, a single service cable is connected to four different outlet cables. Although an interconnect configuration is shown in this figure, it could be used in a cross-connect configuration also. Blocks prewired for bridging can be purchased, or a regular block can be bridged in the field, typically by daisy-chaining the punch-down connectors together.

258 • Chapter 11

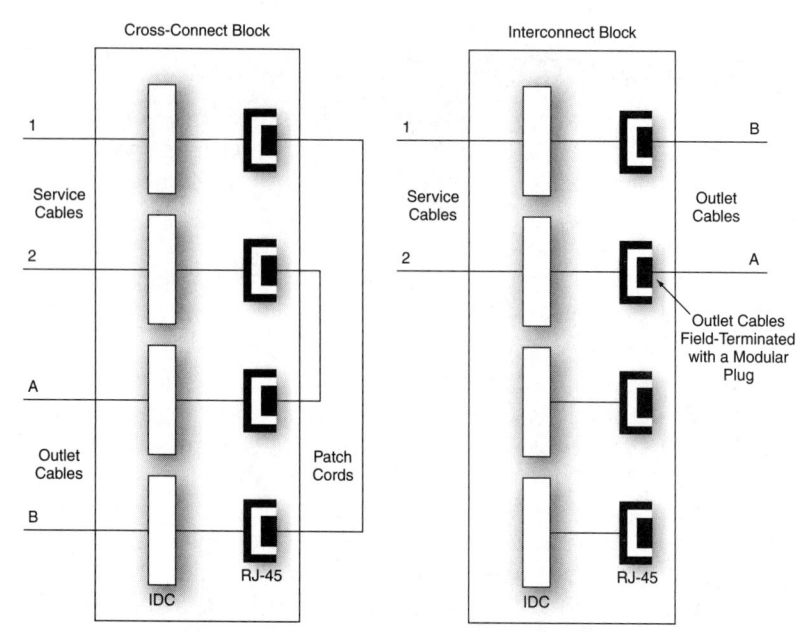

FIGURE 11–8 Cross-Connect Versus Interconnect

There are two main areas where the grade of cabling affects the requirements of the standard. The first concerns the choice of cabling media, and the second involves space allocation for the DD. Table 11–10 summarizes the use of the recognized cabling media.

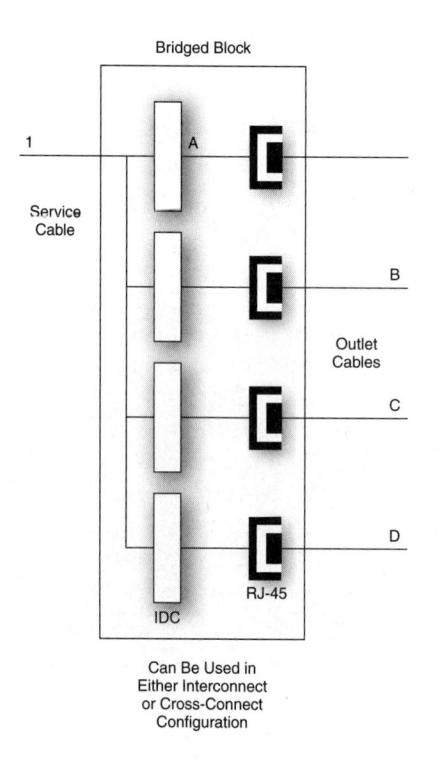

FIGURE 11–9 Bridging

TABLE 11-10
Recognized Cabling Media by Grade*

Cabling	Grade 1	Grade 2
UTP—minimum	Category 5e	Category 5e
UTP—recommended	Category 6	Category 6
Coax	Series 6	Series 6
Fiber		Optional

*In TIA-570-A, the minimum UTP specification was Category 3 and the recommended UTP was Category 5e, so there was a little difference between the grades. In TIA-570-B, the only difference in media is that Grade 2 can optionally use fiber.

For practical purposes, the distinction between the cabling media for the two grades is not particularly important, since we recommend Category 6 UTP in any case, and fiber is optional but seldom used.

Space allocation for the DD is shown in Table 11–11. The spaces are 14.35" wide to allow for mounting the DD between wall studs, if desired.

The additional space for Grade 2 is in anticipation of the need for additional equipment (LAN hubs, video amplifiers, modulators, etc.) at the DD. However, the distinction between Grade 1 and Grade 2 does not make much difference in this case either. The table specifies allocation of wall space that is reserved for the DD. Since there is usually empty wall space above and below the DD, this requirement seems rather inconsequential.

Outlet Locations and Pathways

TIA-570-B requires that at least one telecommunications outlet be placed in each of the following rooms: kitchen, bedrooms, family/great room, and den/study. Additional TOs are recommended (but not required) on unbroken wall spaces of 12' or more, and to ensure that no point in the room is more than 25' (measured horizontally along the wall) from the nearest TO. For example, a room that measures 16' by 10' should have at least two TOs located on opposite walls.

TABLE 11-11
Space Allocation for the DD

Number of TOs	Grade 1	Grade 2
1 to 8	10" to 18" high	18" to 36" high
9 to 16	28" to 36" high	28" to 42" high
17 to 24	28" to 42" high	Multiple interconnected units 28" to 42" high
>24	Multiple interconnected units 28" to 42" high	Multiple interconnected units 28" to 42" high

Note: All DD spaces are 14.35 inches (364 mm) wide.

Cable and Connecting Hardware Specifications

For UTP cable, connecting hardware, and cords, TIA-570-B requires that they meet the specifications of TIA-568-B.2. Similarly, optical fiber cable, connecting hardware, and cords must meet the specifications of TIA-568-B.3.

There is a more substantive discussion of coaxial cable specifications. Coaxial cable specifications are written by the Society of Cable Telecommunications Engineers (SCTE[5]). Series 6 and Series 11 coax are covered by SCTE IPS-SP-001. TIA-570-B lists specifications for attenuation, SRL, and characteristic impedance, as well as some hysical and dimensional requirements. In a nutshell, the following types of coax are covered:

- Series 6 coax is used for outlet cables and backbone cables.
- Series 11 coax is used for backbone cables. Series 11 coax has a thicker conductor and lower attenuation than Series 6. It is not recommended for outlet cables because it has a larger bend radius and is difficult to pull through walls.
- Series 59 coax (which has a thinner conductor and higher attenuation than Series 6) is used only for patch cords and equipment cords.

Installation and Testing Requirements

TIA-570-B gives some requirements and guidelines for the installation of UTP, coax, and fiber cabling. Requirements for field testing are also given. For UTP, field testing consists of three phases—a visual inspection and verification of continuity before the walls are closed up, followed by a performance characterization after trim-out. For fiber, testing should be done as per TIA-568-B.3.

Environmental and Safety Issues

In addition to the architecture and transmission performance, a number of environmental and safety issues must be considered. These include flammability ratings of cable, grounding and bonding, and electromagnetic emissions. Each of these topics is discussed in the following sections.

Flammability Classes

Since cables extend throughout a building, their behavior in the event of a fire is very important. A poorly designed or improperly installed cabling system can spread fire throughout the building and give off noxious smoke that is harmful or fatal to the building occupants. For these reasons, a fire-rating system for communication cables has been established by *Article 800* of the *National Electrical Code®* (*NEC®*), which is officially known as ANSI/NFPA 70.

The fire rating needed for a cable depends on the type of space in which it will be installed:

- Cable that is installed in an ordinary room space must meet the communications cable rating.

---

[5] Until very recently, the SCTE was the Society of Cable Television Engineers.

TABLE 11–12
NFPA Markings for Copper Cables

Marking	Type
MPP	Multipurpose plenum
CMP	Communications plenum
MPR	Multipurpose riser
CMR	Communications riser
MPG	Multipurpose general purpose
CMG	Communications general purpose
MP	Multipurpose general purpose
CM	Communications general purpose
CMX	Communications cable, limited use
CMUC	Undercarpet communications wire and cable

> The NFPA has defined ten flammability ratings for copper cable and eight for fiber cable. The appropriate rating must be used for each application.

- Cable that is installed in a vertical shaft between floors must be riser-rated. Riser-rated cable has fire-resistant characteristics to help prevent carrying flames from one floor to another.
- Cable that is installed in environmental air-handling spaces (ducts, plenums, etc.) must be plenum-rated. Plenum-rated cable is fire resistant and also has low-smoke characteristics.

In general, different materials for the insulation and jacket are used in riser- and plenum-rated cables. The fire rating of a cable is established by testing at an independent laboratory. The rating of the cable is clearly marked on the jacket.

Note that flammability requirements may be different in other areas of the world. In Europe, for example, the requirements and testing procedures are quite different.[6] Be sure to follow the local codes and regulations for your area.

The NFPA ratings for copper-based cables are given in Table 11–12.

Multipurpose cables are similar to communications cables but must meet additional requirements regarding minimum conductor size (26 AWG for multiconductor cables and 18 AWG for single-conductor cables).

For one- and two-family dwellings, CMX cable or better is required by *Article 800.53* of the *NEC®*. For multifamily (that is more than two-family) dwellings,

---

[6] In the United States, the insulation and jacket material for plenum cable often contains halogens, which are forbidden by European codes. As a result, plenum cable, which has the highest fire rating in the United States, cannot be used at all in Europe.

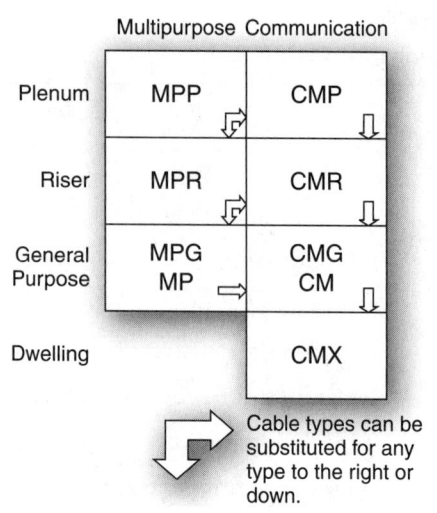

FIGURE 11-10 Permissible Substitutions for Copper Cables

CMX cable is permitted only in nonconcealed locations, such as cords. Concealed cable in multifamily dwellings must be rated CM/CMG or higher. CMUC cable is permitted for undercarpet installations.[7]

In general, higher-rated cables can be substituted for lower-rated cables, with the stipulation that CM cables cannot be substituted for MP cables. Figure 11-10 illustrates the types of cable substitutions that are permitted by the *NEC*®.[8]

Fire ratings for optical fiber cables are established in *Article 770* of the *NEC*®. The fiber ratings use the same riser and plenum concepts, as indicated in Table 11-13.

For fiber cables, one of the important parameters is whether or not the cable contains a (metallic) conductor. When present, such conductors are usually either strength members or part of an armored jacket. As with copper cables, higher-rated cables can be substituted for lower-rated cables with the exception that conductive cables cannot be substituted for nonconductive cables. The substitutions permitted by the *NEC*® are illustrated in Figure 11-11.

Any of the fiber cables listed in Table 11-13 are suitable for use in a residence. Type OFN would normally be used in a one- or two-family dwelling.

Grounding and Bonding

Grounding refers to providing a path to earth for undesired (and potentially hazardous) currents. **Bonding** refers to permanently joining conductors together to provide a low-impedance path to ground.

Proper grounding and bonding are very important to ensure protection against electrical shocks, lightning strikes, and similar hazards. In commercial buildings, the grounding and bonding requirements are given by J-STD-607-A, as well as in various sections of the *NEC*® The 607-A approach provides a telecommunications bonding backbone (TBB) throughout the building so that equipment

---

[7] In general, the use of undercarpet cable is not recommended except in retrofit situations where there is no other feasible means to get a TO to the desired location.

[8] While these substitutions are permitted, keep in mind that higher-rated cables are more expensive. It is generally desirable to use the lowest rating permitted by the *NEC*®.

TABLE 11–13
NFPA Markings for Optical Fiber Cable

Marking	Type
OFNP	Nonconductive optical fiber plenum
OFCP	Conductive optical fiber plenum
OFNR	Nonconductive optical fiber riser
OFCR	Conductive optical fiber riser
OFNG	Nonconductive optical fiber general purpose
OFCG	Conductive optical fiber general purpose
OFN	Nonconductive optical fiber general purpose
OFC	Conductive optical fiber general purpose

on any floor can be easily grounded. This type of structure is overkill for a typical single-unit residence.

For single-unit residences, a telecommunications ground is required for the metal sheath on any cables entering the residence and for all electrical protectors on incoming cables. The grounding requirements for communications cables in a residence are:

- Metallic shields of any cables entering the residence and primary protectors must be grounded.
- The grounding conductor must be 14 AWG or larger and may not be longer than 6.0 m (20′).
- The grounding conductor typically terminates on a grounding electrode or a metal water pipe (as per *Article 250.52* of the *NEC*®).
- The grounding conductor must be bonded to the grounding electrode by exothermic welding or a lug, connector, or clamp that meets the requirements of *Article 250.70* of the *NEC*®. The ground connection may not be soldered.

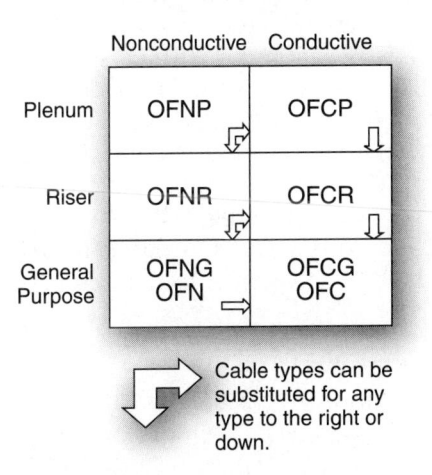

	Nonconductive	Conductive
Plenum	OFNP	OFCP
Riser	OFNR	OFCR
General Purpose	OFNG / OFN	OFCG / OFC

Cable types can be substituted for any type to the right or down.

FIGURE 11–11   Permissible Substitutions for Fiber Cables

Electromagnetic Emissions

FCC Part 15 gives regulations for the emission of electromagnetic radiation (either intentional or accidental) by electronic equipment. Part 15 defines two classes of digital equipment, as follows:

- *Class A digital device.* A digital device that is marketed for use in a commercial, industrial, or business environment, exclusive of a device that is marketed for use by the general public or is intended to be used in the home.

- *Class B digital device.* A digital device that is marketed for use in a residential environment notwithstanding use in commercial, business, and industrial environments. Examples of such devices include but are not limited to personal computers, calculators, and similar electronic devices that are marketed for use by the general public.

The purpose of the Part 15 regulations is to prevent electromagnetic interference (EMI) from digital equipment that could interfere with the operation of other equipment, such as radios, televisions, and so forth. The Class B requirements are more stringent than the Class A requirements. It is the responsibility of the manufacturers of digital equipment (such as PCs, DVD players, LAN hubs, etc.) to test their equipment and make sure it does not generate excessive radiation. If a piece of equipment that conforms to Part 15 is hooked up to an improperly installed cabling system (such as an untwisted UTP system or an improperly shielded STP system), it may generate excessive EMI due to unbalanced currents flowing in the cabling system. It is important to properly install and maintain the cabling system to ensure that this does not happen.

International Cabling Standards

International cabling standards are written by a joint committee of the International Standards Organization (ISO) and the International Electrotechnical Commission (IEC), which is known as ISO/IEC JTC 1/SC 25. The primary document from this committee (similar to the TIA-568 specification) is known as ISO/IEC 11801 (Ed. 2:2002), Information Technology—Generic Cabling for Customer Premises. Although the initial versions of 11801 and TIA-568 had substantial differences, they have been largely harmonized in subsequent editions. Some differences in terminology remain, however, such as the use of the term distributor in the international document, which is synonymous with cross-connect in TIA-568.

The ISO/IEC committee has also issued a residential cabling standard. It is known as ISO/IEC 15018 and is fairly similar to TIA-570 in many respects, such as the use of star architecture, and so forth. The main elements of ISO/IEC 15018 include a home distributor (HD), primary home cables, and application connection points (ACPs). These are very similar to the DD, outlet cables, and TO in TIA-570.

However, ISO/IEC 15018 differs substantially from TIA-570 in terminology and a number of other areas. For example, ISO/IEC 15018 defines three groups of applications:

- Information and communications technologies (ICT)
- Broadcast and communications technologies (BCT)
- Control/command communications in buildings (CCCB)

Different cabling requirements are given for each of these types of applications. ISO/IEC 15018 gives detailed specifications for cables and connecting hardware, unlike TIA-570, which references component specifications in other documents (such as TIA-568).

A detailed examination of ISO/IEC 15018 is beyond the scope of this book. Most of the principles discussed in this book apply to ISO/IEC 15018 as well as TIA-570, but the specific details of ISO/IEC 15018 should be consulted in countries where it is applicable.

Summary

This chapter gives a brief overview of the cabling standards that are relevant to residential networks. The major points covered in this chapter are:

- TIA-568-B (Commercial Building Telecommunications Cabling Standard) defines the architecture and technology of structured cabling systems.
- Structured cabling uses a hierarchical star topology.
- Two test configurations are defined for UTP: the channel and the permanent link.
- A passive link configuration is defined for optical fiber testing.
- Two categories of UTP cable (5e and 6) are defined. Category 6 is the preferred UTP cable for all new installations.
- TIA-570-B is the Residential Telecommunications Cabling Standard.
- The major components of a residential cabling system are the end network interface device (NID), the auxiliary disconnect outlet (ADO), the distribution device (DD), telecommunications outlets (TOs), outlet cables (OCs), and equipment/patch cords.
- Interconnects and cross-connects can both be used at the DD to provide flexible connections between service and outlet cables.
- Residential cabling uses three types of cable: UTP, Series 6 coax, and multimode fiber.
- TIA-570-B defines two grades of cabling (1 and 2), which specify the category of UTP cable to use and the amount of wall space to allocate for the distribution device.
- At least one TO is required in the kitchen, bedrooms, family/great room, and den/study.
- The installation of additional TOs is recommended so that no point is more than 25 feet from the nearest TO.
- TIA-570-B also gives guidelines for cabling security, home-control, and wholehouse audio systems.
- The *NEC®* establishes a fire-rating system for copper and fiber cables. For one- and two-family dwellings, copper cable must be rated CMX or better.
- Proper grounding of the telecommunications cabling is required to ensure safe operation.
- FCC Part 15 sets limits on the amount of EMI that may be generated by digital equipment.
- International cabling standards are written by a joint committee of the ISO and the IEC.
- ISO/IEC 11801 is the international document similar to TIA-568 and ISO/IEC 15018 is the document similar to TIA-570.

Key Terms

Access lines A telecommunications circuit provided by a service provider at the demarcation point.

Bonding The permanent joining of metallic parts to form an electrically conductive path that will ensure electrical continuity and the capacity to conduct safely any current likely to be imposed.

Channel The end-to-end transmission path between two points at which application-specific equipment is connected.

Demarcation point A point at which operational control or ownership changes.

Grounding Establishing a conducting connection between an electrical circuit or equipment and the earth.

Network interface device (NID) The point of connection between the networks. In a residence, the location of the demarcation point.

Passive link An optical fiber test configuration consisting of a duplex run of fiber and the pair of connectors on each end.

Permanent link A test configuration for a link that excludes test cords and patch cords.

Telecommunications closet Obsolete (though still frequently used) terminology, replaced by *telecommunications room*.

Telecommunications outlet (TO) A connecting device (usually a modular jack or coaxial connector) in the home on which the outlet cable terminates.

Telecommunications room (TR) An enclosed space for housing telecommunications equipment, cable terminations, and cross-connect cabling; the location of the horizontal cross-connect. Formerly known as the *telecommunication closet*.

Review Questions

1. What standards specify the requirements for telecommunications cabling in commercial buildings and residences?
2. What topology is used by structured cabling systems?
3. What is the difference between the permanent link and channel test configurations?
4. When would a permanent link test normally be performed?
5. When would a channel test normally be performed?
6. What categories of UTP cable are currently recognized by the TIA standards?
7. To what frequency is the performance of each UTP category specified?
8. What are the two transmission parameters specified for Category 3 channels and permanent links?
9. What additional six parameters are specified for Category 5e and Category 6 channels and permanent links?
10. What parameter is specified only for Category 6 channels and permanent links?
11. What application is driving the development of the Augmented Category 6 UTP specification?
12. What new parameter will be specified for Augmented Category 6 permanent links and channels?
13. What are three preferred cabling media for residential networks?
14. What test configuration is used for fiber-optic links?
15. What is the only test required for a fiber link?
16. What are the major elements of a residential cabling system?
17. What is the primary function of the network interface device (NID)?
18. What is the primary function of the auxiliary disconnect outlet (ADO)?
19. What is the primary function of the distribution device (DD)?
20. What is the maximum length of an outlet cable?
21. Where is the NID usually located?
22. Where is the ADO usually located?
23. Where is the DD usually located?
24. What are the two main differences between Grade 1 and Grade 2 cabling?
25. What types of room must have at least one TO?
26. Additional TOs are recommended on unbroken wall spaces of what length?
27. What is the difference between a cross-connect and an interconnect?
28. What is bridging, and for what application is it most commonly used?
29. What type of coax is recommended for outlet cables? For patch cords and equipment cords?
30. What are the three phases of field testing of installed cable?

31. What residential applications are addressed in the TIA-570-B?
32. Why are fire ratings necessary for communications cable?
33. Are fire ratings uniform in all countries?
34. What are the minimum ratings of copper and fiber cables that are required for a single-family residence?
35. What elements of a residential wiring system must be grounded?
36. Explain the difference between grounding and bonding.
37. Are the FCC Part 15 EMI requirements more stringent for residential or commercial equipment?
38. Can the cabling system influence the amount of EMI generated by a piece of digital equipment? Why or why not?
39. What international specifications are similar to TIA-568 and TIA-570?
40. Name three differences in terminology between United States and international residential cabling standards.

Optical Fiber and Cable

OBJECTIVES *After studying this chapter, you should be able to:*
- Identify the types of fiber-optic cable.
- Identify the physical properties of fiber-optic cable.
- Explain the use of various types of fiber-optic cable.
- Explain fiber attenuation and the results of scattering and absorption of light.
- Describe modal dispersion and its relationship to bandwidth and spectrum of light.
- Define the purpose and use of cable varieties such as simplex cable, distribution cable, breakout cable, and loose tube cable.
- Understand *Article 770* of the *NEC®*.

OUTLINE
Introduction
Fiber Attenuation
Fiber Bandwidth
Bending Losses
Fiber-Optic Cable
Fiber-Optic Cable Installation

Introduction

Optical fiber is composed of a light-carrying core and a **cladding** that traps the light in the core, causing total internal reflection (Figure 12–1). Most fiber is composed of a solid glass core and cladding, with a plastic buffer coating for protection from physical damage and moisture. The plastic buffer is stripped from glass fiber for terminating or splicing. Other fibers may have a glass core and plastic cladding, and some are all plastic.

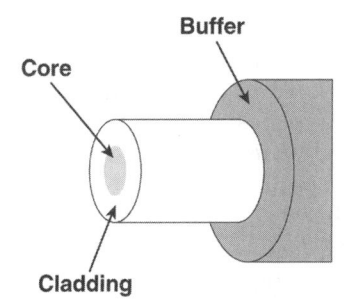

FIGURE 12–1 Optical fiber consists of a core, cladding, and a protective buffer coating.

There are two basic types of fiber: multimode and single-mode (Figure 12–2). **Multimode fiber** means that light can travel many different paths (called modes) through the core of the fiber, which enter and leave the fiber at various angles. The highest angle at which light is accepted into the core of the fiber defines the **numerical aperture (NA).**

Two types of multimode fiber exist, distinguished by the index profile of their cores and how light travels in them. Step index multimode fiber has a core composed of one type of glass. Light traveling in the fiber travels in straight lines, reflecting off the core/cladding interface. The NA is determined by the differences in the indices of refraction of the core and cladding and can be calculated by Snell's law. Because each mode or angle of light travels a different path link, a pulse of light is dispersed while traveling through the fiber, limiting the bandwidth of step index fiber.

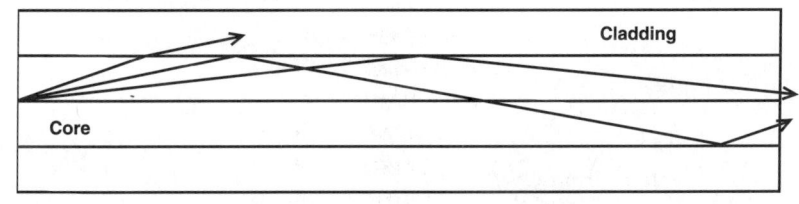

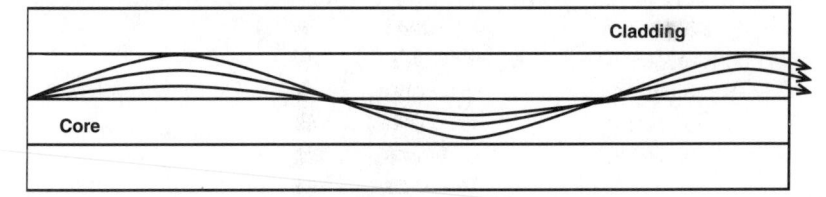

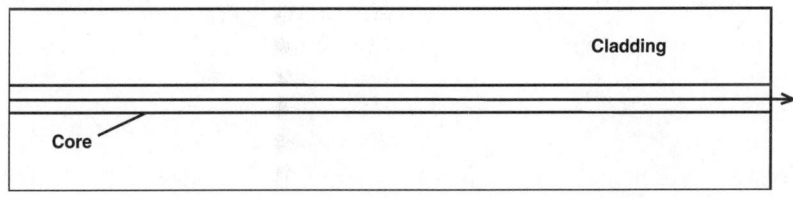

FIGURE 12–2 The Three Types of Optical Fiber

In graded index multimode fiber, the core is composed of many different layers of glass, chosen with indices of refraction to produce an index profile approximating a parabola. Because the light travels faster in lower index of refraction glass, the light will travel faster as it approaches the outside of the core. Likewise, the light traveling closest to the core center will travel the slowest. A properly constructed index profile will compensate for the different path lengths of each mode, increasing the bandwidth capacity of the fiber by as much as one hundred times that of step index fiber.

Single-mode fiber just shrinks the core size to a dimension about six times the wavelength of the light, causing all the light to travel in only one mode. Thus, modal dispersion disappears, and the bandwidth of the fiber increases by at least another factor of 100 over graded index fiber.

Each type of fiber has its specific application, and its performance characteristics are tailored to that application (Table 12–1). **Step index fiber** is used where large core size and efficient coupling of source power is more important than low loss and high bandwidth. It is commonly used in short, low-speed data links with LED sources. It may also be used in applications where radiation is a concern because it can be made with a pure silica core that is not readily affected by radiation.

Most multimode fiber-optic links use 850- or 1,300-nm LEDs because the larger core readily accepts the broad output pattern of the LED. LEDs are limited to speeds of up to 200 MB/s, so they are not appropriate for very high-speed links. A new type of laser, called a VCSEL, has been developed to allow gigabit transmission over multimode fiber and is specified for both Gigabit Ethernet and 10 Gigabit Ethernet.

All single-mode links have very high bit rate and long-distance applications. They use laser sources at either 1,300 nm or 1,500 nm if longer distances

TABLE 12–1
Fiber Types and Typical Specifications

Fiber Type	Core/Cladding Diameter (μm)	Attenuation Coefficient (dBkm) 850 nm	1,300 nm	1,550 nm	Bandwidth (MHz-km)
Step Index	200/240	6			50 at 850 nm
Multimode	50/125	3	1		500 at 850 nm
Graded Index					500 at 1,300 nm
	50/125 Laser rated	3	1		2,000 at 850 nm 500 at 1,300 nm
	62.5/125	3	1		160 at 850 nm 500 at 1,300 nm
	85/125	3	1		(Obsolete)
	100/140	3	1		(Obsolete)
Singlemode	8–9/125		0.5	0.3	Very high (terahertz)
Plastic (POF)	1 mm	(1 dB/m at 665 nm)			(Low for most types)

are expected. Multimode fiber is used for short links, like LANs or security systems, and at lower bit rates.

Although there have been four graded index multimode fibers used in fiber-optic communications, one fiber is by far the most widely used, 62.5/125. Virtually all multimode datacom networks use this fiber. The first multimode fiber widely used by the telephone companies was 50/125, for its greater bandwidth for long-distance phone lines. The 50/125 fiber had the highest bandwidth of all multimode fibers and is compatible with laser sources, so it is now being reconsidered for specialized high-speed applications like Gigabit and 10 Gigabit Ethernet.

However, the small core and low NA of 50/125 fiber made it difficult to couple to the early LED sources, so many data links switched to 100/140 fiber. The 100/140 fiber worked well with these data links, but its large core made it costly to manufacture, and its unique cladding diameter required connector manufacturers to make connectors specifically for it. These factors led to its declining use. The final multimode fiber, 85/125 was designed by Corning to provide efficient coupling to LED sources and use the same connectors as other fibers. Both these fibers should be considered obsolete, but some systems may still be using them.

The telcos switched to single-mode fiber for its better performance at higher bit rates and its lower loss, allowing faster and longer unrepeated links for long-distance telecommunications. Virtually all telecom applications use single-mode fiber. It is also used in CATV, because analog CATV networks use laser sources designed for single-mode fiber. Other high-speed networks are using single-mode fiber, either to support gigabit data rates or long-distance links.

Fiber Attenuation

The most important characteristic of the fiber is the attenuation, or loss of light, as it travels down the fiber. The attenuation of the optical fiber is a result of two factors: absorption and scattering (Figure 12–3). The **absorption** is caused by the absorption of the light and conversion to heat by molecules in the glass. Primary absorbers are residual OH+ and dopants used to modify the refractive index of the glass. This absorption occurs at discrete wavelengths, determined by the elements absorbing the light. The OH+ absorption is predominant and occurs most strongly around 1,000 nm, 1,400 nm, and above 1,600 nm.

The largest cause of attenuation is **scattering.** Scattering occurs when light collides with individual atoms in the glass and is anisotrophic. Light that is scattered at angles outside the NA of the fiber will be absorbed into the cladding or transmitted back toward the source. Scattering is also a function of wavelength, proportional to the inverse fourth power of the wavelength of the light. Thus, if you double the wavelength of the light, you reduce the scattering losses by twenty-four or sixteen times. Therefore, for long-distance transmission, it is advantageous to use the longest practical wavelength for minimal attenuation and maximum distance between repeaters. Together, absorption and scattering produce the attenuation curve for a typical glass optical fiber shown in Figure 12–3.

Fiber-optic systems transmit in the "windows" created between the absorption bands at 850 nm, 1,300 nm, and 1,550 nm, where physics also allows one to fabricate lasers and detectors easily. Plastic fiber has a more limited wavelength band that limits practical use to 660-nm LED sources.

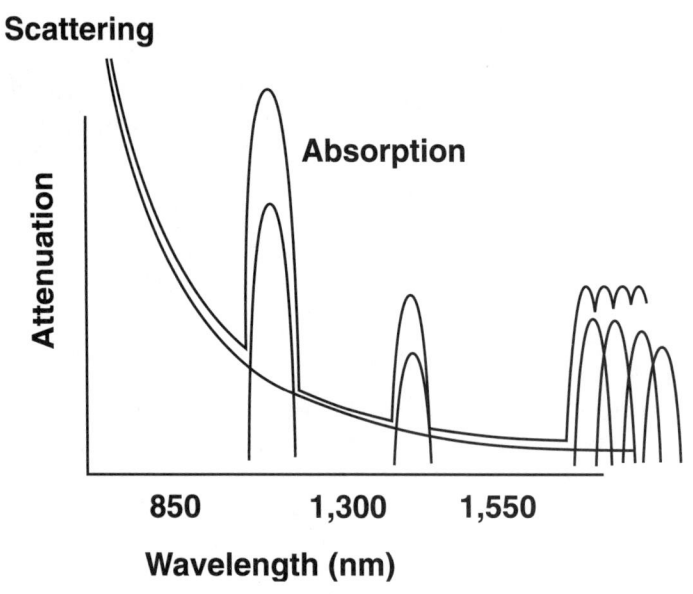

FIGURE 12–3 Fiber attenuation is caused by a combination of scattering and absorption.

Fiber Bandwidth

The information transmission capacity of fiber is limited by two separate components of dispersion: modal and chromatic. Modal dispersion (Figure 12–4) occurs in step index multimode fiber where the paths of different modes are of varying lengths. Modal dispersion also comes from the fact that the index profile of **graded index (GI)** multimode fiber is not perfect. The GI profile was chosen to theoretically allow all modes to have the same group velocity or transit times.

By making the outer parts of the core a lower index of refraction than the inner parts of the core, the higher-order modes speed up as they go away from the

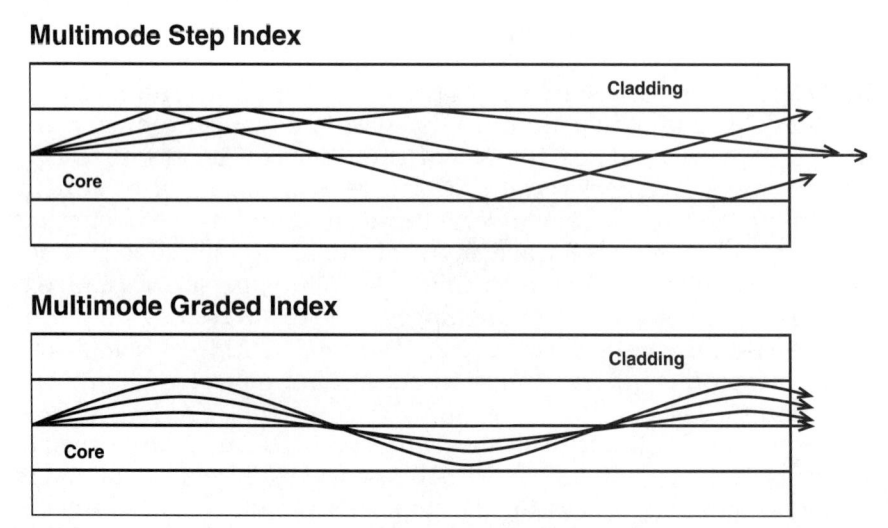

FIGURE 12–4 Modal dispersion caused by different path lengths in the core of the fiber is one aspect of fiber bandwidth.

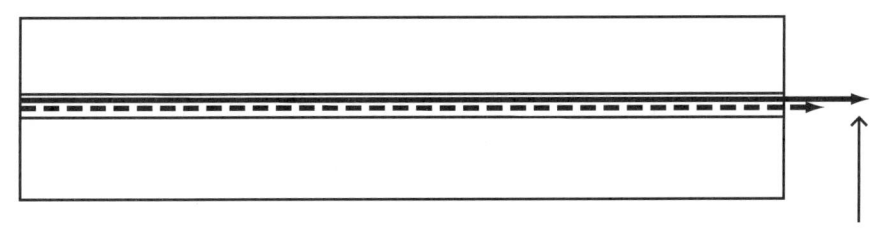

Longer wavelength light goes faster

FIGURE 12–5 Chromatic dispersion occurs because light of different colors (wavelengths) travels at different speeds in the core of the fiber.

center of the core, compensating for their longer path lengths. The index of refraction is a measure of the speed of light in the glass, so the light actually travels faster farther from the core of the fiber. This effect causes the higher-order modes to follow a curved path that is longer than the axial ray (the "zero-order mode"), but by virtue of the lower index of refraction away from the axis, light speeds up as it approaches the cladding and takes approximately the same time to travel through the fiber. Thus, the "dispersion" or variations in transit time for various modes is minimized and the bandwidth of the fiber is maximized.

Chromatic dispersion is caused by the light of different wavelengths traveling at different speeds. The index of refraction of glass is also a function of the wavelength of light, as is shown by the dispersion of sunlight into a spectrum by a prism. The wavelength and spectral width of the LED or laser source can affect the bandwidth, because longer-wavelength light travels faster through glass (Figure 12–5), so the light will be dispersed when traveling through the fiber. The amount of dispersion is determined by the characteristics of the glass in the core of the fiber and the spectral characteristics of the source. LEDs are comprised of more wavelengths of light and therefore are more affected by chromatic dispersion than lasers.

Single-mode fiber does not have modal dispersion, but its ultimate bandwidth is limited by the spectral characteristics of the laser source. Thus, the actual bandwidth of a fiber is determined by the characteristics of the fiber itself and the source used in the data link. Link designers must consider all these factors when designing systems that use multimode fiber.

Bending Losses

Fiber is subject to additional losses as a result of stress. In fact, fiber makes a very good stress sensor. Fiber-optic cables are specifically designed to prevent fiber from being stressed or damaged by the environment in which it is installed. It is also mandatory to minimize stress and/or stress changes on the fiber when manufacturing the cable, installing it, and making measurements.

Fiber-Optic Cable

The main role of fiber-optic cable is to protect the fiber. Cable comes in many different types, depending on the number and types of fibers and the environment where it will be installed. One must choose fiber-optic cable carefully, because the choice will affect how easy it is to install, splice, or terminate, and most important, what it costs.

Choosing a Cable

Because the job of the cable is to protect the fibers from the hazards encountered in an installation, there are many types from which to choose. Cable choice depends on where the cable will be run. Inside buildings, cables do not have to be as strong to protect the fibers, but they have to meet all fire code provisions. Outside buildings, cable type depends on whether the cable is buried directly, put in conduit, strung aerially, or even placed underwater.

The best source of cable information is cable manufacturers. Contact several of them (two minimum, three preferred) and give them the details of the installation. They will want to know where the cable is going, how many fibers you need, and what kind you need (single-mode or multimode, or both in hybrid cables). Some cables have metal strength members or even metal signal or power cables; they are called composite cables. The cable companies will evaluate your requirements and make suggestions, and then you can get competitive bids.

Because the application will call for a certain number of fibers, consider adding spare fibers to the cable. That way, spares will be available if you break a fiber or two when splicing, breaking out, or terminating fibers. And always consider future expansion. Most users install many more fibers than needed, especially by adding single-mode fiber to multimode fiber cables for campus or backbone applications. It is not uncommon to install more than twice as many fibers as needed to allow for future expansion.

Cable Types

All cables share some common characteristics. They all include various plastic coatings to protect the fiber, from the buffer coating on the fiber itself to the outside jacket. All include some strength members, usually a high-strength "aramid" yarn often called "Kevlar," which is the duPont trade name, to use in pulling the cable without harming the fibers. Larger cables with more fibers usually have a fiberglass rod down the middle for more strength and to limit the bend radius. The following are the standard cable types, although the cable makers sometimes have slightly different names for them.

Simplex cable and zip cord (Figure 12–6). A simplex cable consists of one fiber, with a 900-micron buffer coating, Kevlar strength member, and PVC jacket. The jacket is usually 3 mm (1/8 inch) in diameter. Zip cord is simply two of these cables joined with a thin web. It is used mostly for patchcord and backplane applications, but zip cord can also be used for desktop connections.

Distribution cables (Figure 12–7). They contain several 900-micron-buffered fibers bundled under the same jacket with Kevlar® or fiberglass rod reinforcement.

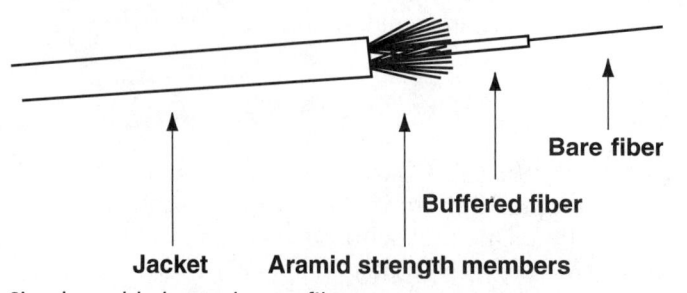

FIGURE 12–6 Simplex cable has only one fiber.

Optical Fiber and Cable • 275

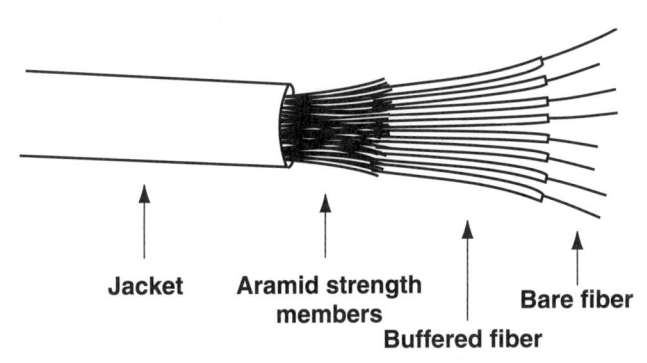

FIGURE 12-7 Distribution cables are similar to simplex cables, except they have several fibers in the center of the cable.

These cables are small in size and used for short, dry conduit runs, and for riser and plenum applications. The fibers are double-buffered and can be directly terminated, but because their fibers are not individually reinforced, these cables need to be broken out with a "breakout box" or terminated inside a patch panel or junction box.

Breakout cables (Figure 12–8). These cables are made of several simplex cables bundled together. This is a strong, rugged design, but they are larger and more expensive than the tightpack or distribution cables. This cable is suitable for conduit runs and for riser and plenum applications. Because each fiber is individually reinforced, this design allows for quick termination to connectors. Breakout cable can be more economical where fiber count is not too large and distances are not too long because it requires so much less labor to terminate.

Loose tube cables (Figure 12–9). These cables are composed of several fibers together inside a small plastic tube, which are, in turn, wound around a central strength member and jacketed, providing a small, high fiber count cable. This type of cable is ideal for outside plant trunking applications because it can be made with the loose tubes filled with gel or dry water-blocking material to prevent harm to the fibers from water. It can be used in conduits, strung overhead, or buried directly into the ground. Because the fibers have only a thin buffer coating, they must be carefully handled and protected to prevent damage.

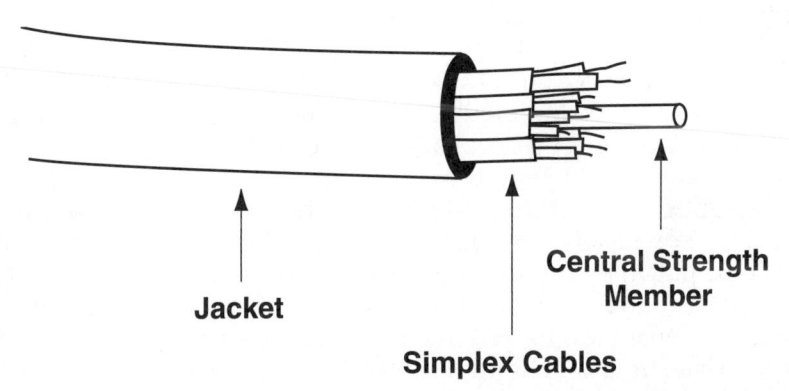

FIGURE 12-8 Breakout cables are a number of simplex cables wound around a central strength member and jacketed.

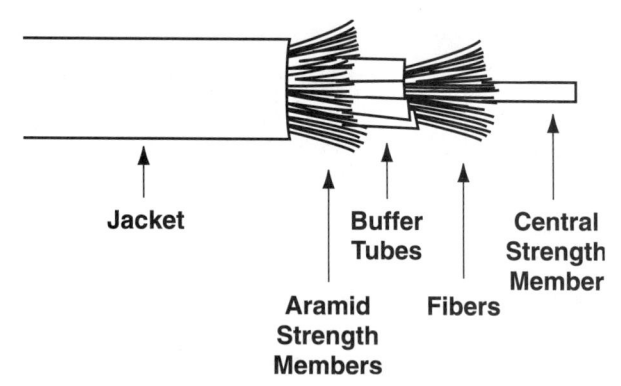

FIGURE 12–9 Loose tube cables have many small tubes that can carry up to twelve fibers each.

Other types. There are other cable types like ribbon cable, and there are different names given to the types already discussed. Every manufacturer has its own favorites, so it is a good idea to get literature from as many cable makers as possible. And do not overlook the smaller manufacturers—often they can help you save costs by making special cable for you.

Cable Ratings and Markings

All indoor cables must carry identification and ratings per *Chapter 770* of the *NEC®*. Cables without markings should never be installed, because they will not pass inspections. The ratings are:

OFN	optical fiber nonconductive
OFC	optical fiber conductive
OFNG or OFCG	general purpose
OFNR or OFCR	riser rated cable for vertical runs
OFNP or OFCP	plenum rated cables for use in air-handling areas
OFN-LS	low smoke density

Fiber-Optic Cable Installation

Installing fiber-optic cable is simplified by the hardy construction of the cable itself. Although the fiber is usually glass, which is perceived as fragile, it is actually stronger than steel. However, if it is bent over too tight a radius, especially under tension, it may fracture. Manufacturers of cable, therefore, design the cable to protect the fiber under stress. Although UTP is limited to a 25-pound pulling tension, most indoor fiber-optic cables are rated at over 100 pounds, and outside plant cables may be rated to 600 pounds or more.

All fiber-optic cables must be pulled by the strength members unless the cable has been specifically designed to be pulled by the jacket. Most cables are designed to be pulled by the strength members included in the cable. Typically, duPont Kevlar® or another aramid fiber will be included as a strength member for pulling. In preparation for pulling, the cable jacket should be stripped, fibers and any internal stiffeners cut off, and the pulling eye attached to the strength members only.

Never pull a cable by the fibers or harm will be done to them. Pulling fibers by the jacket usually results in the cable stretching under tension, then retracting, causing the fibers to be put under great stress. Only specialized cables with double jackets or armoring can normally be safely pulled by the jacket.

Under circumstances where the tension is not too large, small fiber count cables can be pulled by wrapping several turns around a large diameter mandrel to distribute the tension along a length of the cable to reduce the stress on any part of the jacket. The cable must not overlap on the mandrel and must be kept snugly wound on the mandrel. Fiber spools make excellent mandrels for pulling smaller cables.

Although fiber-optic cable can withstand great tension, it still requires care in installation. Twisting the cable is potentially harmful, so it should be unspooled by rolling directly off the reel, not off the ends. If it is unspooled for pulling, it can be laid on the ground in a "figure 8" pattern, which prevents twisting. Even when pulling, a swivel eye should be used to prevent twisting from the pulling rope or tape.

If the fiber is to be pulled around a corner, care should be taken to minimize both the pulling tension and bending radius. Observe the manufacturers' recommendations, or if they are not known, assume a bending radius under tension of twenty times the cable diameter.

Fiber-optic cable should not be left unsupported, nor should it be covered by heavier copper cables. If cable trays are used, fiber should be installed last, on top of the copper cables or lashed to the bottom of the tray. If "J hooks" are used with fiber, use the wide ones specifically designed for Cat 5E/6.

Outside plant cable can be direct buried or pulled in conduit. Long cable runs, up to several kilometers, can be installed with proper cable lubrication and pullers that monitor and limit tension. Lubricants should be chosen for compatibility to the cable jacket to prevent long-term damage to the cable. If a single run is desired but it is too long or has too many bends for a single pull, the cable can be pulled from an intermediate point, despooled into a figure 8, and pulled in the opposite direction.

Aerial installations require a self-supporting cable or attachment to a messenger. All dielectric aerial cable is available, as well as cable with an attached metal messenger for support, but many outside cables can be lashed for aerial installation.

It is wise to try to pull cable between locations without splicing to reduce installation complexity and cost. The cost is not only the cost of making a splice but also that of making the enclosures and providing proper mounting for the enclosures. Connectors and patch panels or boxes prove to be more cost-effective in most installations.

Fiber-optic cable is typically installed, then terminated on-site. Cable can sometimes be installed with connectors already installed if proper precautions are taken to protect the connectors. A protective boot must be installed over the connectors and attached to the strength members of the cable before pulling. Because this makes the cable much more bulky, pulling becomes more difficult.

Two other alternatives to field termination are available. You can terminate one end of the cable then pull the unterminated end, reducing the number of field terminations by one-half. Or you might consider some of the new multifiber connectors that have up to twelve fibers in a connector that is smaller than some single-fiber connectors.

Whenever pulling preterminated cable, remember that accurate length calculations are mandatory to prevent wasting cable or being too short. Consider vertical runs (such as from a floor outlet to above a ceiling) and service loops as part of the length.

In fact, service loops should be included on all cable installations. This extra length could be critical if splicing for restoration or retermination ever becomes necessary. Coil up the excess fiber where it will not be harmed but where its location will be obvious when it is needed.

Summary

- Fiber-optic cable is composed of a light-carrying core and a cladding that transmits light through internal reflection.
- Fiber core may be glass or plastic surrounded by a layer of cladding, which helps to transmit the light through the cable.
- Two basic types of fiber include multimode and single-mode.
- Multimode fiber transmits light over many different pathways called modes.
- The highest angle that the light is accepted into the core defines the numerical aperture of the cable.
- Two types of multimode fiber exist: step index and grade index.
- Step index is composed of one type of glass.
- Light travels in straight lines.
- Grade index is composed of many different layers of glass having different indices of refraction, the index profile of which approximates a parabola.
- Grade index accommodates increased bandwidth and capacity over that of step index.
- Fiber attenuation is a result of two factors: absorption and scattering.
- Fiber-optic cable is specifically designed to prevent damage by the environment.
- The types of manufactured cable include simplex, zipcord, distribution cables, breakout cables, and loose tube cables.
- The type of job determines the type of cable to be used.
- Loose tube cables are made for outdoor use.
- Care must be taken when installing fiber optical cable so as to not stress or overly bend the fiber.
- Indoor fiber-optic cable are rated to 100 lb. of pull tension, and outdoor cable to 600 lbs. or more.

Key Terms

Absorption That portion of fiber-optic attenuation resulting from conversion of optical power to heat.

Cladding The lower refractive index optical coating over the core of the fiber that "traps" light into the core.

Graded index (GI) A type of multimode fiber that uses a graded profile of refractive index in the core material to correct for dispersion.

Mulitmode fiber A device that removes optical power in higher-order modes in fiber.

Numerical aperture (NA) A measure of the angular acceptance of an optical fiber.

Scattering The change of direction of light after striking small particles that causes loss in optical fibers.

Single-mode fiber A fiber with a small core, only a few times the wavelength of light transmitted, that allows only one mode of light to propagate; commonly used with laser sources for high-speed, long-distance links.

Step index fiber A multimode fiber where the core is all the same index of refraction.

Review Questions

1. What part of the fiber carries the light? What part traps light in the core?
2. What is defined by the highest angle at which the light is accepted into the core of the fiber?
3. In multimode fiber, does the light travel faster near the center or the outside of the core? Why?
4. How many times more is the bandwidth capacity of a multimode graded index cable than a step index cable?
5. What fiber can support gigabit data rates over long distances?
6. What two factors cause attenuation?
7. In what wavelength windows do fiber-optic systems transmit?
8. List two types of dispersion and define them.
9. What is the main purpose of the fiber-optic cable?
10. What standards must inside cables meet?
11. What fibers does a hybrid cable contain?
12. What part of the cable do you use to pull the fiber cable?
13. What cable rating is used in an environmental air area?
14. Which fiber cable is easy to terminate?
15. The cable that has a gel to prevent water getting to the fiber is called what?
16. Name the cable rating or marking of the fiber cable used in vertical runs.
17. When you do not know the bend radius, what should you use as a "rule of thumb"?
18. How may you accomplish a long single-pull run?
19. How can you reduce the number of field terminations?

Classification of Circuits, *Article 725* of the *NEC®*

OBJECTIVES *After studying this chapter, you should be able to:*

- Explain the differences between remote-control, signaling, and power-limited circuits.
- Define Class 1, Class 2, and Class 3 circuits.
- Identify Class 1, Class 2, and Class 3 circuit identifications and markings.
- Identify Class 1 Power-Limited Circuits, including conductor size, use, and insulation rating. List their power source requirements.
- Describe Class 1 circuit overcurrent protection requirements and designate the location of such protection devices.
- Identify acceptable practices for placing conductors of different Class 1 circuits in the same cable, enclosure, or raceway.
- Identify Class 2 and Class 3 circuits including conductor size, use, and insulation rating. List their power source requirements.
- Explain wiring methods for Class 2 and Class 3 circuits for both supply-side and load-side applications.
- Explain reclassification guidelines of Class 2 and Class 3 circuits.
- Define the separation requirements between Class 1 circuits and Class 2 and Class 3 circuits.
- Outline the installation requirements for conductors of different Class 2 and Class 3 circuits in the same cable, enclosure, or raceway.
- Calculate the number and size of conductors in a raceway.

OUTLINE Types of Electrical Circuits

Article 725 of the *NEC®*

Classification of Circuits and Class 1

Class 2 and Class 3 Circuits

Wiring Methods for Class 2 and Class 3 Circuits, Supply-Side and Load-Side Applications, *Sections 725.51* and *725.52*

Power-Limited Tray Cable and Instrumentation Tray Cable

Reclassification of Class 2 and Class 3 Circuits, Markings, and Separation Requirements

The Installation Requirements for Multiple Class 2 and Class 3 Circuits and Communications Circuits, *Article 725.56*

Support of Conductors and Cables

Calculate the Number and Size of Conductors in a Raceway

Types of Electrical Circuits

Electrical wiring in most modern office buildings, shopping malls, schools, colleges, hospitals, and industrial manufacturing plants may often include any or all of the following types of circuits and systems:

- Electric light and power circuits
- Motor control circuits
- Instrumentation signaling circuits
- Process control circuits
- Thermocouple circuits—for measuring temperature
- Distributed Control Systems (DCS)—for controlling large-scale automation processes of complex building systems
- Programmable Logic Controllers (PLCs)—to control and automate repetitive manufacturing processes
- Communication systems—including audio, video, radio, television, CCTV, computer networking, and wireless.

Except for the electric light and power circuits, all others are examples of specialized circuits, which as the *NEC®* states, are not considered an integral part of a device or appliance (*NEC®, Section 752.1*). Wiring not considered an integral part of a device or appliance includes all circuit conductors that are connected and run externally, to and from devices, as required by the design of a system and the specifics of a location. In comparison, integral circuit wiring is typically factory installed and not subject to change or alteration by the user. Integral circuit wiring is essentially the internal functional wiring of a device. In most cases, integral circuit wiring is concealed within a device, and is inaccessible to the user or system installer. An example is the internal wiring of a washing machine. Except for the external power cord, there are no electrical connections to be made externally to the device since all the control circuits and system wiring is self-contained within the unit.

The connection of a coaxial cable from a parking lot video camera to a guard station video monitor would be an example of a communication circuit. The wiring of the circuit conductors between the camera and the monitor is not considered an integral part of the two devices since it must be externally applied and installed according to the unique physical parameters and restrictions of the parking lot and building. Also, the external cable does not actually make the camera or monitor function; it simply transfers a communication signal between the two devices.

Obviously, each installation will be different, and in some cases, specialized hardware and wiring techniques may be required under special circumstances; special circumstances may include hazardous locations, as outlined in Chapter 5 of the *NEC®*, or issues of life-safety.

The specific definitions, classifications, and requirements of specialized circuits, other than electric light and power, which are not considered an integral part of a device or appliance, are given in *Article 725* of the *NEC®*.

Article 725 of the *NEC®*

Article 725 of *The National Electrical Code®* (*NEC*) outlines the requirements of four general categories of circuits that are considered separate from those of electric light and power. They are remote-control circuits, signaling circuits, power-limited circuits, and communication circuits. Such circuits are typically designed and designated for a specific use and purpose. In addition, the requirements of their power supplies are often made more restrictive to those of electric light and power circuits. As an example, **Power-limited circuits** typically operate from lower levels of voltage and current than the rest, and often implement some form of internal, active current limiting.

It is important to note that the wiring methods and installation requirements of electrical circuits, as laid out in Chapters 1 through 4 of the *NEC®* still apply in all cases, except where they are amended by *Article 725*. *Article 725* does not replace Chapters 1 though 4 of the *NEC®*, but instead only amends certain requirements, as determined by the type and classification of a circuit.

When designing and laying out an installation, remote-control circuits, signaling circuits, power-limited, and communication circuits must each be further classified as either a Class 1, Class 2, or a Class 3 circuit, depending on the voltage and current requirements of the connecting loads and the type of power supplies being used. Additional considerations will also include the size and placement of circuit conductors, type of conductors, pipe-fill, overcurrent protection, and the requirements of power supplies.

Let us first define the four types of circuits and then the specific ratings and variables of Class 1, Class 2, and Class 3 circuits.

Remote-Control Circuits

Remote-control circuits are circuits that control other circuits. Control voltages can often vary from as high as 600 volts to as low as 5 volts, depending on the type of system. As an example, motor-control circuits will typically use 120 volts on starter coils, but many other types of control circuits will operate from 24 volts or less. Low-voltage, low-current circuits are commonly used to control high-voltage, high-current systems. Coiled relays, transistors, and SCRs will often be used to accomplish the task. The inputs to these devices will typically require control voltages ranging from 5 to 24 volts, with current levels in the milliamps. Such minimal input specifications have the ability to control large output loads due to the electrical isolation and high-impedance values measured between the inputs and outputs. Common uses for remote-control circuits are motor controls, elevators, conveyor systems, automated processes, and garage door openers.

Signaling Circuits

Signaling circuits are circuits that activate notification devices. Examples may include lights, doorbells, buzzers, sirens, annunciators, or alarm devices.

Power-Limited Circuits

Power-limited circuits are limited in output and capacity by the use of any or all of the following types of devices: overcurrent devices, overvoltage devices, or by internal, active electronic circuitry. The power supplies of such circuits may be used on remote-control or signaling circuits, as well as in various other types of electronic circuits, including communications circuits, audio, video, computers, or any other type of specialty low-power application. Chapter 9, *Table 11(A)* and *(B)* of the *NEC®* outlines the limitations of Class 2 and Class 3 power sources for AC and DC applications. In addition, Chapter 9, *Table 12 (A)* and *(B)* outlines the power source limitations for AC and DC power-limited fire alarm circuits (PLFA). We will be discussing these tables in more detail later in the chapter.

Communication Circuits

Communication circuits include telegraph and telephone, alarm systems, radio and television systems, community antenna television (CATV) and radio distribution, computer networks, and network-powered broadband systems. While *Article 725* briefly discusses communication systems, as related to the classification of circuits, the installation specifics of such systems are covered in far greater detail in *Articles 800, 810, 820,* and *830* of the *NEC®*.

Classification of Circuits and Class 1

Once a circuit has been designated as either remote-control, signaling, or power-limited, it must then be further classified as a Class 1, Class 2, or Class 3 for cabling and installation purposes. The circuit classification then defines the type of power supply to be used, as well as the voltage, current, and insulation requirements of the conductors.

Class 1 Circuit

Class 1 circuits typically operate from higher voltage and current levels than those of Class 2 or Class 3 circuits. For this reason, they are considered to be more of a shock hazard and danger to unknowing individuals. The installation of such circuits must therefore be performed by licensed electricians, in accordance with Chapters 1 though 4 of the *NEC®*, as if they were electric light and power circuits. Licensed power-limited technicians and low-voltage installers are not allowed to service or install such circuits. Two types of Class 1 circuits exist; they are Class 1 remote-control and signaling circuits and Class 1 power-limited circuits.

Class 1 Remote-Control and Signaling Circuits

Class 1 remote-control and signaling circuits include all wiring connected between the load side of an overcurrent device and the connecting equipment. In most cases, overcurrent devices will be either fuses or circuit breakers.

    Class 1 remote-control and signaling circuits do not require power limiting and shall operate at levels up to but not exceeding 600 volts. In the majority of cases, however, such circuits will rarely exceed 120 volts.

Due to potentially high levels of voltage and current, the wiring methods for all Class 1 circuits shall be installed in accordance with Chapter 3 of the *NEC®*, just as electric light and power circuits. But be careful not to mistakenly jump to the next conclusion, as most people do. Class 1 circuits are not to be intermixed in the same cable, raceway, box, or enclosure with electric light and power circuits.

Two exceptions, however, do exist. Power conductors and Class 1 circuit conductors may be installed in the same cable, raceway, box, or enclosure when they are *functionally associated,* or when they are installed in factory or field-assembled control centers.

The term *functionally associated* means that the power conductors and the Class 1 remote-control or signaling circuits are connecting to the same piece of equipment. Factory or field-assembled control centers are considered a low risk for intermixing power conductors with those of Class 1 since they are typically in locations staffed by qualified personnel who are routinely servicing and maintaining the facility.

Separation of Unrelated Circuits

To prevent the intermixing of unrelated power conductors and Class 1 circuits, power conductors must be separated from those of Class 1 by a solid and firmly fixed barrier made from a material compatible with the cable tray. As an alternative, if a barrier is not used, either the power conductors *or* the Class 1 circuit conductors must be placed inside of a metal-enclosed cable as a means of providing the necessary isolation. Two or more Class 1 circuits are allowed to occupy the same cable, cable tray, enclosure, or raceway, provided that all the conductors are insulated for the maximum voltage present.

In practice, it makes good sense to separate control circuit wiring from electric light and power conductors to help minimize the risk of unwanted electrostatic or electromagnetic noise interference on the lines. Metal conduits, while expensive and time-consuming to install, will help to act as a noise shield to high-current-carrying conductors, helping to greatly reduce the effects of unwanted interference and noise spikes on sensitive circuits.

Class 1 Power-Limited

Class 1 power-limited circuits were designed and intended to handle the demands of higher power loads where the use of Class 2 or Class 3 would prove to be insufficient or impractical. While not as common as Class 1 remote-control and signaling circuits, Class 1 power-limited circuits are used in a variety of low-voltage applications, most commonly on the damper controls of environmental air systems, typically found in commercial and industrial building spaces.

Class 1 Power-Limited Circuit Specifications

Class 1 power-limited circuits shall be supplied by a source having a limited output of 30 volts (AC or DC) and 1,000 volt-amps.

Class 1 circuits are not allowed in the same cable with communication circuits.

Class 1 transformers must comply with *Article 450* of the *NEC®*.

Class 1 power-limited circuits should only be supplied through listed and labeled, Class 1, power-limited sources. *Section 725.21* A 1 and 2 outlines the specific details and specifications of these supplies.

Class 1 power supplies, not including transformers, shall have a maximum output of 2,500 VA. The product of the maximum voltage and maximum current output shall not exceed 10,000 VA; these ratings are determined when overcurrent devices are bypassed.

Class 1 Overcurrent Protection Devices

Transformer supplies that feed Class 1 circuits must be protected by an overcurrent device rated to 167 percent of the power supply rating. The overcurrent device can be built into the power supply, but it cannot be interchangeable with devices of a higher rating; interchangeable fuses are not allowed.

Class 1 **overcurrent protection** devices must be located at the point of supply, where the conductors receive their power.

Single-phase, single-voltage transformers may have their overcurrent devices placed on the primary side of the circuit, provided that the maximum rating does not exceed the value determined by multiplying the secondary load current by the secondary-to-primary voltage ratio of the transformer. Consult *Section 450.3* for further details on transformer overcurrent protection.

Transformers having multiple secondary taps or windings must have their overcurrent devices placed on the secondary side of the circuit, at the point of supply. While it is true that a device on the primary side would be able to protect the transformer from a total power overload, it would not be able to differentiate or protect individual secondary outputs to their required power-limited maximums.

Likewise, each output of a multi-output electronic power source must be protected from shorts and overloads individually, through separately connected overcurrent protection devices. While a fuse or breaker on the input side would be able to protect the source from a total power overload, it would also not be able to differentiate and protect the individual outputs to their required power-limited maximums.

Class 1 Conductor Ampacity Ratings

The load capacity for 18 AWG conductors shall not exceed 6 amps; 16 AWG conductors shall not exceed 8 amps, as listed in *NEC®, Section 402.5*.

Ampacity levels for conductors larger than 16 AWG must be taken and calculated from *Section 310.15* and *Tables 310.16* through *310.21*. These calculations should be left to licensed electricians and are beyond the scope of this book.

Insulation Requirements

Insulation requirements for Class 1 conductors 18 AWG and 16 AWG shall be suitable for 600 volts. Conductors larger than 16 AWG shall comply with *Article 310*. All conductors must be listed for use in Class 1 circuits.

As stated earlier, all Class 1 circuit calculations and installations must be performed by licensed electricians. The Class 1 circuit ratings and regulations presented in this book have been included for general informational purposes only, as a comparison to those of Class 2 and Class 3 circuits. Table 13–1 has been included for such a comparison.

TABLE 13–1
Comparison of Class 1, Class 2, and Class 3 Circuits

Class of Circuit	Insulation Rating of Conductors	Circuit Voltage Limit	Power Limit	Load Current Maximum
Class 1 Remote-Control and Signaling	600 Volts Conductors larger than 16 AWG shall comply with NEC®, Article 310	600 Volts	No limit	See NEC®, Table 402.5 16 AWG not more than 8 Amps 18 AWG not more than 6 amps The ampacity of conductors larger than 16 AWG shall be calculated according to NEC®, 310.15 Flexible cords shall comply with NEC®, Article 400
Class 1 Power-Limited	Same as above	30 Volts	1000 VA	Same as above
Class 2	150 Volts	150 Volts See Chapter 9, Table 11 A and B of the NEC®	100 VA Exceptions: AC between 30 and 150 volts is limited to 5 mA × Vmax DC between 60 and 150 volts is limited to 5 mA × Vmax See Chapter 9, Table 11 A and B of the NEC®	See Chapter 9, Table 11 A and B of the NEC®
Class 3	300 Volts	150 Volts See Chapter 9, Table 11 A and B of the NEC®	100 VA	See Chapter 9, Table 11 A and B of the NEC® Conductors shall not be smaller than 18 AWG

Class 2 and Class 3 Circuits

Class 2 Circuit

Class 2 circuits include all wiring connected between the load side of a listed and labeled Class 2 power source and the connecting load or equipment. Class 2 circuits are power-limited, and considered safe from a shock hazard and fire initiation point of view.

Class 3 Circuit

Class 3 circuits include all wiring connected between the load side of a listed and labeled Class 3 power source and the connecting load or equipment. Class 3 circuits are also power-limited, and considered safe from a shock hazard and fire initiation point of view; but they often operate at higher voltage and current levels, as compared to Class 2 circuits, and must therefore include additional safeguards to help protect individuals from possible electric shock.

Class 2 and Class 3 Power Sources

Class 2 and Class 3 circuits can derive their power from any of the following sources:

- A listed and labeled Class 2 or Class 3 transformer
- A listed and labeled Class 2 or Class 3 power supply
- Any alternative, listed equipment marked and identified as a Class 2 or Class 3 power source.

Alternative power sources may include computer circuit cards that supply Class 2 or Class 3 power to circuits, stored energy sources, batteries, and thermocouples. Dry cell batteries are considered to be inherently limited Class 2 power sources, provided that they are less than 30 volts and have a maximum current capacity equal to or less than series-connected #6 zinc cells.

A power source designed to be **inherently limited** is internally clamped and unable to deliver more than a specific amount of energy to a load. Any attempt to push an inherently limited source past its maximum limit will cause it to either shut down or self-destruct, in a safe manner. Both Class 2 and Class 3 power sources are available as either inherently limited or non-inherently limited. Power sources that are inherently limited do not require overcurrent protection, to do so would be fine, but redundant.

An example of an inherently limited power source is an inherently limited Class 2 transformer. Such a supply often contains an internal fusable link, buried deep within the windings of the secondary. The link, however, is not user serviceable, thus forcing the operator or technician to replace the entire device if load currents were to ever rise above maximum levels. The actual replacement of the link would require a total rewinding of the transformer secondary, a job most people would find highly impractical and not easy to accomplish.

Chapter 9, *Table 11* (*A* and *B*) of the *NEC*® outlines the specifics of AC and DC, Class 2 and Class 3 power sources. Specifications for both inherently limited and not inherently limited sources are given in the table. In general, it should be noted that the power output maximum for almost all categories of Class 2 and Class 3 sources is 100 VA. The only exception is that a type AC, Class 2, inherently limited source, between 30 and 150 volts must be limited to 5 mA; and a type DC, Class 2, inherently limited source, between 60 and 150 volts must also be limited to 5 mA.

Let's now look at the specifications for current limitation, I_{max}, and maximum current, in *Table 11* (*A* and *B*). At first glance, it may appear a bit confusing. You will notice that I_{max} in the 0 to 20 volt, inherently limited column, is listed as 8 amps; but the maximum current specification listed at the bottom of the table states 5 amps. To understand the difference you must first read note number 1.

Note number 1 explains that the I_{max} specification is a measured current limitation when using a noncapacitive load, with the current-limiting protection of the power source disabled or bypassed. The 8 amp I_{max} rating, therefore, represents a worst case scenario rating for the power source if the current-limiting protection were to ever fail or become bypassed. The 5-amp rating is the maximum operating current of the power source under normal circuit conditions.

The power calculation for the above example can then be shown to be 20 volts × 5 amps = 100 VA; this is correct. (Remember that we were in the 0–20 volt column of the table.)

Notice too that VA_{max} for power sources not inherently limited will be limited to 250 VA, noticeably higher than the 100 VA circuit limit. As in the previous example for maximum current, note 1 explains that the VA_{max} measurements are to be taken when the current-limiting protection has been bypassed or disabled.

Be careful not to confuse the current-limiting protection with the current-limiting impedance. Note number 1 also states that the current-limiting impedance is *not* to be bypassed. The current-limiting impedance of a device is a series-connected output impedance, used to help match the supply to a desired load impedance, possibly 50, 75, or 600 ohms. Impedance matching helps to achieve maximum power transfer from the source to the load, making the circuit more efficient. The current-limiting protection, on the other hand, is used to limit load currents to some maximum level.

The Interconnection of Power Sources

As a final note, Class 2 and/or Class 3 power sources must never have their outputs connected in parallel across a single circuit or load, or interconnected in any manner to boost the over-all current output or capacity of the sources. Since the rated outputs of individual Class 2 and Class 3 power sources are required to be limited to 100 VA, for purposes of safety, they must be kept separate. Connecting two Class 2 or two Class 3 power sources in parallel would ultimately increase the total output power capacity of the source above the 100-VA limit. The circuit would then need to be reclassified as Class 1 and approved by the authorities having jurisdiction.

Wiring Methods for Class 2 and Class 3 Circuits, Supply-Side and Load-Side Applications, *Sections 725.51* and *725.52*

The supply side of all Class 2 and Class 3 power sources must be wired in accordance with Chapters 1 through 4 of the *NEC®*; overcurrent devices on the supply side must not be rated above 20 amps. Remember too that supply side connections are considered electric light and power circuits; for this reason, they must be installed by a licensed electrician. A power-limited technician can only work on the load side of a power-limited source or circuit, while the supply side is reserved for an electrician.

Wiring methods of Class 2 and Class 3 power sources on the load side of the circuit can be accomplished in one of two ways:

1. The circuits can be reclassified as Class 1 and installed in accordance with other Class 1 circuits. In such cases, the Class 2 and Class 3 circuit markings must be eliminated, and the derating factors of the conductors, as given in *Section 310.15 (B)(2)(a)*, shall not apply. Reclassification as a Class 1 circuit will also require the installation and future servicing to be conducted by a licensed electrician. To reclassify a Class 2 or Class 3 circuit would ultimately mean that the circuit is now off limits to power-limited technicians.
2. The Class 2 and Class 3 circuits can be installed under the requirements of *Sections 725.54, 725.61,* and *725.71,* which lay out the listing and marking of conductors, the installation and separation requirements from non-power-limited circuits, and the application requirement of cables with regard to plenum, riser, general-purpose, and dwelling, and their permitted substitutions.

Voltage and Insulation Requirements for Class 2 and Class 3 Circuits

Class 2 circuits shall be rated for not less than 150 volts.

Class 3 circuits shall be rated for not less than 300 volts.

Single conductor, Class 3 wires shall not be smaller than 18 AWG, and shall be listed and labeled as CL3, CL3P, CL3R, or CL3X cable.

Class 2 cables shall be listed and labeled as CL2, CL2P, CL2R, or CL2X.

The use and type of cable needed for an installation shall be determined by the location. Examples include general purpose location (CL2 or CL3), plenum space (CL2P or CL3P), riser (CL2R or CL3R), or dwelling (CL2X or CL3X).

Cable Substitutions

As an installer of electrical wiring, it is important to know which types of cable may be used for the various types of installations. In many cases, certain types of cable may be substituted for others, as permitted by location. The examples include plenum, riser, general-purpose, or dwelling. *Table 725.61* of the *NEC®* details the specific use, classification, reference number, and permitted substitutions of all types of Class 2, Class 3, and **Power-Limited Tray Cable (PLTC).**

Based on the permitted substitution chart, *Table 725.61* of the *NEC®*, Communication Plenum (CMP) is shown to be suitable for use in any location. The logic then follows that plenum-grade cable can be used as a substitute for riser cable, riser cable can be substituted for general-purpose cable, and general-purpose cables can be used in one- and two-family dwellings. The allowable substitutions drop down through the list, but are not backward compatible.

Power-Limited Tray Cable and Instrumentation Tray Cable

Power-Limited Tray Cable (PLTC)

Power-limited tray cable (PLTC) is a special type of listed and labeled, nonmetallic-sheathed cabling, intended for use in cable trays of factories or industrial establishments. *Table 725.61* of the *NEC®* also states that PLTC may be used as an approved substitute for Class 2 or Class 3 wiring in general-purpose or dwelling locations.

PLTC is rated for 300 volts and can be purchased in sizes 22 AWG through 12 AWG. For industrial or factory settings, another alternative to PLTC would be to use **Instrumentation Tray Cable (ITC)** (*Article 727* of the *NEC®*). The main difference is that ITC can only be used and installed in industrial establishments that are maintained and supervised by qualified personnel. That being the case, a further benefit instrumentation tray cable has over PLTC is that it can be installed under the raised floors of control rooms or equipment rooms, and without the use of cable trays or raceways. The only stipulation is that cables must be securely mounted and protected against physical damage. Instrumentation tray cables installed in open settings without the use of conduits or raceways, between cable trays and equipment racks, may extend to lengths of 50 feet, maximum; in such cases the cables must be mounted and secured at intervals of not more than 6 feet.

Since instrumentation tray cable circuits do not fall under the regulations of *Article 725*, they are not considered to be Class 1, Class 2, or Class 3 circuits; instead they are governed solely by *Article 727* of the *NEC®*. For this reason, ITC shall not be used as a replacement or substitute for Class 2 or Class 3 wiring in a non-industrial setting; only PLTC can be used as an approved substitute for Class 2 or Class 3 wiring.

ITC specifications are limited to 150 volts and 5 amps, for sizes 22 AWG to 12 AWG. Size 22 AWG is limited to 150 volts and 3 amps. Although the specifications of ITC and Class 2 cable may look similar, they are not the same, and should not be confused with each other since ITC can only be installed in industrial settings.

Some manufacturers do, however, offer a brand of cable that can be used as PLTC or ICT, having a dual rating; in such cases, the cable will be labeled appropriately for use as PLTC/ITC.

Neither PLTC nor ITC can be used on Class 1 circuits.

Reclassification of Class 2 and Class 3 Circuits, Markings, and Separation Requirements

Reclassification Guidelines of Class 2 and Class 3 Circuits

Section 725.11 (A) of the *NEC®* states that remote control circuits of safety control equipment shall be reclassified as Class 1 if an equipment failure would introduce a direct fire or life hazard to the environment. Controls such as thermostats, heating and air-conditioning circuits, water temperature devices, or electrical household control devices are not to be included or regarded as safety-control equipment. The main designation would be the failure of the equipment to introduce a *direct* fire or life hazard.

One example of circuit reclassification may include the exhaust blower fans of a paint booth. The failure of the fans to remove paint fumes from the internal atmosphere of the booth would result in the accumulation of toxic vapors; such a situation would be considered a definite safety hazard from an explosive gas point of view, as well as life threatening to the operator. In such instances, the control circuits on the blower exhaust fans would need to be reclassified as Class 1 circuits.

Part B of *NEC® Section 725.11* also discusses the issue of possible damage to the remote-control circuits of safety equipment. In such cases, where the possibility exists for potential damage to control circuits of safety equipment, all conductors shall be installed inside of either rigid metal conduits, intermediate metal conduits, rigid nonmetallic conduits, electrical metallic tubing, type MI cable, type MC cable, or be otherwise suitably protected from the environment and the elements. The control circuit conductors must be protected from possible abuse or damage to ensure that the functional use of safety equipment is not compromised or made inoperable.

Class 1, Class 2, and Class 3 Circuit Identification and Markings, *NEC® Sections 725.10, 725.61, 725.71*

Class 1, Class 2, and Class 3 circuits shall be identified at terminals and junctions in a manner that will prevent the unintentional interference by service technicians. When servicing a system there should be no confusion on the part of the technician as to which wires and cables are to be tested. The unwanted interference to other circuits can be easily avoided by simply marking and labeling conductors as to their type and proper designation.

Separation Requirements for Power-Limited Class 2 and Class 3 Circuits from Those of Class 1 and All Other Non-Power-Limited Circuit Conductors, as Covered in *Sections 725.54, 725.55 (A through J)*

A common thread exists through all of *Section 725.55, Subsections A* through *J*—the requirement that Class 2 and Class 3 circuits must be separated from electric light and power circuits, Class 1 circuits, non-power-limited fire alarm, or medium powered network-powered broadband communication circuits.

The general requirement of *725.55, Subsection A,* states that Class 2 and Class 3 circuits shall not be placed inside of any cables, cable trays, compartments, enclosures, manholes, outlet boxes, device boxes, raceway, or any other similar fitting with Class 1 or any other non-power-limited circuit conductors.

Non-power-limited circuit conductors include electric light and power conductors, non-power-limited fire alarm, or conductors of medium powered network-powered broadband communication circuits.

But, provided that separation does exist between power-limited Class 2 and Class 3 circuits and those conductors of Class 1 and non-power-limited circuits, certain allowances are permissible, as referenced by *Subsections B* through *J*.

There are a number of reasons why the separation ruling by the *NEC®* is required, and also why it simply makes good sense. The reasons include protecting conductors of power-limited, Class 1, and non-power-limited circuits from internally shorting to each other; isolating circuits based on the rights and permissions of individuals to work on and maintain them; the safety and protection of individuals from their own technical ignorance; and also to electrically isolate power-limited circuits from possible electrostatic and electromagnetic interference.

First, if power-limited Class 2 or Class 3 circuits were to ever short out across a Class 1 circuit, or to other non-power-limited circuits, the surrounding environment could very easily develop into a potential shock or fire hazard. Remember too, that the insulation requirements for Class 2 or Class 3 circuit conductors are significantly lower than those of Class 1 or electric light and power circuits. Imagine what would happen if a shorted Class 2 circuit conductor, having a maximum insulation rating of 150 volts, were to be exposed to the higher voltage and current specifications of a 600-volt rated, electric light and power conductor.

Ultimately, the Class 2 cable would not be able to hold back the potential threat, and over time the buildup of heat from the additional current flow, resulting from the short, would start to deteriorate the internal and external insulation and eventually burn the conductors. In order to prevent such an occurrence, the separation of power-limited conductors from those of Class 1 and non-power-limited conductors is required.

A second reason for the need to separate power-limited circuits from those of Class 1 and non-power-limited circuits has more to do with the rights and restrictions of those individuals intending to work on and maintain such circuits. High-power circuits are intended to be installed and maintained by trained, licensed electricians. Master electricians are allowed to work on any type of circuits ranging from electric light and power to those of power-limited and communication circuits. Master electricians are also allowed to install any type of circuit inside of a hazardous location, as stipulated by *Article 500* of the *NEC*®. Power-limited technicians, however, are far more restricted with regard to the type of work they can perform; they also are not allowed to work in hazardous locations, as is the case of the master electrician. The permissible types of circuits that power-limited technicians can install include Class 2 and Class 3 circuits, audio circuits (as referenced by *Article 640* of the *NEC*®), power-limited fire alarm circuits (as referenced by *Article 760* of the *NEC*®), fiber-optic circuits (as referenced by *Article 770* of the *NEC*®), and communication circuits (as described in *Articles 800, 810, 820,* and *830* of the *NEC*®).

For obvious reasons, power-limited circuits should be logically separated and isolated from those of Class 1 and non-power-limited circuits as a guarantee that the power-limited technicians will not accidentally interfere with conductors and circuits they are restricted from working on. Separation also helps to lower the risk of potential shorts on the lines of Class 1 or electric light and power circuits because of accidents, misunderstandings, or misconnections by service technicians.

From a safety point of view, Class 1 and non-power-limited circuit conductors should also be separated from those of power-limited circuits as a way to help protect unwitting individuals from a potential threat or shock hazard. I call this the idiot factor. Most individuals seem to believe that simply because they were able to install their own car stereo system, this somehow makes them knowledgeable enough to install all types of communication and electrical circuits. In certain situations, such beliefs can be very dangerous. To help protect such individuals from themselves and their own ignorance, having power-limited circuits separated and installed inside of isolated compartments and away from electric light and power circuits is a wise and useful practice.

The last reason why circuit separation simply makes good sense is to help protect low-voltage conductors from the unwanted electrostatic or electromagnetic interference of high-power conductors. Very often, low-voltage circuits having a high degree of sensitivity are unable to tolerate even small levels of electrical noise or interference on their lines; in some cases, noise levels as low as 10 millivolts can often generate unwanted intermittencies or circuit failures.

One example may be the need to separate audio conductors from those of electric light and power conductors. Grounded, unshielded audio lines, running long distances, will ultimately pick up an annoying 60-Hz power hum as a result of their proximity to high-current power lines. The unwanted interference will then be permanently imposed on the intended audio signal and amplified, to the great annoyance of listeners. To help prevent such unwanted interference, unbalanced audio conductors should be placed inside of insolated raceways, separate and away from high-power circuits and conductors.

The permitted subparts to *Section 725.55* of *NEC®* allow some deviation from *Subpart A*, which insists on maintaining the absolute separation and isolation of Class 1 and non-power-limited circuit conductors from those of power-limited Class 2 and Class 3 circuits. The permissions, as stated in *Subparts B* through *J*, all include some form of separation and isolation, either by barriers, isolated compartments, or by placing raceways within enclosures. Let us now look at the specifics of *Subparts B* through *J*.

Subpart B, Separated by Barriers

Class 2 and Class 3 circuits may be installed with conductors of Class 1 and non-power-limited circuit conductors where they are separated by a barrier. *Subpart C*, Raceways within enclosures—Inside of enclosures, Class 2 and Class 3 circuits may be installed inside of isolated raceways as a means or separating them from electric light and power, Class 1, and all other types of non-power-limited circuit conductors.

Subpart D, Associated Systems within Enclosures

Class 2 and Class 3 circuits may be installed inside of compartments, enclosures, device boxes, or outlet boxes with those of electric light and power circuits, Class 1 circuits or all other types of non-power-limited circuit conductors, provided that the circuits are functionally associated, and that they enter the enclosure through separate opening.

Functionally associated implies that the non-power-limited conductors are providing the means of power to the equipment within the enclosure, and the Class 2 or Class 3 conductors are being used for remote control or signaling purposes. In such a situation the Class 1 or non-power-limited conductors must be routed and secured within the enclosure to maintain a 1/4 in. separation from all power-limited conductors.

An alternative would be to install all circuits operating at less than 150 volts to ground as Class 3 circuits by using CL3, CL3R, or CL3P cabling. In such cases, the conductors extending beyond the jacketing must maintain a ¼ in. separation from all other conductors, or the Class 3 connections can be placed inside of a non-conductive barrier or non-conductive sleeve, such as a flexible tubing.

A last alternative would be to install the Class 2 or Class 3 circuits as Class 1, in accordance with all the rules and regulations of *Section 725.21* of the *NEC®*.

Subpart E, Enclosures with Single Openings

Class 2 and Class 3 circuit conductors can be included inside of enclosures with single openings, with those of Class 1 circuits and non-power-limited circuit conductors, provided that they are functionally associated; and that the Class 2 or Class 3 conductors are permanently separated from all others by a continuous, non-conductive type of insulator, like flexible tubing. When entering the enclosure, the insulator is required to be placed over the outer jacketing of the power-limited cables as an added layer of protection against possible shorts and contact with the non-power-limited conductors. Fire alarm sprinkler heads are a perfect example of this type of assembly; the high power conductors are used to power the water flow valves, and the power-limited conductors are used to control the initiating relays and also to establish the supervisory circuit for the fire alarm control panel.

Subpart F, Manholes

Inside of manholes, one of three possibilities exists when combining conductors and cabling from various types of circuits.

Either all of the electric light and power, Class 1, non-power-limited fire alarm, and/or medium power network-powered broadband communication circuits are to be installed inside of a metal enclosed cable or type UF, underground feeder cable.

Or, the Class 2 or Class 3 circuit conductors are installed inside of a firmly fixed, nonconductive barrier, such as flexible tubing, as a means to permanently separate the power-limited conductors from all other Class 1 and nonpower-limited conductors.

Or, the Class 2 or Class 3 circuit conductors are to be securely fastened to rack mounts, insulators, or any other approved supports meant to permanently separate them from all Class 1 and nonpower-limited circuit conductors.

Subpart G, Closed-Loop and Programmed Power Distribution, as Covered by *NEC®, Article 780.*

Article 780 is a special case that covers the installation of closed-loop and programmed power distribution systems, found in what are now commonly termed "smart houses." The cabling for such systems often requires a special hybrid, which combines multiple conductors, including 120-volt ac power, 24-volt dc UPS, telephone, remote-control and signaling, and coax. The hybrid cables are connected to receptacle outlets known as convenience centers, which often supply different types of signals and energy to various appliances and utilization equipment.

The closed-loop control provides a reduced shock hazard, since receptacles are only energized when the attachment plugs have been inserted into the outlet. The presence of the connected load at the convenience centers is supervised by the service center, allowing for outlets to be energized only when loads are present and de-energized when they are removed. The UPS is used by the service center to help maintain the system electronics during a transient or power outage.

Subpart H, Cable Trays

Power-limited circuits may be installed inside of cable trays with conductors of non-power-limited circuit conductors, provided that they are separated by a solid and firmly fixed barrier, of a material compatible with the cable tray.

A second alternative would be to place the power-limited conductors inside of type MC cable (metal-clad cable, as described in *Article 330* of the *NEC®*).

Subpart I, Hoistways

Inside of hoistways, designed for the purposes of elevators or dumbwaiters, power-limited circuit conductors are to be installed inside of rigid metal conduit, rigid nonmetallic conduit, intermediate metal conduit, liquid-tight flexible nonmetallic conduit, or electrical metallic tubing, as a means of separating them from all Class 1 and non-power-limited circuit conductors.

Subpart J, All Other Applications

In all other cases, Class 2 or Class 3 conductors must maintain a minimum separation of 2 inches from all Class 1 circuits and non-power-limited circuits.

The 2-inch separation, however, is not required when the Class 1 and non-power-limited conductors are place inside of isolated raceways, or are in metal-sheathed, metal-clad, non-metalic-sheathed, or type UF underground feeder cable.

The 2-inch separation is also not required if, instead, the Class 2 and Class 3 circuit conductors are placed inside of the isolated raceways, or are in metal-sheathed, metal-clad, non-metalic-sheathed, or type UF underground feeder cable.

A third and last possibility allows for the Class 2 and Class 3 circuit conductors to be separated from those of Class 1 and non-power-limited circuit conductors by placing them inside of a firmly fixed non-conductor, such as a porcelain tube or flexible tube. In all the previously listed examples of *Subparts A* through *J*, the basic rule is simple: non-power-limited conductors are required to be separated from power-limited conductors.

Electric light and power circuits, Class 1 circuits, non-power-limited fire alarm, and medium power network-powered broadband communication circuits are all considered to be non-power-limited circuits; and for the purposes of safety and fire prevention they must maintain a fixed and continuous separation from all Class 2 and Class 3 circuit conductors. Other reasons for the separation may include:

- The prevention of possible short circuits between power limited and non-power-limited conductors
- The isolating of circuits based on the rights and permissions of individuals to work on and maintain them
- As a means to help protect individuals from their own technical ignorance
- To electrically isolate power-limited circuits from electrostatic and electromagnetic interference.

The Installation Requirements for Multiple Class 2 and Class 3 Circuits and Communications Circuits, *Section 725.56*

Class 2 and Class 3 Circuits in the Same Cable, Enclosure, or Raceway

When intermixing multiple conductors of Class 2 and Class 3 circuits within the same cable, enclosure or raceway, the following rules apply:

Two or more Class 2 circuits may share the same cable, enclosure, or raceway.

Two or more Class 3 circuits may share the same cable, enclosure, or raceway.

Class 2 circuit conductors may be included inside of the same cable, enclosure, or raceway with conductors of Class 3 circuits, provided that the Class 2 conductors are insulated to the minimum requirements of the Class 3 circuit; 300 volts or better.

Class 2 and Class 3 Conductors with Communications Circuits

Class 2 and Class 3 circuit conductors are permitted to be installed in multi-stranded cables along with communication circuits, provided that they are reclassified as communication circuits, and are installed in accordance with *Article 800* of the *NEC*®. In such cases, the cabling will need to be a listed communication cable or multipurpose cable.

Composite Cables

Compared to the previous example, a composite cable is somewhat different; this is not the same as having multiple circuits running through a single multiconductor cable. Instead, cable vendors will often engineer a composite cable made up of individually listed Class 2 and Class 3 cables and combined with those of communication cables, all under a common outer jacket. The fire resistance of the entire composite cable is then tested and rated based on its own individual performance. In such cases, all conductors within the composite cable are then permitted to be reclassified as communication cables, and installed in accordance with *Article 800* of the *NEC*®.

Class 2 or Class 3 Cables and Other Circuits

Jacketed Class 2 or Class 3 circuit cables are allowed to be installed in the same enclosure or raceway with jacketed cables of:

- Power-limited fire alarm circuits
- Nonconductive and conductive optical fiber cables, in accordance with *Article 770*
- Communications circuits, in accordance with *Article 800*
- Community antenna television (CATV) and radio distribution systems, in accordance with *Article 820*
- Low-power, network-powered broadband communications, in accordance with *Article 830*.

Class 2 or Class 3 Conductors or Cables and Audio Circuits

A new ruling in the 2005 edition of the *NEC*® requires that audio circuit conductors, as described in *NEC*® *640.9 (C)*, installed as Class 2 or Class 3 wiring from an amplifier, shall not be permitted to occupy the same cable or raceway with other Class 2 or Class 3 conductors or cables. Essentially, this ruling requires that audio cables must maintain a permanent separation from remote-control or signaling conductors, even though they may both be considered as Class 2 or Class 3.

Support of Conductors and Cables

Section 725.58 of the *NEC*® states that conductors of Class 2 or Class 3 circuits are not to be strapped, taped, or attached by any means to the exterior of any conduit or raceway as a means of support. The same applies to fire alarm cables, *Section 760.58;* communication cables, *Section 800.133 (C);* coaxial cables, *Section 820.133(C);* and network cables, *Section 830.133 (B)*. One permitted case, however, does exist, as referenced by *Section 300.11(B)(2)*.

Section 300.11(B)(2) allows for a Class 2 cable to be attached to the exterior of a conduit when the conduit internally contains functionally associated power supply conductors. This specific example exists in almost every house across America: the Class 2 thermostat control wires tied to the exterior of the main electrical power conduit of the furnace. In such cases, the internal conduit wires supply power to the furnace, while the external thermostat wires provide the Class 2 control circuit; both are functionally associated to the furnace.

Except for HVAC service technicians and installers, there are very few additional cases, if any, where Class 2 conductors or cables are allowed to be attached to the exterior of a conduit or raceway. Notice too, that Class 3 conductors or cables were never even mentioned by *Section 300.11(B)(2)*. Even if they are functionally associated, Class 3 circuit conductors do not have the same option available to them as those of Class 2.

Not attaching cables to the exterior of any conduit or raceway is usually a wise and good practice. To do so would make future changes or remodeling efforts far more difficult to achieve, often requiring entire systems to be pulled out altogether, replaced, or rewired. In such instances, it will make life a lot easier to keep dissimilar systems isolated from each other and installed on separately supported mounting hardware.

As a related topic, *Section 725.8,* entitled the Mechanical Execution of Work, states that circuits shall be installed in a neat and workmanlike manner; cables installed on the outer surfaces of walls and ceilings shall be structurally supported so as not to subject them to possible damage or abuse from normal building use. Support hardware, such as hangers, straps, and staples should also be designed and used appropriately and not cause any undue stress or damage to cables.

An example of normal building use would be the opening and closing of a door. Obviously, cables should not be hanging down in such a manner so as to be trapped or pulled by the movement of a door throughout the course of a day.

A last topic related to the installation practices of cables involves the access to electrical equipment behind panels, *Section 725.7*. The section states that the accumulation of wires and cables shall not prevent the removal of access panels, including suspended ceiling tiles.

When installing cables above suspended ceilings, cables must never be allowed to rest on the top of tiles, since the accumulation and weight of cables will make it nearly impossible to remove tiles at a later date. Over time, the excessive weight of cable bundles may also cause tiles to warp or crack.

When installing cables above a suspended ceiling, cables must also never be tied to the suspended ceiling support wires. Instead, cables and conductors should be installed and supported by an approved means. Approved means indicates that the installer is using an appropriate mounting hardware acceptable to the electrical inspector. At the end of the day, conductors must be mounted up and out of the way of removable tiles or panels.

Regulation by the *NEC®* regarding the support of cables and the mechanical execution of work, simply makes good sense. In many instances, the regulations are intended to help installation technicians assume a reasonable amount of pride in their work. A job performed and installed in a neat, organized and workmanlike manner will result in a safer environment, and will most likely not develop into a potential safety or fire hazard.

Calculate the Number and Size of Conductors in a Raceway

When installing electrical wiring within a raceway, the number of conductors to be run should be limited to a certain percentage of fill. Limiting the number of conductors within a raceway will allow for the easy addition and removal of future wires, so as not to damage those already existing. In addition, not stuffing the raceway to the maximum fill will allow room for the dissipation of heat by the

EXAMPLE

Problem

What size Flexible Metal Conduit (FMT) is required for the following conductors?

5— # 14 THW   6— # 12 TW   7— # 10 THHN   8— # 8 THWN

conductors. In fact, the *NEC®* requires the installation of electrical wiring to conform to **pipe fill** restrictions.

When determining the number of conductors permissible in a particular conduit, the starting point should be the type of raceway to be used. From there you need to identify the number of conductors that will be placed inside the raceway, remembering to take into account the *NEC®* fill restrictions; the pipe must then be sized accordingly. Let's take a closer look at the steps involved.

To solve this problem, first find the article specifically dedicated to flexible metal conduit (*Article 348*). As you read *Article 348*, be aware that there are some additional details you will need to know. First, where to find the number of conductors allowed in this type of conduit, and where to find the percentage fill requirements for FMC.

Notice that *Section 348.22* refers to *Chapter 9, Table 1* of the *NEC®*. In looking at *Table 1*, you should determine that since there are more than two conductors running through the raceway, the allowable fill requirements should be 40 percent. We will use this percentage later when trying to determine the right size of pipe to use in the installation. Now, even though *Table 1* is quite small, we still cannot ignore the notes supplementing the table. *Table 1, note 6*, for example, tells exactly where to find the dimensions of different types and sizes of conductors; it also points to the appropriate table for sizing out conduit once the cable dimensions are known. To find the total area of all conductors, in square inches, we must first go to *NEC®*, Chapter 9, *Table 5*. First, locate the types and sizes of wire that will run through the raceway. Then, multiply the area of each, in square inches, by the number of conductors. In this example, the total area in square inches, for each type of conductor, will be:

14 THW   = .0209 × 5 CONDUCTORS = .1045 in.^{2}
12 TW   = .0181 × 6 CONDUCTORS = .1086 in.^{2}
10 THHN = .0211 × 7 CONDUCTORS = .1477 in.^{2}
8 THWN   = .0366 × 8 CONDUCTORS = .2928 in.^{2}

The totals for all conductors in the above problem are then added together to give a grand total of 0.6499 in.^{2}

Now, turn back to *NEC®*, Chapter 9, *Table 4*, and find Flexible Metal Conduit (*Article 348*). Look at the 40 percent fill column. Because there are more than two conductors in this problem, compare the total calculated area of all conductors, 0.6499 in.^{2}, to the trade size dimensions of various conduits listed in the table. Choose the conduit size that is large enough to fit all of the conductors. The answer should be a 1½" Flexible Metal Conduit. Why? Because the conduit one size smaller measures only .511 square inches, and this would be too small to fit the required total of .6499 square inches. As a result, the next size up, 0.743 in.^{2}, is the obvious choice.

In the previous example, none of the cables were multiconductor, and all were listed in Chapter 9, *Table 5*. If multiconductor cables are to be used, refer to Chapter 9, *Table 1, Note 5* and *Note 9*. *Note 5* states that the actual dimension of the multiconductor cable can be used to calculate pipe fill. *Note 9* also states that for munticonductor cables having an elliptical dimension, the widest or major diameter of the cable shall be used when calculating the cross-sectional area. All that is required then, is to simply measure the widest outer diameter of the cable manually, and then calculate the cross-sectional area by using the formula, area = πr^2. Be sure to divide the diameter by 2 to obtain the radius or the calculation will be off by a factor of four.

The last point to make has to do with the 60 percent fill column in *NEC®*, Chapter 9, *Table 4*. When can you use 60 percent fill? The answer can be found in *NEC®*, Chapter 9, *Table 1, Note 4*. *Note 4* states that 60 percent fill can be used when a conduit connects between two boxes, cabinets, or enclosures, not to exceed 24 inches in length. This would be the only case for using 60 percent fill. In this special case, adjustment factors as given in *NEC® 310.15(B)(2)(A)*, need not apply.

Summary

- *Article 725* outlines the requirements of remote-control, signaling, and power-limited circuits.
- Remote-control circuits control other circuits.
- Signaling circuits activate notification devices such as lights and buzzers.
- Power-limited circuits are limited in output and capacity. Such circuits may be inherently limited or non-inherently limited.
- Circuits are classified as Class 1, Class 2, or Class 3.
- Class 1 circuits operate at higher voltage and current levels than those of Class 2 or Class 3.
- Power-limited technicians are prohibited from installing or servicing Class 1 circuits.
- Inherently limited Class 2 or Class 3 circuits will internally clamp or self-destruct in a safe manner when pushed past their limit.
- *NEC® Table 11A* and *11B* outline the specifics of Class 2 and Class 3 power supplies.
- Class 2 or Class 3 must never be interconnected.
- In general, Class 2 and Class 3 circuits shall not share the cable, conduit, box, or raceway with those of Class 1 circuits or electric light and power circuits.
- *Section 725.55* of the *NEC®*, Parts B–J list all permissible applications for installing Class 1, Class 2, and Class 3 circuits. In all cases some form of separation and isolation either by barriers, compartments, or by raceways is required.
- Class 2 or Class 3 circuits are not to be strapped, taped, or attached by any means to the exterior of any conduit or raceway as a means of support.
- When installing electrical wiring in raceways, the number of conductors to be run should be limited to a percentage of fill based on *Chapter 9, Table 1* of the *NEC®*.

Key Terms

Class 1 circuit Class 1 circuits are defined by *Article 725* of the *NEC®*. There are two types of Class 1 circuits: Class 1 remote-control and signaling, and Class 1 power-limited. Class 1 remote-control and signaling has no power limit and can operate up to 600 volts. Class 1 power-limited is limited to 30 volts and 1000 volt-amps.

Class 2 circuit Class 2 circuits are defined by *Article 725* of the *NEC®*. In general, Class 2 circuits are power limited up to 100 volt-amps. Class 2 circuit conductors must be rated up to 150 volts. Current limitations do change based on voltage levels. See Chapter 9, *Table 11* (*A*) and (*B*) of the *NEC®*, which defines the specific voltage ranges and current limitations of inherently limited and non-inherently limited, AC and DC, Class 2 circuits and power supplies.

Class 3 circuit Class 3 circuits are defined by *Article 725* of the *NEC®*. In general, Class 3 circuits are power limited up to 100 volt-amps. Class 3 circuit conductors must also be rated up to 300 volts and must not be smaller than 18 AWG. Current limitations do change based on voltage levels. See Chapter 9, *Table 11* (*A*) and (*B*) of the *NEC®*, which defines the specific voltage ranges and current limitations of inherently limited and non-inherently limited, AC and DC, Class 3 circuits and power supplies.

Communication circuits and systems are covered by *Articles 800, 810, 820,* and *830* of the *NEC®*. They include telecommunication circuits, security and alarm circuits, radio and TV antenna systems, CCTV and radio distribution systems, and network-powered broadband.

Inherently limited An inherently limited power supply is clamped internally and unable to deliver more than a specific amount of energy to a load. Any attempt to push an inherently limited source past its maximum limit will cause it to either shut down or self-destruct in a safe manner.

Instrumentation Tray Cable (ITC) is covered by *Article 727* of the *NEC®*. ITC specifications are limited to 150 V and 5 A, for sizes 22 to 12 AWG wire. Size 22 AWG is limited to 150 V and 3 A. ITC can only be used and installed in industrial establishments that are maintained and supervised by qualified personnel; it is not to be used for general-purpose installations or as a substitute for Class 2 wiring (Class 2 wiring is covered by *Article 725* of the *NEC®*).

Overcurrent protection The purpose of overcurrent protection is to protect cables and circuits from excessive levels of current flow that could result in a potential fire hazard. Examples of overcurrent protection include fuses, circuit breakers, or active electronic feedback circuits.

Pipe fill When installing electrical wiring within a raceway, the number of conductors to be run should be limited to a certain percentage of fill. Limiting the number of conductors within a raceway will allow for the easy addition and removal of future wires to avoid damaging those already existing. In addition, not stuffing the raceway to the maximum fill will allow room for the dissipation of heat. To calculate pipe fill, refer to Chapter 9, *Tables 1* through *5* of the *NEC®*, Pipe Fill.

Power-limited circuits Power-limited circuits are limited in output and capacity. Power-limited falls into 3 varieties, Class 1, Class 2, and Class 3. Class 1 power-limited is limited to a maximum output power of 1000 VA and 30 volts. The limitations of Class 2 and Class 3 power-limited circuits are defined by Chapter 9, *Table 11 A* and *B* of the *NEC®*. In general, Class 2 and Class 3 power-limited circuits must not exceed 100 VA. The allowable voltages for Class 2 and Class 3 circuits may vary as high as 150 volts, with maximum current levels calculated based on the level of supply voltage.

Power-Limited Tray Cable (PLTC) A special type of listed and labeled, nonmetallic-sheathed cabling that is intended for use in cable trays of factories or industrial establishments. *Table 725.61* of the *NEC®* also states that PLTC may be used as an approved substitute for Class 2 or 3 wiring in general-purpose or dwelling locations. PLTC is rated for 300 V, 100 VA, and can be purchased in sizes 22 AWG through 12 AWG.

Remote-control circuits Circuits that control other circuits. Control voltages often can vary from as high as 600 V to as low as 5 V, depending on the type of system. As an example, motor-control circuits typically will use 120 V on starter coils, but many other types of control circuits will operate from 24 V or less. Coiled relays, transistors, and silicon-controlled rectifiers (SCRs) often will be used to accomplish the task. Common uses for remote-control circuits are motor controls, elevators, conveyor systems, automated processes, and garage door openers.

Signaling circuit A signaling circuit is defined in *Article 725* of the *NEC®*. Signaling circuits are circuits that activate notification devices. Examples may include lights, doorbells, buzzers, sirens, annunciators, and alarm devices.

Review Questions

1. Define integral circuit wiring.
2. Which article of the *NEC*® outlines the requirements of four general categories of circuits that are considered separate from those of electric light and power?
3. Give an example of a remote-control circuit.
4. What devices are commonly used to control high-voltage, high-current systems?
5. A doorbell would be classified as a _____ circuit.
6. List three commonly used methods to limit current flow in a power-limited circuit.
7. List three examples of a communication circuit.
8. Define the term *functionally associated*.
9. Why is it good practice to separate control circuit wiring from electric light and power conductors?
10. What is the maximum allowable output voltage for a non-limited, Class 1 power supply?
11. An 18 AWG conductor shall not exceed _____ amps, and that of a 16 AWG, _____ amps.
12. The voltage ratings of 18 AWG, Class 1 conductors shall be _____.
13. The voltage rating of a Class 2 conductor shall be limited to _____, and that of Class 3, _____.
14. List three acceptable power sources for Class 2 and Class 3 circuits.
15. Where in the *NEC*® can you find specifications for Class 2 and Class 3, AC and DC power supplies?
16. Conductors for a Class 3 circuit shall not be smaller than _____ AWG.
17. What types of environments utilize power-limited tray cable?
18. The minimum voltage rating of PLTC shall be _____.
19. Explain the meaning of reclassification when combining Class 2 and Class 3 circuits within the same raceway.
20. How should Class 1, Class 2, and Class 3 cables be marked and identified?
21. Why must power-limited circuits be separated from non-power-limited circuits?
22. When can power-limited and non-power-limited circuits be installed inside of the same cable tray or raceway?
23. Define a composite cable.
24. In general, cables and conductors of Class 2 or Class 3 circuits shall not be placed inside of the same cable, enclosure, raceway, compartment, manhole, outlet box, or device box with those of Class 1, or non-power-limited circuits, unless of course they are _____. List 5 permitted cases.
25. Conductors of Class 2 or Class 3 circuits shall be separated by at least _____ inches from conductors of Class 1 circuits, when they are not being installed inside of isolated raceways, or cable trays.
26. Can Class 2 or Class 3 conductors be strapped or taped to the outside of any conduit? If so, give an example of an allowable case.
27. What size Electrical Metallic Tubing (EMT) would you need to accommodate 20, 18 AWG XFF and 18, 16 AWG XFF conductors?

Telephone Wiring

OBJECTIVES After studying this chapter, you should be able to:

- Define the network interface and describe its typical location.
- Compare the modular plug style to the four prong telephone plug.
- Apply the *NEC®, Article 800* to telecommunication wiring.
- Explain the use and location of protectors and surge arrestors.
- Describe the proper grounding methods for a telephone system.
- Categorize the minimize gauge requirements and color codes for telephone wire.
- Describe the differences between RJ-11 and RJ-14 and the required pinouts.
- List the spacing requirements for fastening hardware.
- List the required safety issues when installing telephone systems.
- Identify the differences between residential and small office telephone systems.
- List the types of residential networks and define their specifications.

OUTLINE Residential or Commercial?

Inside Wiring

Installation and Code Requirements

Circuit Protection

Interior Communications Conductors

Telephone Wiring Components

Second Line Installations

Small Office Installations

Planning the Installation

Separation and Physical Protection for Premises Wiring

Installation Safety

Installation Steps

Testing

Installation Checklist

Residential Networking

Residential or Commercial?

You are undoubtedly familiar with residential phone wiring. Most everyone has by now wired a new phone jack, or at least has seen how the phone wiring runs (often chaotically) throughout houses. New construction tends to be much more logical, with most wiring done in a star pattern out from a central location in the house to facilitate troubleshooting and expansion.

Small businesses, with more lines and phones, often use small modular systems with microprocessor technology to offer services similar to large phone systems but on a smaller scale. Such installations will mount on a wall in an office or closet area, connecting to a large copper cable for outside trunk connections and multiple UTP cables to each work area.

The advent of structured cabling systems and general acceptance of TIA/EIA 568 standards for communications wiring in commercial installations have changed the nature of commercial telephone wiring. The simple 2- or 4-conductor wire, plugs or modular outlets, and daisy-chain wiring typical of telephone installations is now common only in residential installations. Although this section describes the typical telephone installation, bear in mind that in larger commercial installations, a structured wiring system using UTP and following 568 guidelines is more typical.

Inside Wiring

Inside wiring is all telephone wire that is inside a telephone company customer's premises and is located on the customer's side of the **network interface (NI)**. This wiring comprises the vast majority of telephone wiring.

The NI is the physical and electrical boundary between the inside wiring and the telecommunications network. The NI can be any type of telephone company–provided connecting point. Usually it is some type of small box with connectors and/or modular jacks inside. The NI is almost always mounted on an exterior wall, and may be either inside or outside of the structure. For single-family dwellings, the NI is most commonly installed outside, and for commercial or multifamily dwellings, it is most commonly installed indoors in a closet or basement.

A telephone circuit runs from a home or business to the local telephone company switching office. At the local office it is connected to equipment that hooks up to the national telephone network.

Generally, the telephone wiring entry to a structure will be located at or near the same place as the entry for electrical wiring. The NI will be placed near this entry. If an NI is not in place, any existing telephone company–provided modular jack may be used to connect newly installed customer-provided inside wire to existing inside wire.

The NI must be located inside the customer's premises at an accessible point for several reasons:

- Connection through a telephone company–provided modular jack is required by the FCC's registration program.
- Utilization of a jack makes it easier to connect or disconnect customer equipment or wire to the telecommunications network.
- Having the jack inside the customer's premises is not required but helps ensure the customer's privacy of communication and helps to prevent unauthorized use.
- Utilization of a jack forms a boundary for the ending of the network service and the beginning of the inside wiring and equipment. This creates a dividing point for the wiring and for deciding who is responsible for it, should a problem develop.

The point of location for the NI will be determined by the telephone company, although on new construction, you will have to verify the location with phone company representatives before beginning the work.

When you complete the wiring, you will plug your wiring directly to the NI or other telephone company–provided **modular jack.** For residential wiring, the end of your wire should have a **modular plug** on it to enable you to connect to the NI or telephone company–provided modular jack. For commercial premises, you will more often than not have to punch your connections down at the NI.

For existing installations, connecting points between your inside wire and your telephones may be of several types, depending on when your phones were installed. The following connecting points must be considered:

- *Modular:* Most recently installed telephones are connected to the inside wire via a modular system, which, for desk-type phones (Figure 14–1), consists of a miniature plug at the end of the telephone cord and a matching jack on the wall or baseboard.
- *Connecting block:* Wall-mounted phones have a pair of slots and a sliding modular plug on the back (Figure 14–2). The phone is attached to a "connecting block" on the wall, which has two rivets that fit into the slots and a modular jack that accepts the plug located on the back of the phone.
- *Permanently wired:* The telephone is connected directly to the inside wire and cannot be unplugged. The connection point is usually a small, square plastic box near or on the baseboard by the floor.

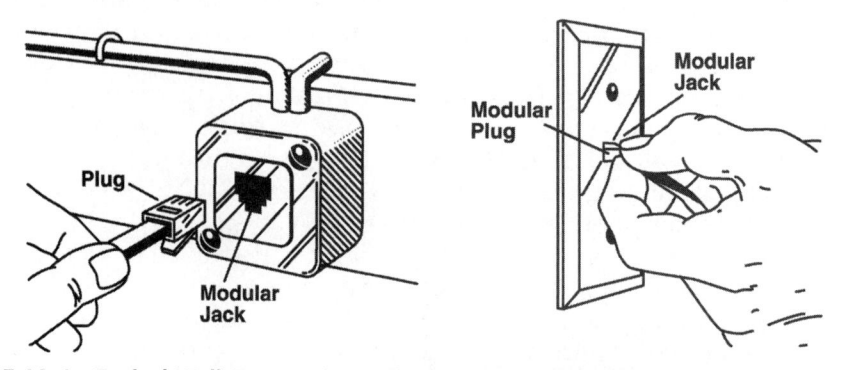

FIGURE 14–1 Typical Wall-Mounted and Flush-Mounted Modular Phone Jacks

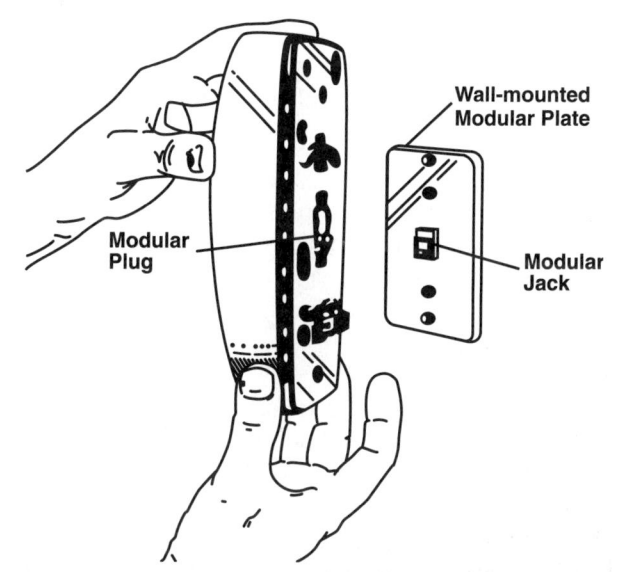

FIGURE 14–2 Modular Jack and Plate for Mounting Wall Phones

- *Four-prong:* On some desk-type telephones, there is a round or rectangular four-prong plug at the end of the telephone cord. The four-prong plug plugs into a jack with four holes. Telephones equipped with such plugs may be plugged in and unplugged easily, enabling you to move them from room to room as needed.

To convert permanently wired phones or those with four-prong plugs to a modular system, you will need an appropriate converter kit, available at most stores that sell telephone accessories.

Installation and Code Requirements

All building and electrical codes applicable in your state to telephone wiring must be complied with. *Article 800* of the *NEC®* covers communication circuits, such as telephone systems and outside wiring for fire- and burglar-alarm systems. Generally, these circuits must be separated from power circuits and grounded. In addition, all such circuits that run out of doors (even if only partially) must be provided with circuit protectors (surge or voltage suppressors). The requirements for these installations are as follows.

For conductors entering buildings, if communications and power conductors are supported by the same pole, or run parallel in span, the following three conditions must be met:

1. Wherever possible, communications conductors should be located below power conductors.
2. Communications conductors cannot be connected to crossarms.
3. Power service drops must be separated from communications service drops by at least 12 inches.

Above roofs, communications conductors must have the following clearances:

- *Flat roofs:* 8 feet
- *Garages and other auxiliary buildings:* None required

- *Overhangs*, where no more than 4 feet of communications cable will run over the area: 18 inches
- *Roof slope*, where the roof slope is 4 inches rise for every 12 inches horizontally: 3 feet.

Underground communications conductors must be separated from power conductors in a manhole or near handholes by brick, concrete, or tile partitions. Communications conductors should be kept at least 6 feet away from lightning protection system conductors.

Circuit Protection

Protectors are surge arresters designed for the specific requirements of communication circuits. They are required for all aerial circuits not confined with a block. ("Block" here means city block.) They must be installed on all circuits within a block that could accidentally contact power circuits over 300 volts to ground. They must also be listed for the type of installation.

Metal sheaths of any communication cables must be grounded or interrupted with an insulating joint as close as practicable to the point where they enter any building (such point of entrance being the place where the communications cable emerges through an exterior wall or concrete floor slab, or from a grounded rigid or intermediate metal conduit).

Grounding conductors for communication circuits must be copper or some other corrosion-resistant material, and they must have insulation suitable for the area in which they are installed. Communications grounding conductors may be no smaller than No. 14. The grounding conductor must be run as directly as possible to the grounding electrode, and it must be protected if necessary.

If the grounding conductor is protected by metal raceway, it must be bonded to the grounding conductor on both ends. Grounding electrodes for communications ground may be any of the following:

1. The grounding electrode of an electrical power system
2. A grounded interior metal piping system (avoid gas piping systems for obvious reasons)
3. Metal power service raceway
4. Power service equipment enclosures
5. A separate grounding electrode.

If the building being served has no grounding electrode system, the following can be used as a grounding electrode:

1. Any acceptable power system grounding electrode (see *NEC*®, *Section 250.81*)
2. A grounded metal structure
3. A ground rod or pipe at least 5 feet long and 1/2 inch in diameter. (This rod should be driven into damp earth, if possible, and kept separate from any lightning protection system grounds or conductors.)

Connections to grounding electrodes must be made with approved means. If the power and communications systems use separate grounding electrodes, they must be bonded together with a No. 6 copper conductor. Other electrodes may be bonded also. This is not required for mobile homes.

For mobile homes, if there is no service equipment or disconnect within 30 feet of the mobile home wall, the communications circuit must have its own grounding electrode. In this case, or if the mobile home is connected with cord-and-plug, the communications circuit protector must be bonded to the mobile home frame or grounding terminal with a copper conductor no smaller than No. 12.

Interior Communications Conductors

Communications conductors must be kept at least 2 inches away from power or Class 1 conductors, unless they are permanently separated from them or unless the power or Class 1 conductors are enclosed in one of the following:

1. Raceway
2. Type AC, MC, UF, NM, or NM cable, or metal-sheathed cable.

Communication cables are allowed in the same raceway, box, or cable with any of the following:

1. Class 2 and 3 remote-control, signaling, and power-limited circuits
2. Power-limited fire protective signaling systems
3. Conductive or nonconductive optical fiber cables
4. Community antenna television and radio distribution systems.

Communications conductors are not allowed to be in the same raceway or fitting with power or Class 1 circuits, and they are not allowed to be supported by raceways unless the raceway runs directly to the piece of equipment the communications circuit serves.

Openings through fire-resistant floors, walls, and so on must be sealed with an appropriate fire-stopping material.

Any communication cables used in plenums or environmental air-handling spaces must be listed for such use.

Communication and multipurpose cables can be installed in cable trays.

Any communication cables used in risers must be listed for such use.

Telephone Wiring Components

Before beginning any wiring job, you must plan ahead. Determine what types of components you will need. There are several types of standard components associated with telephone wiring.

Telephone Wire

Conductors in telephone wires shall be solid copper, 22 AWG minimum, and have at least four insulated conductor wires, which may be colored red, green, black, and yellow, or which may follow standard color coding. The conductors shall have an outer plastic coating protecting all conductors with a 1500-volt minimum breakdown rating. Although one phone line only needs two conductors (for "Tip" and "Ring"), the other two conductors are provided for powering dial lighting on some phones or to allow a second phone line to be easily installed (Table 14–1 and Table 14–2).

TABLE 14–1
Typical Inside Wire

Type of Wire	Pair Number	Pair Color Matches	
2-pair wire	1	Green	Red
	2	Black	Yellow
3-pair wire	1	White/Blue	Blue/White
	2	White/Orange	Orange/White
	3	White/Green	Green/White

Bridges or Cross-Connects

The purpose of a bridge is to connect two or more sets of telephone wires. Some bridges include a cord with a modular plug on the end, which can serve as an entrance plug in connecting your wire to the telephone company–provided NI or modular jack. Other bridges are designed to be placed at a junction where several telephone wires meet. Proper use of bridges will minimize the amount of wire required for the job.

Modular Outlets

Modular outlets are the jacks or connecting blocks into which modular phones are plugged. These jacks are known as **RJ-11** for two-wire connections; RJ-14 for four wires (two lines); and RJ-45, which may have four, six, or eight connections. There are also sixwire versions of RJ-11 and RJ-14 jacks. There are two basic types: jacks for desk telephone sets and jacks for wall telephone sets. In shopping for wiring components, you may find several variations for modular jacks. Some attach to the surface of the baseboard or wall, whereas others are flush-mounted, requiring a hole in the wall. Some also provide a spring-loaded door to cover the jack opening when nothing is plugged into it. This protects the inside of the jack from dust or dirt, which can damage the electronic contacts. These outlets must meet the FCC's registration program requirements.

TABLE 14–2
Inside Wire Connecting Terminations

Wire Color		Wire Function	
2-Pair	3-Pair	Service w/o Dial Light	Service w/ Dial Light
Green	White/Blue	Tip	Tip
Red	Blue/White	Ring	Ring
Black	White/Orange	Not used (2nd line—Tip)	Transformer
Yellow	Orange/White	Ground (2nd line—Ring)	Transformer

Telephone Wiring • 309

TABLE 14–3
Fastener Spacing Guidelines

Fasteners	Horizontal	Vertical	From Corner
Wire clamp	16 inches	16 inches	2 inches
Staples (wire)	7.5 inches	7.5 inches	2 inches
Bridle rings**	4 feet		2 to 8.5 inches*
Drive rings**	4 feet	8 feet	2 to 8.5 inches*

* When changing direction, the fasteners should be spaced to hold the wire at approximately a 45-degree angle.

** To avoid possible injury do not use drive rings below a 6-foot clearance level; instead use bridle rings.

Typical Fasteners and Recommended Spacing Distances

Wire staples generally are used to secure the cables to structural surfaces. Other types of fasteners are sometimes used and should be installed using the spacing guidelines in Table 14–3.

Second Line Installations

Residences often have two phone lines installed today, either for the convenience of family members or for the use of a modem or fax machine in a home office. The addition of a second line is simplified if the wiring in the house is already four-wire. Ordering another phone line from the local telephone company will result in a second line added to the NI. If four wires are available in the house, determine that one pair (black/yellow) is not attached to a transformer for powering dial lights (or remove the transformer if it is). Then the black wire attaches to **"Tip"** and the yellow wire to **"Ring"** on the second line.

Most modular outlets are wired with all four wires, so if the second pair is connected, an RJ-14 (four-wire) connector (Figure 14–3) can be used to connect two-line phones directly to the outlet. If two separate phones, or a phone and modem

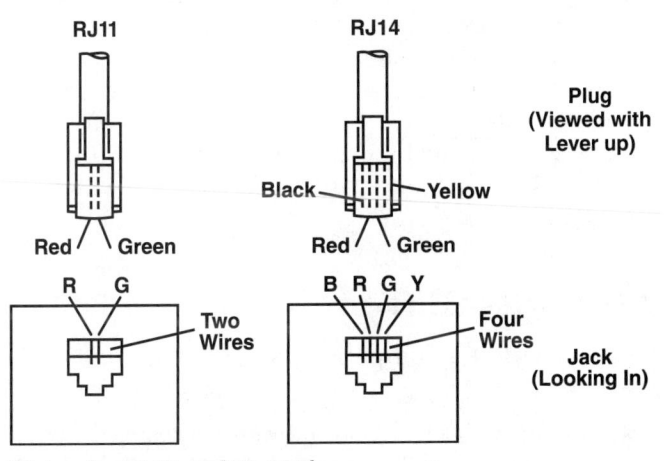

FIGURE 14–3 Wiring for RJ-11 and RJ-14 Plugs

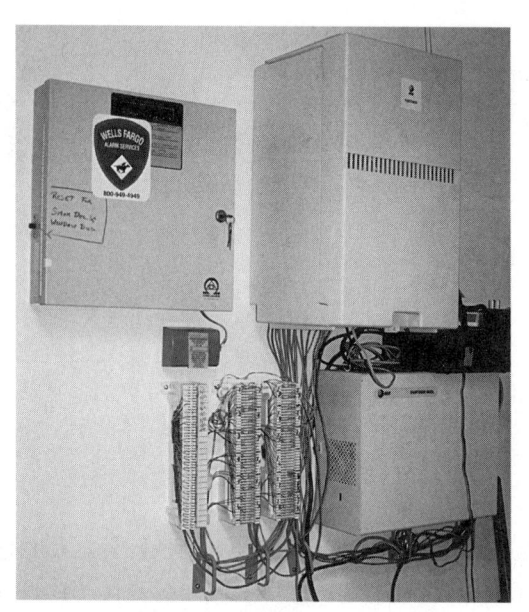

FIGURE 14–4 Typical Small Office Phone System

or fax machine, are desired from the single outlet, a special breakout adapter is available to provide two separate phone lines.

Small Office Installations

Small office installations have a larger number of lines, and probably more telephones, than residential installations. The phone company will bring to the customer a large multiple-pair cable and terminate it on a punchdown block on the customer premises. The phone system usually includes a switch box (KSU) that allows all phones to access outside lines, plus boxes for other functions like voice mail, music on hold, alarm call, and so on. A typical installation is shown in Figure 14–4.

The block on the far left is the phone company drop, with a 25-pair cable hidden in the wall and individual line cables terminated in modular plugs connected to the switch (upper right corner). The two 66 blocks are where the switch is connected to the individual phone lines, running out to modular jacks at each work area. By using a terminal block, the lines can be switched easily by changing cross-connect wires.

The box on the middle right is the voice mail option, and the box on the upper left is the alarm system dialer. Note that the cables are neatly routed and held in place by guides below the blocks. The phone outlet just below the switch has its cable protected in a stick-on snap-closed raceway, which is used in most areas where wires are not snaked through the walls.

Planning the Installation

The general rules for planning and performing a telephone installation are the following:

1. Determine where you want to place the modular outlets. This will likely be determined by the owner of the structure itself, because there are no *Code* guidelines for telephone wiring.

2. Determine which type of outlet is best for each location. If the jack is likely to be exposed to excessive dust or dirt, use jacks with protective covers.
3. Determine the best path to run the wiring from the NI or other existing telephone company–provided modular jacks to each of the new outlets. Place bridges where two or more paths come together.
4. Inventory the tools you will need to do the wiring job, such as
    a. screwdriver with insulated handle
    b. a pair of diagonal cutters, with insulated grips, to cut wire
    c. a tool to strip the wire coating off without damaging any of the four conductors
    d. hammer or staple gun for staples used to attach wire to wall or baseboard
    e. drill, with appropriate-sized bits, to drill holes for screws, anchors, and toggle bolts
    f. keyhole saw, if a hole in the wall is necessary, and a drill with a large enough bit to make a hole for the saw blade.
5. **Do not** place connections to wiring in outlet or junction boxes containing other electrical wiring.
6. Avoid the following if possible:
    a. damp locations
    b. locations not easily accessible
    c. temporary structures
    d. wire runs that support lighting, decorative objects, and so on
    e. hot locations, such as steam pipes, furnaces, and so on
    f. locations that subject wire and cable to abrasion.
7. Place telephone wire at least 6 feet from lightning rods and associated wires, and at least 6 inches from other kinds of conductors (e.g., antenna wires, wires from transformers to neon signs, and so on), steam or hot water pipes, and heating ducts.
8. Do not connect an external power source to inside wire or outlets.
9. Do not run conductors between separate buildings
10. Do not expose conductors to mechanical stress, such as being pinched when doors or windows close on them.
11. Do not place wire where it would allow a person to use the phone while in a bathtub, shower, swimming pool, or other hazardous location. Telephone ringing signals are a shock hazard!
12. Do not try to pull or push wire behind walls when electric wiring is already present in the wall area.
13. Use only bridged connections if it is necessary to establish a splice of two or more wires.
14. Place connecting blocks and jacks high enough to remain moisture-free during normal floor cleaning.
15. Do not attach jacks so that the opening faces upward. This increases the potential for damage from dirt and dust.
16. Wires should run horizontally and vertically in straight lines and should be kept as short as possible between bridges and other connections.

17. Run exposed wiring along door and window casings, baseboards, trim, and the underside of moldings so it will not be conspicuous or unsightly.
18. Wood surfaces are better for fastening wire and attaching connecting blocks, jacks, and bridges. When attaching hardware to walls, place fasteners in studs (wooden beams behind the walls) whenever possible.
19. If drilling through walls, floors, or ceilings, be careful to avoid contacts with concealed hazards, such as electric wiring, gas pipes, steam or hot water pipes, and so on.
20. If installing cables next to grating, metal grillwork, and so on, use a wire guard or other protective barrier to resist abrasion.
21. Always fasten cables to cement or cinder blocks with screw anchors, drive anchors, or masonry fasteners.
22. Avoid running cables outside whenever possible. If exterior wiring is necessary, drill holes through wooden window or door frames and slope entrance holes upward from the outside. Try to use rear and side walls so the wire will not be as noticeable; place horizontal runs out of reach of children; and avoid placing wiring in front of signs, doors, windows, fire escapes, "drop wires," and across flat roofs.
23. When fastening wire to metal siding, the type of fastener used depends on the type of siding and the method used to install it. Extra caution should be used when working on mobile homes. Mobile homes should be properly grounded. Line voltages present an extreme danger when working on metal. Therefore, proper grounding is very important.

Small Office Installations

In a small office with several lines, the phone company will usually bring a 25-pair cable into the building and terminate it on a punchdown block. The lines will be tested and marked on the block or nearby. The connection to the telephone equipment will involve punching down wires to connect from the block to the equipment. If the installation involves a small switch that allows all phones to share the lines, the switch will connect every phone in the system to the equipment. Most of these systems now use 4-pair UTP (Cat 3 or 5) cable and 8-pin modular outlets, just like structured cabling systems.

Each phone will connect to a modular outlet connected back to the switching equipment. Many small offices now install networking cabling along with telephone cables to support both services with one installation. For installations like this, refer to the sections of this book on structured cabling.

See Table 14–4 for the standard color coding used for 25-pair cables. Note that pairs are coded for pair number and "Tip" and "Ring," which is important for correct installation.

Separation and Physical Protection for Premises Wiring

Table 14–4 applies only to telephone wiring from the network interface or other telephone company–provided modular jacks to telephone equipment. Minimum separations between telephone wiring, whether located inside or attached to the outside of buildings, and other types of wiring involved are shown in Table 14–5. Separations apply to crossing and to parallel runs (minimum separations).

TABLE 14–4
25-Pair Backbone Cable Color Code

Pair	Tip Base Color/Stripe	Ring Base Color/Stripe
1	White/Blue	Blue/White
2	White/Orange	Orange/White
3	White/Green	Green/White
4	White/Brown	Brown/White
5	White/Slate	Slate/White
6	Red/Blue	Blue/Red
7	Red/Orange	Orange/Red
8	Red/Green	Green/Red
9	Red/Brown	Brown/Red
10	Red/Slate	Slate/Red
11	Black/Blue	Blue/Black
12	Black/Orange	Orange/Black
13	Black/Green	Green/Black
14	Black/Brown	Brown/Black
15	Black/Slate	Slate/Black
16	Yellow/Blue	Blue/Yellow
17	Yellow/Orange	Orange/Yellow
18	Yellow/Green	Green/Yellow
19	Yellow/Brown	Brown/Yellow
20	Yellow/Slate	Slate/Yellow
21	Violet/Blue	Blue/Violet
22	Violet/Orange	Orange/Violet
23	Violet/Green	Green/Violet
24	Violet/Brown	Brown/Violet
25	Violet/Slate	Slate/Violet

Installation Safety

Telephone connections may have varying amounts of voltage in the bare wires and terminal screws. Therefore, before you begin an installation, make sure the entrance point of any existing wire is unplugged from the NI or telephone company–provided modular jack while you are working. This will disconnect any wiring from the telephone network. If you are just connecting a new modular outlet to existing wiring that you cannot disconnect, take the handset of one telephone off the hook.

TABLE 14–5
Separation of Phone Cable from Other Types of Cabling

Types of Wire Involved		Minimum Separation	Wire Crossing Alternatives
Electric supply	Bare light or power wire of any voltage	5 feet	None
	Open wiring not over 300 volts	2 inches	Note 1
	Wire in conduit or in armored or nonmetallic-sheathed cable or power ground wires	None	NA
Radio and TV	Antenna lead-in or ground wires	4 inches	Note 1
Signal or control	Open wiring or wires in conduit or cable	None	NA
Communication	CATV coax with grounded shielding	None	NA
Telephone drop	Using fused protectors	2 inches	Note 1
	Using fuseless protectors or where there is no protector wiring from transformer	None	NA
Sign	Neon signs and associated wiring from transformer	6 inches	None
Lightning systems	Lightning rods and wires	6 feet	See wiring separations

Note 1: If minimum separations cannot be obtained, additional protection of a plastic tube, wire guard, or two layers of vinyl tape extending 2 inches beyond each side of object being crossed must be provided.

This will prevent the phone from ringing and reduce the possibility of electrical shock. Disregard messages or tones coming from the handset signaling you to hang up. In addition:

- Use a screwdriver with an insulated handle.
- Do not touch screw terminals or bare conductors with your hands.
- Do not work on telephone wiring while a thunderstorm is in the vicinity.

Installation Steps

1. Install a bridge or some other component to act as an entrance plug for your wire. This plug will connect to the NI or telephone company–provided modular jack. The bridge should have a modular-type cord with a plug at the end to insert into the NI or modular jack. Another acceptable type of entrance plug is a length of telephone wire with a modular plug on the end. Do not insert the entrance plug into the NI or modular jack until your wiring is completed.

2. Install all modular jacks in or on walls or baseboards. Use wood screws on wooden surfaces. Drill holes slightly smaller than the diameter of the screws being used to make installation easier. To fasten components to plasterboard walls, use screw anchors or toggle bolts.

3. Run wire to each modular jack, stapling it to the wall or baseboard about every 8 inches. Be sure you do not pierce or pinch the wire with staples. Allow enough wire to make the electric connections to the modular jack attached to the wall or baseboard. (In new installations, the wiring will be in place before the walls are completed.)

4. Strip the plastic coating on the phone wire as needed and connect the colored conductors to the terminals for each modular jack. Trim excess wire and attach the modular jack cover (if any) to the base.

5. When finished, place the plug on the end of your bridge into the NI or telephone company–provided jack.

Testing

After installing the wiring, the first step in testing it is to lift the handset of a phone plugged into one of the new outlets, listen for dial tone, then dial any single number other than "0." Listen. If you hear a lot of excessive noise, or if the dial tone cannot be interrupted, you have a problem. Attempt to locate it by using the following "troubleshooting" guidelines. If you cannot locate or repair the trouble yourself, disconnect the defective wiring until you can get the problem repaired.

Troubleshooting Guidelines

If testing indicates problems in the wiring you have installed, or if problems develop with the phone service later, try to determine if the problems are being caused by your own wire and equipment or by the telephone line. Here are some of the things you can do to try to identify the nature of the problem.

1. Unplug the wire you installed from the NI or telephone company–provided modular jack. Plug any phone (other than the one used when you detected the problem) directly into either of these jacks. If the problem persists, the telephone company lines or equipment may be faulty and you should proceed to step 2; otherwise, see step 3.

2. Dial the telephone company's repair service bureau listed in your directory. Describe the problem you are experiencing; be sure to state that you have installed your own wiring.

3. If the problem no longer exists when you plug another phone into the NI or telephone company–provided modular jack, it probably is being caused by your wiring or equipment. You may be able to localize the source of the problem by plugging the working phone into different outlets and testing each separately as before. Among the possible sources of trouble are broken wires, worn insulation, incorrect (e.g., red and green conductors reversed) or loose connections, and staples put through the wire.

4. If you have Touch-Tone® Service and, after lifting the handset of a phone plugged into the new outlet you installed, you hear the dial tone but the Touch-Tone® dial does not operate, unplug the wire from the NI or other telephone company–provided modular jack, reverse the red and green conductors at the outlet, then plug it back into the NI and check the phone again (reversing the polarity). If you still cannot locate the problem, call the telephone company's repair service bureau.

Installation Checklist

1. Be sure the entrance plug is unplugged from the NI or telephone company–provided modular jack.
2. Attach each component securely to the wall or baseboard.
3. Run wire to each component, allowing enough extra wire to make electric connections.
4. Make electrical connections and put covers on components.
5. Plug the entrance plug into the NI or telephone company–provided modular jack.
6. Plug in telephones and test (see "Testing" instructions).
7. See "Troubleshooting Guidelines" if problems occur.

Residential Networking

In October 1999, the TIA/EIA approved a formal standard for residential networks. The title is *ANSI/TIA/EIA-570-A Residential Telecommunications Cabling Standard*. This standard is derived from the usual TIA/EIA 568 standard for structured cabling systems. The basic 570 specifications are shown in Figure 14–5.

Following are several of the 570 standard's key points:

- Daisy-chaining of telephone circuits, long the standard, is out. Instead, each outlet must have its own home run. This is called "star topography." The 100-meter link length is carried over from EIA 569.
- Two grades of cabling, jacks, and distribution devices are specified: Grades 1 and 2.
- Grade 1 cabling may be Cat 3, which likely will not be used. Grade 2 cabling must be Cat 5, with Cat 5E recommended.
- Grade 2 distributive devices are required to be larger than Grade 1 devices.
- Grade 1 outlets terminate one 4-pair UTP cable and one 75-ohm coax cable. Grade 2 outlets terminate two 4-pair UTP cables, two 75-ohm coax cables, and provide for an optional optical fiber termination.
- At least one outlet must be provided in each kitchen, bedroom, family/great room, and den/study. It is recommended that one outlet be provided for each 12 feet of unbroken wall space.
- The 8-position modular jack is the only UTP jack allowed for the outlet and it must be wired in the "A" configuration. The 6-position RJ-11 is not allowed. Additionally, splitting of pairs is only allowed with an external adapter and not behind the outlet.
- The standard specifies a **distribution device (DD)** for each residence. This device is a panel of sorts, functioning as a type of service-entrance panel for the telephone, cable TV, and broadband services to the home.
- Location, space, and electrical power requirements are provided in 570. The DD must be located in a centralized, accessible location in the tenant space, if practical. This is to minimize the length of outlet cables and to allow for easy maintenance and configuration of the DD.

Typical Single Residential Unit

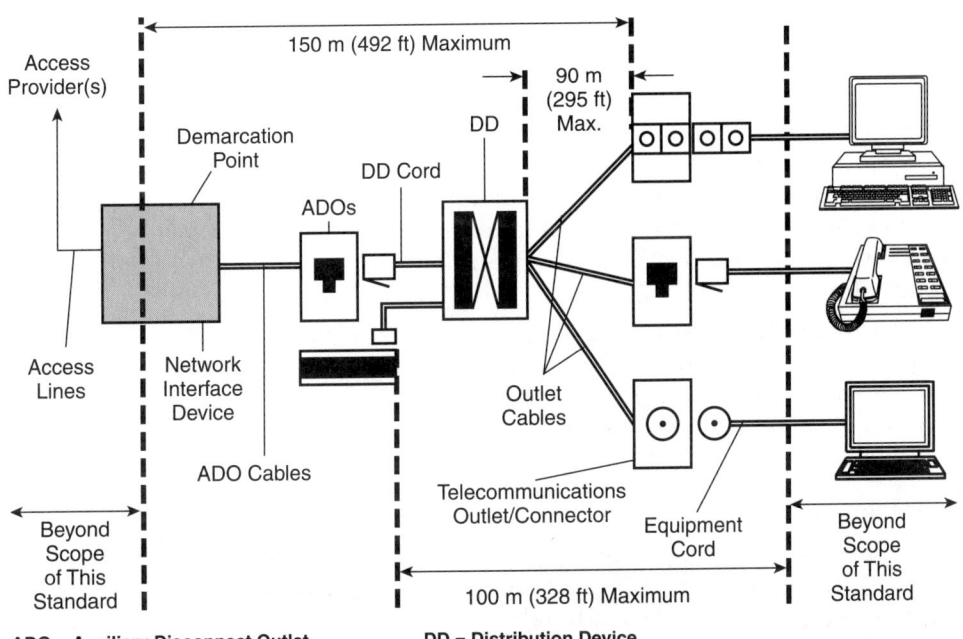

Recognized Cabling by Grade

Cabling	Grade 1	Grade 2
4-pair UTP	1 cable per outlet	2 cables per outlet
	Cat 3 minimum, Cat 5 recommended	Cat 5 minimum, Cat 5E recommended
75-ohm Coax	1 cable per outlet	2 cables per outlet
	Series 6	Series 6
Fiber	Not recommended	Optional
		50- or 62.5-micron multimode

Space Allocation for the Distribution Device

Number of Outlets	Grade 1 (height by width)	Grade 2 (height by width)
1 to 8	24" by 16"	36" by 32"
9 to 6	36" by 16"	36" by 32"
17 to 24	48" by 16"	48" by 32"
Greater than 24	60" by 16"	60" by 32"

FIGURE 14–5 Recognized Cabling by Grade

- Space allocations for the DD are provided based on grade and number of outlets served. The recommendations are provided based on the spacing between wall studs. A nonswitchable 15-amp duplex outlet is required at the DD for Grade 2 systems, and it is recommended for Grade 1. The standard also provides recommendations for multitenant dwellings and the associated backbone-cabling infrastructure.

- The required testing for residential networks is not as rigorous as that for commercial networks. Commercial systems go through a difficult testing process called "certification." Home cabling systems go through a less difficult process called "verification." Verification ensures that the cabling system is continuous (i.e., it has no shorts or open circuits) and that the correct terminations have been made. Verification, unlike certification, does not measure the information-carrying capacity of the link. This is considered unnecessary because residential links are nearly always considerably shorter than commercial links and, thus, suffer much less from attenuation losses—a significant factor in a link's capacity. In the shorter links, NEXT and far end cross-talk (FEXT) are a much reduced concern.

- The primary test for residential links is the wire map test, verifying the pin connections on both ends of the link. This is not to say that there is anything wrong with doing a complete certification with the much more expensive Cat 5 tester, only that it is not necessary for normal residential links.

FireWire®

Although the standard for residential networking is now EIA 570, this is not the only broadband system available. We recommend that you use 570, but you should be aware that there are others.

Two of these other systems are generally called *FireWire®* and *iLink®*. These are registered trademarks of Apple and Sony, respectively. *FireWire®* can transfer a lot of signal (some say up to 400 Mbps). It does this for both audio and video over the IEEE 1394 standard system. IEEE 1394 runs over plastic or glass optical fiber or over short distances of high-quality twisted copper pair cabling, such as Cat 5 or better. IEEE 1394 is typically used as an audio/video bus for interconnecting devices in a high-end entertainment race, or for digital cameras and computers to form an entertainment area network.

Universal Serial Bus (USB) is also occasionally mentioned as a networking technology to be used in homes. USB devices can transmit data at 1.5 and 12 Mbps speeds over twisted-copper-pair cables at lengths of up to 3 meters (9 feet 10 inches) at the lower speed, and up to 5 meters (16 feet 4 inches) for the higher-speed devices. USB is frequently used for connecting personal computer peripherals and pairs of computers to each other, but it is not generally applicable for use as a home area network.

Summary

- The network interface represents the physical and electrical boundary between the inside wiring and the telecommunications network.
- The advent of structured cabling systems and the general acceptance of TIA/EIA 568 standards for communications wiring in commercial installations have changed the nature of commercial telephone wiring.
- Residential telephone wiring is typically represented by two- or four-conductor modular outlets that daisy-chain throughout the residential installation.
- Small businesses, with more lines and phones, use small modular systems with microprocessor technology.
- The entry-point residential telephone wiring is generally located at or near the entry point for electrical wiring.
- *Article 800* of the *NEC®* covers communication circuits, such as telephone systems and

- outside wiring for fire- and burglar-alarm systems.
- All metal sheaths of any communication cable that must be grounded are interrupted with an insulated joint as close as practical to the point of entry.
- Conductors of telephone wire should be solid copper 22 AWG minimum, and have at least two insulated twisted pairs.
- Residential telephone connectors use modular RJ-11 or RJ-14 connectors.
- The second pair of a telephone wire is reserved for the second line.
- Small office installations will typically be installed through large multipair cable on type 66 or 110 punchdown blocks.
- When installing telephone wiring, always avoid damp locations and wire runs that support lighting, high voltage, lightning rods, and antenna systems.
- Most residential broadband networks are specified by EIA 570.
- Two other types of systems are generally called *FireWire* and *iLink*.

Key Terms

Distribution device (DD) A facility within the residence that contains the main cross-connect or interconnect where one end of each of the outlet cables terminates.

Modular jack A female connector for wall or panel installation; mates with modular plugs.

Modular plug A standard connector used with wire, with four to ten contacts, to mate cables with modular jacks.

Network interface (NI) The point of interconnection between a user terminal and a private or public network.

Ring One conductor in a phone line, connected to the "Ring" of the contact on old-fashioned phone plugs; a network where computers are connected in series to form a ring—each computer in turn has an opportunity to use the network.

RJ-11 6 position modular jack/plug.

Tip One conductor in a phone line, connected to the "Tip" of the old-fashioned phone plug.

Review Questions

1. What is NI?
2. True or False: The point of location of the NI is always inside the residence.
3. What type of modular plugs and jacks does the telephone system use?
4. What must you have if you have any circuits outside?
5. How far apart should power service drops be from communication service drops?
6. How much clearance must communication conductors have in a roof with a 4/12 slope (4-inch rise and 12 inches horizontally)?
7. A surge arrester must be put on all circuits with a block that may accidentally contact power circuits over _____ volts.
8. What material must ground conductors be made from?
9. What copper wire gauge is the minimum used in telephone wires?
10. What colors are the insulated wires used in old-style 4-conductor telephone cables?
11. What is the function of a bridge in telephone circuits?
12. How far apart should you staple the wire in the vertical or horizontal direction?
13. What angle is suggested for cables when changing direction?
14. Besides voice communications, what additional functions may a small office phone system have?

15. Where should you *not* place connections to telephone wiring?
16. Why could the telephone be a problem in the shower?
17. What should one be careful of when drilling into a wall, floor, or ceiling?
18. What should one make sure of when working on mobile homes?
19. What can one do if minimum separations cannot be arranged?
20. List four sources of problems one might encounter in telephone cabling installations.
21. What is the standard for residential networks?
22. Name an alternate standard for residential networks.

Video System Installations

After studying this chapter, you should be able to: **OBJECTIVES**

- Identify the types for video applications including cabling and connectivity.
- Demonstrate the process for terminating coaxial cable.
- List the specific code requirements for cable television and security monitoring systems as covered by *Article 820* of the *NEC®*.
- Explain the grounding and installation requirements for coaxial cable.
- List the designations and classifications of the types of coaxial cable and their uses including substitution hierarchy.

OUTLINE

Introduction
Video Cabling
Installation
Termination
Code Requirements

Introduction

Cabling for video applications covers two distinct uses: **CCTV**, or closed-circuit TV for security surveillance; and **CATV**, which stands for community antenna TV (not cable TV) and is used to distribute television signals. Both applications use similar cabling and installation techniques.

Video Cabling

The high bandwidth of video signals requires the performance of coaxial cables for transmission. Coaxial cable has always been the industry standard for high-bandwidth applications, although it is being replaced by fiber optics for long-distance or higher-bandwidth applications. Microwave transmission is used only where required because its cost is significantly more than the other two methods.

Using a high-quality coaxial cable is essential. For CCTV installations of less than 1,000 feet, RG-59U cable is fine. But for distances of 1,000 to 2,000 feet, RG-11U should be used. Installations of more than 2,000 feet in length require the use of amplifiers to keep the signals at usable levels, or the use of optical fiber as the communication media. Most CCTV installations will use BNC connectors, a bayonet-mount connector, or occasionally an "F" connector like CATV.

CATV uses large RG-11 or RG-8 for its trunk lines and the smaller RG-59 or RG-6 for drops to the home and within it. Larger buildings with longer cable runs will use Series 7 or RG-11 cable for its lower loss. The standard connector for CATV is the **F connector.**

Perhaps most important with CATV cables is proper cable selection and termination. CATV cables are directly connected to the public CATV network. **FCC** rules limit signal leakage, so it is important to use good cable with proper shielding and to terminate properly to prevent signal interference with other electronic devices. In addition, poor termination can cause reflections in the cable that affect the return path, or connection back to the system.

Because more networks now use or plan for cable modems for Internet connections, proper CATV installation becomes more important. A poorly installed installation in a building can cause trouble for thousands of CATV subscribers. The **Society of Cable Television Engineers (SCTE)** has published standards for the installation of cabling for CATV systems.

Installation

Installing video cabling is relatively simple. Coax cables must be installed with care; that is, they may not be pulled beyond their tension limits and may not be sharply bent. In addition, they must remain safe from physical damage and from environmental hazards. Care must also be taken when strapping or (especially) stapling cables to structural surfaces (walls, ceilings, and so on). If the staple or strap is cinched too tightly, it will deform the cable and alter its transmission characteristics; in fact, the system may not work properly, or even at all.

As mentioned earlier, surveillance cameras must be firmly mounted. In addition to standard wall-mounting brackets, cameras may also be mounted in vandal-resistant cases, which are recommended in trouble-prone areas. Mountings are also available with built-in panning and tilting mechanisms. Obviously, these will cost a bit more than the standard mountings but will provide an additional benefit. However, they cannot be used in all installations, particularly not with video motion detector devices.

Termination

Coaxial cable connections (Figure 15–1) include a center contact and an outer shell that connects to the shield on the cable. CATV systems use F connectors, a screw thread connector, whereas CATV may also use **BNC,** bayonet-mount connectors. The N connector is also used on larger coax cables. Note that the F connector uses the center conductor of the cable as the center contact, minimizing installation time and cost.

Coaxial connectors are so simple to install that we all can and do install them all the time—for connecting our VCRs—using simple screw-on F connectors. For

Video System Installations • 323

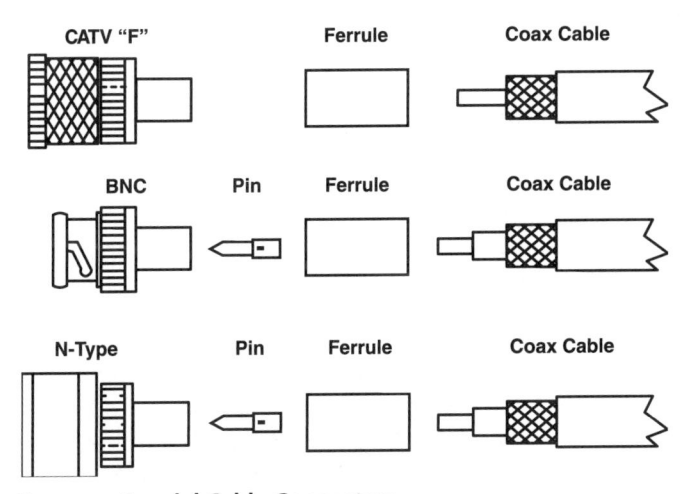

FIGURE 15–1 Common Coaxial Cable Connectors

building wiring with coax, especially for CATV, we need to be more careful to ensure high performance and low signal leakage. Crimp connectors are preferred, and cable should be top quality. When making up a coaxial connector, it is important to make sure that the connector type matches the cable type and that the connection is made up securely.

The process of terminating coaxial cables, shown in Figure 15–2, is as follows:

1. The outer insulating jacket must be stripped away, exposing the braided shield.
2. The braided shield must be pulled back over the outer jacket, leaving the inner insulation and its foil shield exposed.
3. The end of the inner insulation should then be removed, exposing the center conductor.
4. The connector is then placed on the end of the cable and crimped or tightened down. There are any number of different connectors available, with slightly different termination styles. Follow the directions for the given connector precisely.

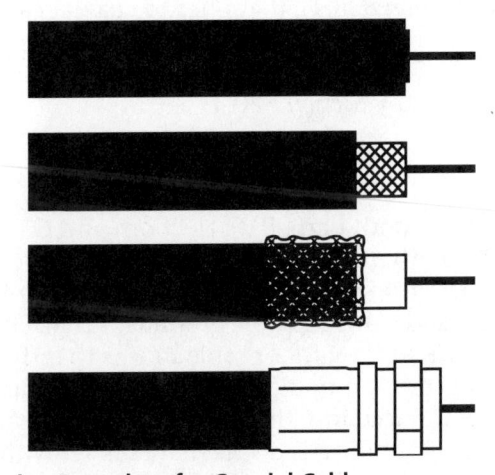

FIGURE 15–2 Termination Procedure for Coaxial Cable

Code Requirements

Cable television and security monitoring circuits are covered by *Article 820* of the *NEC®*. As we all know, the use of such circuits has expanded dramatically in recent years, and is likely to expand much further as the "information highway" continues to develop. The services that will probably do more to promote high-tech telecommunication services than any other are Internet connections by "cable modems" and "video on demand."

You will notice that the title of *Article 820* is "Community Antenna Television." Because very few of us have ever done a community antenna installation, and because we are relating this article to cable television and security monitoring, the following provides a brief explanation of CATV.

CATV started decades ago as a means of providing television signals to communities that could not receive broadcast stations, either because of distance or shadow areas, where the signal was too weak. Community antennas were installed at remote locations (such as on top of a nearby hill), and signals from them were fed to the homes in the area.

Later, when television signals transmitted via satellite became common, cable TV systems were able to provide a much wider variety of programming than was available via broadcast. Once programming from all over the world was available, the demand for the services became enormous, and cable television companies began to provide them. The new services were based on the same standards and methods as the community antenna systems from which they evolved. Today, cable television has developed into a huge system that serves at least 40 million homes in the United States; and it is growing steadily.

Security monitoring has been used ever since it became possible. At first, when the technology was new and costs were relatively high, security monitoring was used only for more vital uses. Later, as costs came down, its use became widespread.

Article 820 is, therefore, very broadly defined, covering all radio frequency signals sent through coaxial cables. Although the term *radio frequency* (often abbreviated as *RF*) is not defined in the *NEC®*, it would generally include every frequency from several kilohertz to hundreds of megahertz. This would include all types of radio signals, television signals, and computer network signals.

Article 820 does not, however, cover television cabling that is not coaxial (see *Fine Print Note* [FPN] to *Section 820.1*); it applies to coaxial cable only.

The primary safety requirements of this article are: the voltage applied to coaxial cables cannot exceed 60 volts, and the power source must be energy limited. Energy limitation is defined by *Section 820.4*.

Grounding

Grounding is mentioned in both parts *B* (Protection) and *C* (Grounding) of *Article 820*. Grounding is particularly required for coaxial cables run outside of buildings. Although the *NEC®* does not specifically state that all outdoor cables must be grounded, almost all outdoor runs must be grounded to meet the requirements of *Section 820.33*. The concern with outdoor cable runs is that they will be exposed to lightning strikes. And, in addition to direct strikes, outdoor runs of conductors can have substantial voltages induced into the conductors from nearby lightning strikes.

Another concern for grounding is accidental contact with power conductors. Whenever an outdoor coaxial cable could accidentally come into contact with power conductors operating at over 300 volts, grounding is required.

It is always the outer conductor of the coaxial cable that is grounded. The rules are generally as follows:

1. The cable must be grounded as close to the cable's entrance to the building as possible.
2. The grounding conductor must be insulated.
3. The grounding conductor must be at least #14 copper, and must have an ampacity at least equal to that of the coaxial sheath.
4. The grounding conductor must be run in a more or less straight line to the grounding electrode. The run must be protected if subject to damage.
5. The grounding electrode can be any suitable type. *Section 820.40* specifically mentions the following:
   a. the building grounding electrode
   b. water piping
   c. service raceway
   d. the grounding electrode conductor, or its metal enclosure.

The connection to the grounding electrode can be with any suitable means, as detailed in *Section 250.70*.

If the coaxial cable is grounded to an electrode other than the building electrode, the two electrodes must be bonded together with a #6 or larger copper conductor.

The grounding connection may not be made to any lightning protection conductor, whether it is a grounded conductor or not. The coaxial grounding electrode may, however, be bonded to a lightning protection grounding conductor.

Pieces of unpowered metallic equipment such as amplifiers and splitters that are connected to the outer conductor of grounded coaxial cables are considered to be grounded.

Surge Suppressors

Although not required by the *Code,* surge suppressors are a practical necessity for virtually all outdoor runs of coaxial cable. Because the pieces of equipment connected to these cables are very sensitive, they are easily damaged by voltage spikes. The most commonly used type of surge suppressor for communication circuits is the **metal oxide varistor (MOV).** These suppressors are usually called *protectors* in the communications industry. These devices, which are made of sintered zinc oxide particles pressed into a wafer and equipped with connecting leads or terminals, have a more gradual clamping action (clamping is the act of connecting a conductor to ground when the voltage rises too high) than either spark-gap arresters or gas tubes. As the surge voltages increase, these devices conduct more heavily and provide clamping action. And unlike spark-gap arresters and gas tubes, these protectors absorb energy during surge conditions. They also tend to wear out over time.

Routing of Outdoor Circuits

The *Code's* rules for the routing of coaxial communication circuits are essentially the same as those for other communication circuits. Specifically, coaxial cables must:

- Be run below power conductors on poles
- Remain separated from power conductors at the attachment point to a building
- Have a vertical clearance of 8 feet above roofs (there are several exceptions)

- If run on the outside of buildings, be kept at least 4 inches from power cables (but not conduits)
- Be installed so that they do not interfere with other communication circuits, which generally means that they must be kept far enough away from the other circuits
- Be kept at least 6 feet from all lightning protection conductors, except if such spacing is very impractical.

Messenger Cables

The vast majority of outdoor runs of coaxial cables are run with **messenger cables** (or are especially designed to be messenger cable assemblies themselves). In *Article 820*, the only requirement made for such runs is that the runs be attached to a messenger cable that is acceptable for the purpose and has enough strength for the load to which it will be subjected (such as the weight of ice or snow, or wind tensions). Messenger cables are covered in more detail in *Article 321* of the *NEC*®.

Indoor Circuits

Coax cables indoors are subject to the following requirements:

- They must be kept away from power or Class 1 circuits, unless the circuits are in a raceway, metal-sheathed cables, or UF cables.
- Coaxial cables can be run in the same raceway (or enclosure) with Class 2 or 3 circuits, power-limited fire protective signaling circuits, communication circuits, or optical cables.
- They may not be run in the same raceway or enclosure with Class 1 or power conductors. Exceptions are made if there are permanent dividers in the raceway or enclosure, or in junction boxes used solely as power feeds to the cables.
- They are allowed to be run in the same shaft as power and Class 1 conductors, but in these cases, they must remain at least 2 inches away. (But, as noted earlier, this applies to open conductors, not to conductors in raceways, metal-sheathed cables, or UF cables.)

Cable Types

The *NEC*® goes into great detail on designated cable types. Many of these requirements apply more to the cable manufacturer than to the installer. Nonetheless, the proper cable type must be used for the installation. The *NEC*® designations and their uses are as follows:

- **Type CATVP.** CATVP is plenum cable (hence the "P" designation) and may be used in plenums, ducts, or other spaces for environmental air.
- **Type CATVR.** CATVR is riser cable (extremely fireresistant) and is suitable for run installation in shafts or from floor to floor in buildings.
- **Type CATV.** CATV is general-use cable. It can be used in almost any location, except for risers and plenums.
- **Type CATVX.** CATVX is a limited-use cable and is allowed only in dwellings and in raceways.

You will find in practice that many coax cables are multiple rated. In other words, their jacket is tested and suitable for several different applications. In such cases, they will be stamped with all of the applicable markings, such as both CATV and CATVR. (Notice that all the cable types start with "CATV," which refers to community antenna television, *not cable television*.)

Trade designations generally refer to the cable's electrical characteristics; specifically, the impedance of the cable. This is why different cable types (RG-59U, RG-58U, and so on) should not be mixed. Even though they appear to be virtually identical they have differing levels of impedance, and mixing them may degrade system performance.

Substitutions

The *Code* defines a clear hierarchy for cable substitutions. The highest of the cable types is plenum cable (CATVP); it can be used anywhere at all. The next highest is riser cable (CATVR); it can be used anywhere, except in plenums. Third on the list is CATV, which can be used anywhere except in plenums or risers. Last is CATVX, which can be used only in dwellings and in raceways.

As always, only plenum types of cables can be used in plenums, riser types in risers, and so on. The formal listing is shown in *Table 820.53* in the *NEC*®.

Summary

- Two distinct type of video applications include CCTV and CATV.
- High-bandwidth video signals require performance of coaxial cables for transmission.
- CATV is generally connected through F connectors, and CCTV is connected through BNC connectors.
- Care must be used when installing coaxial cable especially when strapping or stapling cable so as not to alter the characteristic impedance of the transmission line.
- *Article 820* of *NEC*® covers television and security monitoring systems.
- *Article 820* also covers all radio frequency signals sent over coaxial cable.
- The outer shield of coaxial cables must be grounded.
- Outdoor cable runs must contain surge suppression devices.
- CATVP may be used in plenum spaces, CATVR in riser locations, CATV as general use, CATVX in residences and dwellings.

Key Terms

BNC Bayonet CXC connectors.

CATV An abbreviation for community antenna television, usually delivered by coax cable or hybrid fiber coax (HFC) networks, or cable TV.

CCTV Closed-circuit television, commonly used for security.

F connector The standard connector for community antenna television (CATV).

FCC Federal Communications Commission; oversees all communications issues in the United States.

Messenger cable The aerial cable used to attach communications cable that has no strength member of its own.

Metal oxide varistor (MOV) The most commonly used type of surge suppressor for communications circuits.

SCTE Society of Cable and Telecommunications Engineers.

Review Questions

1. What do the acronyms CCTV and CATV stand for?
2. What type of copper cable is used for the high-bandwidth CATV and CCTV signals?
3. What needs to be done to install cables of more than 2,000 feet?
4. What cables and connector does CATV use?
5. How does one prevent signal interference with other electrical devices?
6. What installation problems can alter a cable's transmission characteristics?
7. What type of mounting cannot be used with built-in motion detector devices?
8. Why was CATV created?
9. What made CATV grow to the point of providing service to over half the homes in the United States?
10. *NEC®, Article 820* refers to what type of cable?
11. Why is grounding important to outdoor cables?
12. What gauge conductor is used for outdoor grounding?
13. To protect outdoor runs of coaxial cable, what type of surge suppressors are commonly used?
14. The coaxial communication circuits follow the same code in the *NEC®* as which of the following?
    a. Electrical wiring
    b. Communication circuits
    c. Fiber-optic cables
    d. Security cables
15. *Matching:* Match each cable type with its use.
    a. CATVR
    b. CATVX
    c. CATVP
    d. CATV

    (1) Any location except risers and plenums
    (2) Spaces that have environmental air, plenums, or ducts
    (3) Very fire resistant runs in installation shafts or floors in buildings
    (4) Limited to only dwellings and raceways

Network Cabling

After studying this chapter, you should be able to:

OBJECTIVES

- Identify the types of cabling used on networks.
- Define terms associated with cable specifications and performance.
- Identify the types, specs, and uses of twisted-pair cable.
- Define the basics and standards of wireless transmission.
- List the parameters and specifications of power line carrier transmission.
- Understand the methods for installing commercial network cabling.

OUTLINE

Introduction
Cabling Requirements
Network Cabling Types
Wireless Transmission
Infrared Transmission
Power Line Carrier Networks
Other Transmission Means
Network Cable Handling
Typical Installations
Pulling Cables
Mounting Hardware in Closets or Equipment Rooms
Cabling in the Closet or Equipment Room

Introduction

Unlike telephone and video cabling, which have used one basic type of cable all along, computer networks have used many different types of cables and, in fact, several other means of communication. All of these methods have been tried as solutions to providing low-cost, high-performance cabling. They include:

- Unshielded twisted-pair cables (UTP)
- Shielded twisted-pair cables (STP)
- Coaxial cables
- Optical fiber cables
- Radio signals
- Infrared lights, reflected off ceilings or focused on line-of-sight
- Electronic signals sent through power lines.

Although most network cabling has migrated to UTP cabling following the structured cabling guidelines in TIA/EIA 568, there are still applications that use other cabling schemes. Therefore, a review of the entire spectrum of cables is needed to understand network cabling.

Cabling Requirements

With power work, we are most concerned about the path the current will take, and less concerned about the quality of the power going from one point to another. In data wiring, however, we must consider both qualities of the transmission.

First, we must have a clear path from one machine to the next. Here we are concerned with the signal's strength; it must arrive at the far end of the line with enough strength to be useful.

Second, we are concerned with the quality of the signal. For instance, if we send a square-wave digital signal into one end of a cable, we want a good square-wave signal coming out of the far end. If this signal is distorted, it is unusable, even if it is still strong.

The requirements are determined by a number of performance parameters of the cable, including its attenuation, characteristic impedance, **capacitance,** and **cross-talk.** Because these terms are used many times in describing cables, we will take time to define them now.

Attenuation

Attenuation (Figure 16–1) is the loss of signal power over the length of a cable. This attenuation is normally measured as a number of decibels per 100 feet, at a given frequency. For fiber-optic work, attenuation is measured in decibels per kilometer.

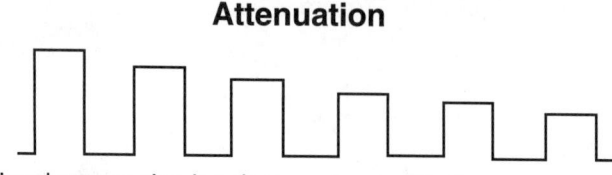

FIGURE 16–1 Signal attenuation is a decrease in amplitude.

Signals sent over copper wires deteriorate differently at different frequencies—the higher the frequency, the greater the attenuation. Attenuation is a problem, since a weakened signal can only be picked up by a very sensitive receiver. Such very sensitive receivers are quite expensive; therefore, low attenuation is desirable.

Signal attenuation also depends on the construction of the cables, particularly the size of the wire and the dielectric characteristics of the cable insulation. For example, you could have two 100-ohm, 24-gauge cables and one of the cables might have a lower attenuation than the other, strictly because of the construction characteristics of the cables. This would allow the cable with lower attenuation to be used over longer distances with better results.

Impedance

Impedance, which is the total opposition to current flow, is an important consideration for coaxial and twisted-pair cables for computer systems. The most important consideration for impedance is that it remain consistent throughout the entire system. If it does not remain consistent, a portion of the data signals can be reflected back down the cable, often causing errors in the data transmissions.

Cable Capacitance

As mentioned, capacitance is the most common element of attenuation. This capacitance usually comes in the form of *mutual capacitance*, which is capacitance between the conductors within the cable. Capacitance is a big factor because it causes the cable to filter out high-frequency signals. Data transmissions are square-wave signals. When the capacitance is too great, it has a tendency to round off the digital data signal, making it more difficult to receive.

When data signals have a clear square wave, they are easily intelligible to the receiver. If, however, the signals become rounded, they can be confusing to the receiver, causing a high **bit error rate (BER)** or causing data to be received in error.

Cross-talk

Cross-talk (Figure 16–2) is the amount of signal that is picked up by a quiet conductor (a conductor with no signal being transmitted over it at the moment) from other conductors that are conducting data. This signal is picked up through electromagnetic induction, the same principle by which transformers operate. Cross-talk contaminates adjacent lines and can cause interference, overloaded circuits, and other similar problems.

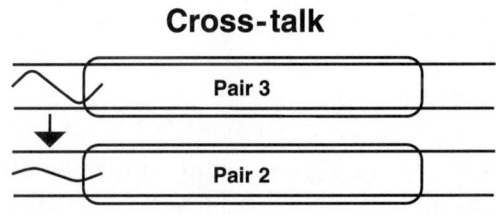

FIGURE 16–2 Cross-talk is signal coupling from one pair to another.

Preventing Cross-talk

The traditional method to prevent cross-talk and electromagnetic interference is to place a shield around the conductors. The three principal types of shields are these:

1. Longitudinally applied metallic tape
2. Braided conductors, such as are commonly used in coaxial cables
3. Foil laminated to plastic sheets.

Spirally applied metallic tapes can also be used, but they are not generally good for data transmission.

Shielding stops interference and cross-talk by absorbing magnetic fields. Because the shielding is conductive itself, when the magnetic field crosses through it, it is absorbed into the shield. It does induce a current into the shield, but because this current is spread over the wide, flat surface of the shield, the field is diffused and is usually not strong enough at any one point to cause a problem.

Foil shields work better at higher frequencies (>10 MHz), reducing electromagnetic interference by 35 decibels; wire braid shields generally reduce interference by as much as 55 decibels at lower frequencies (<10 MHz); and a combination of the two types of shields reduces interference by over 100 decibels. It is also possible to use electronics to reduce interference, although they must be closely matched to the network's data transmission rate.

Most of today's cables use twisted pairs and balanced transmission to minimize cross-talk. If the pairs are twisted at different rates, their antenna characteristics are different enough to greatly reduce cross-talk without any additional shielding.

Network Cabling Types

Unshielded Twisted-Pair (UTP) Cables, 22–24 Gauge

Applications: Most current networks. UTP cable is the primary cable used for networks, as specified in the TIA/EIA 568 standard.

Advantages: Inexpensive; may be in place in some places; familiar and simple to install.

Disadvantages: Subject to interference, both internal and external; limited bandwidth, which translates into slower transmissions unless multiple pairs or encoding electronics are used.

Screened Twisted-Pair (ScTP) Cables

Applications: Same as UTP. Although not currently specified for any networks or covered in the TIA/EIA 568 standard, used in many networks in Europe, where EMI is a greater concern.

Advantages: A foil screen enclosing all four pairs helps electromagnetic interference (EMI) emissions and interference but has no effect on cross-talk.

Disadvantages: More expensive; harder to terminate; require special plugs and jacks.

Shielded Twisted-Pair (STP) Cables, 22–24 Gauge

Applications: IBM Token Ring. STP is covered in TIA/EIA 568 but is no longer widely used.

Advantages: Easy installation; reasonable cost; resistant to interference; better electrical characteristics than unshielded cables; better data security.

Disadvantages: Not easily terminated with modular connectors; may become obsolete due to technical advances.

Coaxial Cables

Applications: Original Ethernet, Thinnet.

Advantages: Familiar and fairly easy to install; better electrical characteristics (lower attenuation and greater bandwidth) than shielded or unshielded cables; highly resistant to interference; generally good data security.

Disadvantages: More expensive; bulky; may become obsolete due to technical advances.

Optical Fiber Cables

Applications: FDDI, ESCON, Fibre Channel, all versions of Ethernet, optional for most networks.

Advantages: Top performance; excellent bandwidth (high in the gigabit range, and theoretically higher); very long life span; excellent security; allow for very high rates of data transmission; cause no interference and are not subject to electromagnetic interference; smaller and lighter than other cable types.

Disadvantages: Slightly higher installed cost than twisted-pair cables because more expensive electronics interface to them.

Wireless Transmission

In the past several years, there has been a good deal of interest in wireless networks. Although part of the reason for this interest is due to the expense of installing cables (the installation expenses are usually far higher than the material expenses), and the frequent moving of terminals within a large office, the most compelling reason is the need to have mobile terminals connected to a network.

With a wireless network, no data transmission cables are required to connect any individual terminal. Within the range of the radio signals, a terminal can be moved anywhere. Wireless networks do require cabling, however, to connect the antennas to the network backbone, so they are not completely without cabling requirements.

Wireless networks are usually more expensive than cabled networks. However, the real money savings occur when terminals must be mobile, such as for delivery services, warehouses, or hospitals. Money can also be saved in locations where it would be especially difficult to install cables.

Wireless networks have three major problems: bandwidth, security, and standards. Current wireless networks have a maximum throughput of 11 MHz, well below the rates of cabling-based Ethernet, although techniques are being developed to expand this to over 50 MHz. Because the signals are broadcast, anyone with a compatible receiver can enter any network that has sufficient signal strength. This is a boon to hackers and requires extreme measures of security to prevent unauthorized usage of the network.

Standards continue to be elusive. The IEEE Ethernet committee (the "802" committee) has already had four wireless standard projects: 802.11, 802.11a, 802.11b (Wi-Fi, the current most popular version), and 802.11g. In addition, there are two more, Bluetooth® and Home RF, that have proponents. Standards are changing so fast that it is impossible to predict current conditions in a textbook. Refer to appropriate current information for the status of wireless standards.

A potential hazard with wireless networks is the health effects of radio emissions. Although no definitive studies have been completed, there have been indications that radio waves in these frequency ranges may be harmful.

Infrared Transmission

Another method of transmitting data signals without the use of wires is by using infrared (IR) light. By sending pulses of IR light in the same patterns as electronic pulses sent over cables, it is possible to send data from one place to another. Networks based on IR transmission have been developed for use in offices and for line-of-sight transmissions between buildings.

IR light is used for this purpose because it is invisible to the naked eye, and because it is inexpensive to implement. This is a variation on the same technology used for TV remote controls. The distance between terminals is normally limited to around 80 feet, although newer systems exceed that figure. Line-of-sight systems between buildings can go thousands of feet but are vulnerable to heavy rainstorms, fog, and other natural phenomena (including big birds).

This technology works fairly well, although there are problems that develop in offices with numerous walls. Just like normal light, IR light cannot pass through walls. In open offices, this is not much of a problem, but in walled-off offices, remote transmitter/receivers are often required. A transmitter/receiver can be mounted in an area where it will easily receive data signals and be connected to the computer with a cable. Not all of these systems use combination transmitter/receivers, however; some of them use separate transmitters and receivers.

As with other wireless signal transmission, IR networks require all of the parts that conventional networks require. Where the data cables would normally connect to the back of the computer, however, transmitter/receivers are installed instead. IR signals can be used to send and receive through these devices, usually at a rate of between 4 and 16 mbps. Depending on the design, the units can be rather large and awkward.

One great advantage of the wireless networks is that they can be set up anywhere, almost instantly. This can be a terrific feature for people whose work location is constantly shifting from one place to another. Some people in the trade have come to call these systems "networks in a box."

Power Line Carrier Networks

Another method of sending data signals involves sending them through regular power lines. This method requires a special network adapter that modulates the computer data and transmits it over the power lines in an office or home. Other devices or computers connected the same way may receive data over the power lines. This system is very attractive from the standpoint of installation costs. No cables are required, and the installation is very quick and easy.

This system imposes data signals right over the power line current. The voltage and frequency differences are easily separated from each other by tuned receivers. Power lines operate at 60 cycles per second and 120 volts, whereas data signals are normally in the 5-volt range and have a frequency of hundreds of thousands or millions of cycles per second.

Power line carrier systems will never meet the technical performance of optical fiber cables, or even other types of cables. They are, however, a very appealing option for home networks, for example, where computers in the future may control heating and air-conditioning systems, turn lights on and off, and manage security systems.

The installation of these systems is not completely hassle-free, however. Power lines do not travel unbroken throughout whole buildings, or even through parts of buildings. To bridge the gaps in wiring systems, special devices called "signal bridges" are required. These devices connect to two separate wiring systems and transfer the data signals from one to another without allowing current to be transferred from one system to another (which would cause major problems and, most likely, injuries). Signal amplifiers are also frequently required when wiring systems cover long distances. There can also be filters and other devices required for these systems, depending on the location of the installation.

Other Transmission Means

Although the methods of transmission that we have discussed are by far the most common methods, there are other methods that are sometimes used. The chief among these are microwave and laser signal transmission for line-of-sight connections between nearby buildings.

The typical method is to set up transmitter/receivers on the roofs of both buildings and to connect both ends to the networks in the buildings. The buildings can be up to about 1 km apart for most systems, and even farther apart for others. Signal transmission rates of 1.5 Mb/s or greater are not uncommon.

It is important to remember in planning such a system that the management of any buildings that are under the transmission area should be notified, and that any appropriate permissions are granted before any work is begun. Most microwave systems require FCC licenses, which may be difficult to obtain in metropolitan areas.

Network Cable Handling

The performance of the cabling network is heavily dependent on the installation. The Category 5E or 6 components used in most structured cabling installations have been carefully designed and exhaustively tested to meet or exceed the requirements of TIA/EIA 568 for performance at 100–200 MHz. If the cable is not properly installed, performance will be degraded. You cannot mix components either; every component in the cable plant must be rated to the same category and properly installed or the cabling will fail testing.

As general guidelines, remember the following:

- All components must be similarly rated for performance.
- Cable must be pulled from the reel or box without kinking.

- Cable must be pulled with less than 25 pounds of tension.
- Use cable lubricant in conduit if necessary.
- Cable must not be pulled around sharp corners or kinked.
- Inspect the cable routes for surfaces that may abrade the cable.
- On riser installations, try to lower the cable down, not to pull it up.
- Cables must be supported to prevent stress. Cable supports should not have sharp edges that may distort the cable.
- Cable ties must not be so tight as to distort the jacket of the cable. They are only used to prevent unnecessary movement of the cable, so snug is tight enough.

Typical Installations

The typical installation of communications cabling is in an office building. There are communications wiring closets on every floor, and several per floor if the floor covers a large area. A main communications closet will have the primary communications equipment (PBX, network routers, and so on). Cables will run from the wiring closet to work areas overhead above a suspended ceiling. Wiring closets will be connected by multiple cables run overhead or in risers.

Each work area (Figure 16–3) will be connected to the closet with at least two cables, one for telephone and one for data. Although the telephone cable can be Cat 3, it is more common to install two Cat 5E cables for simplicity and easier cabling management. If the telephone is already installed and only a data cable is being installed, only one cable may be needed. Cables will be terminated in 8-pin modular jacks at the outlet and either in patch panels (Figure 16–4) or on punch-down blocks in the closet.

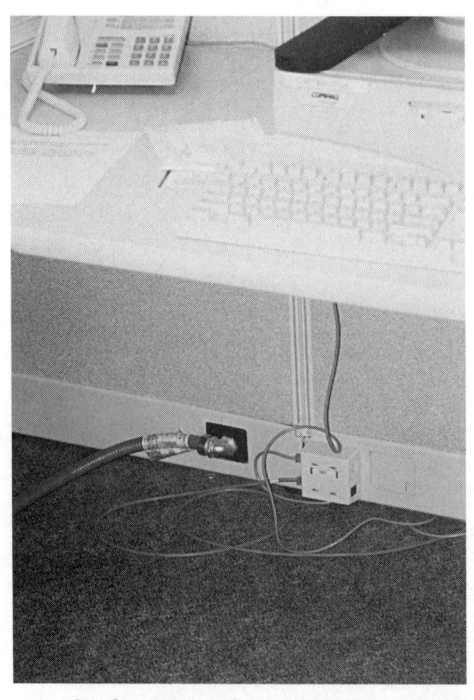

FIGURE 16–3 Work Area Outlet for Network and Telephone in Modular Office

FIGURE 16–4 Terminated Cat 5 Cables in Modular Patch Panels

Closets may have numerous cables running between them, unless the backbone cabling is fiber-optic, which will be only one cable with several fibers. Numerous cables will be needed to connect wiring closets to the equipment room. Terminations may be either punchdowns or patch panels or both, depending on the destination of the cables.

Some applications will have cables in conduit to protect cables in areas where damage is possible to prevent EMI where that is a problem, or to assist in firestopping. Cable trays (Figure 16–5) may be used to keep all cables neatly together and out of harm's way. Some installations may have false floors to allow interconnection cables to be run below the floor.

Pulling Cables

Horizontal Pulls

In a typical installation, the horizontal cables will be pulled between the wiring closet and the work area above a suspended ceiling. This pull will involve routing the cables around everything else already there, including other cables; heating, ventilating, and air-conditioning (HVAC) systems; light fixtures; and so on. It is usually easier in new construction because the tiles are not in place. The procedure involves gaining access, determining the best route for the cables, running a handline, affixing cables to the handline, and pulling the cables.

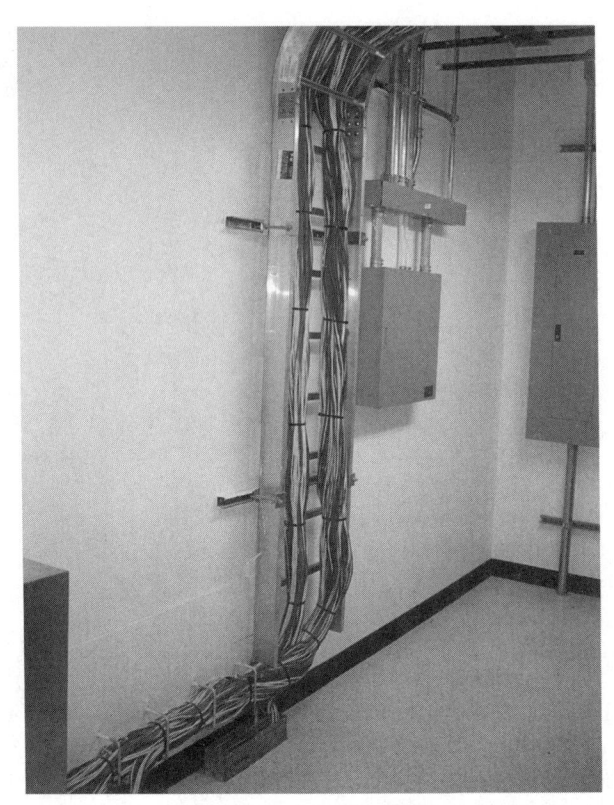

FIGURE 16–5 Large numbers of cables should be neatly bundled and placed in cable trays.

The most efficient way to pull horizontal cables is to pull bundles. Using drawings of the area, determine where all work areas are located and establish several consolidation points or clusters from which to pull. This is easy with open areas and modular furniture but may be much more difficult in walled offices. Sometimes the best way to pull cables is from the work area (sometimes called the "drop") to the closet, but sometimes pulling from the closet is more efficient. It all depends on the actual installation situation.

The first step is to gain access to the area. Clear as many obstacles as possible and secure the area. Put caution tape around the ladder you are working on, and keep everyone away for safety (theirs as well as yours). In the path you will be taking with the pull, carefully remove every other ceiling tile along the route, but leave them above the ceiling to prevent damage.

After examining the area and finding the most direct route, tie a weight to the handline or pull string (typically a thin nylon rope) and, starting at the drop end, toss it from ceiling opening to opening until you reach the closet. Some installers even use a slingshot to shoot the handline over ceiling tiles.

Cables come in several styles of boxes designed to allow easy pulling directly from the box (Figure 16–6). Note the instructions on the box regarding the proper placement of the box for pulling to ensure that the cable is not twisted or kinked when pulled from it. Place all the cable boxes in the drop area, and mark each as to the final location. Post-it® notes are good temporary labels for the boxes.

If the cables are on reels, set up a stand with an axle to allow the cable to unroll. Do not feed the cable off the sides of a spool because it will twist and kink. Two chairs and a short length of conduit or a broomstick can be a makeshift axle,

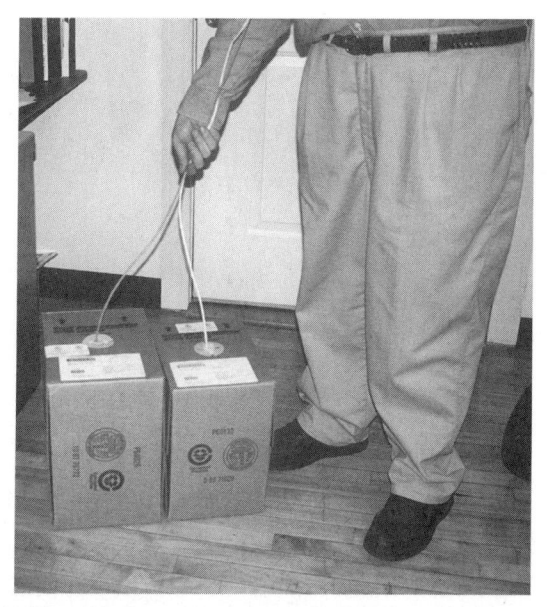

FIGURE 16–6 Boxed UTP cable is made to be pulled directly from the box.

but some contractors use jack stands and a piece of conduit. Cable "trees" are also available at reasonable prices.

Mark the cables with a permanent marker with the same identification at both ends (Figure 16–7). This will make later identification of the cable easier and make tracing unnecessary.

Although cables can be pulled individually, it is more efficient to pull as many cables as possible at one time. Bundle the cables (Figure 16–8) as follows:

- Gather all the cables to be pulled at once and align the ends. Tape them together with electrician's tape for 8 to 10 inches, including the ends.

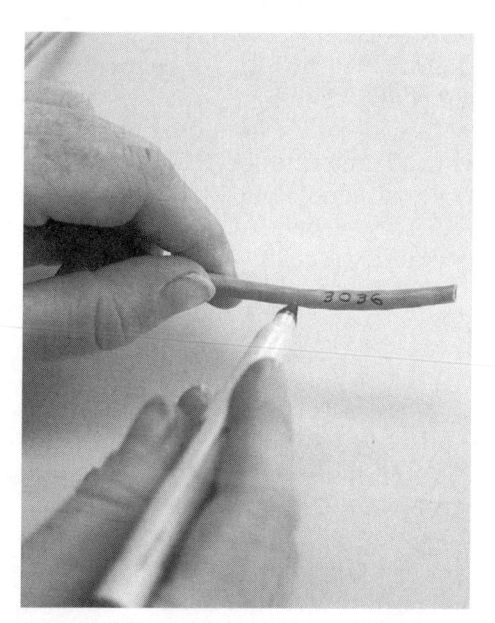

FIGURE 16–7 Mark the cables with a permanent marker to facilitate identification.

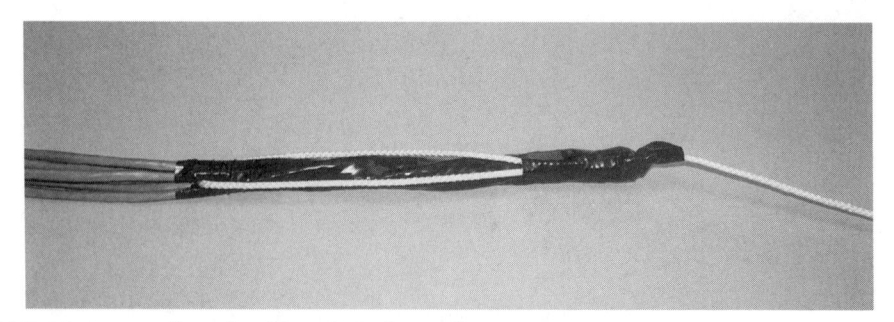

FIGURE 16–8 Bundles of Cable Ready for Pulling

- Split the bundle in half.
- Run the handline between the two bundles and tie it to the handline. Tape the handline to the tip. This makes the bundle less resistant to pulling.
- Some installers like to make loops on the end by taping the bundle about a foot back from the end, dividing the cables, tying them in a knot to form a loop on the end, and taping it. The handline can be attached to that loop.

If you are pulling only one cable, make a loop in the end, twist the cable back over itself, tape it, and tie the handline to the loop (Figure 16–9).

Pull the cable, watching for snags, as you pull it around sharp corners or other obstacles. If you hit a snag, stop and free the cable. Never jerk the cable, because that may overstress it and cause performance degradation. Find the cause of the snag and fix it!

Horizontal Cable Supports

After pulling the cable, it must be supported. It is not safe to leave it lying on the ceiling; it must be supported and kept away from the AC power cables and lighting fixtures. Numerous hangers, cable trays, and other cable-handling solutions are available from vendors of cabling hardware.

Above suspended ceilings, **"J" hooks** can be used to separate the cables from most other hardware above the ceiling, especially power cables and lighting fixtures. Special J hooks with wide bases should be used for communication cables (Figure 16–10). Hangers can sometimes be attached to the ceiling where the suspended ceiling hangers are attached, but J hooks should not be attached to the ceiling suspension wires. Using suspended ceiling tile wires is not allowed in most localities; such wires may cause the ceiling to become uneven. In new or renovated

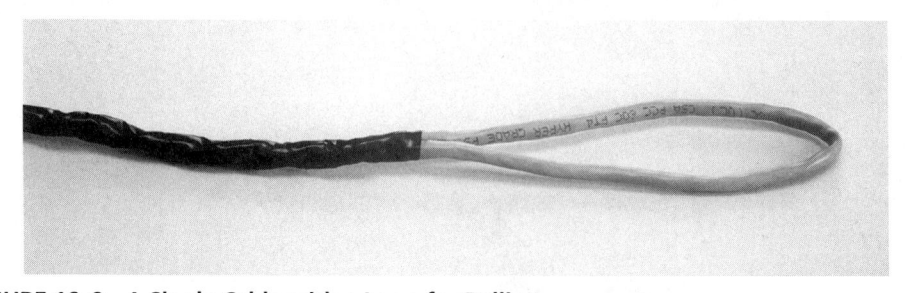

FIGURE 16–9 A Single Cable with a Loop for Pulling

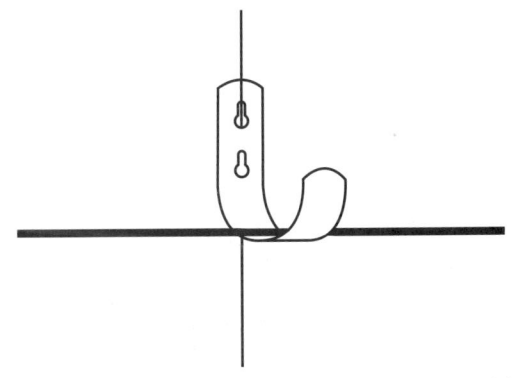

FIGURE 16–10 Special wide "J hooks" should be used for Cat 5 cables.

construction, it may be possible to have separate wires installed for cable J hooks at the same time as the ceiling support wires are installed.

Closet-to-Closet Pulls

Pulls from closet to closet or equipment rooms can be simple pulls of cable bundles like horizontal pulls or may involve conduit, cable trays, or riser installation. The first consideration is the same as that in all installations—examine the route closely and plan the pull accordingly.

Conduit Pulls

Make sure you know how large and long the conduit is and how many bends are involved. The diameter of the conduit will limit the number of cables that can be installed, and the current length and bends of the cables will dictate whether lubrication is necessary. For long runs, use special cable lubricants only, because other lubricants may damage the cables.

Conduit pulls follow this procedure:

1. Open the conduit on each end. Examine the ends for roughness, and use a special "leaderguard" (Figure 16–11) if necessary to prevent cable damage.
2. Feed a snake into the conduit from the end you will be pulling the cables into.

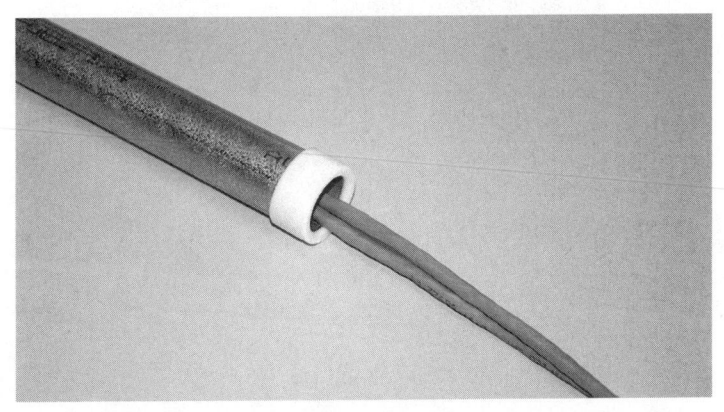

FIGURE 16–11 Plastic guards protect cable in conduit.

3. At the far end, attach a handline to the snake and pull back through the conduit.
4. Set out all cable boxes and mark each by number or location. Arrange so the cables will pull smoothly without kinking or twisting.
5. Bundle the cables as described in the section on horizontal pulls.
6. Attach the handline to the cable bundle.
7. With one operator at the beginning feeding cables and one or more pulling on the far end, pull the cables slowly and smoothly through the conduit.
8. Use mechanical pullers if the tension is too high. They can pull more smoothly and at higher tensions but should have tension monitors to prevent overstress.

Cable Trays and Raceways

Cable trays (Figure 16–12) are used to make installations neater and to protect cables from sharp edges and abrasive surfaces and from having other cables or hardware put stress on them. In order to protect the cables, it is necessary to arrange the cables properly in the trays. Lay the cables flat in the trays, starting from the sides and avoiding sharp bends at corners. If fiber-optic and copper cables are placed in the same trays, the fiber cables should be on top.

Riser Installation

Often closets or equipment rooms will be on different floors. For short runs, the cable can be supported from the top with a mesh grip or similar hardware available from cabling hardware vendors. For longer runs, special cables may be used, or a "messenger," a steel cable to which the cable can be attached, is installed with anchors at the top and bottom (and along the way if needed) to which the cables are attached.

FIGURE 16–12 Overhead cable trays organize and protect cables.

Although we call it a riser "pull," it is much easier to drop the cable down from the top and let gravity help you. Like all other pulls, make sure you know the exact path of the cables and ensure that the path is clear. Make sure you know where and how the cable will be attached. You do not want to lose the cables!

If needed, use a cable pulley above the riser opening. This is sometimes necessary if the cable is being dropped through a large opening and needs guidance.

Riser installation procedures are as follows:

1. Set up the cable or cable bundle about 30 feet from the riser opening.
2. Have as many people as needed holding the cables and helping feed them into the riser opening.
3. Have personnel at each location where the cable needs guiding, such as each floor in a multifloor drop or where the cable will be attached to a messenger.
4. Slowly feed the cable into the opening, guide it through each floor, and proceed until the drop is complete.

Mounting Hardware in Closets or Equipment Rooms

The closet or equipment room can be anything from a broom closet with communications hardware mounted on the wall to giant rooms with raised floors full of racks of equipment and cable trays connecting them. This makes describing this part of the installation difficult, so we will offer some generalities.

The layout of the room will be dictated by the space available and the equipment to be installed. Someone will lay out the area and specify where all the hardware will be installed. The installation of cable trays, racks, and so on is so vendor-specific that the best advice is to read the directions.

All hardware should be installed before any cabling begins. Once all the hardware is installed, the cable can be pulled in and terminated at the proper locations.

Firestopping

All penetrations of firewalls require firestopping to meet fire codes. This can be done with permanent foam-in firestopping material or removable material available in bags if more cables are to be installed in the future (Figure 16–13).

Cabling in the Closet or Equipment Room

First and foremost, be neat. If this is a new installation, remember that others will be inspecting your work and later will be doing moves and changes. Run cables neatly and carefully. Avoid sharp bends that hurt cable performance. If you have excess cables, coil them up out of the way and tie them loosely. Mark everything and document where everything goes.

Cable ties are often used to hold bundles of cables neatly. Normal nylon cable ties are acceptable as long as they are not tied too tightly, which can cause

FIGURE 16–13 Penetrations of firewalls or floors must be firestopped to meet building codes.

problems with cross-talk or attenuation in Cat 5E or Cat 6 cables or loss in fiber-optic cables.

If they are used, they should be hand-tightened only until they are snug, but can still be easily slid along the cable, then cut off. A better choice, although more expensive, are ties made from hook-and-loop fabric, which cannot harm the cables and offer the added advantage of being easily opened for adding additional cables (Figure 16–14).

If you are installing the cable hardware, you will usually be attaching all the punchdowns on plywood backboards attached to the wall. Again, cables should be routed neatly, avoiding sharp bends or stresses, and held in guides or wire saddles. Follow the guidelines on terminating to preserve cable performance.

Patch panels are mostly rack-mounted. Cables must be brought to the racks, terminated, and dressed neatly. Patchcords will be used by the equipment installers to attach communications hardware to each location. Obviously, neatness is preferred, but it is sometimes difficult as more patchcords are added to the racks (Figure 16–15). The hook-and-loop type of cable tie can be used to hold coiled excess patchcord cable for neatness. Good documentation is mandatory to speed the setup and moves and changes.

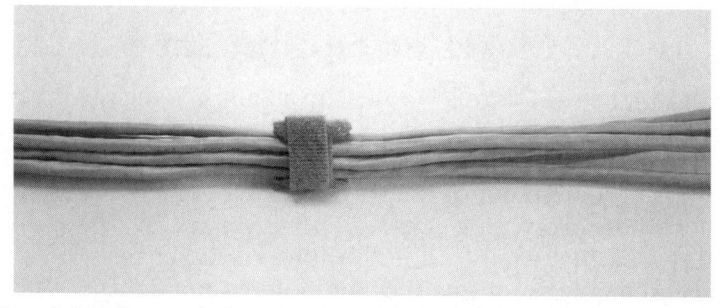

FIGURE 16–14 Cable ties made from hook-and-loop fasteners keep cables neat.

FIGURE 16–15 After many moves and changes, even the neatest patch panel will probably look like this!

Summary

- The type of network cable is determined by the type of network.
- Two considerations when choosing a network cable are the pathways from machine to machine and the signal strength.
- Signal performance is determined by the type of cable used.
- Wireless installations, while more expensive than cabled networks, are used for difficult installation where running cables is impractical.
- Problems with wireless networks include bandwidth, security, and compatibility of standards.
- Infrared uses wireless light transmission as a means of networking equipment.
- Infrared cannot pass through walls or barriers and is slow compared to other methods of transmission.
- Power line carrier networks involve the sending of signals over power lines.
- The performance of network cabling is greatly dependant on the quality of the installation.
- As a guideline, all network components must be similarly rated for performance.
- Cables must not be kinked or pulled too tightly.
- Cable supports must not stress cables unnecessarily.
- The termination of network cabling typically runs through communication closets.
- The communication closets include punch-down blocks, equipment racks, and modular patch panels to facilitate the interconnection of network devices.

Key Terms

Attenuation The reduction in optical power as it passes along a fiber, usually expressed in decibels (dB); the reduction of signal strength over distance.

Bit error rate (BER) The fraction of data bits transmitted that are received in error.

Capacitance The ability of a conductor to store charge.

Cross-talk Signal coupling from one pair to another.

Impedance The AC resistance.

J Hook A hook shaped like the letter "J" used to suspend cables.

Screened twisted-pair (ScTP) UTP cable with an outer shield under the jacket to prevent interference.

Shielding Stops interference and cross-talk by absorbing magnetic fields.

Review Questions

1. What two qualities does one concern oneself with regarding cable requirements of network cabling?
2. What is attenuation, and how is it specified in copper and fiber?
3. What affects attenuation in copper cables?
4. If the opposition of current flow impedance is not consistent, what happens to the signal?
5. What does capacitance filter out?
6. What does a distorted or rounded data signal cause?
7. Electromagnetic induction from one conductor to another causes what type of problem?
8. List the three types of shields used to prevent cross-talk.
9. What type of transmission is used to minimize cross-talk in unshielded cables?
10. *Matching:* Match each wire with the appropriate application, advantage, and disadvantage. Each of the wires listed will have three numbers indicating these elements.
    a. Coaxial cables
    b. UTP cables
    c. Optical fiber cables
    d. ScTP cables
    e. STP cables

    Applications:
    (1) Most current networks
    (2) Original Ethernet
    (3) Same as UTP
    (4) Optional for most networks
    (5) IBM Token Ring

    Advantages:
    (6) Inexpensive
    (7) Easy installation
    (8) Better data security
    (9) Excellent bandwidth
    (10) Has no effect on cross-talk

    Disadvantages:
    (11) Hard to terminate
    (12) Limited bandwidth
    (13) Higher cost
    (14) Bulky
    (15) Breaching security

11. Explain why wireless transmission is not "wireless."
12. What is the advantage of wireless networks?
13. True or False: You can use Cat 3 and Cat 5E components together because the wires are so similar in color and size.
14. What type of terminations are used for cross-connects in wiring closets?
15. What is the step-by-step procedure in pulling horizontal cables?
16. Once you have each box of cable in the drop area, what should you do to each box?
17. How do you support the wires above the ceiling?

18. When you run cable through a conduit, what can you use to protect the cable from the rough area at the end where the conduit was cut?

19. When placing both fiber and copper cables in a cable tray, which type of cable is placed on the bottom of a cable tray? Describe why.

20. For long-run riser installations, you would use special cables or install a _____.

21. What will help in a riser "pull" to get the cable to the end destination?

22. What should be done in the closets or equipment room before pulling the cable?

23. What will help speed the process of a setup in a closet or equipment room?

Cabling for Wireless Networks

OBJECTIVES *After studying this chapter, you should be able to:*

- Define the components of a wireless network.
- Explain the fundamentals of 802.11.
- Categorize the differences between 802.11a, 802.11b, and 802.11g.
- Describe the design issues of setting up a wireless network.

OUTLINE
Introduction

Wireless Is Not Wireless

Installation: The Site Survey

Introduction

Wireless networks are becoming increasingly important because of their unique properties. Wireless computer networks allow users to connect to the network without patchcords by using wireless cards in their PC, thus making them popular with mobile users such as workers in hospitals and warehouses or traveling salespersons.

Travelers can have network access in airports, hotels, and even restaurants and coffee shops. Structures where installing cabling is difficult or disruptive can often benefit from the use of wireless networks.

Wireless network standards are still evolving. Most wireless computer networks use radio frequency (RF) transmission and are based in the IEEE 802.11 standards, which are under constant development and updating to improve service and speed. An alternative RF network, Bluetooth®, is used for short-distance connections, mostly for consumer devices. Several wireless networks have been developed using IR light, but none have become widely used because of their limited range and line-of-sight transmission requirements.

The biggest concern about wireless is security. Unless carefully set up, the security of wireless computer networks is minimal. Practically anyone with a laptop and wireless card can access any RF wireless network, creating severe security problems.

Anyone considering a wireless network, commercial or residential, should thoroughly understand and carefully set up security to limit unauthorized use.

One should not be concerned that wireless will make cabling obsolete. Wireless is limited in connection speed, harder to manage, subject to interference from other RF devices such as remote controls, and more expensive than cabling. And, of course, wireless is not wireless.

Wireless Is Not Wireless

Wireless is not entirely wireless. The easiest way to understand wireless is to think of it as a link that replaces the patchcord that connects a PC to the network. A wireless network diagram looks like zone cabling with the final drop to the desktop replaced with a wireless link.

The wireless "antenna," officially called an **access point (AP)** but also referred to as a "hot spot," is a lot more than that. It is a radio transceiver and network adapter matching the PCMCIA card you insert in your laptop to access the network with some logic that implements part of the network protocols, allowing access to the network and sharing among several users. The transceiver in the AP has limited power, so the distance from the antenna to your laptop is limited. The connection between PCs and APs can be affected by metal in a building that reflects or attenuates signals. A typical office building may need four to twelve APs per floor to get consistent connections throughout the area.

The AP is connected to a wired network just like a PC using UTP or fiber-optic cable to a local hub, which connects it into the network backbone. Not only does the wireless AP require a network cable to connect to the network, but it needs power—uninterruptible power just like any network hub or switch—to operate. In some systems, that power is provided over the two spare pairs of a UTP cable using the Power Over Ethernet Standard (IEEE 802.af).

So, replacing a wired network with a wireless one does not mean you do not need cabling; you may, in fact, need more when you consider the power needs of the APs. Any advantage of a wireless network is not in the installation but in the flexibility of users roaming and maintaining connections.

System Layout

Any LAN application, network operating system, or protocol will run on an 802.11 compliant wireless LAN as easily as it will run over Ethernet, but it will run more slowly. Because IEEE 802.11b wireless LANs communicate using radio waves that can penetrate or reflect around many indoor structures, how much more slowly they will run is primarily determined by the quality of the design of access point locations.

To establish a wireless LAN, you must install and configure the access points and NICs. In addition, you will probably need to install cable from the existing network infrastructure to the access points because they are rarely located near existing outlets. Generally, access points are located in or near the ceiling to extend range and reduce interference from room furnishings. The number of access points depends on the coverage area, number of users, and types of services required.

The single most important part of the installation is placement of the access points to ensure proper coverage. Most manufacturers provide a site-survey tool to measure local signal output. Place the access points and use the site-survey tool to record signal strength and quality while roving within the intended coverage area.

Once the access points are installed, access points and NICs must be configured. Configuration options vary according to the manufacturer. When you are designing a wireless LAN, the cost of the equipment is just one concern; you must also consider installation and maintenance expenses.

Although all 802.11a, b, or g-compliant products are based on a standard, the standard offers no guarantee that access points, NICs, and other equipment from various manufacturers will interoperate. Choose equipment from compatible vendors or talk to equipment suppliers to ensure that you will have no problems. In a few years, everything will probably be seamlessly interoperable, but for the moment, the responsibility lies with the network designer.

Interoperability and Interference

Because the acceptance of any network depends on multivendor interoperability, industry leaders have formed a group, the Wireless Ethernet Compatibility Alliance (WECA), to certify cross-vendor interoperability and compatibility of 802.11b wireless networking products. This group has created tests to certify interoperability and announced the "WiFi" standard, which is an awarded seal for those wireless LAN products that have successfully completed prescribed interoperability testing. The seal provides assurance that products bearing the logo will work together. Many of the industry's leading wireless LAN manufacturers belong to the group.

The **WiFi** effort is effective to the extent it reaches, but understand its limitations. For example, both IEEE 802.11 and Bluetooth share common spectrum in the 2.4-GHz ISM band. IEEE 802.11 devices can provide a mobile extension to wired networks in large enterprise installations, and even replace a cabling infrastructure in small office and home office environments. Meanwhile, Bluetooth is available to the mobile worker and business traveler by facilitating e-mail to a laptop using a cellular telephone, synchronization with palm-top devices, and access to local printers.

It is inevitable that 802.11b and Bluetooth will bump into each other at some point. Currently, studies are underway to see exactly what the effects will be. But it is known that IEEE 802.11b's susceptibility to Bluetooth interference increases as the distance between wireless nodes and access points increases.

Installation: The Site Survey

Before a wireless network can be installed, the installer or designer must first do a thorough site survey. Essentially, setting up a wireless network is the same as setting up a cellular telephone system, except on a smaller scale. If you do not do a thorough site survey, the network may end up with many dead zones where no signal can be transmitted. Elevator shafts, kitchens with metal cabinets, thick concrete walls, steel studs, and even large metal fans can seriously restrict the effectiveness of a wireless installation.

A good site survey addresses two primary concerns: first, it examines the signal coverage requirements for the site. Second, the site survey enables you to understand the overall networking and system requirements. Does the existing backbone have adequate bandwidth to handle the overlaid wireless network? Are there sufficient resources to support the wireless capabilities needed?

When performing the site survey, use the tools and equipment provided by the wireless equipment vendor. If you are installing, designing, or selling a wireless LAN for the first time, it is advisable to have a specialist come in and walk you through the process.

There are seven major areas to consider when designing the site:

1. **Future use:** The customer may want to expand or modify the system at a later date. Build in plenty of excess capacity if at all possible.
2. **Coverage:** Define the area in which you want wireless coverage and its characteristics.

    Indoors or outdoors? Do you need coverage to be isolated to certain spaces? A wireless access point's typical range is spherical, and each area will generally need four access points with cells overlapping by about 30 percent to achieve optimum coverage. Access point and antenna ranges differ by vendor specification; make sure you know the ranges for your system.
3. **Capacity:** How many users are you expecting? What sorts of applications will they run? What types of equipment do you plan to use? These factors will affect the number of access points you need. Site capacity can be increased by adding more access points. Networks operating at higher data rates have a much shorter range, and such installations generally require more access points.
4. **Interference:** What are your potential sources of interference now and in the future?

    These can include sensitive equipment, previously installed systems, and Bluetooth networks for very-short-range mobile communications. In outdoor installations, moving vehicles such as trucks, planes, or other equipment may also be large enough to temporarily block signals. In such difficult locations, you may need to place access points at very high locations for vertical signal penetration. Access points may be placed at opposite sides of the covered area so that, if one is shadowed, the signal may still be transmitted through the other. You should also examine potential antenna performance patterns to decide on an omnidirectional or directional antenna. Interference avoidance design requires a combination of sufficient signal strength, the appropriate technology, intelligent access point and antenna placement, and a thorough site survey.
5. **Connectivity and power requirements:** What are your environment's networking constraints? All your networks must be based on the same standards, usually Ethernet. One cannot assume that the current environment will be able to handle the increased workload, and you do not want new access points suddenly overstraining the system. You should also calculate the installation based on the network device with the least range so that every device is sure to transmit clearly. Connecting to the wired infrastructure may also require different cabling alternatives. Wireless access points are typically installed in ceilings for better coverage, which means that data cabling and AC power (if power over Ethernet is not used) will need to be run in some potentially difficult locations.
6. **Cost and ease of installation:** Performing a site survey provides a realistic understanding of the wireless installation's cost before finalizing the design. Perhaps the site has unusually high interference issues to resolve, or capacity is greater than you anticipated. Based on a good site survey, you can put together a reasonable cost estimate so there are no surprises. Trying to build a wireless LAN in stages is a complex undertaking. Many installers have found themselves in the midst of a communications disaster as sessions drop out and cell locations are lost. Because wireless systems simply fit in on top of existing environments, however, you can anticipate limited work interruptions during the installation process.

7. **Setting up temporary antennas and access points:** The most important task when installing a wireless network is to do a site survey. This process includes assessing every part of the site with a software survey tool installed on a handheld computer or laptop that monitors the signal and identifies failure zones. Such site-survey tools measure performance between access points, identify sources of interference, and help determine access-point placement.

Site surveys differ in their complexity and level of effort, based on technology and space. Small facilities may not require one at all, and you can often use only product specifications and a good set of blueprints. However, in larger installations the site survey offers a level of security in an arena in which, unlike wired networks, there are many variables to consider and few fixed rules. A thoughtful, accurate, and effective assessment of coverage and system requirements will make the wireless LAN installation more straightforward and less expensive while laying the groundwork for ongoing expansion.

Summary

- Wireless standards are still evolving and most are based on the IEEE 802.11 standard.
- A short distance alternative to 802.11 is Bluetooth.
- The biggest concern about wireless is security.
- Wireless is limited in speed, hard to manage, and subject to interference.
- The wireless antenna is officially called an access point or a "hot spot."
- Any LAN application, network operating system, or protocol will run on an 802.11 wireless LAN, but it will run slower than over Ethernet.
- To establish a wireless LAN you must configure access points with cables back to an existing network infrastructure.
- The single most important part of an installation is placing the access points to ensure proper coverage.
- The Wireless Ethernet Compatibility Alliance (WECA) was established to certify the cross-vendor interoperability of systems and 802.11.
- Site capacity can be increased by adding more access points.

Key Terms

Access point (AP) A device that connects wireless communication devices together to form a wireless network.

WiFi Set of product compatibility standards for wireless local area networks (WLAN) based on the IEEE 802.11 specifications.

Review Questions

1. What is the most common standard wireless data signal?
2. What is an access point?
3. What is the wireless technology designed for low-cost, low-power, shortrange (<10-meter) links?
4. How does Bluetooth compare to 802.11?
5. What does multivendor interoperability require?

Testing Voice, Data, and Video Wiring

After studying this chapter, you should be able to:

- Describe the procedure for testing coaxial cable and unshielded twisted-pair.
- List the test instruments required for testing coaxial cable.
- Identify the tests for wiremapping a voice data cable.
- Explain the pinout for TIA/EIA 568a and 568b as well as FCC USOC.
- List a variety of problems that occur when wiremapping.
- Explain the relationship of structural return loss and impedance.
- Define attenuation, NEXT, Powersum NEXT, ACR, FEXT, propagation delay, and delay skew.

OBJECTIVES

Introduction
Coax Cable Testing
Unshielded Twisted-Pair (UTP) Testing
Wire Mapping
Impedance, Resistance, and Return Loss
Cable Length
Attenuation
Near End Cross-talk (NEXT)
Power Sum NEXT
Attenuation to Cross-talk Ratio (ACR)
FEXT and ELFEXT

OUTLINE

Propagation Delay and Delay Skew

Cable Plant "Certification" to Standards

Introduction

All wiring installed for voice, data, or video applications needs testing to ensure proper operation. Data and voice cabling are similar in testing requirements and are covered in one section. Coax testing can be either easier (if it works) or much harder (if it does not), so it is covered first.

Coax Cable Testing

Because coax has only two conductors, the inner conductor and the shield, you want continuity and no shorts. This can be tested by a simple coax tester or a **digital multimeter (DMM).** With the DMM, testing the cable with no terminations will tell if the cable is shorted, and testing with it terminated will tell whether or not it is open. If the cable passes these tests, it should work with most applications.

Sometimes coax can be damaged in installation and will have unusual attenuation characteristics over its frequency range. If a coax cable has problems transmitting a signal but shows neither an open nor a short, look for damage like kinks that may be causing the problem.

Unshielded Twisted-Pair (UTP) Testing

Testing voice and data UTP wiring is a function of use. Unless the network running on the wire is high speed, the wire only needs testing for correct connections by using a low-cost wiring verifier (wire mapper) or a "toner." As the network speed increases, the need for testing bandwidth, cross-talk, and so on at appropriate network speeds becomes more important.

Voice cabling requires a simple verification of the connections, or wire mapping. This is done with an instrument that plugs into the cables at either end and tests for proper pin-to-pin connections. A toner is a device that puts a tone on the wires and allows checking at any point with a small receiver. Toners can be used to trace and identify wires in termination blocks also.

Even though voice wiring may not require more than a continuity check, many structured wiring installations pull the same Cat 5E cabling for both voice and data applications because the incremental cost is low and the installation of Cat 5E everywhere gives the most future flexibility. Therefore, it may be appropriate that all new wiring be fully tested as if any link may eventually have to carry network data.

Most testing today, with the exception of simple wire mapping, is done with automated loss testers that are preprogrammed to perform tests to the requirements of TIA/EIA 568. These instruments test wire mapping, length, attenuation, and **NEXT.** As long as the instruments meet the 568 requirements and are designed for testing the networks proposed for the cable plant you are installing, you can expect valid test results and can generally accept "pass/fail" results.

This discussion focuses on the tests used, what causes problems, and how to troubleshoot testing failures. If you are using an automated tester, read the directions and practice with the instrument before using it in the field because that will make mistakes in interpreting data much less likely.

TABLE 18–1
Network Cabling Test Requirements

Cable/Network	Cat 3	Cat 5E	Cat6
Analog phone	W	W	W
Digital phone/PBX	W	W	W
Ethernet	L, W, X, A	L, W, X, A	L, W, X, A
4 MB Token Ring	L, W, X, A	L, W, X, A	L, W, X, A
16 MB Token Ring	L, W, X, A	L, W, X, A	L, W, X, A
100Base-T4	NR	L, W, X, A, PD, S	L, W, X, A, PD, S
100Base-TX	NR	L, W, X, A, PD	L, W, X, A, PD
100VG AnyLAN	L, W, X, A	L, W, X, A, S	L, W, X, A, S
TP-PMD/FDDI	NR	L, W, X, A	L, W, X, A
155 MB/s ATM	NR	L, W, X, A	L, W, X, A
1000Base-T	NR	L, W, X, PSX, A, PD, E, PSE	L, W, X, PSX, A, PD, E, PSE
1000Base-TX	NR	NR	L, W, X, PSX, A, PD, E, PSE

Tests: L = length, W = wire map of connections, A = attenuation, X = near end cross-talk (NEXT), PSX = power sum NEXT, E = ELFEXT (equal level far end cross-talk), PSE = PSELFEXT (power sum equal level far end cross-talk), S = delay skew, PD = propagation delay, NR = not recommended.

Table 18–1 shows the recommended testing requirements of different cables when used for different communications applications.

Wire Mapping

Wire mapping includes all tests for correct connections. The correct pairs of wires must be connected to the correct pins, according to the color codes defined by the standards (either TIA T568A or T568B mapping). There must be no transposing of wires or pairs and no shorts or opens.

Figure 18–1 shows TIA/EIA T568A, and Table 18–2 gives a wire map for T568A pin assignments. Figure 18–2 is another way to look at the wiring diagram for the 568A modular jack.

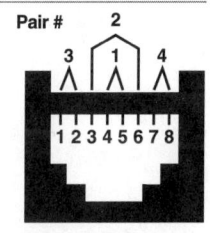

FIGURE 18–2 T568A Wiring at the Jack, Looking into the Jack

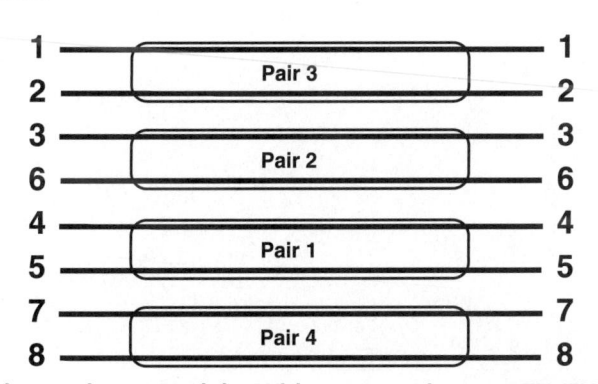

FIGURE 18–1 Eight-Conductor Modular Wiring Connections per TIA/EIA 568

356 • Chapter 18

TABLE 18–2
TIA T568A Modular Connector Wiring

Pin Number	Pair Number	Color Codes
1	3-Tip	White/Green
2	3-Ring	Green
3	2-Tip	White/Orange
4	1-Ring	Blue
5	1-Tip	White/Blue
6	2-Ring	Orange
7	4-Tip	White/Brown
8	4-Ring	Brown

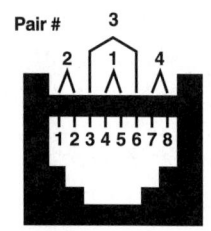

FIGURE 18–3
T568B wiring switches the position of pairs 2 and 3.

The T568A or T568B standard may be used for building wiring. The T568B standard (Figure 18–3) is more commonly specified by AT&T. The only difference between T568A and T568B wiring is the swapping of pairs 2 and 3, which makes no difference in performance—it is simply a different color-code convention adopted by various vendors. (Table 18–3 gives a wire map for T568B pin assignments.) Whichever version is used, it must be used throughout an installation to prevent problems. Some plugs and jacks include both T568A and T568B wiring guides on the part, so be careful that one end is not wired to a different standard. If such a problem is detected, it can be solved by an equally miswired jumper, but this invites future problems. Better to correct it when detected!

There is yet another variation of the 8-pin pinout, the FCC USOC (Universal Service Order Code) RJ-61X (Figure 18–4) used for voice only. Table 18–4 gives a wire map for USOC pin assignments. Although this pinout works for voice and

FIGURE 18–4
USOC wiring for the 8-pin modular jack is used for voice wiring only.

TABLE 18–3
TIA T568A Modular Connector Wiring

Pin Number	Pair Number	Color Codes
1	3-Tip	White/Orange
2	2-Ring	Orange
3	3-Tip	White/Green
4	1-Ring	Blue
5	1-Tip	White/Blue
6	3-Ring	Green
7	4-Tip	White/Brown
8	4-Ring	Brown

TABLE 18–4
USOC Modular Connector Wiring

Pin Number	Pair Number	Color Codes
1	4-Ring	Brown
2	3-Tip	White/Green
3	2-Tip	White/Orange
4	1-Ring	Blue
5	1-Tip	White/Blue
6	2-Ring	Orange
7	3-Ring	Green
8	4-Tip	White/Brown

some low-speed data applications, the pair layout causes too much cross-talk for higher speed networks, so it is seldom used for data installations and is not covered in the 568 standard. It does allow a modular 6-pin plug to attach to an 8-pin plug, although that is not recommended.

Open DECconnect is a variation of T568A that leaves out pair 1. (Table 18–5 gives a wire map for DEC connect pin assignments.) It has only been used in Digital Equipment Corporation's proprietary DECconnect networks.

When EAI/TIA 568 was published, the T568A convention was recommended because its pin/pair assignments, with pairs 1 and 2 located on the center four pins, were compatible with a wide variety of existing 2-pair voice and data applications, including some data systems that used the old USOC pair configuration for 6- and 8-conductor jacks. T568B was an accepted alternative, for it was a convention that AT&T had already widely established for 4-pair data wiring. T568B is the same as

TABLE 18–5
Open DECconnect Wiring

Pin Number	Pair Number	Color Codes
1	3-Tip	White/Green
2	3-Ring	Green
3	2-Tip	White/Orange
4	NC	
5	NC	
6	2-Ring	Orange
7	4-Tip	White/Brown
8	4-Ring	Brown

TABLE 18–6
Network Conductor Use

Network	Pins Used
10Base-T	1-2, 3-6
Token Ring	4-5, 3-6
TP-PMD (FDDI)	1-2, 7-8
ATM	1-2, 7-8
100Base-TX	1-2, 3-6
100Base-T4	1-2, 3-6, 4-5, 7-8
100VG-AnyLAN	1-2, 3-6, 4-5, 7-8
1000Base-T or 1000Base-TX	1-2, 3-6, 4-5, 7-8

AT&T or WECO 268A. Federal government publication FIPS 174 only recognizes T568A.

Most important for any cable is correct connections—that is, the proper conductors must be connected to the proper pins on the plugs and jacks. But first you must be certain whether you are dealing with T568A, T568B, USOC, or even DECconnect terminations, as opposed to nonstandard connections or older installations where less than four pairs are terminated, to get the correct test results.

Networks may not use all four pairs. In fact, most use only two pairs chosen to allow proper performance for that network. Table 18–6 gives a rundown of what pins are used by most common networks. You must translate that into "pairs," using Tables 18–2 through 18–5, depending on the pin configuration.

Wire Map Problems

Most wire map problems occur at the connections. Physical examination of the connections should find the fault. Wire map errors fall into several basic categories, which are illustrated with T568A connections in the following text.

Shorts and opens. A short occurs when two conductors are accidentally connected, and an open exists when one or more wires are not connected to the pins on the plug or jack (Figure 18–5). Opens can also occur due to cable damage. A **time domain reflectometer (TDR)** test, which shows the distance to the fault, can assist you in locating the fault.

Reversed pairs. Reversed pairs occur when the conductors (Tip and Ring) are reversed in the pair (Figure 18–6).

Transposed or crossed pairs. Transposed or crossed pairs occur when both conductors of one pair are swapped with both conductors of another pair at one end (Figure 18–7). The most common cause of crossed pairs is termination of one end of the cable with T568A and the other end with T568B, thus reversing pairs 2 and 3.

Split pairs. A wiring verifier will provide basic connection information, but some faults like split pairs (Figure 18–8) may not show up in a wire map. Split

Testing Voice, Data, and Video Wiring • 359

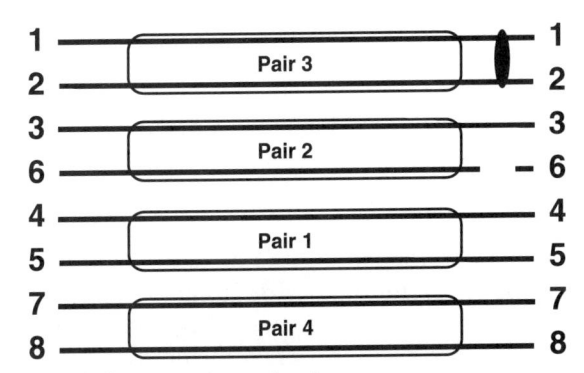

FIGURE 18–5 Opens and Shorts at a Termination

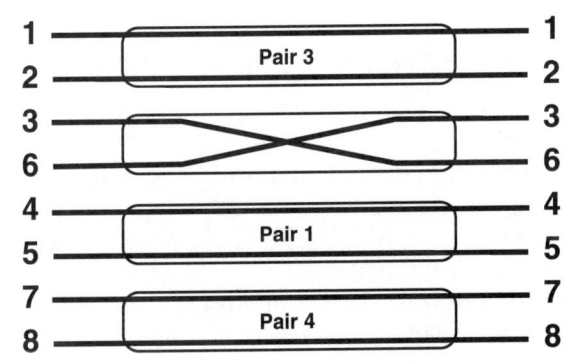

FIGURE 18–6 Reversed pair has conductors crossed at one connection.

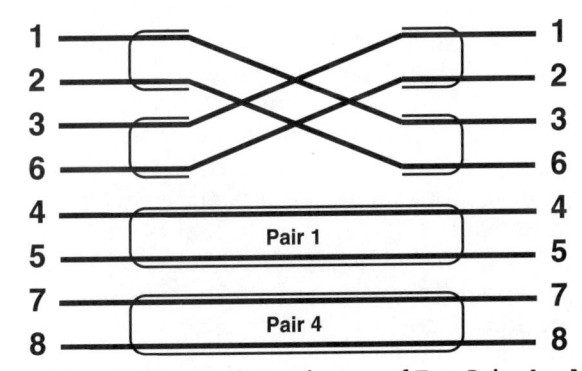

FIGURE 18–7 Crossed Pairs Where Both Conductors of Two Pairs Are Miswired

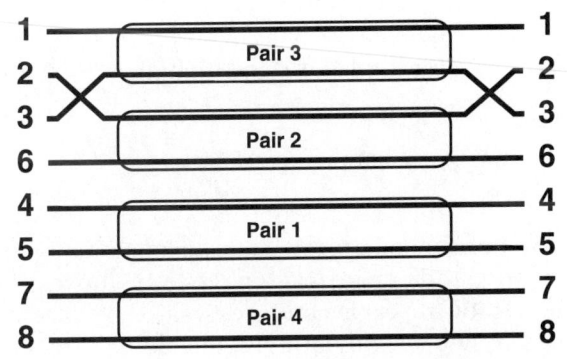

FIGURE 18–8 Split pairs have proper wire mapping but will cause cross-talk problems.

pairs occur when one wire on each of two pairs is crossed on *both* ends. This fault is impossible to find with a normal wiring verifier because the wire map is correct—that is, the pin connections are correct, but the wires are not in proper pairs. Split pairs can be detected only in a cross-talk (NEXT) test, where the unbalanced pairs can be detected. Split pairs are usually caused by terminating a jack with punchdown color codes (pairs 1-2-3-4) instead of T568A or T568B, which splits pairs 1 and 2.

Conclusion. Remember that these miswirings can occur in any combination. A common error is simply mistaking the color coding and mixing up wires. Another common error is terminating as T568A on one end and T568B or USOC on the other. The wire mapper should detect these errors.

Impedance, Resistance, and Return Loss

Impedance is the "resistance" of the cable at the frequency of signals transmitted. Return loss refers to reflections that occur at changes in impedance. These reflections can cause errors in signal transmission if they are too large. UTP cable is specified to have a nominal impedance of 100 ± 15 ohms. For high-speed data, both impedance and return loss are functions of the signal frequency, and impedance tends to decrease with frequency.

Cables, connectors, and other hardware for high-speed networks are designed to have very consistent impedance to prevent reflections. For cable, this means having the size of the conductors, twist of the pairs, and insulation materials carefully controlled. The term *structural return loss* is used to refer to the reflections caused by variations of the impedance of the cable itself along its length. Even variations in production of cable can result in varying impedance, so consistently using the same cable in a network can minimize problems.

At connectors, the twist of each pair must be maintained to within 1/2 inch, or 13 mm, of the connection points to prevent cross-talk and undesirable reflections. If one expects consistent high-speed network performance, every component in the cable plant must be rated for Cat 5 and installed precisely.

The impedance typically goes down with higher frequency, and return loss will vary significantly with frequency. Both should therefore be measured with a tester that tests at a frequency consistent with the network planned for the cable plant. At the present time, there is no industry agreement on field-testing requirements for return loss, although this is expected to change.

Resistance is the DC component of the impedance and can be measured with a DMM. Every cable has a rated resistance in ohms/foot. If you know the length of the cable, you can test resistance with your DMM; if you know the resistance in ohms/foot, you can measure the total resistance of a cable and determine the length.

Cable Length

Cable length needs to be known to verify that the length is within the limitations of the design standards and for future reference in moves, changes, or troubleshooting. The length of the cable can be estimated by measuring resistance as described in the preceding text or by using a time domain reflectometer (TDR).

Time domain reflectometry works like radar, sending an electrical pulse down the cable to an open end where the signal is reflected back to the transmitting end. By knowing the characteristic speed of the signal in the cable, called the nominal velocity of propagation (NVP), and the round-trip transit time, you can calculate the length of the cable.

The NVP is an average value; actual cable samples may vary as much as 10 percent among production lots. Even each pair, which has different twist rates to reduce cross-talk, will have a different velocity of propogation (varying as much as 3 to 4 percent) and a different physical length. Add to this the inherent inaccuracy of the instrument, another 2 to 3 percent, and you see the measurement is a good approximation of the length, but not an exact measurement. All testers will measure and report the length of each pair.

TDRs offer another valuable piece of information—they find opens, shorts, and terminations. If you see no reflection, the end of the cable is properly terminated. If your return pulse is the same polarity as the transmitted pulse, the cable is open at the end. And if the return pulse is of opposite polarity to the transmitted pulse, the cable is shorted. Thus the TDR can find cable faults by type and location, enhancing its use as a troubleshooting tool.

Attenuation

As an electrical signal travels down the cable, the impedance causes attenuation (Figure 18–9). At the far end, the signal will be smaller than at the transmitter. It is important that the attenuation be less than a specified value so the received signal is of adequate strength for proper data transmission.

Attenuation is expressed in dB (decibels), where 20 dB is an attenuation factor of ten. Like everything else we have discussed, attenuation is a function of frequency, increasing at higher frequencies, and must be tested at operating frequencies. Testers will test at the frequencies specified in the category of cable being tested, up to 16 MHz for Cat 3, 100 MHz for Cat 5E, and 250 MHz for Cat 6.

Measuring attenuation requires an instrument at each end of the cable: one to transmit a known value signal and one at the far end to measure the signal level and calculate the attenuation. Pass/fail criteria are set by the 568 standard.

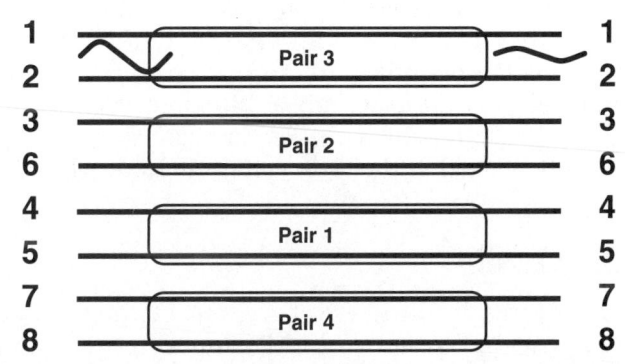

FIGURE 18–9 Attenuation in the cabling causes a lower signal amplitude at the receiver end.

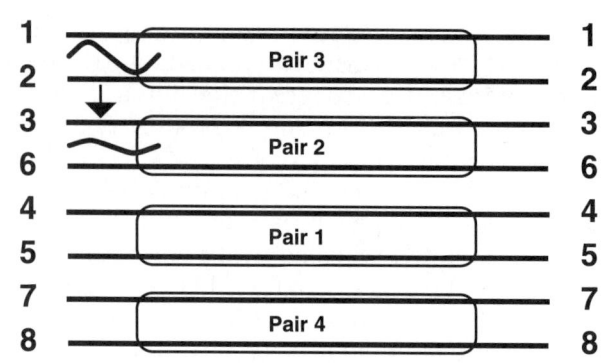

FIGURE 18–10 Near end cross-talk (NEXT) is signal coupling at the transmitter end.

Near End Cross-talk (NEXT)

In a cable that has four pairs of electrical conductors, whenever one pair is carrying signals it may couple some of its energy into an adjacent pair (Figure 18–10). If a signal is being transmitted from the other end simultaneously, the proper signal may be compromised by interference by the cross-talk.

Each pair works like an antenna; the pair carrying a signal is the transmitter, and every other pair is a receiver. The cable construction, including the variation in twist rates in the pairs, is designed to minimize cross-talk. Like everything else, cross-talk is frequency dependent, so it must be tested over the full frequency range specified for the category of cable being installed.

As stated earlier, at the connectors, where cross-talk is most critical, the twist of each pair must be maintained to within 1/2 inch, or 13 mm, of the connection points to prevent undesirable reflections. If one expects consistent high-speed network performance, every component in the cable plant must be rated the same and installed precisely.

Testing cross-talk is quite simple. Terminate the far end of each pair to prevent reflections that would interfere with cross-talk measurements, transmit a signal on one pair, measure the coupled signal on another pair, and calculate the cross-talk in dB, just like attenuation. As with attenuation, pass/fail criteria are set by TSB-67 and are given in the discussion on certification in the last section of this chapter.

Note that each pair must be tested against all the other pairs for a total of six tests, and the test must be repeated at both ends of the cables for it to be valid.

The most common failure for NEXT is improper termination. If the twists are not maintained to within 1/2 inch (13 mm) of the termination, NEXT will fail. Poor quality components, including patchcords, termination blocks, plugs, jacks, or cables, will cause failures also, especially at high frequencies.

As mentioned earlier, split pairs will cause unbalanced transmission and high cross-talk. It is with NEXT testing that you will likely find these problems.

Power Sum NEXT

Ethernet and Token Ring use only two pairs out of the four-pair UTP cable. Some of the new higher-speed networks like Fast Ethernet, ATM, and Gigabit Ethernet use all four pairs. Because they transmit over multiple pairs simultaneously, cross-talk occurs from several pairs at once. It can be a bigger problem

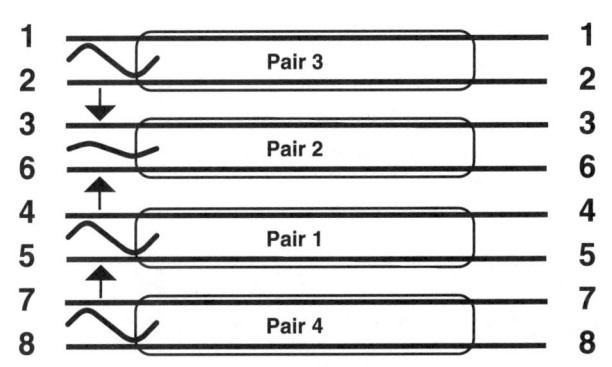

FIGURE 18–11 Power sum NEXT measures the coupling effect of all other pairs into one of the four pairs in the cable.

and needs testing. **Power sum NEXT** measures cross-talk on one pair while all other pairs are transmitting (Figure 18–11). This test has already been implemented by several manufacturers in their testers but has not yet been included in TSB-67.

Attenuation to Cross-talk Ratio (ACR)

In a duplex communication link, like networks running on UTP cabling, signals can be traveling in both directions simultaneously. Thus, at the receiver end of the cable, one can have a signal from the other end attenuated by the cable and cross-talk coupled from the local transmitter simultaneously arriving at the receiver. Because both ends transmit signals of approximately the same amplitude, the received signal needs to be higher than the interfering cross-talk.

Attenuation-to-cross-talk ratio **(ACR)** is the measure of this situation (Figure 18–12) and an excellent indication of the overall quality of the cable link. For example, if the cross-talk is 35 dB and the attenuation is 15 dB, the received signal will be 20 dB larger (a factor of ten) than the cross-talk. If the cross-talk is 28 dB and the attenuation is 22 dB, the received signal will be 6 dB larger (a factor of two) than the cross-talk. Thus, the higher the ACR, the better the cabling performance.

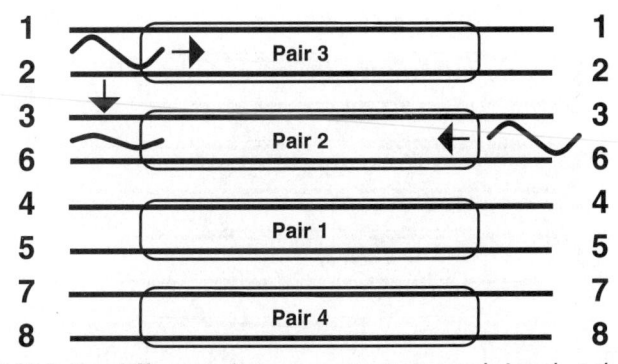

FIGURE 18–12 ACR is the difference between an attenuated signal at the receiver and cross-talk from an adjacent transmitter.

FEXT and ELFEXT

FEXT is just like NEXT except it is measured at the end opposite the test signal, the far end. Because the test signal is attenuated more on longer links, FEXT may be greater on short links. To normalize the measurement, ELFEXT compensates for the attenuation of the signal as in ACR.

Propagation Delay and Delay Skew

Propagation delay (or simply delay) is a measure of the time it takes for an electrical signal (traveling at about two-thirds the speed of light) to reach the far end of the cable. Testers measure delay as part of the process of measuring length. Because each pair has a different length due to the twist rate and a different NVP, there will be variations among the four pairs. This is not a problem with most networks because the signals are carried on only one pair in each direction. However, many of the high-speed networks (100 MB/s and higher) are using two or even four pairs in each direction, so if the parallel signals arrive too far apart, data errors will occur.

The design of the cable includes having a different twist rate on each pair of wires; a more tightly twisted pair will have slightly longer wires for a given cable length. Thus, a signal will take different times for end-to-end transmission for each pair; and NVP can also vary with wire insulation, also causing differences in transit times. Some cables have been built with different (cheaper) insulation on the pairs not normally used with Ethernet or Token Ring, and the variation in transit times may be more than is tolerable.

The maximum difference in transit times between all four pairs is called **"delay skew"** (Figure 18–13). The delay skew must be less than a specified amount to allow the cable to work with the high-speed networks. Because delay skew is a function of the cable itself, it is important to buy only cable that meets the required specifications as outlined in TIA/EIA 568. However, delay skew is not likely to be a field-test requirement because it cannot be easily affected by installation practices.

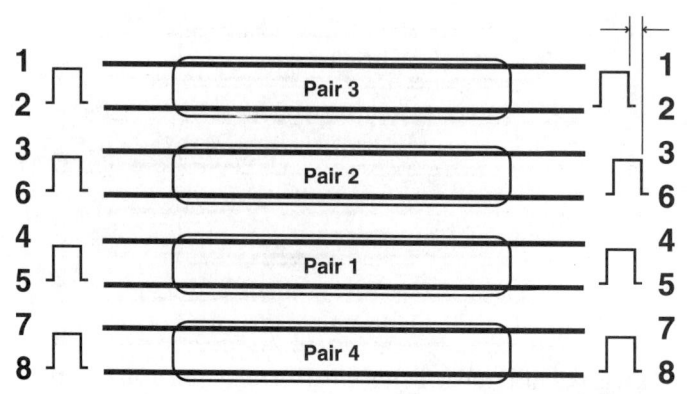

FIGURE 18–13 Delay skew measures the timing differences between pairs.

Cable Plant "Certification" to Standards

The industry has agreed on a set of test standards for UTP cabling included in TIA/EIA 568. The automatic testers will give you a "pass/fail" result, and perhaps even an indication if the result is uncertain, as when the measured result is closer to the limit than the actual instrument accuracy.

Length testing requires the link be less than 90 meters and the channel to be less than 100 meters, including the test equipment patchcords. Wire mapping requires connections per 568A or 568B. Other connections, including USOC and DECconnect, although perhaps valid for the installation, will not meet 568 standards.

Summary

- All wiring installed for voice, data, or video applications needs testing to ensure proper operation.
- Coaxial cable can be tested by using a digital multimeter.
- Unshielded twisted-pair requires a low-cost wire verifier or toner.
- A wire mapping instrument plugs into each end of the cable to test proper pin-to-pin connection.
- Pinout requirements vary depending on the equipment standard, including TIA/EIA 568a and 568b and USOC.
- Wiring errors can occur as shorts and opens, reversed pairs, transposed pairs, or split pairs.
- Structural return loss refers to the reflections caused by variations of impedance along the cable.
- Cable testers will perform a variety of tests and give a pass/fail result for each test.

Key Terms

ACR Attenuation to cross-talk ratio; a measure of how much more signal than noise exists in the link by comparing the attenuated signal from one pair at the receiver to the cross-talk induced in the same pair.

Delay Scew The maximum difference of propagation time in all pairs of cable.

Digital multimeter (DMM) Modern multimeters are exclusively digital, and identified by the term DMM or digital multimeter. In such an instrument, the signal under test is converted to a digital voltage and an amplifier with an electronically controlled gain preconditions the signal. Since the digital display directly indicates a quantity as a number, there is no risk of *parallax* causing an error when viewing a reading.

NEXT Near end cross-talk; a measure of interference between pairs in UTP cable.

Power Sum NEXT Near end cross-talk tested with all pairs but one energized to find the total amount of cross-talk caused by simultaneous use of all pairs for communication.

Time domain reflectometer (TDR) A testing device used for copper cable that operates like radar to find length, shorts or opens, and impedance mismatches.

Review Questions

1. What can be used as a simple coax tester?
2. What should you look for if there is a problem with video transmission in a coax cable?
3. What UTP test instruments do you use to test either low-speed or Cat 5E/6 high-speed cable wire for correct connections?

4. Why is Cat 5E a better choice to install than Cat 3?
5. What is used to test wire mapping, length, attenuation, and NEXT?
6. What are the color codes used in the 4-pair UTP cable? List the pairs together.
7. Other than the TIA/EIA 568A or TIA/EIA 568B, what variations are used but not recommended for structured cabling?
8. What company's standard established T568B wiring?
9. What is most important and the first thing to test in any cable?
10. What instrument is used to test the length of the cable or the distance to the fault?
11. What fault is a split pair, and what happens to signals transmitted on these pairs?
12. What is impedance?
13. What are reflections that occur at changes in impedance?
14. What is it when the cable has a reflection from a variation in impedance in the cable?
15. How much untwisting is allowable before you lose performance in Cat 5E or Cat 6 cable?
16. What do you need to know to use a TDR to calculate the length of a cable?
17. What is reduced by the different twist rates in the separate pairs of UTP cable?
18. *Matching:* Match the fault with the return pulse indication on a TDR.

    a. Open at the end

    b. Properly terminated

    c. Shorted

    (1) Opposite polarity to the transmitted pulse

    (2) Same polarity as the transmitted pulse

    (3) Little return pulse
19. True or False: NEXT is when the transmitting pair couples part of its signal to other pairs of wires.
20. What does power sum NEXT measure?
21. What occurs in a delay skew and why is it important?
22. True or False: Different insulation in a Cat 5 wire does not affect the performance of the wire.
23. What are the specifications for a cable plant length for a link and a channel?
24. What does a channel include that a link does not include when dealing with length?

Communications Infrastructure Design

OBJECTIVES

After studying this chapter, you should be able to:

- Understand the steps in the infrastructure design process.
- Explain the considerations involved in specifying the location of the major infrastructure components, such as the DD, the NID, and the ADO.
- Discuss the requirements and recommendations for the number and location of TOs.
- Explain the design considerations for wireless networks.
- Understand the infrastructure design requirements of entertainment networks and home automation systems.
- List several ways of providing future flexibility for the infrastructure.
- Discuss the types of documentation produced during the design process.
- Identify the unique requirements of retrofit installations.

OUTLINE

Introduction

The Infrastructure Design Process

Retrofit Installations

Example with Floor Plan

Introduction

This chapter covers the process of designing the **communications infrastructure** for a single-dwelling unit (SDU). This material presents a general overview of the design process and the main items to be considered. It is not intended to be a substitute for the more detailed material in the Building Industry Consulting Service International manual (BICSI 2002) or the instructions provided by the manufacturers of the various infrastructure products.

Several parties have a vested interest in the communications infrastructure design, including the homeowner, the architect, the builder, and the installer. Each of them is likely to have slightly different interests and priorities. The homeowner is likely to be interested primarily in the functionality of the system, placement of the **telecommunication outlets** (TOs), aesthetics of the installation, the cost, and the flexibility of the system to meet future requirements.

The architect is likely to be interested primarily in the functionality and aesthetics. The builder will be concerned with the cost, the schedule for installation, and how standard the design is. Standard designs are preferred because it is easier to procure the components and install them without problems. The installer is likely to be interested in the details of the design (such as the brand of components that are selected), ease of installation, and the schedule.

A successful communications infrastructure design should meet the needs of all the relevant parties that are involved.

The Infrastructure Design Process

Before the design can begin, the user requirements for the system must be determined. Two key questions are:

1. What systems will the infrastructure support—voice and data, entertainment network, home automation systems, and so on?
2. What allowance should be made for future flexibility in the system?

The installation also must meet all relevant codes and standards, including local building codes, national codes such as the *NEC®*, and FCC regulations. Compliance with industry standards, such as TIA-570B, is optional but highly desirable. The design process can be thought of as proceeding in six stages.

1. Placement of the major infrastructure components
2. Voice and data network
3. Entertainment network
4. Home automation systems
5. Future flexibility
6. Documentation

Each of these items is discussed in more detail in the following sections.

Placement of the Major Infrastructure Components

The Network Interface Device (NID), the **Auxiliary Disconnect Outlet (ADO)**, and the **Distribution Device (DD)** are major infrastructure components that must be located within the house. The NID is usually located on an exterior wall, often at a location selected by the service provider. The ADO is placed at a convenient point between the NID and the DD.

The location of the DD is very important. While theoretically it could be almost anywhere within the house, several principles should be followed to determine an optimum location, as indicated in Table 19–1. If the house contains a utility closet, that is often a good location for the DD. Another frequently used location is either underneath or behind the basement stairs.

TABLE 19–1
Requirements for Location of the Distribution Device

The DD should:	*The DD should not:*
Be located in an area with sufficient room to hold all the necessary equipment, including room for future expansion.	Be located on an outside wall that is subject to extremes of temperature and/or possible water leakage.
Provide adequate access for servicing and administering the cabling and equipment.	Be located near sources of EMI such as large motors, power transformers, and so forth.
Be in a fairly central location to minimize the length of the cable runs.	Be located in a fire-rated wall, as most DD cabinets are not rated for firewall installation. If the DD must be installed on a fire-rated wall, be sure to get approval from the local building inspector before the installation.
Be in a secure location that can be locked if desired.	
Be provided with at least one alternating-current power outlet.	Take up prime floor space in the living area of the house.

Once the location is specified, the make and model of DD must be chosen.[1] Most manufacturers have a number of different sizes of DD. It is important to select a model that meets the current requirements of the house and provides some room for future expansion.

The DD housing contains different types of modules (such as voice, LAN hubs, whole-house audio, etc.) for terminating and cross-connecting (or interconnecting) cables providing the various applications within the house. Figure 19–1 and Figure 19–2 illustrate a typical LAN hub and video module that may be mounted within a DD. The types and number of modules within the DD will be selected based on the service requirements of the house.

FIGURE 19–1 A LAN Hub with Brackets for Mounting in the DD (Courtesy of Leviton Manufacturing Co., Inc.)

---

[1] In this book, we do not discuss the pros and cons of specific manufacturers. Many different companies make residential cabling products. The choice of a particular company's products is often based on previous experience of the builder or installer and is beyond the scope of this book.

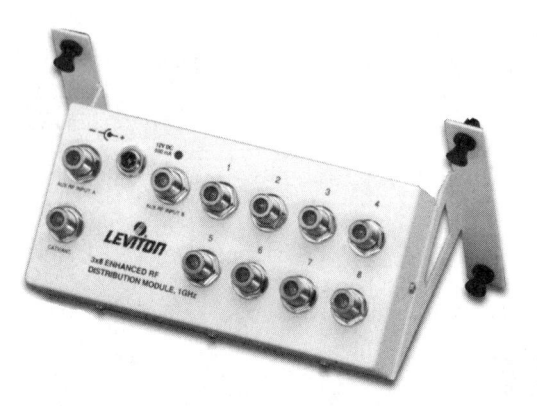

FIGURE 19–2   A Cable TV Distribution Module   (Courtesy of Leviton Manufacturing Co., Inc.)

Voice and Data Network

Several points must be considered when designing the voice and data network, including the number and location of TOs, the type of cable to use, support for the various broadband access technologies, and provisioning of WAPs.

Telecommunications Outlets

Each room must have a sufficient number of telecommunications outlets (TOs) to allow equipment to be connected without using long extension cords. The only requirement in TIA-570B for TO placement is that at least one TO shall be placed in each of the following rooms: kitchen, bedroom, family/great room, and den/study. TIA-570B also recommends that additional TOs be placed:

- On any unbroken wall space of 12 feet or more in length.[2]
- So that no point along the floor line is more that 25 feet (measured along the floor line) from the nearest TO.

These are the absolute minimums. While the growth of wireless devices (including cordless phones, cell phones, and Wi-Fi) has reduced the need for TOs somewhat, what is most important is that there is a TO available where you need it. In most houses, many additional TOs beyond those required by TIA-570B will be provided. To ensure that the house contains an adequate number of TOs for the voice and data network, the following guidelines are recommended.

1. Provide at least one TO in every room, including bathrooms, large entry halls, and other rooms not covered by the TIA-570B requirement.
2. Provide at least one TO in each basement and garage area.
3. In rooms that are likely to have extensive data or entertainment networking, provide at least one TO for every 12 linear feet of wall. In such areas, consideration should be given to installing dual TOs.
4. Provide at least one TO on all unbroken wall spaces where it is likely that telecommunications equipment will be needed, even if the wall space is less than 12 feet in length. This ensures that telecommunications cords will never have to be run across a doorway.

---

[2] An *unbroken wall space* is a continuous expanse of wall without doorways. It can go around a corner but cannot cross a doorway.

5. In bedrooms, provide at least two TOs on opposite walls so that a telecommunications cord never has to be run across the room. Most people will want a telephone near their bed, so place a TO near where the bed(s) will likely be.
6. Rooms that may be used as a home office should follow Recommendation 3.

The intent of these guidelines is to ensure that sufficient TOs are installed in the house to accommodate a wide variety of equipment configurations, at locations desired by the homeowner, without running long telecommunications cords in areas where they present tripping hazards.

Cable Selection

Unshielded twisted-pair cabling is universally used for voice and data networks. TIA-570B requires Category 5e or better UTP cable and recommends Category 6 UTP. Due to the large improvement in performance at a relatively small additional cost, Category 6 UTP should be used whenever possible. Only low-end installations, when minimizing the cost is critical, should use Category 5e cabling.

As an intermediate-cost solution, it sometimes makes sense to pull Category 6 cable and install Category 5e connecting hardware. This only gives Category 5e performance, but the cabling system can be easily upgraded to Category 6 performance in the future by replacing the connecting hardware.

Keep in mind that cabling typically represents no more than 1 percent of the price of a new home, so it is not possible to save a significant amount of money by going to a cheaper cabling system. Attempting to cut corners on the cabling system often winds up requiring a much more expensive upgrade later.

Use Category 6 UTP cabling instead of Category 5e whenever possible; it provides much better performance for a slight increase in cost. If cost is an issue, consider pulling Category 6 cable and using Category 5e connecting hardware.

Fiber is rarely needed for current applications but is sometimes touted as a way to "future-proof" an installation. However, unless the homeowner has plans to install a system that needs fiber in the near future, it is difficult to justify the investment. There are several kinds of multimode, single-mode, and plastic fiber. It is not clear which of these will be needed in future years. Installing conduit so that the appropriate type of fiber can be easily installed when it is needed is often a better solution.

Access Technologies

Any or all of the broadband access technologies may be used in the home. The communications infrastructure should provide support for any access technologies that are available at the time of construction. Typically, this would include coaxial cable for cable television and cable modem Internet access, and UTP for voice and DSL service.

Provisions should also be made for connection to satellite and wireless access networks. This may involve determining likely locations for antennas or WAPs and running cable or conduit to those locations.

Wireless Access Points

Wireless networks have become extremely popular, and their market penetration is growing rapidly. Almost all networks installed in new homes will include some wireless equipment. Different types of wireless equipment commonly found in the home include WLANs, WPANs, and cordless phones.

Wireless personal area networks (WPANs) are generally very short-range networks (typically 10 m or less) and are usually set up on an ad hoc basis without any impact on the permanent infrastructure in the house. Cordless phones also have little or no impact on the communications infrastructure. The base unit is simply plugged into any convenient TO. At the DD, it is cross-connected to a phone line. The wireless handsets may then be used in any part of the house or yard.

Wireless local area networks (WLANs), however, do require some planning. The wireless access point (WAP) must be installed in a suitable location and cabled to the DD for connectivity to the rest of the home network. For most homes, a single WAP is adequate, but a large home might require more than one. Design of multi-WAP wireless networks is beyond the scope of this book.

The location of the WAP is important to ensure good coverage of the home. When determining the location of the WAP, the following principles should be followed.

- The WAP should be located as centrally as possible, both to minimize the distance to the wireless equipment and to reduce emissions outside the house.
- The WAP should not be located adjacent to large metal objects.
- The WAP should be located away from sources of interference such as other transmitters.
- Changing the orientation of the antenna(s) may improve the coverage in different areas of the house.

In many cases, the WAP can simply be placed on a shelf near the DD. If the DD is in the basement, it may be desirable to locate the WAP on the first or second floor to get better coverage of a multistory house. Some WAPs look like smoke detectors and are specifically designed for ceiling mounting. In this case, a UTP cable must be run to the WAP location during construction of the house. Wireless networks may result in a decrease in the number of TOs needed, but be sure to provide UTP connectivity to likely WAP sites.

Wherever the WAP is located, it is very important that the security features of the wireless network are enabled. Wi-Fi networks, for example, should, at a minimum, have the WEP enabled. This is particularly critical in dense housing arrangements like townhouses or high-rise apartment buildings.

Entertainment Networks

Entertainment networks add a number of additional requirements to the communications infrastructure, including:

- Coaxial cable for television signals
- Additional TO locations for entertainment equipment
- Audio cabling for in-wall/ceiling speakers
- Special considerations for home theater rooms.

Coaxial Cable for Television Signals

Although coaxial cable will someday be supplanted by a combination of UTP and wireless networks, it is currently the preferred media for transporting television signals. As per TIA-570B, Series 6 coaxial cable with F connectors should be used. At the option of the designer, hybrid cable (which includes both UTP and coax)

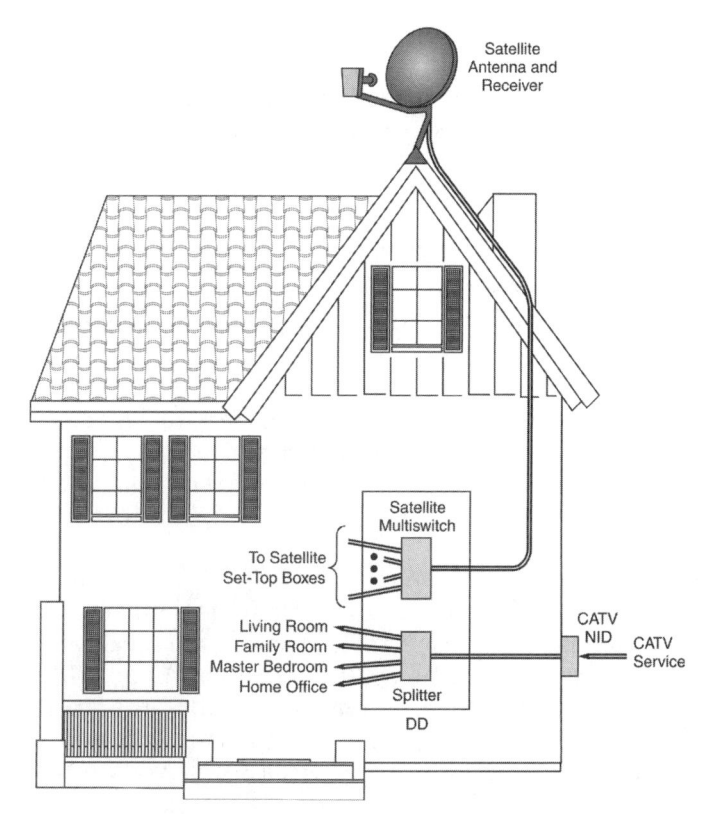

FIGURE 19–3 Coaxial Cabling in a Typical Home

may be used when coax and UTP outlets are collocated. The decision to use a single hybrid cable instead of multiple UTP and coax cables is often based on the cost of the cables and the difference in labor cost to install them.

In a typical residence with CATV service, coaxial cable is run from the CATV NID to the DD. At the DD, the coax connects to either a splitter or an amplifier or to a video distribution module like the one illustrated in Figure 19–2. After the splitter or video module, coax cables are run to each site where it is likely that a television will be installed. This configuration is illustrated in Figure 19–3.

If the house has satellite television service, then multiple coaxial cables should be run to the attic or roof where the satellite antenna is located. Different satellite service providers specify different antenna cabling requirements. Satellite receivers often require a better grade of coaxial cable than is required by TIA-570B, typically RG6 quad shield cable with a copper signal conductor (as opposed to dual shield with a copper-clad steel center conductor). The installer should determine these requirements from the satellite service providers. Install the required type and number of coaxial cables with sufficient cable length to reach from the DD to a good satellite antenna location on a southerly exposure of the roof or eve of the house. Some satellite service providers include a satellite multi-switch on the antenna itself; others provide it as a separate device that can mount in the DD. The output of the multiswitch must be cabled directly to the satellite set-top boxes, without a splitter. The direct-current voltage necessary to control the antenna low-noise-blocking receivers does not pass through CATV splitters. Both satellite and CATV configurations are illustrated in Figure 19–3.

Since coax is not a general-purpose medium like UTP, relatively few coaxial outlets are required. It is recommended that a coax outlet be placed in each room on the particular wall where it is likely that a television will be used,[3] including living/family rooms, kitchen, master bedroom, and so on. Also note that many satellite and CATV subscription services require a telephone line connection near the set-top box for video-on-demand or other pay-television services.

Additional Telecommunications Outlet Locations for Entertainment Equipment

For homes that have (or might someday have) a digital entertainment network, additional UTP TOs should be installed as per Recommendation 3 in the section on TOs earlier in this chapter. An entertainment server presents some unique connectivity needs. As shown in Figure 6–12, the entertainment server needs connectivity to the CATV network, the digital entertainment network, and the LAN (if different from the entertainment network). Potential locations for the entertainment server should have sufficient UTP and coax cabling and outlets to support these connections.

Audio Cabling for In-Wall/Ceiling Speakers

Whole-house audio systems are included in many new homes. If a whole-house system is being installed at the time of construction, the speaker locations should be determined as per the manufacturer's specifications. Audio cable is then pulled from the DD (or wherever the audio system is located) to the speaker locations.

If a whole-house audio system is not being installed, it may be desirable to pull some audio cables for future installation of speakers. We come back to this point in the section on Future Flexibility.

Special Considerations for Home Theater Rooms

If a small home theater system will be installed in a living room or family room, no additions to the communications infrastructure are normally necessary. Such systems are normally connected with point-to-point cords and speaker wire in the room.

On the other hand, if the house has a special theater room, then it is usually desirable to permanently install the cabling within the walls of the room. In this case, the cabling should be designed as specified by the manufacturer of the home theater system. The theater equipment will need connectivity to the CATV network and possibly the LAN and entertainment network. Such deluxe home theater installations often include special lighting systems as well.

Home Automation Systems

As discussed in the previous chapter, many home automation systems are not compatible with structured cabling. It is therefore important to find out what types of HASs (lighting control, climate control, security, and fire alarm) are being installed in the house and whether the cabling for them is provided by the HAS vendor or is to be installed as part of the communications infrastructure.

If cabling for any of the HAS equipment is being provided as part of the communications infrastructure, it is necessary to look at the manufacturer's installation instructions for the particular systems (such as lighting control, door locks, etc.). If the system is compatible with structured cabling, then it can be included with

---

[3] Note that some television displays, especially flat screens, use wall-mounting brackets that are higher than a typical coaxial TO mounting height.

the design, material ordering, and installation of the communications infrastructure. If the particular HAS equipment is not compatible with structured cabling, then the cabling for that system must be designed separately using the appropriate cables and design rules as specified by the manufacturer.

A couple of applications—security cameras and smart appliances—are usually compatible with structured cabling and should be considered during the communications infrastructure design.

Security Cameras

Security cameras may use coax, UTP, or wireless as their transmission medium. Using UTP is often advantageous because it allows power for the camera to be sent over the same cable. Coax or wireless cameras must have another source of power. For this reason, in addition to a coaxial TO, it is good practice to place UTP TOs at places where security cameras will likely be installed. Both the UTP and coax TOs can be mounted in a single-gang box.

If the house is being built for a specific client, the homeowner can be consulted about the number and location of security cameras. If the homeowner is not known or is not available, the following are the usual locations for security cameras.

- At each exterior door
- On the garage with a view of the driveway
- On the back of the house with a view of the deck, patio, pool area, or backyard
- In infant's room(s) if a baby monitor is desired
- If the house will be unoccupied for extended periods (such as a vacation home), it may be desirable to equip additional camera locations inside the house so it can be remotely monitored for unauthorized access. Possible locations would include the living room and major interior hallways.

Smart Appliances

Although smart appliances are not currently widely deployed, it is likely that they will be more common in the future. To provide for easy installation of smart appliances, UTP-based TOs may be installed at the locations of major appliances such as the refrigerator, stove, and washing machine.

Future Flexibility

In general, it is much easier and cheaper to install cabling during construction when the walls are open than it is to add more cabling later after the walls are finished. In all but the most cost-sensitive tract homes, it is desirable to consider making a few minor additions to the communications infrastructure to accommodate items likely to be needed or desired in the future. Some specific examples include:

- Running extra conduit from the DD. A spare conduit from the DD to the attic facilitates the future connection of a satellite antenna and can also be used to get additional cabling to the top floor of the house. Depending on the design of the house and the location of the DD, it may also be advantageous to run a conduit to the basement or crawl space, if any.
- Installing wiring for the future addition of in-wall or in-ceiling speakers. Audio cable for this purpose can be run from the DD to the future location of the speakers. The audio cable is usually left unterminated inside the wall or ceiling.

A few feet of spare cable should be left at each end. The location of the cable must be marked on the floor plan so that it can be easily located when needed.

- Installing UTP and coax wiring and TOs for the future addition of security cameras: For this purpose, UTP/coax TOs are usually installed in locations where cameras are likely to be placed, as detailed earlier in this chapter.
- Adding a TO or two for ceiling-mount WAPs.
- Adding additional TOs in rooms (such as a spare bedroom) that are potential home office locations.
- Adding additional TOs in locations where smart appliances are likely to be installed.

A little planning for future flexibility can save a lot of expense down the road by installing cable for equipment that is desirable but not affordable when the house is being constructed.

Documentation

The output of the design process is a set of documentation for the communications infrastructure of the house. At a minimum, this should include four items:

1. Floor plans that show the location of all the major components of the infrastructure, including:
   - The NID, the ADO, and the DD
   - Any conduits (or other cable pathways, such as cable trays) that were installed
   - The location of all TOs and a notation indicating the type of media (UTP, coax, fiber, etc.)
   - The location and type (UTP, audio, coax, etc.) of any spare cables that are left unterminated inside the walls or ceilings.
2. A wiring list for all the cables in the installation.[4] The wiring list shows where both ends of each cable are connected (for example, from Port 1 of Voice Module 1 at the DD to the TO on the east wall of the dining room).
3. A materials list that gives the manufacturer, part number, description, and quantity of all material needed for the installation. The cost of the job will be calculated from the materials list and the associated labor needed for installation.
4. A schedule for the various steps of the installation.

Retrofit Installations

In both new and older homes without structured wiring,[5] it is common for the entire communications infrastructure to consist of one or two daisy-chained telephone outlets and maybe one or two cable television jacks. With the much heavier demand for voice, data, and entertainment equipment in homes today, it is

---

[4] If desired, the wiring list may be more generic at this point and merely indicate whether each TO is connected to a voice module, LAN hub, and so on. During the installation the details, such as specific port numbers on the DD, can be recorded.

[5] About 50% of the homes currently being built do not have structured cabling.

often necessary to add additional communications infrastructure to existing homes. This process is called **retrofit,** as distinguished from new construction.

A retrofit installation may be done specifically to upgrade the communications infrastructure of the home, or it may occur in conjunction with an addition to or remodeling of the home. If the remodeling is extensive enough that the walls are opened up, then the installation is very much like new construction and the design process discussed earlier can be used.

On the other hand, if the walls are not opened up, then the basic design philosophy still applies but the implementation will be tempered by the realities of the installation process. Although it is easier and cheaper to install cable during the initial construction, a skilled installer with the proper tools can put a cable anywhere it needs to go without damaging existing walls or ceilings. Contractors that install fire and security alarm systems, for example, quite often work in vintage homes.

Retrofit installations use the same basic design philosophy as new construction but may use some slightly different design considerations and materials to make the installation easier.

When designing infrastructure for a retrofit installation, the design process starts with the connectivity requirements of the systems or applications being installed (as it does for installations in new construction). However, unlike new construction, the design of the infrastructure is influenced by the ease or difficulty of installation that is presented by the home.

Several techniques and products can be used to make retrofit installations easier. They generally fall into three categories: design, cables and pathways, and tools and techniques.

Retrofit Design Considerations

The first design consideration is to minimize the amount of new cable that needs to be installed. The easiest way to do this is to use wireless or other equipment that does not need communications wiring (such as PLC) wherever possible. This avoids the need to run new cables. However, do not hesitate to use a wired solution if the performance of wireless equipment is not adequate for the intended application.

Where new cabling is run, it is best to allow some flexibility in repositioning TOs a few feet one way or the other if it makes the installation easier.

Retrofit Cables and Pathways

There are a couple of additional options for cable pathways that are usually considered undesirable in new construction but may be used where needed for retrofit applications. When routing cable within a room, a common retrofit technique is the use of channel molding or surface-mount raceways.

Undercarpet cable may sometimes be used as an alternative to channel molding. Be sure to use cable specifically designed and designated as undercarpet cable; ordinary UTP or coax cable will not hold up under the mechanical pounding that undercarpet cable is subjected to.

Getting cables to the upper floors in a multistory house is sometimes difficult. One way to do this, which is usually used only as a last resort, is to run the cables through a conduit up the outside of the house. This obviously has aesthetic considerations and should only be done with the concurrence of the homeowner. Another possibility is to run a vertical conduit or raceway in closets if they are stacked one above the other.

Retrofit Tools and Techniques

Many tools and techniques have been developed specifically for installing cable in existing construction. However, it is important to keep these tools and techniques in mind when designing the infrastructure.

Example with Floor Plan

In this section, we consider the design of the communications infrastructure for a typical single-family house. The house consists of two floors and a basement.

- The basement (Figure 19–4) is divided into three areas—a workshop, a laundry/utilities area, and a storage area.
- The first floor (Figure 19–5) contains a living room, a family room, a kitchen, a dining room, a bathroom, an enclosed porch, an entry hall, and a deck.
- The second floor (Figure 19–6) contains a master bedroom, two other bedrooms, a home office, and a bathroom.

Requirements

The basic requirements for the infrastructure are:

- Voice, data, and CATV cabling will be provided throughout the house. Category 6 UTP and Series 6 coaxial cable will be used.

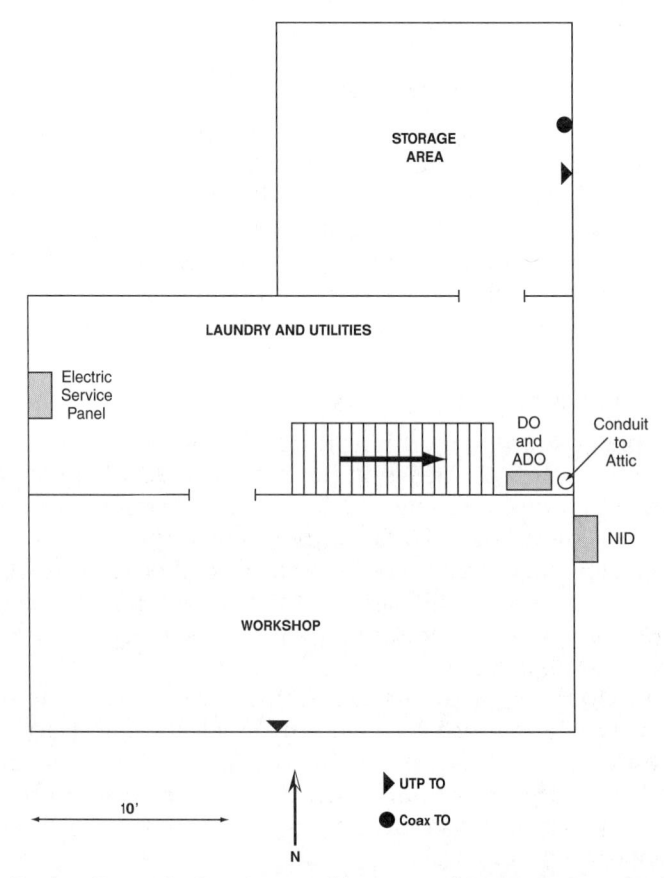

FIGURE 19–4 Design Example, Basement (Courtesy of Baxter Enterprises)

Communications Infrastructure Design • 379

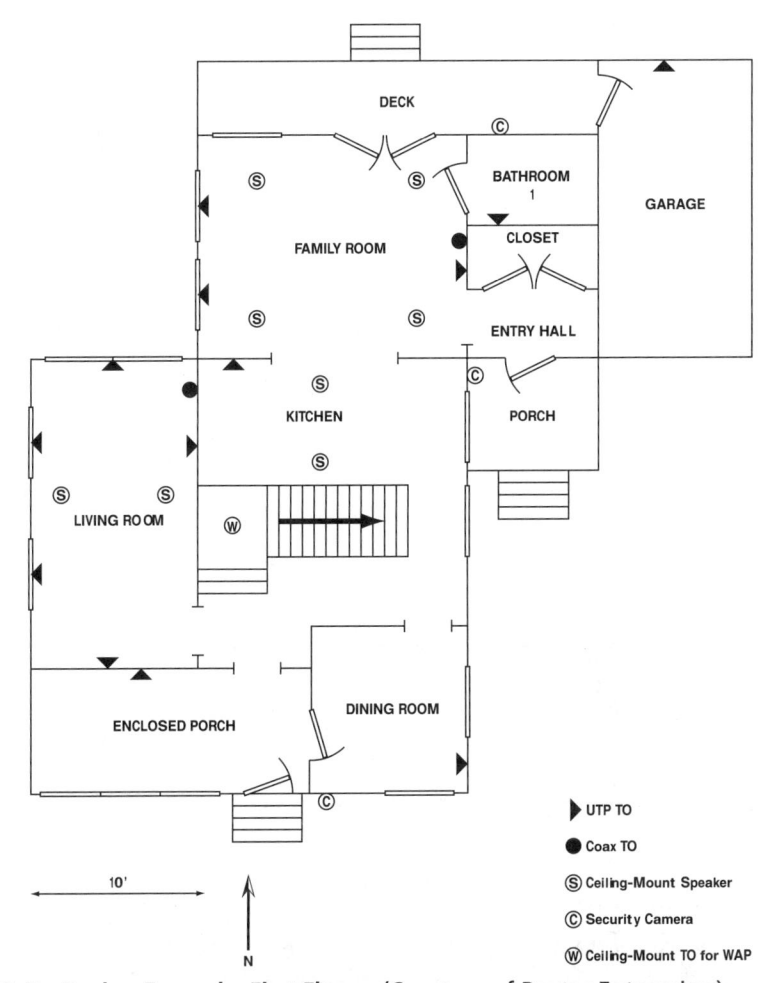

FIGURE 19–5 Design Example, First Floor (Courtesy of Baxter Enterprises)

- A whole-house audio system will be installed with speakers in the kitchen, living room, family room, and master bedroom.
- No HAS wiring is required except for security cameras located at each exterior door.
- Provision should be made for a single, centrally located WAP.

Infrastructure Design Process

Having defined the requirements, let us step through the six stages of the design process.

1. Placement of the major infrastructure components

Since the house has a full basement, it is convenient to locate the DD there. It will be placed on the wall near or under the stairs. The DD cabinet should be centered at a convenient height for working and far enough off the floor so that some wiring can enter at the bottom and power receptacles, if needed, can be located at the bottom of the DD cabinet. The NID will be located on the east exterior basement wall near the DD. Since the DD and the NID are so close together, the ADO will be collocated with the DD.

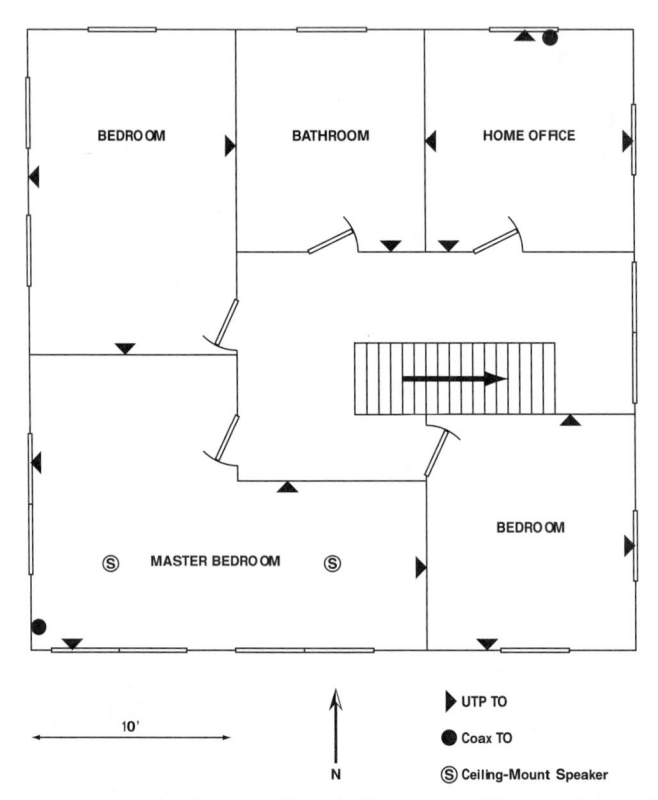

FIGURE 19–6 Design Example, Second Floor (Courtesy of Baxter Enterprises)

2. Voice and data network

Unshielded twisted-pair connectivity will be provided to every room in the house. In the basement, a TO is provided at a central location in the workshop. Another TO is provided in the storage area because it could easily be converted into a play room.

On the ground floor, a single TO is provided in the kitchen, bathroom, dining room, enclosed porch, and garage as per the recommendations earlier in this chapter. The family room and living room are much more likely to have a lot of communication and entertainment equipment in them. The living room is provided with a total of five TOs, one on each short wall and two on the long (west) wall. Despite its larger size, the family room is provided with only three TOs because there is relatively little wall space on which to locate them. A covered floor TO is another possibility if more TOs are needed near or under the stairs.

On the second floor, a single TO is provided in the bathroom. Three TOs are provided in each of the two small bedrooms so that long cords can be avoided. Due to its larger size, the master bedroom has a TO on each wall. The home office also has a TO on each wall to accommodate telephones, fax machines, LANs, and so on.

To support a WLAN, a ceiling-mounted TO is placed over the foot of the stairs on the first floor so that a WAP can be installed there. The WAP will be powered over the UTP cable, so no power outlet is necessary.

3. Entertainment network

Unshielded twisted-pair TOs have been generously allocated to the living room, family room, and master bedroom, so there is no need to place additional outlets for entertainment equipment.

Coax TOs are installed in areas of the living room (northeast corner) and family room (east wall) where it is likely that televisions will be located. Additional coax TOs are placed in the master bedroom and the basement storage room (for future use).

In-ceiling speakers for the whole-house audio system are located in the living room, family room, kitchen, and master bedroom. Be sure to consult the manufacturer's instructions for the optimal placement of the speakers to ensure good sound distribution in the room. In most cases, there will be an amplifier and volume control in each zone, as discussed in Chapter 6. For simplicity, in this example we are using a centralized configuration with the speakers wired back to the DD area with high-quality audio cable (as illustrated in Figure 6–6).

4. Home automation systems

As previously mentioned, HAS wiring is not required except for security cameras. Unshielded twisted-pair TOs for security cameras are placed outside the front, side, and back doors so that security cameras can be installed. The cameras will be powered over the UTP so no power outlets at the camera locations are necessary.

5. Future flexibility

To facilitate pulling more cables to the second floor or installing a satellite antenna at a later date, a conduit is run from the DD area to the attic. As previously mentioned, UTP and coax TOs were placed in the basement storage area for future use.

6. Documentation

At this point, the design is largely finished and the documentation remains to be done. Floor plan information is shown in Figure 19–4 through Figure 19–6. For this example, we will not produce the detailed documentation but will merely illustrate what needs to be done.

Table 19–2 summarizes the components required for this installation. This installation includes 33 UTP TOs (including the ones to support the WAP and security cameras), five coax TOs, and ten ceiling-mounted speakers. To produce a materials list, specific products are selected and totaled up, including the TOs, DD and associated modules, patchcords, faceplates, audio equipment, and so on.

The amount of each type of cable (Category 6 UTP, Series 6 coax, and audio cable) must be calculated. If we roughly estimate (by eyeballing the floor plan) that the basement runs are 30 feet long, the first floor runs are 50 feet long, and the second floor runs are 60 feet long, then we can calculate the amount of cable required, as shown in Table 19–3.

The cable quantities are then rounded up to an integral number of packages. For example, UTP cable comes in 1,000-foot boxes, so two boxes will be needed. If hybrid cable is being used, then all five of the coax runs will be converted to hybrid cable and the UTP quantity will be reduced by 250 feet.

The materials list can be generated with a simple spreadsheet, or it can be automated with configurator or estimator software. Finally, a wiring list is generated and a schedule for the work is specified.

TABLE 19–2
Design Example Component Summary

	Quantity				
Component	*Basement*	*1st Floor*	*2nd Floor*	*Total*	*Cable Type*
UTP TO	2	17	15	34	Cat. 6 UTP
Coax TO	1	2	2	5	Series 6 coax
In-ceiling speaker	0	8	2	10	Audio
DD	1	0	0	0	
ADO, NID	1 per line	0	0	0	

TABLE 19–3
Design Example Cable Quantity Calculation

	Qty. × Length (feet)			
Cable Type	*Basement*	*1st Floor*	*2nd Floor*	*Total Length (ft.)*
Cat. 6 UTP	2 × 30	17 × 50	15 × 60	1,810
Series 6 coax	1 × 30	2 × 50	2 × 60	250
Audio	0	8 × 50	2 × 60	520

Summary

This chapter presents design guidelines for communication infrastructure to support voice, data, entertainment, and home automation equipment in an SOU. The major points covered in the chapter are:

- There are four major stakeholders in the communications infrastructure design process: the homeowner, the architect, the builder, and the installer. Each has slightly different interests and priorities.
- The infrastructure design process has six stages.
- The location of the major infrastructure components—such as the DD, the NID, and the ADO—is a key consideration.
- Specifying the number and location of TOs is another very important decision.
- Category 6 UTP cabling should be used whenever possible instead of Category 5e UTP.
- Optical fiber is sometimes installed for future use, but it is difficult to know what type of fiber will eventually be needed.
- The use of wireless LANs may reduce the number of TOs somewhat. Consideration should be given to the location of the WAP.
- Entertainment networks usually require coaxial cable for CATV signals, additional TOs for entertainment equipment, and possibly audio cabling for in-wall/ceiling speakers and volume controls.
- Home automation systems are not always compatible with structured cabling and must be considered on a case-by-case basis. Security

- cameras and smart appliances, however, can usually be served with UTP or wireless infrastructure.
- Consideration should be given to installing a small amount of additional infrastructure to ensure flexibility for future applications, such as running a spare conduit from the DD to the attic.
- Documentation for the design process should consist of at least four items: floor plans showing the location of all major infrastructure components, a wiring list, a materials list, and an installation schedule.
- Although retrofit installations use the same basic design philosophy, design tradeoffs are made to ease the installation process.

Key Terms

Auxiliary Disconnect Outlet (ADO) A connector that allows incoming services to be easily disconnected for troubleshooting or other reasons.

Communications infrastructure The cabling systems and their associated pathways that support the distribution of information throughout the home.

Distribution device (DD) A facility within the residence that contains the main cross-connect or interconnect where one end of each of the outlet cables terminates.

Retrofit The installation of communications infrastructure into an existing residence.

Telecommunication outlet A connecting device (usually a modular jack or coaxial connector) in the home on which the outlet cable terminates.

Review Questions

1. Name the four major stakeholders in the communications infrastructure design and state one or two main interests for each of them.
2. List the six steps in the infrastructure design process.
3. State at least six principles that should be met when specifying the location of the DD.
4. Summarize the one requirement and two recommendations that TIA-570B makes for the placement of TOs.
5. List at least four other recommendations for TO placement that should be considered in addition to the TIA-570B recommendations.
6. Why is Category 6 UTP cabling preferred over Category 5e?
7. Name at least two places where a WAP is commonly located.
8. How do satellite system installations differ from CATV system installations? Can you use a CATV splitter to connect a satellite signal to multiple set-top boxes?
9. From the DD, to what locations should coaxial cable be run?
10. What is unique about the connectivity requirements of entertainment servers?
11. What two home automation applications are most often compatible with structured cabling?
12. Name at least four locations in the home that would normally be wired for security cameras.
13. Give at least four examples of additional infrastructure that is added to ensure future flexibility.
14. What four types of documentation should be provided?
15. List at least two design considerations and two types of cable or pathways that are used only for retrofit installations.

Television

OBJECTIVES *After studying this chapter, you should be able to:*

- Install residential television wiring, antennas, and CATV cables, in conformance to *NEC®* requirements.
- Describe the basic operation of satellite antennas.

OUTLINE Installing the Wiring for Home Television

Satellite Antennas

Code Rules for the Installation of Antennas and Lead-In Wires (*Article 810*)

Installing the Wiring for Home Television

This chapter discusses the basics of installing outlets, cables, receivers, antennas, amplifiers, multiset couplers, and some of the *NEC®* rules for home television. Television is a highly technical and complex field. To ensure a trouble-free installation of a home system, it should be done by a competent television technician. These individuals receive training and certification through the Electronic Technicians Association in much the same way that electricians receive their training through apprenticeship and journeyman training programs.

Some cities and states require that, when more than three television outlets are to be installed in a residence, the television technician making the installation be licensed and certified. Just as electricians can experience problems because of poor connections, terminations, and splices, poor reception problems can arise as a result of poor terminations. In most cases, poor crimping is the culprit. Electrical shock hazards are present when components are improperly grounded and bonded.

According to the plans and for this residence, television outlets are installed in the following rooms:

Front Bedroom	2
Kitchen	1
Laundry	1
Living Room	3

Master Bedroom	2
Recreation Room	3
Study/Bedroom	2
Workshop	<u>1</u>
Total	15

Because TV is a low-voltage system, metallic or nonmetallic standard single-gang device boxes, 4-in. square outlet boxes with a plaster ring, plaster rings only, or special mounting brackets can be installed during the rough-in stage at each likely location of a television set. (More on this later.) See Figure 20–1. For new construction, shielded coaxial cables are installed concealed in the walls. For remodel work, cables can be concealed in the walls by fishing the cables through the walls and installing mounting brackets that are inserted and snapped into place through a hole cut into the wall.

Shielded 75-ohm RG-6 coaxial cable is most often used to hook up television sets to minimize interference and keep the color signal strong. There are different kinds of shielded coaxial cable. Double-shielded cable that has a 100 percent foil shield covered with a 40 percent or greater woven braid is recommended. This coaxial cable has a PVC outer jacket. The older style, flat 300-ohm twin-lead cable, can still be found on existing installations, but the reception might be poor. To improve reception, 300-ohm cables can be replaced with shielded 75-ohm coaxial cables

For this residence, 15 television outlets are to be installed. A shielded 75-ohm RG-6 coaxial cable will be run from each outlet location back to one central point, as in Figure 20–2.

This central point could be where the incoming Community Antenna Television (**CATV**) cable comes in, where the cable from an antenna comes in, or where the TV output of a satellite receiver is carried. For rooftop or similar antenna mounting, the cable is run down the outside of the house, then into the basement, garage, or other convenient location where the proper connections are made. CATV companies generally run their incoming coaxial cables underground to some convenient point just inside of the house. Here, the technician hooks up all of the coaxial cables coming from the TV outlets to an amplifier that boosts the signal and improves reception, Figure 20–2, or to a multiset splitter, Figure 20–3. An amplifier may not be necessary if only those coaxial cables that will be used are hooked up.

Cable television does not require an antenna on the house because generally the cable company has the proper antenna to receive signals from satellites. The cable company then distributes these signals throughout the community they have contractually agreed to serve. These contracts usually require the cable company to run their coaxial cable to a point just inside of the house. Inside the house,

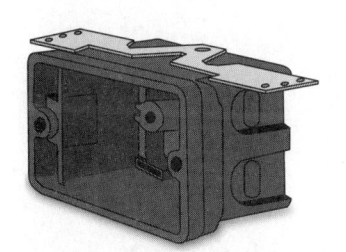

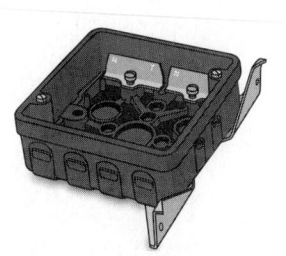

FIGURE 20–1 Nonmetallic Boxes and a Nonmetallic Raised Cover

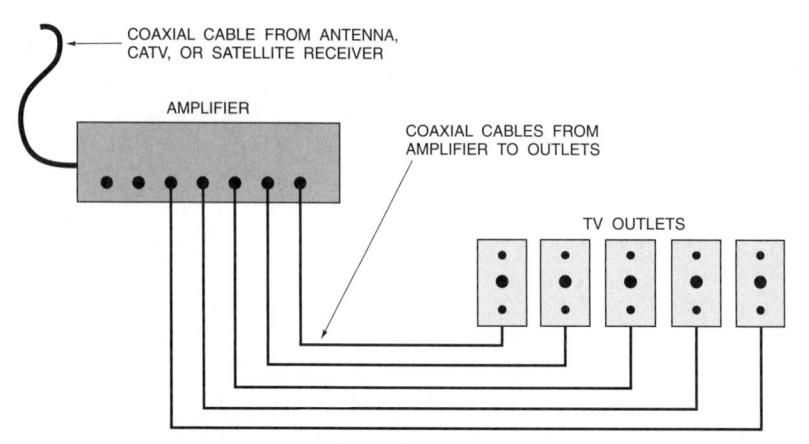

FIGURE 20–2 A television master amplifier distribution system may be needed where many TV outlets are to be installed. This will minimize the signal loss. In simple installations, a multiset coupler can be used, as shown in Figure 20–3.

they will complete the installation by furnishing cable boxes, controls, and the necessary wiring. In geographical areas where cable television is not available, antennas as shown in Figure 20–4 and Figure 20–6 are needed.

Faceplates for the TV outlets are available in styles to match most types of electrical faceplates in the home.

Hazards of Mixing Different Voltages

For safety reasons, video/voice/data wiring must be separated from the 120-volt wiring, *810.18(C)*. If both 120-volt and low-voltage circuits are run into the same box, then a permanent barrier must be provided in the box to separate the two systems. A better choice is to keep the two systems totally separate using two wall boxes.

Another popular choice is where video/voice/data cables are run to a location right next to a 120-volt receptacle. Mounting brackets that are commercially available can be installed around the electrical device box during the rough-in stage. This lets you trim out a 120-volt receptacle and video/voice/data jacks with one faceplate, as shown in Figure 20–5. These mounting brackets keep the center-to-center measurements of the device mounting holes for the electrical box and the video/voice/data wiring precisely in alignment. With this method, a wall box is not provided for the video/voice/data wiring. The video/voice/data cables merely come out of the drywall next to the electrical box, and the faceplate takes care of trimming out the receptacle and the video/voice/data wiring jacks.

When wiring a new house, it certainly makes sense to run at least one coaxial cable and one Category 5 cable to a single wall box wherever TV, telephone, and/or

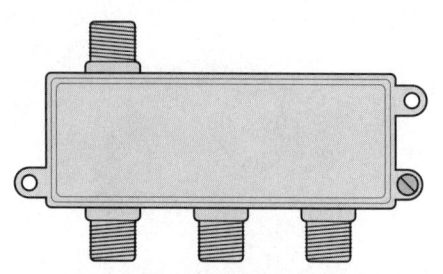

FIGURE 20–3 A Three-Way CATV Cable Splitter: One In, Three Out Two-way and four-way splitters are also commonly available.

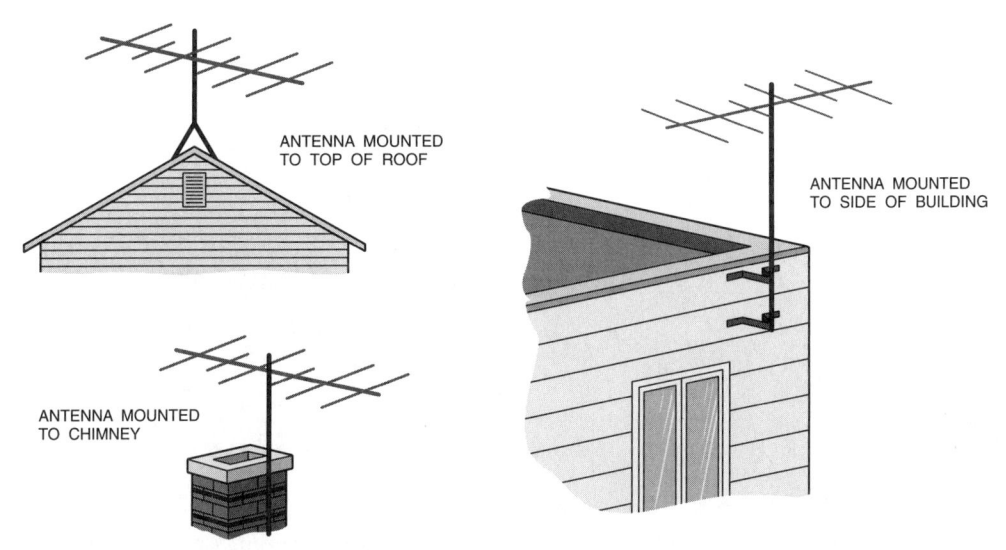

FIGURE 20–4 Older Style Antennas Although still available, they have given way to the more popular digital satellite "dish," Figure 20–6(A).

computers might be used. This could be a single box located close to a 120-volt receptacle outlet, or it could be a mounting bracket as shown in Figure 20–5. Run a separate coaxial cable and a separate Category 5 cable from each location to a common point in the house. You will use a lot of cable, but this will enable you to interconnect the cables as required, such as for a local area network (LAN) to serve more than one computer located in different rooms, or other home automation systems.

For adding video/voice/data wiring to an existing home, many types of nonmetallic brackets are available that snap into a hole cut in the drywall and lock in place.

More *NEC*® rules are discussed later in this chapter.

Although shielded coaxial cable minimizes interference, it is good practice to keep the coaxial cables on one side of the stud space and to keep the light and power cables on the other side of the stud space.

FIGURE 20–5 A Special Mounting Bracket Fastened to the Wood Stud The bracket fits nicely over the electrical wall box to accommodate a 120-volt receptacle, two coaxial outlets, and one telephone outlet. A single two-gang faceplate is used for this installation. Many other combinations are possible.

388 • Chapter 20

FIGURE 20–6 **(A) A Digital Satellite System 18-in. (450-mm) Antenna, a Receiver, and a Remote Control** (Courtesy of Thomson Consumer Electronics). **(B) A Large Satellite Antenna Securely Mounted to a Post that Is Anchored in the Ground.** These large antennas are rarely used today. (Courtesy of Hi-Tech Advisors, www.advisors.com)

Code Rules for Cable Television (CATV) (*Article 820*)

CATV systems are installed both overhead and underground in a community. Then, to supply an individual customer, the CATV company runs coaxial cables through the wall of a residence at some convenient point. Up to this point of entry (*Article 820, Part II*) and inside the building (*Article 820, Part V*), the cable company must conform to the requirements of *Article 820*, plus local codes if applicable.

Coaxial cable generally used for residential installations is Type CATV or CATVX. Other types are listed in *Table 820.50* of the *NEC®*.

Here are some of the key rules to follow when making a coaxial cable installation.

1. The outer conductive shield of the coaxial cables must be grounded as close to the point of entry as possible, *820.33*.
2. Coaxial cables shall not be run in the same conduits or box with electric light and power conductors, *820.50(A)(1)(2)*.
3. Do not support coaxial cables from raceways that contain electrical light and power conductors, *820.55(C)*.
4. Keep the coaxial cable at least 2 in. (50 mm) from light and power conductors unless the conductors are in a raceway, nonmetallic-sheathed cable, armored cable, or UF cable, *820.55(A)(2)*. This clearance requirement really pertains to old knob-and-tube wiring. The 2-in. (50-mm) clearance is not required if the coaxial cable is in a raceway.
5. Where underground coaxial cables are run, they must be separated by at least 12 in. (300 mm) from underground light and power conductors, unless the underground conductors are in a raceway, Type UF cable, or Type USE cable, *820.12(B)*. The 12-in. (300-mm) clearance is not required if the coaxial cable has a metal cable armor.
6. The grounding conductor (*Article 820, Part III* and *Part IV*):
   a. outer conductive shield shall be grounded as close as possible to the coaxial cable entrance or attachment to the building, *820.33*.

> **CAUTION**
>
> **Do Not** simply drive a separate ground rod to ground the metal shielding tape on the coax cable. Difference of potential between the coax cable shield and the electrical system ground during lightning strikes could result in a shock hazard as well as damage to the electronic equipment. If for any reason one grounding electrode is installed for the CATV shield grounding and another grounding electrode for the electrical system, bonding these two electrodes together with no smaller than a 6 AWG copper conductor will minimize the possibility that a difference of potential voltage might exist between the two electrodes.

b. shall not be smaller than a 14 AWG copper or other corrosion-resistant conductive material. It need not be larger than 6 AWG copper. The grounding conductor must be insulated. It shall have a current-carrying capacity not less than that of the coaxial cable's outer metal shield. See *NEC® 810.40(A)*.

c. length shall be as short as possible, but not longer than 20 ft (6.0 m). If impossible to keep the grounding conductor to this maximum permitted length, then the Exception to *820.40(A)(4)* permits installing a separate grounding electrode. These two electrodes shall be bonded together with a bonding jumper not smaller than 6 AWG copper.

d. may be solid or stranded, *820.40(A)(2)*.

e. shall be guarded from physical damage, *820.40(A)(6)*.

f. shall be run in a line as straight as practical, *820.40(A)(5)*.

g. shall be connected to the nearest accessible location on one of the following, *820.40(B)*:
  - the building grounding electrode.
  - the grounded interior metal water piping pipe, within 5 ft (1.5 m) of where the pipe enters the building. Refer to *250.52*.
  - the power service accessible means external to the enclosure.
  - the metallic power service raceway.
  - the service equipment enclosure.
  - the grounding electrode conductor or its metal enclosure.
  - the grounding conductor or grounding electrode of a disconnecting means that is grounded to an electrode according to *250.32*. This pertains to a second building on the same property served by a feeder or branch-circuit from the main electrical service in the first building.

h. If none of the options in g is available, then ground to any of the electrodes per *250.52*, such as metal underground water pipes, the metal frame of the building, a concrete encased electrode, or a ground ring. Watch out for the maximum length permitted for the grounding conductor, as mentioned above.

Satellite Antennas

A satellite antenna is often referred to as a "dish." See Figure 20–6. A satellite dish has a parabolic shape that concentrates and reflects the signal beamed down from one of the many stationary satellites orbiting 22,245 miles (35,800 kilometers) above the equator. Stationary means that the satellite is traveling at the same

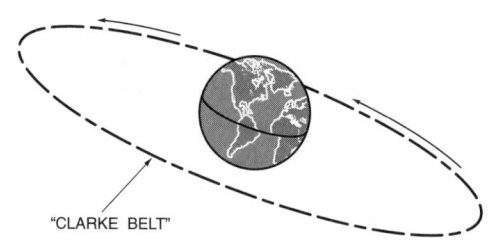

FIGURE 20–7 All satellites travel around the earth in the same orbit called the "Clarke Belt." They appear to be stationary in space because they are rotating at the same speed at which the earth rotates. This is called "geosynchronous orbit (geostationary)." The satellite receives the uplink signal from earth, amplifies the signal, and transmits it back to earth. The downlink signal is picked up by the satellite antenna.

speed that the earth rotates. The orbit in which satellites travel is called the "Clarke Belt," as shown in Figure 20–7.

The latest satellite technology is the digital satellite system (DSS). A digital system allows the use of small antennas, approximately 18 in. (450 mm) in diameter, that can easily be mounted on a roof, on a chimney, on the side of the house, on a pipe, or on a pedestal, using the proper mounting hardware.

Although rarely used today for residential TV reception, large antennas are still in use. Because of their huge size and weight, proper installation is important.

The basic operation of a satellite system is for a transmitter on earth to beam an "uplink" signal to a satellite in space. Electronic devices on the satellite reamplify and convert this signal to a "downlink" signal, then retransmit the downlink signal back to earth, as shown in Figure 20–8.

Regardless of the type of antenna used, the line-of-sight from the antenna to the satellite must not be obstructed by trees, buildings, utility poles, or other structures. Instructions furnished with an antenna provide the necessary data for direction and up angle based on ZIP codes.

Before installing an antenna, check with your local inspection department. Local codes might have restrictions and requirements in addition to the *NEC*®.

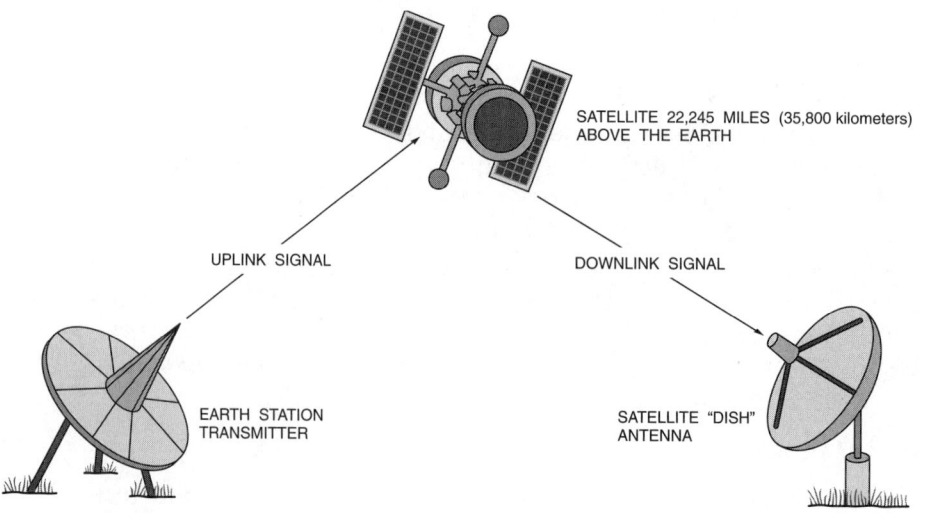

FIGURE 20–8 The earth station transmitter beams the signal up to the satellite where the signal is amplified, converted, and beamed back to earth.

Television • 391

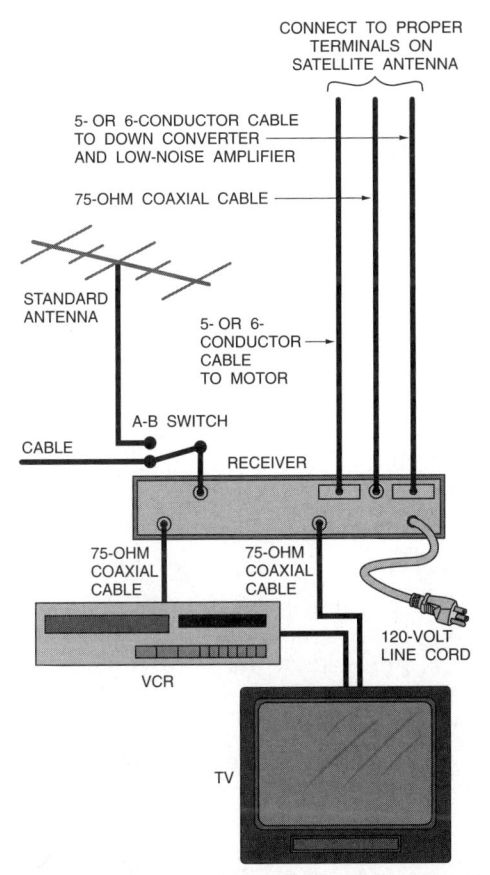

FIGURE 20–9   Typical Satellite Antenna/Cable/Standard Antenna Wiring   Connection to a standard antenna cable, CATV cable, or to a satellite cable enables the homeowner to watch one channel while recording another channel. The installation manual for the receiver usually has a number of different hookup diagrams. In this diagram, the A-B switch allows a choice of connecting to a standard antenna or to the CATV cable. This hookup can be desirable should the CATV or satellite reception fail.

Figure 20–9 and Figure 20–10 show typical connections between a television set, receiver, VCR, and antenna. The manufacturer's installation instructions are always to be followed.

Figure 20–11 illustrates one method of installing a large antenna post in the ground in accordance to a manufacturer's instruction.

Code Rules for the Installation of Antennas and Lead-In Wires (*Article 810*)

Home television and AM and FM radios generally come complete with built-in antennas. For those locations in outlying areas on the fringe or out of reach of strong signals, it is quite common to install a separate antenna system or a satellite antenna system.

Indoor or outdoor antennas may be used with televisions. The front of an outdoor antenna is aimed at the television transmitting station. When there is more

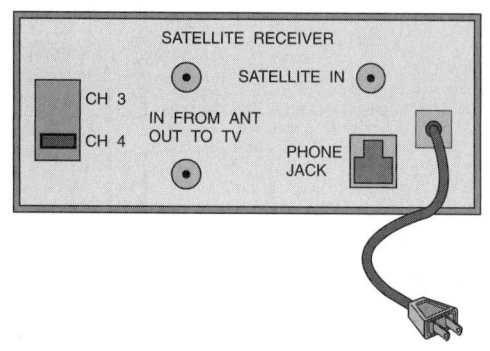

FIGURE 20–10 The Terminals on the Back of a Digital Satellite Receiver The connections are similar to those in Figure 20–9. Note the difference in that the digital satellite receiver has a modular telephone plug, a feature that uses a toll-free number to update the access card inside the receiver to ensure continuous program service. The telephone can also handle program billing.

than one transmitting station and they are located in different directions, a rotor is installed. A rotor turns the antenna on its mast so that it can face in the direction of each transmitter. The rotor is controlled from inside the building. The rotor controller's cord is plugged into a regular 120-volt receptacle to obtain power. A three-wire or four-conductor cable is usually installed between the rotor motor and the control unit. The wiring for a rotor may be installed during the roughing-in stage of construction, running the rotor's cable into a device wall

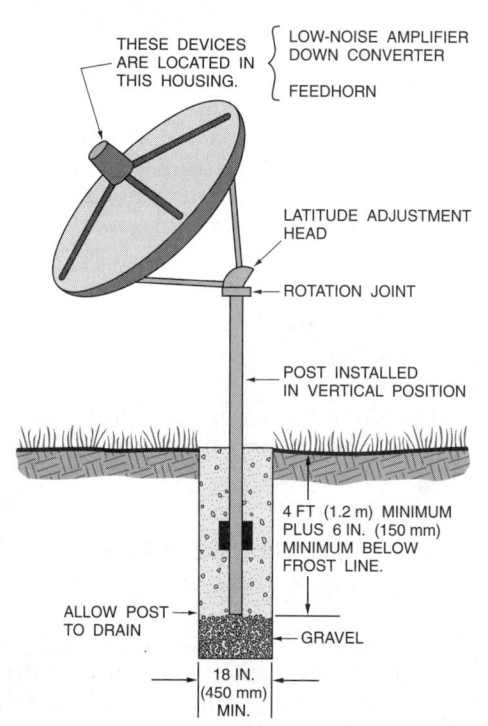

FIGURE 20–11 Satellite Antenna Solidly Installed in the Ground According to the Manufacturer's Instructions The popularity of these large antennas has given way to the small digital satellite system antenna, as shown in Figure 20–6.

> **CAUTION**
>
> **Do Not** simply drive a separate ground rod to ground the metal mast, structure, or antenna. Difference of potential between the metal mast, structure, or antenna and the electrical system ground during a lightning strike could result in a shock hazard as well as damage to the electronic equipment. If any reason permitted by the *Code* results in one grounding electrode for the antenna and another grounding electrode for the electrical system, bond the two electrodes together with a bonding jumper not smaller than 6 AWG copper or equivalent, *810.21(J)*.

box, allowing 5 to 6 ft (1.5 to 1.8 m) of extra cable, then installing a regular single-gang switchplate when finishing.

Article 810 covers radio and television equipment. Although instructions are supplied with antennas, the following key points of the *Code* regarding the installation of antennas and lead-in wires should be followed.

1. Antennas and lead-in conductors shall be securely supported, *810.12*.
2. Antennas and lead-in conductors shall not be attached to the electric service mast, *810.12*.
3. Antennas and lead-in conductors shall be kept away from all light and power conductors to avoid accidental contact with the light and power conductors, *810.13*.
4. Antennas and lead-in conductors shall not be attached to any poles that carry light and power wires over 250 volts between conductors, *810.12*.
5. Lead-in conductors shall be securely attached to the antenna, *810.12*.
6. Outdoor antennas and lead-in conductors shall not cross over light and power conductors, *810.13*.
7. Outdoor antennas and lead-in conductors shall be kept at least 24 in. (600 mm) away from open light and power conductors, *810.13*.
8. Where practicable, antenna conductors shall not be run under open light and power conductors, *810.13*.
9. On the outside of a building:
   a. Position and fasten lead-in conductors so they cannot swing closer than 24 in. (600 mm) to light and power conductors having not over 250 volts between conductors; 10 ft (3.0 m) if over 250 volts between conductors, *810.18(A)*.
   b. Keep lead-in conductors at least 6 ft (1.8 m) away from a lightning rod system, *810.18(A)*, or bonded together according to *250.60*.

   Note: The clearances in (a) and (b) are not required if the light and power conductors or the lead-in conductors are in a metal raceway or metal cable armor.
10. On the inside of a building
    a. Keep the antenna and lead-in conductors at least 2 in. (50 mm) from other open wiring (as in old houses) unless the other wiring is in a metal raceway or cable, *810.18(B)*.

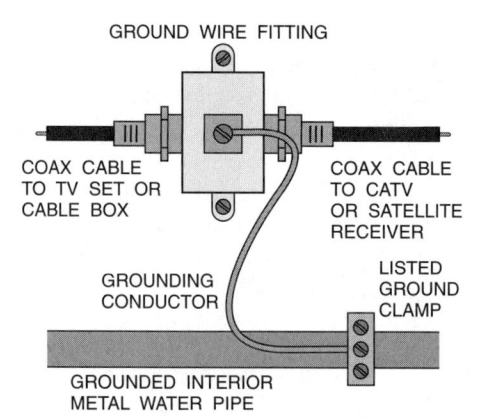

FIGURE 20–12 Typical Connection for the Required Grounding of the Metal Shield on Coaxial Cable to the Interior Metal Water Piping The connection must be made within 5 feet of where the water pipe enters the building.

    b. Keep lead-in conductors out of electric boxes unless there is an effective, permanently installed barrier to separate the light and power wires from the lead-in wire, *810.18(C)*.

11. Grounding:

    a. All metal masts and metal structures that support antennas shall be grounded, *810.21*.

    b. The grounding conductor must be copper, aluminum, copper-clad steel, bronze, or a similar corrosion-resistant material, *810.21(A)*.

    c. The grounding conductor need not be insulated. It must be securely fastened in place, may be attached directly to a surface without the need for insulating supports, shall be protected from physical damage or be large enough to compensate for lack of protection, and shall be run in as straight a line as is practicable, *810.21(B), (C), (D),* and *(E)*.

    d. The grounding conductor shall be connected to the nearest accessible location on one of the following:

- the building or structure grounding electrode. Refer to *250.50* for more details.
- the grounded interior metal water pipe (as shown in Figure 20–12), within 5 ft (1.5 m) of where the water pipe enters the building. Refer to *250.52* for more details.
- the metallic power service raceway.
- the service equipment enclosure.
- the grounding electrode conductor or its metal enclosure.

    e. If none of d is available, then ground to any one of the electrodes per *250.52*, such as metal underground water pipe, metal frame of building, concrete-encased electrode, or ground ring, *810.21(F)(2)*.

    f. If neither d nor e is available, then ground to an effectively grounded metal structure or to any one of the electrodes per *250.52*, a driven rod or pipe, or a metal plate, *810.21(F)(3)*.

g. The grounding conductor may be run inside or outside of the building, *810.21(G)*.
h. The grounding conductor shall not be smaller than 10 AWG copper or 8 AWG aluminum, *810.21(H)*.

The objective of grounding and bonding to the same grounding electrode as the main service ground is to reduce the possibility of having a difference of voltage potential between the two systems.

Wiring for Computers and Internet Access

Wiring for telephones, high-speed Internet access, computers, television, printers, modems, security systems, intercoms, and similar home automation equipment could be thought of as one big project. Planning ahead is critical. Deciding on where this equipment is likely to be located will help design an entire video/voice/data system.

You will want to install Category 5 or Category 5e cables between each telephone outlet and a central distribution point so as to have the flexibility to connect the video/voice/data equipment in any configuration.

In addition to the previously mentioned equipment, there very well might be a scanner, an answering machine, desk lamps, floor lamps, an adding machine, a calculator, a radio, a CD player, a CD "burner," audio/videocassette tape players, VCR and DVD players, and ZIP drives. They draw little current, but all need to be plugged into a 120-volt receptacle! Managing and plugging in that many power cords is a major problem. Give consideration to installing two or three duplex or quadplex 120-volt receptacles at these locations. Multioutlet plug-in strips with surge protection will most likely be needed.

Bringing Technology into Your Home

In some areas, the telephone company provides the entire "package" for homes. This might include digital local and long-distance telephone service, Internet access, cable modems for personal computers, cable television, digital subscriber lines (DSL) (through existing copper telephone lines), and whatever other great new things may come next.

Don't Conceal Boxes!

It may seem obvious, but the issue of accessibility bears repeating. Electrical equipment junction boxes and conduit bodies must be accessible, *314.29*. It is permitted to have certain electrical equipment such as a chime transformer, junction boxes, and conduit bodies above a dropped lay-in ceiling because these are accessible by dropping out a ceiling panel. Never install electrical equipment, junction boxes, or conduit boxes above a permanently closed-in ceiling or within a wall. There may come a time when the electrical equipment and junction boxes need to be accessed. This includes chime transformers.

Summary

- Poor connections, terminations, and splices will result in poor television, CATV, and satellite reception.
- Electrical shock hazards are present when television, CATV, and satellite systems are improperly grounded.

- Sheilded 75-ohm RG-6 coaxial cable is most often used to hook up television, CATV, and satellite systems.
- A signal amplifier may be necessary to boost the signal level, especially when signal splitters are being used to drive multiple outlets.
- For safety reasons, voice/data/video wiring must be separate from 120-volt wiring.
- When high-voltage and low-voltage systems are run in the same box, they must be separated by a permanent protective barrier.
- Television, CATV, and satellite systems must be grounded to the main electrical system.
- A second ground rod is not recommended since a difference of potential can exist between it and the main system ground rod during a lightning strike. If a second ground rod is installed, then it must be bonded back to the main electrical ground with a 6 AWG copper conductor.
- *Article 810* of the *NEC®* covers the installation requirements of radio and television equipment.
- *Article 820* of the *NEC®* covers the installation requirements of CATV and radio distribution.

Key Terms

CATV An abbreviation for community antenna television, usually delivered by coax cable or hybrid fiber coax (HFC) networks, or cable TV.

Review Questions

1. Which type of television cable is commonly used and recommended?
2. What must be provided when installing a television outlet and receptacle outlet in one wall box?
3. From a cost standpoint, which system is more economical to install: a master amplifier distribution system or a multiset coupler? Explain the basic differences between these two systems.
4. Digital satellite systems use an antenna that is approximately (18 in. [450 mm]), (36 in. [900 mm]), or (72 in. [1.8 mm]) in diameter. Circle the correct answer.
5. Which article of the *Code* references the requirements for cable community television installation?
6. It is generally understood that grounding and bonding together all metal parts of an electrical system and the metal shield of the cable television cable to the same grounding reference point in a residence will keep both systems at the same voltage level should a surge, such as lightning, occur. Therefore, if the incoming cable that has been installed by the CATV cable company installer has the metal shield grounded to a driven ground rod, does this installation conform to the *NEC®*?
7. All television satellites rotate above the earth in (the same orbit) or (different orbits). Circle the correct answer.
8. Television satellites are set in orbit (10,000), (18,000), or (22,245) miles above the earth, which results in their rotating around the earth at (precisely the same) or (different) rotational speed as the earth rotates. This is done so that the satellite "dish" can be focused on a specific satellite (once), (one time each month), or (whenever the television set is used). Circle the correct answers.
9. Which section of the *Code* prohibits supporting coaxial cables from raceways that contain light or power conductors?
10. When hooking up one CATV cable and another cable from an outdoor antenna to a receiver that has only one antenna input terminal, an _____ switch is usually installed.

Smoke, Heat, and Carbon Monoxide Alarms, and Security Systems

After studying this chapter, you should be able to:

- Understand the basics of *NFPA 72 (2002)*, the *National Fire Alarm Code, NFPA 720 (2003)*, the *Recommended Practice for the Installation of Household Carbon Monoxide Warning Equipment,* and *NFPA 70 (2005)*, the *National Electrical Code®*.
- Understand the basics of smoke, heat, and carbon monoxide alarms.
- Understand the location requirements for the installation of smoke, heat, and carbon monoxide alarms for minimum acceptable levels of protection.
- Understand the location requirements for the installation of smoke and heat alarms that exceed the minimum acceptable levels of protection.
- Discuss general requirements for the installation of security systems.
- Be aware of important UL Standards covering fire warning equipment.

OBJECTIVES

National Fire Alarm Code

Smoke, Heat, and Carbon Monoxide Alarms

Detector Types

Types of Smoke Alarms

Types of Heat Alarms

Installation Requirements

Maintenance and Testing

Carbon Monoxide Alarms

Fire Alarm Systems

Security Systems

OUTLINE

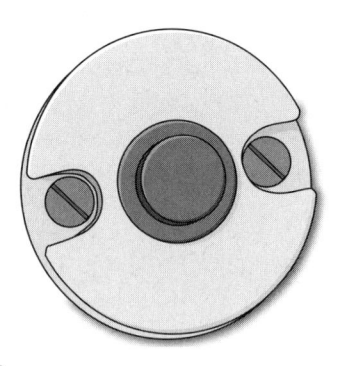

FIGURE 21–1 Heat Detector

National Fire Alarm Code

NFPA 72 and *NFPA 720* tell us "where to install" smoke and heat alarms.

NFPA 70 in *Article 760* of the *NEC®* tells us "how to" install the wiring for smoke and heat alarms.

NFPA 72 is the *National Fire Alarm Code®*. It presents the minimum requirements for the proper selection, installation, operation, and maintenance of fire warning equipment that will provide reasonable fire safety. The primary function of fire warning equipment is to provide a reliable means to notify the occupants of the presence of a threatening fire and the need to escape to a place of safety before such escape might be obstructed by smoke or other delaying conditions in the normal path of the way out. *Chapter 11* specifically covers household fire warning equipment. Smoke alarms are required by *NFPA 72*. **Heat detectors** are not required but are recommended by *NFPA 72*.

Fire warning devices commonly used in a residence are heat detectors, Figure 21–1, and combination **smoke alarms/detectors,** Figure 21–2.

The more elaborate systems connect to a central monitoring customer service center through a telephone line. These systems offer instant contact with police or fire departments, a panic button for emergency medical problems, low-temperature detection, flood (high-water) detection, perimeter protection, interior motion detection, and other features.

All detections are transmitted first to the company's customer service center where personnel on duty monitor the system 24 hours a day, 7 days a week. They will verify that the signal is valid. After verification, they will contact the police department, the fire department, and individuals whose names appear on

FIGURE 21–2 Combination Smoke Alarm/Detector

a previously agreed-upon list that the homeowner prepared and submitted to the company's customer service center.

NFPA Standards are not law until adopted by a city, state, or other governmental body. Some NFPA Standards are adopted totally, and others are adopted in part. Installers must be aware of all requirements of the locality in which they are doing work. Most communities require a special permit for the installation of fire warning equipment and security systems in homes and require registration of the system with the police and/or fire departments. In most instances, fire protection requirements are found in the building codes of a community and are not necessarily spelled out in the electrical code.

Pay particular attention to *11.5.1.1* of *NFPA 72*, which clearly states in part that smoke alarms shall be installed "*Where required by applicable laws, codes, or standards for the specified occupancy.*" This puts the responsibility on the local building department. It might seem confusing to have so many codes. Without question, you must become familiar with your local applicable laws, codes, or standards.

Most communities adopt building codes published by the **International Code Council (ICC).** These codes contain requirements for the installation of smoke alarms. Your local building department officials can explain what is required for new work and remodel work (alterations, repairs, and additions) where a permit is required. Whereas new work requires interconnected, hardwired, battery-backup smoke alarms, it might be acceptable on remodel jobs to install battery-powered-only smoke alarms for existing areas where interior walls or ceiling finishes are not removed to expose structural framing members. To install, interconnect, and hardwire smoke alarms in these existing areas could result in damage to walls and ceilings, requiring patching, repainting, or other repairs—an uncalled for and tremendous expense. If there is access from an attic, crawl space, or basement, the inspector might require interconnected, hardwired, battery-backup smoke alarms if you can do the job without having to remove interior finishes. Check this out with your local inspector!

In this chapter, rather than going back and forth between the word "alarm" and "detector," we refer to fire warning devices as alarms, knowing full well that before an alarm can sound, the cause must be detected. Most home-type fire warning devices combine the detector and alarm in one device. The location requirements for smoke and heat alarms are pretty much the same.

Smoke, Heat, and Carbon Monoxide Alarms

This unit covers the basic requirements for protection in homes against the hazards of fire, heat, smoke, and carbon monoxide. Fire alarm systems for commercial installations are much more complicated than the typical household system.

Household fire alarm requirements are found in *Chapter 11* of *NFPA 72*, the *National Fire Alarm Code*. Specifics for one- and two-family dwellings are found in *11.5.1*.

Fire is the third leading cause of accidental death. Home fires account for the biggest share of these fatalities, most of which occur at night during sleeping hours. Rapidly developing high-heat fires and slow smoldering fires are the culprits. Both produce smoke and deadly gases.

Smoke, heat, and **carbon monoxide alarms** are installed in a residence to give the occupants early warning of the presence of fire or toxic fumes. Fires produce smoke and toxic gases that can overcome the occupants while they sleep.

Most fatalities result from the inhalation of smoke and toxic gases, rather than from burns. Heavy smoke reduces visibility.

In nearly all home fires, detectable smoke precedes detectable levels of heat. People sleeping are less likely to smell smoke than people who are awake. The smell of smoke probably will not awaken a sleeping person. Therefore, smoke alarms are considered to be the primary devices for protecting lives.

Heat alarms **do not** take the place of smoke alarms. Heat alarms are installed in addition to smoke detectors.

Another lifesaving device is the carbon monoxide (CO) alarm, which senses dangerous carbon monoxide emitted from a malfunctioning furnace or other source. *NFPA 720* is the *Recommended Practice for the Installation of Household Carbon Monoxide Warning Equipment*.

Home-type smoke, heat, and carbon monoxide devices as a rule have both the detector and the alarm in one device. There also are alarms that combine more than one type of sensor into one unit, such as smoke/heat, smoke/carbon monoxide, or carbon monoxide/explosive gases.

In larger, more complex commercial installations, detectors and alarms are generally separate devices that, when triggered, send a signal to a central control panel, which in turn sounds the general alarm.

As with all electrical installations, always install fire warning equipment that has been listed by a nationally recognized testing laboratory (NRTL).

Detector Codes

AC/Battery. Required in new construction. These conform to the *NFPA 72* requirement that alarms have two power supplies—regular household 120-volt ac for the primary source of power and a 9-volt battery for the secondary source. The alarm "chirps" when the battery gets low. See Figure 21–3.

Interconnected. Required in new construction. Should any smoke, heat, or carbon monoxide alarm trigger, that alarm will sound and simultaneously send a

FIGURE 21–3 An ac–dc smoke alarm operates on 120 volts ac as the primary source of power and a 9-volt battery as the secondary source of power. In the event of a power outage, the alarm continues to operate on the battery. (Courtesy of BRK Electronics)

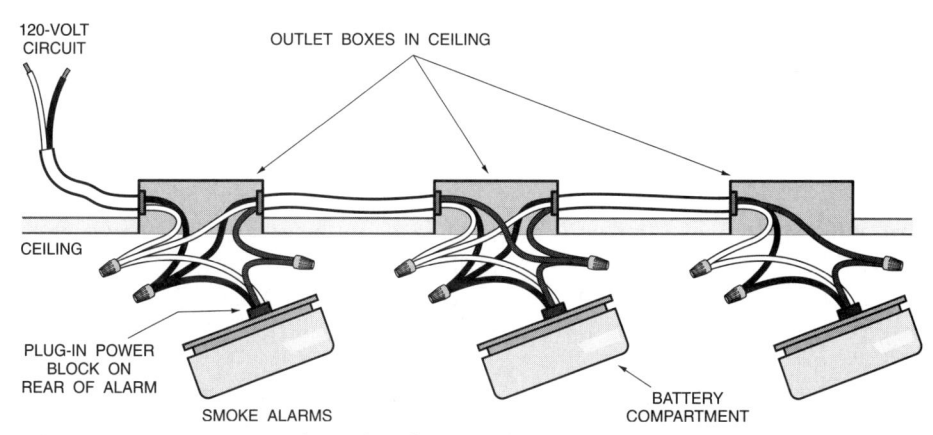

FIGURE 21–4 Interconnected Smoke Alarms The two-wire branch-circuit is run to the first alarm, then three-wire nonmetallic-sheathed cable interconnects the other alarms. If one alarm in the series is triggered, all other alarms will sound off.

signal to set off all other interconnected alarms. Check the manufacturer's instructions to determine how many units may be interconnected. Never connect more units than the number specified by the manufacturer. Never "mix" different manufacturer's devices. Testing laboratories do not perform tests using more than one manufacturer's devices in the test. The alarm "chirps" when the battery gets low. These units usually have a small three-wire power connector—the wires connect to the field wiring and the other end has a connector that plugs into the back of the unit. See Figure 21–4. The pigtail may have a red or orange-colored "interconnect" wire. The field wiring between alarms is three-wire nonmetallic-sheathed cable that contains a red insulated conductor that is used as the "interconnect" wire.

Plug-in. Not permitted in new construction. These alarms plug into a wall outlet that is not controlled by a wall switch. They operate on 120 volts as the primary source of power and on a 9-volt battery as the secondary source. The alarm "chirps" when the battery gets low.

Battery-Operated Only. Not permitted in new construction. These alarms operate on a 9-volt battery only. The alarm "chirps" when the battery gets low. These alarms are commonly used in existing homes.

For the Hearing Impaired. These alarms have a bright strobe light that provides a visual alarm for the hearing impaired. This complies with the requirements found in the Americans with Disabilities Act (ADA).

Types of Smoke Alarms

Two common types of smoke alarms are the **photoelectric** type (sometimes called photoelectronic) and the **ionization** type. They usually contain an indication light to show that the unit is functioning properly. They also may have a test button that simulates smoke so that, when the button is pushed, the detector's smoke-detecting ability as well as its circuitry and alarm are tested.

Some alarms are tested with a magnet. Others are tested with a listed spray from an aerosol can.

Smoke alarms generally do not sense heat, flame, or gas. However, some smoke alarms can be set off by acetylene and propane gas.

Some alarms have a "hush button" that can temporarily silence a nuisance alarm for about 15 minutes.

Photoelectric Type

The photoelectric type of smoke alarm has a light sensor that measures the amount of light in a chamber. When smoke is present, an alarm sounds, indicating a reduction in light due to the obstruction of the smoke. This type of sensor detects smoke from burning materials that produce large quantities of smoke, such as furniture, mattresses, and rags. The photoelectric type of alarm is less effective for gasoline and alcohol fires, which do not produce heavy smoke. This type of alarm can become more sensitive to smoke as it gets older.

Photoelectric alarms are less likely to sound an alarm from normal cooking, unlike the ionization type.

There are two types of photoelectric devices: light obscuration and light scattering.

Ionization Type

The ionization type of alarm contains a low-level radioactive source (less than used in luminescent watch and clock dials), which supplies particles that ionize the air in the detector's smoke chamber. Plates in this chamber are oppositely charged. Because the air is ionized, an extremely small amount of current (millionths of an ampere) flows between the plates. Smoke entering the chamber impedes the movement of the ions, reducing the current flow, which triggers the alarm. This type often sets off a nuisance alarm because cooking routinely gives off small "invisible" smoke particles. This type of alarm can become more sensitive to smoke as it get older.

The ionization type of alarm is effective for detecting small amounts of smoke, as in gasoline and alcohol fires, which are fast-flaming with little or no smoke.

Some smoke alarms are available with both ionization and photoelectric sensors combined. Other smoke alarms are available with smoke and carbon monoxide detection capabilities or smoke and heat detection capabilities in one unit.

Types of Heat Alarms

Heat alarms respond to heat—not smoke!

Three types of heat alarms are described as follows.

- Fixed temperature heat detectors sense a specific fixed temperature, such as 135°F (57°C) or 200°F (93°C). They shall have a temperature rating of at least 25°F (14°C) above the normal temperature expected, but not to exceed 50°F (28°C) above the expected temperature. Fixed temperature detectors are sometimes combined with smoke alarms and carbon monoxide alarms.
- Rate-of-rise heat detectors sense rapid changes in temperature (12°F to 15°F per minute) such as those caused by flash fires.
- Fixed/rate-of-rise temperature detectors are available as a combination unit.
- Fixed/rate-of-rise/smoke combination detectors are available in one unit.

Installation Requirements

The following information includes specific recommendations for installing smoke alarms and heat alarms in homes. Remember that smoke alarms are the primary fire warning devices, and heat detectors are installed in addition to the required smoke alarms. As you continue reading, study Figure 21–5, Figure 21–6, Figure 21–7, Figure 21–8, and Figure 21–9. Complete data are found in *NFPA Standard No. 72* and in the instructions furnished by the manufacturer of the equipment.

The Absolute Minimum Level of Protection

Here is the absolute minimum level of protection for smoke alarms in conformance to *NFPA 72* for new residential construction.

- Install multiple station smoke alarms:
  1. In new construction in all sleeping rooms. This is not a requirement for existing one- and two-family dwellings.
  2. Outside each separate sleeping area, in the immediate vicinity of the sleeping rooms.
  3. On each additional story of the dwelling unit, including basements but excluding crawl spaces and unfinished attics.
  4. For existing dwellings, listed battery-powered smoke alarms that are not interconnected are generally permitted. This provides a reasonable level of protection as opposed to having no smoke alarms at all. However, your local code might require ac/dc dual-powered interconnected alarms.
- Smoke alarms must have at least two independent sources of power. The primary source is the 120-volt ac circuit, and the secondary (standby) source is the integral battery power supply.
- Smoke alarms must be interconnected using multiple station devices so that when one alarm in the series triggers, all other alarms in the house will sound.

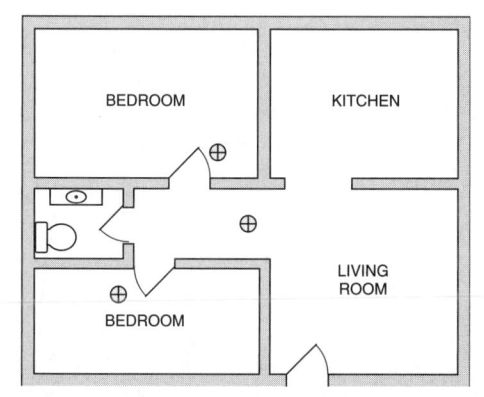

FIGURE 21–5 Smoke alarms are required in all sleeping areas, outside all sleeping areas, and on every story. The logical location is near the center of the bedroom. However, this may be impossible because of a ceiling luminaire (fixture) and/or ceiling paddle fan. An alternate location is on the wall space above the door, not closer than 4 in. (102 mm) nor more than 12 in. (305 mm) from the ceiling.

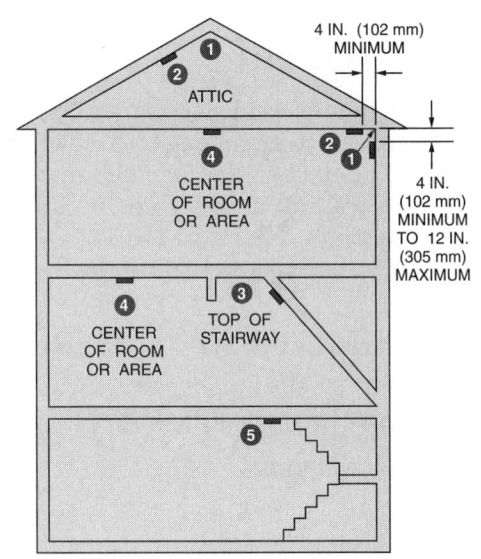

FIGURE 21–6 Recommendations for the Installation of Heat and Smoke Alarms

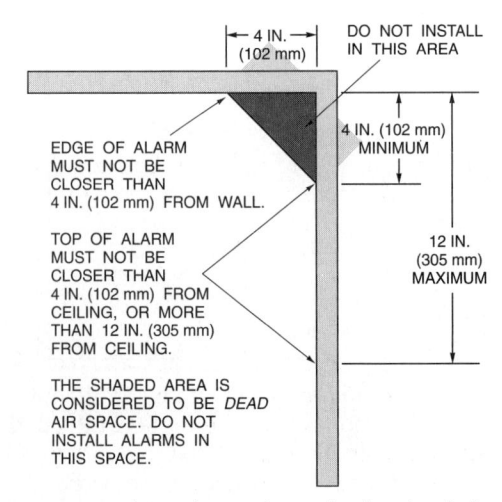

FIGURE 21–7 Do not mount smoke or heat alarms in the dead airspace where the ceiling meets the wall.

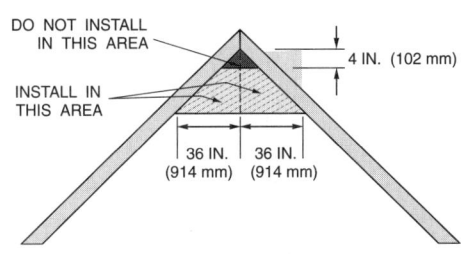

Smoke, Heat, and Carbon Monoxide Alarms, and Security Systems • 405

FIGURE 21–8 Installing Smoke and Heat Detectors/Alarms in Peaked (Cathedral) Ceilings

Do not exceed the maximum number of alarms permitted by the manufacturer to be interconnected.
- The alarms shall provide a sound that is audible in all dwelling areas that can be occupied or used.

Where to Install

In *bedrooms* and *halls outside of sleeping areas*. See Figure 21–5.

In *other rooms* and areas. See Figure 21–6.

On *walls*—not closer than 4 in. (102 mm) but not farther than 12 in. (305 mm) from the adjoining ceiling. Do not install in dead airspace. See Figure 21–6 and Figure 21–7.

On *flat ceilings*—not closer than 4 in. (102 mm) from the adjoining wall. See Figure 21–6 and Figure 21–7.

On *peaked ceilings*—locate with 36 in. (914 mm) horizontally from the peak, but not closer than 4 in. (102 mm) vertically from the peak. A peaked ceiling might also be called a cathedral ceiling. Do not install in dead airspace. See Figure 21–8.

On *sloped ceilings*—a sloped ceiling is defined as having a rise greater than 1 ft in 8 ft (1 m in 8 m). Locate within 36 in. (914 mm) of the high side but not closer than 4 in. (102 mm) from the adjoining wall. Do not install in dead airspace. See Figure 21–9.

How to Wire Smoke and Heat Alarms

In new construction, smoke and heat alarms require both primary (120 volts) and secondary (battery) sources of power. The *National Fire Alarm Code* does not specify where to pick up the 120-volt supply.

NEC® 210.12 requires that all 125-volt, single-phase, 15- and 20-ampere electrical outlets in bedrooms be AFCI protected. This includes smoke alarms.

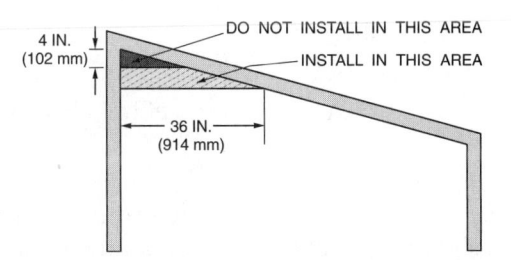

FIGURE 21–9 Installing Smoke and Heat Detectors/Alarms in Sloped Ceilings

Probably the simplest way to hardwire and interconnect smoke alarms and heat alarms when wiring new homes is to pick up one smoke alarm from one of the bedroom branch-circuits, then run the interconnecting wiring to all of the other required and optional smoke alarms. The electrical plans indicate the location of the required and optional smoke and heat alarms.

Some inspectors require a "lock-off" device over the handle of the circuit breaker that serves the alarms so that it will not unintentionally be turned off. This lock-off device is really not necessary. Should the breaker supplying the alarms get turned off intentionally or unintentionally, or if it trips due to trouble in the circuit, the battery backup will power the alarms. When the batteries get "low," the alarms start "chirping."

The usual way to connect alarms is to run the two-wire branch-circuit to the first alarm, then run three-wire nonmetallic-sheathed cable from the first alarm to all other alarms. The third conductor provides the interconnect feature that will sound all alarms should any one of the alarms activate. See Figure 21–4.

Summary of Do's and Do Nots

When installing detectors and alarms, ask yourself, "Where will the smoke and greatly heated air travel?" Because smoke and heated air rise, detectors and alarms must be in the path of the smoke and greatly heated air.

Figure 21–5, Figure 21–6, Figure 21–7, Figure 21–8, and Figure 21–9 illustrate some of the more important issues regarding smoke and heat alarms.

Here is a list of the Do's and Don'ts in no particular order of importance.

The Do's

1. Do install listed smoke alarms, and install them according to the manufacturer's instructions.
2. Do install smoke alarms on the ceiling, as close as possible to the center of the room or hallway. *Note:* Because ceiling paddle fans are often installed in bedrooms, install the alarms on the wall space above the bedroom door, not closer than 4 in (102 mm) nor more than 12 in. (305 mm) from the ceiling.
3. Do make sure that the path of rising smoke will reach the alarm when the alarm is installed in a stairwell. This is usually at the top of the stairway. The alarm must not be located in a dead airspace created by a closed door at the head of the stairway.
4. Do space according to the instructions furnished with the devices.
5. Do install the type that has a "hush" button if located within 20 ft (6.1 m) of a cooking appliance, or install a photoelectric type.
6. Do install smoke alarms at the high end of a room that has a sloped, gabled, or peaked ceiling where the rise is greater than 1 ft (305 mm) per 8 ft (2.6 m). Do not install in the dead airspace.
7. Do mount on the bottom of open joists or beams.
8. Do install a smoke alarm on the basement ceiling close to the stairway to the first floor.
9. Do install smoke alarms in split-level homes. A smoke alarm installed on the ceiling of an upper level can suffice for the protection of an adjacent lower level if the two levels are not separated by a door. Better protection is to install alarms for each level.

10. Do install, in new construction, dual-powered smoke alarms that are hard-wired directly to a 120-volt ac source and also have a battery. In existing homes, battery-powered alarms are the most common, but 120-volt ac alarms or dual-powered alarms can be installed. Remember, battery-powered alarms will not operate with dead batteries. Ac-powered alarms will not operate when the power supply is off.
11. Do install interconnected smoke alarms in new construction so that the operation of any alarm will cause all other alarms that are interconnected to sound.
12. Do install smoke alarms so they are not the only load on the branch-circuit. Connect smoke alarms to a branch-circuit that supplies other lighting outlets in habitable spaces. Because smoke alarms draw such a miniscule amount of current, they can easily be connected to a general lighting branch-circuit, such as a bedroom branch-circuit.
13. Do always consider the fact that doors, beams, joists, walls, partitions, and similar obstructions will interfere with the flow of smoke and heat, and, in most cases, create new areas needing additional smoke alarms and heat alarms.
14. Do make sure that the gap around the ceiling outlet box is sealed to prevent dust from entering the smoke chamber. This is a big problem for ceiling-mounted alarms.
15. Do clean smoke alarms according to the manufacturer's instructions.
16. Do consider that the maximum distance between heat alarms mounted on flat ceilings is 50 ft (15 m) and 25 ft (7.5 m) from the detector to the wall. This information is explained in detail by *NFPA 72*. Where obstructions such as beams and joists will interfere with the flow of heat, the 50-ft distance is reduced to 25 ft (7.5 m), and the 25-ft distance is reduced to 12-½ ft (3.75 m).
17. Do install alarms after construction clean-up of all trades is complete and final.
18. Do install interconnected units of the same manufacturer. Different manufacturers' units may or may not be compatible.
19. Do connect all interconnected units to the same branch-circuit. Different circuits cannot be shared.

The Do Nots

1. Do not install smoke alarms in the dead airspace at the top of a stairway that can be closed off by a door.
2. Do not install within 36 in. (914 mm) of a door to a kitchen or to a bathroom containing a shower or tub.
3. Do not locate where the smoke alarm will be subject to temperature and/or humidity that exceeds the limitations stated by the manufacturer.
4. Do not install within a 36-in. (914-mm) horizontal path from a supply hot air register. Install outside the direct flow of air from these registers.
5. Do not install within a horizontal path 36 in. (914 mm) from the tip of the blade of a ceiling-suspended paddle fan.
6. Do not install where smoke rising in a stairway could be blocked by a closed door or an obstruction.
7. Do not place the edge of a ceiling-mounted smoke alarm or heat alarm closer than 4 in. (102 mm) from the wall.

8. Do not place the top edge of a wall-mounted smoke alarm or heat alarm closer than 4 in. (102 mm) from the ceiling and farther than 12 in. (305 mm) down from the ceiling. Some manufacturers recommend placement not farther down than 6 in. (153 mm).

9. Do not install smoke alarms or heat detectors on an outside wall that is not insulated, or one that is poorly insulated. Instead mount the alarms on an inside wall.

10. Do not install smoke alarms or heat alarms on a ceiling where the ceiling will be excessively cold or hot. Smoke and heat will have difficulty reaching the alarms. This could be the case in older homes that are not insulated or are poorly insulated. Instead, mount the alarms on an inside wall.

11. Do not install smoke alarms or heat alarms where the ceiling meets the wall because this is considered dead airspace where smoke and heat may not reach the detector.

12. Do not connect smoke alarms or heat alarms to wiring that is controlled by a wall switch.

13. Do not install smoke alarms or heat alarms where the relative humidity exceeds 85 percent, such as in bathrooms with showers, laundry areas, or other areas where large amounts of visible water vapor collect. Check the manufacturer's instructions.

14. Do not install smoke alarms or heat alarms in front of air ducts, air conditioners, or any high-draft areas where the moving air will keep the smoke or heat from entering the detector.

15. Do not install smoke alarms or heat alarms in kitchens where the accumulation of household smoke can result in setting off an alarm, even though there is no real hazard. The person in the kitchen will know why the alarm sounded, but other people in the house may panic. This problem exists in multifamily dwelling units where unwanted triggering of the alarm in one dwelling unit might cause people in the other units to panic. The photoelectric type may be installed in kitchens but must not be installed directly over the range or cooking appliance. A better choice in a kitchen is to install a heat alarm.

16. Do not install smoke alarms where the temperature can fall below 32°F (0°C) or rise above 120°F (49°C) unless the detector is specifically identified for this application.

17. Do not install smoke alarms in garages where vehicle exhaust might set off the detector. Instead of a smoke alarm, install a heat detector.

18. Do not install smoke alarms in airstreams that will pass air originating at the kitchen cooking appliances across the alarm. False alarms will result.

19. Do not install smoke alarms or heat detectors on ceilings that employ radiant heating.

20. Do not install smoke alarms or heat alarms in a recessed location.

21. Do not connect smoke alarms or heat alarms to a switched circuit or a circuit controlled by a dimmer.

Maintenance and Testing

Once installed, smoke, heat, and carbon monoxide alarms must be tested periodically and maintained (blow out accumulated dust) to make sure they are operating properly. This is a requirement of *NFPA 72*, which states, "*Homeowners shall inspect and test smoke alarms and all connected appliances in accordance with the manufacturer's instructions at least monthly."*

When to replace. In conformance to *10.4.6* and *11.8.1.4.(5)* in *NFPA 72*, smoke alarms *"shall be replaced when they fail to respond to tests"* and *"shall not remain in service longer than 10 years from the date of installation."*

   Note: The ten-year mandatory replacement requirement is often overlooked. This requirement is based on studies of tens of thousands of smoke alarms in operation to determine the acceptable number of failures over many years. *Code* committees and manufacturers analyzed the data and agreed that ten years in service provided a reasonable lifetime. The ten-year replacement program, along with regular testing per the manufacturer's instruction results in very few homes going unprotected for any extended period of time. Hardwired and battery-operated alarms are equally affected by age.

   The expected life of lithium batteries used in some alarms is said to be ten years.

After You Install the System, Then What?

After installing the system, there is still more to do. *NFPA 72* requires that the installer of the fire warning system provide the homeowner with the following information:

1. An instruction booklet illustrating typical installation layouts
2. Instruction charts describing the operation, method, and frequency of testing and maintenance of fire warning equipment
3. Printed information for establishing an emergency evacuation plan
4. Printed information to inform owners where they can obtain repair or replacement service and where and how parts requiring regular replacement, such as batteries or bulbs, can be obtained within 2 weeks
5. Information that, unless otherwise recommended by the manufacturer, smoke alarms shall be replaced when they fail to respond to tests
6. Instruction that smoke alarms shall not remain in service longer than ten years from the date of installation.

   This information is usually part of the instructions furnished by the manufacturer of the alarms.

Exceeding Minimum Levels of Protection

The following are recommendations for attaining levels of fire warning protection that exceed the minimum level stated previously and include guidelines for installing smoke and heat detectors in a home:

- Install smoke alarms in all rooms, basements, hallways, heated attached garages, and storage areas. Installing alarms in these locations will increase escape time, particularly if the room or area is separated by a door(s). In some instances, smoke alarms are installed in attics and crawl spaces.
- Install heat alarms in kitchens because conventional smoke alarms can nuisance sound an alarm.
- Install heat alarms in garages because gasoline fires give off little smoke.
- Consider special alarms that light up or vibrate for occupants who are hard of hearing.

- Consider smoke alarms that have an "escape" light.
- Consider low-temperature alarms (i.e., 45°F or 7°C) that can detect low temperatures should the heating system fail. The damage caused by frozen water pipes bursting can be extremely costly.

Building Codes

In most instances the various building codes published by the International Code Council (ICC) (formerly BOCA, ICBO, and SBCCI) adopt *NFPA 72* and *NFPA 720* by reference, but it would be wise to check with your local AHJ to find out if there are any differences between these codes that might affect your installation.

Many companies specialize in installing complete fire and security systems. They also offer system monitoring at a central office and can notify the police or fire department when the system gives the alarm. How elaborate the system should be is up to the homeowner.

The ICC *Code* in *1202.5* requires that smoke alarms be connected to a circuit that also supplies lighting outlets in habitable spaces.

Manufacturers' Requirements

Here are a few of the responsibilities of the manufacturers of smoke alarms and heat alarms. Complete data are found in *NFPA 72*.

- The power supply must be capable of operating the signal for at least 4 minutes continuously.
- Battery-powered units must be capable of a low-battery warning "chirp" of at least one chirp per minute for 7 consecutive days.
- Direct-connected 120-volt ac alarms must have a visible indicator that shows "power-on."
- Alarms must not signal when a power loss occurs or when power is restored.

  Note: In *4.3.1* in *NFPA 72*, we find that we must:
  - always install fire alarm equipment that is "Listed for the purpose."
  - always install fire alarm equipment in accordance with the manufacturer's installation and maintenance instructions.

Carbon Monoxide Alarms

Carbon monoxide alarms are installed in addition to smoke and heat alarms.

Carbon monoxide (CO) is referred to as the "silent killer"! It is responsible for more deaths that any other single poison.

When a heat or smoke alarm is triggered, it is generally not difficult to determine the source of the problem. But when a carbon monoxide alarm goes off, it is hard to determine the source.

Carbon monoxide is odorless, colorless, and tasteless—undetectable by any of a person's five senses—taste, smell, sight, touch, hearing; but it is highly poisonous. Carbon monoxide replaces oxygen in the bloodstream, resulting in brain damage or total suffocation.

Carbon monoxide is produced by the incomplete burning of fuels. Common sources of carbon monoxide in a home can come from a malfunctioning furnace, gas appliances, kerosene heaters, automobile or other gas engines, charcoal grills, fireplaces, or a clogged chimney.

Symptoms of carbon monoxide poisoning are similar to flulike illnesses such as dizziness, fatigue, headaches, nausea, diarrhea, stomach pains, irregular breathing, and erratic behavior.

Carbon monoxide has about the same characteristics as air. It moves about just like air.

Carbon monoxide alarms detect carbon monoxide from any source of combustion. They are not designed to detect smoke, heat, or gas. The Consumer Product Safety Commission (CPSC) recommends that at least one carbon monoxide alarm be installed in each home, and preferably one on each floor of a multi-level house. Some cities require this. If only one carbon monoxide alarm is installed, it should be in the area just outside individual bedrooms.

Carbon monoxide alarms monitor the air in the house and sound a loud alarm when carbon monoxide above a predetermined level is detected. They provide early warning before deadly gases build up to dangerous levels.

For residential applications, carbon monoxide alarms are available for hardwiring (direct connection to the branch-circuit wiring), plug-in (the male attachment plug is built into the back of the detector for plugging directly into a 120-volt wall receptacle outlet), and combination 120-volt and battery-operated units. They are available with the interconnect feature for interconnecting with smoke and heat alarms. Some carbon monoxide alarms also have an explosive gas sensor.

NFPA 720 is the *Recommended Practice for the Installation of Household Carbon Monoxide (CO) Warning Equipment*. This standard contains the installation recommendations for carbon monoxide alarms. Here is a summary of these recommendations.

Installation "Do's" for Carbon Monoxide Alarms

1. Do install carbon monoxide alarms that are tested and listed in conformance to UL Standard 2034.
2. Do install carbon monoxide alarms according to the manufacturer's instructions.
3. Do use household electricity as the primary source of power. In existing homes, monitored battery units are permitted.
4. Do install carbon monoxide alarms in or near bedrooms and living areas.
5. Do install carbon monoxide alarms in locations where smoke alarms are installed.
6. Do test carbon monoxide alarms once each month or as recommended by the manufacturer.
7. Do remove carbon monoxide alarms before painting, stripping, wallpapering, or using aerosol sprays. Store in a plastic bag until the project is completed.
8. Do carefully vacuum carbon monoxide alarms once each month or as recommended by the manufacturer.
9. Do interconnect alarms where multiple alarms are installed.

Installation "Do Nots" for Carbon Monoxide Alarms

1. Do not connect carbon monoxide alarms to a switched circuit or a circuit controlled by a dimmer.
2. Do not install carbon monoxide alarms closer than 6 in. (150 mm) from a ceiling. This is considered dead airspace.

3. Do not install carbon monoxide alarms in garages, kitchens, or furnace rooms. This could lead to nuisance alarms, may subject the detector to substances that can damage or contaminate the alarm, or the alarm might not be heard.
4. Do not install carbon monoxide alarms within 15 ft (4.5 m) of a cooking or heating gas appliance.
5. Do not install carbon monoxide alarms in dusty, dirty, or greasy areas. The sensor inside of the alarm can become coated or contaminated.
6. Do not install carbon monoxide alarms where the air will be blocked from reaching the alarm.
7. Do not install carbon monoxide alarms in dead airspaces such as in the peak of a vaulted ceiling or where a ceiling and wall meet.
8. Do not install carbon monoxide alarms where the detector will be in the direct airstream of a fan.
9. Do not install carbon monoxide alarms where temperatures are expected to drop below 40°F (4°C) or get hotter than 100°F (38°C). Their sensitivity will be affected.
10. Do not install carbon monoxide alarms in damp or wet locations such as showers and steamy bathrooms. Their sensitivity will be affected.
11. Do not mount carbon monoxide alarms directly above or near a diaper pail. The high methane gas will cause the detector to register, and this is not carbon monoxide.
12. Do not clean carbon monoxide alarms with detergents or solvents.
13. Do not spray carbon monoxide alarms with hair spray, paint, air fresheners, or other aerosol sprays.

Fire Alarm Systems

Fire-protective signaling systems installed in homes have power-limited fire alarm (PLFA) circuits where the power output is limited by the listed power supply for the system. Here are the key requirements in *Article 760, Part III*:

- Must have power output capabilities limited to the values specified in Chapter 9, *Table 12(A)* and *Table 12(B)*. Refer to *760.41*. Residential fire-protective signaling systems are generally "power-limited."
- Do not connect the equipment to a GFCI-protected branch-circuit, *760.41*.
- Wiring on the supply side of the equipment is installed according to the conventional wiring methods found in Chapter 1 through Chapter 4 of the *NEC®*. Refer to *760.51*.
- The branch-circuit overcurrent protection supplying the fire alarm system shall not exceed 20 amperes, *760.51*.
- Cables may be run on the surface or be concealed and shall be adequately supported and protected against physical damage, *760.52(B)(1)*.
- When run exposed within 7 ft (2.1 m) of the floor, the cable shall be securely fastened at intervals not over 18 in. (450 mm), *760.52(B)(1)*.
- Cables shall not be run in the same raceway, cable, outlet box, device box, or be in the same enclosure as light and power conductors unless separated by a barrier, or introduced solely to connect that particular piece of equipment. Refer to *760.55*. This is also a ruling in *725.55*.

- Cables shall not be supported by taping, strapping, or attaching by any means to any electrical conduits, or other raceways, *760.58*.
- Cables shall be supported by structural components, *760.8*.
- Wiring on the load side shall be insulated solid or stranded copper conductors of the types listed in *Table 760.82(A)*.
- *Table 760.61* shows different types of cable that can be substituted for one another when necessary.
- If installed in a duct or plenum, the cable must be plenum rated, *760.61(A)*.
- Type FLP cable is permitted in one- and two-family dwellings, *760.61(B)(3)*.
  - Conductors in cable shall not be smaller than 26 AWG. Single conductors shall not be smaller than 18 AWG. Refer to *760.57* and *760.82(B)*.
  - Cable minimum voltage rating is 300 volts, *760.82(C)*.
- Remove all abandoned cables. This is extremely important since abandoned cables provide fuel for a fire. Removing abandoned cables is required by *725.2, 725.3(B), 725.61(E), 760.2, 760.3(A), 800.2, 800.3(C), 820.2*, and *820.3*.

It makes a lot of sense to contact your local inspection authority before installing a fire warning system to be sure the system you intend to install meets the requirements of the local codes! Contractors who specialize in this kind of work are well aware of the *Code* requirements and will install the proper types of cables.

Security Systems

It is beyond the scope of this text to cover all types of residential security systems. What follows is a typical system.

Professionally installed and homeowner installed security systems can range from a simple system to a very complex one. Features and options include master control panels, remote controls, perimeter sensors for doors and windows, motion sensors, passive infrared sensors, wireless devices, interior and exterior sirens, bells, electronic buzzers, strobe lights to provide audio (sound) as well as visual detection, heat sensors, carbon monoxide alarms, smoke alarms, glass-break sensors, flood sensors, low-temperature sensors, and lighting modules. Figure 21–10 shows some of these devices. Figure 21–11 shows a schematic one-line diagram for a typical security system.

Monitored security systems are connected by telephone lines to a central station, a 24-hour monitoring facility. If the alarm sounds, security professionals will contact the end user to verify the alarm before sending the appropriate authorities (fire, police, or ambulance) to the home.

The decision about how detailed an individual residential security system should be generally begins with a meeting between the homeowner and the security consultant or licensed electrical contractor before the actual installation. Most security system installers, consultants, and licensed electrical contractors are familiar with a particular manufacturer's system.

The systems are usually explained and even demonstrated during an in-home security presentation. They are then professionally installed by the security system company or licensed electrical contractor.

Do-it-yourself kits that contain many of the features provided by the professionally installed systems can be found at home centers, hardware stores, electrical

414 • Chapter 21

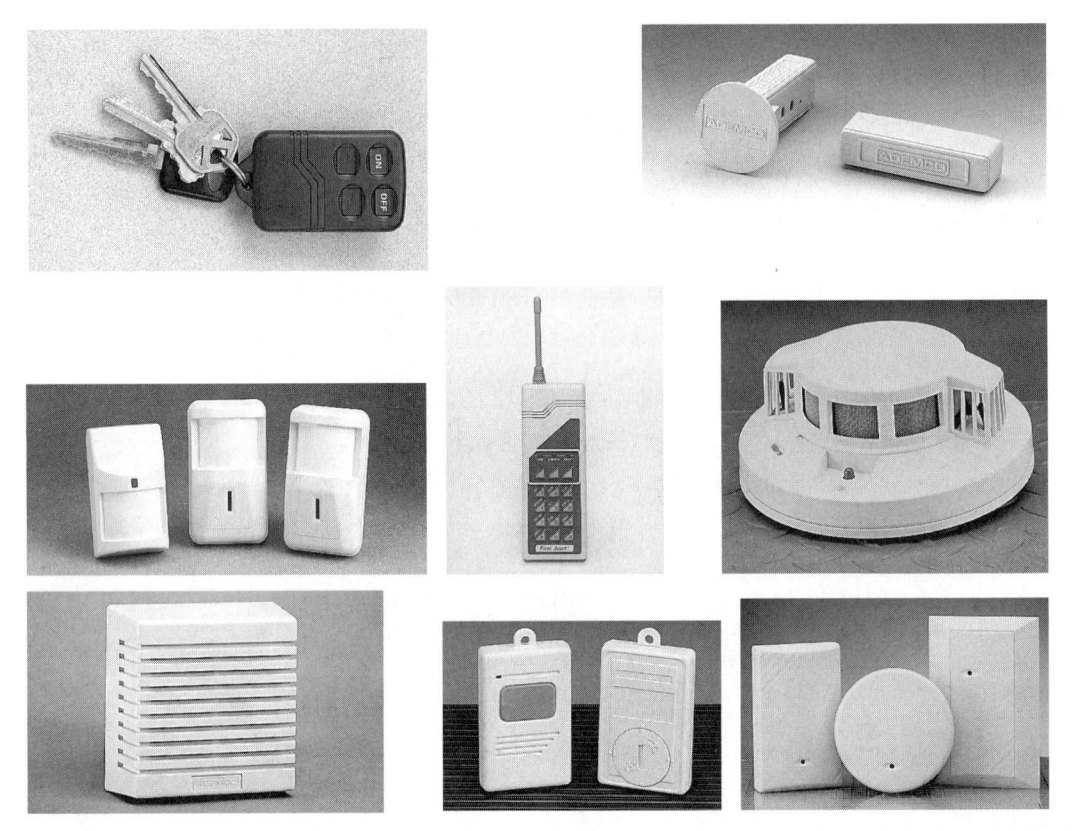

FIGURE 21–10 Photos courtesy of First Alert Professional Security Systems. First Alert is a Registered Trademark of the First Alert Trust.

distributors, and electronics stores. Depending on the system, it may or may not be possible to connect to a central monitoring facility.

The wiring of a security system consists of small, easy-to-install low-voltage, multiconductor cables made up of 18 AWG conductors. These conductors should be installed after the regular house wiring is completed to prevent damage to these smaller cables. Usually the wiring can be done at the same time as the chime wiring is being installed.

Security system wiring comes under the scope of *Article 725*.

When wiring detectors such as door-entry, glass-break, floor-mat, and window-foil detectors, circuits are electrically connected in series so that if any part of the circuit is opened, the security system will detect the open circuit. These circuits are generally referred to as "closed" or "closed loop." Alarms, horns, and other signaling devices are connected in parallel because they will all signal at the same time when the security system is set off. Heat detectors and smoke detectors are generally connected in parallel because any of these devices will "close" the circuit to the security master control unit, setting off the alarms.

The instructions furnished with all security systems cover the installation requirements in detail, alerting the installer to *Code* regulations, clearances, suggested locations, and mounting heights of the systems' components.

Always check with the local electrical inspector to determine if there are any special requirements in your locality relative to the installation of security systems.

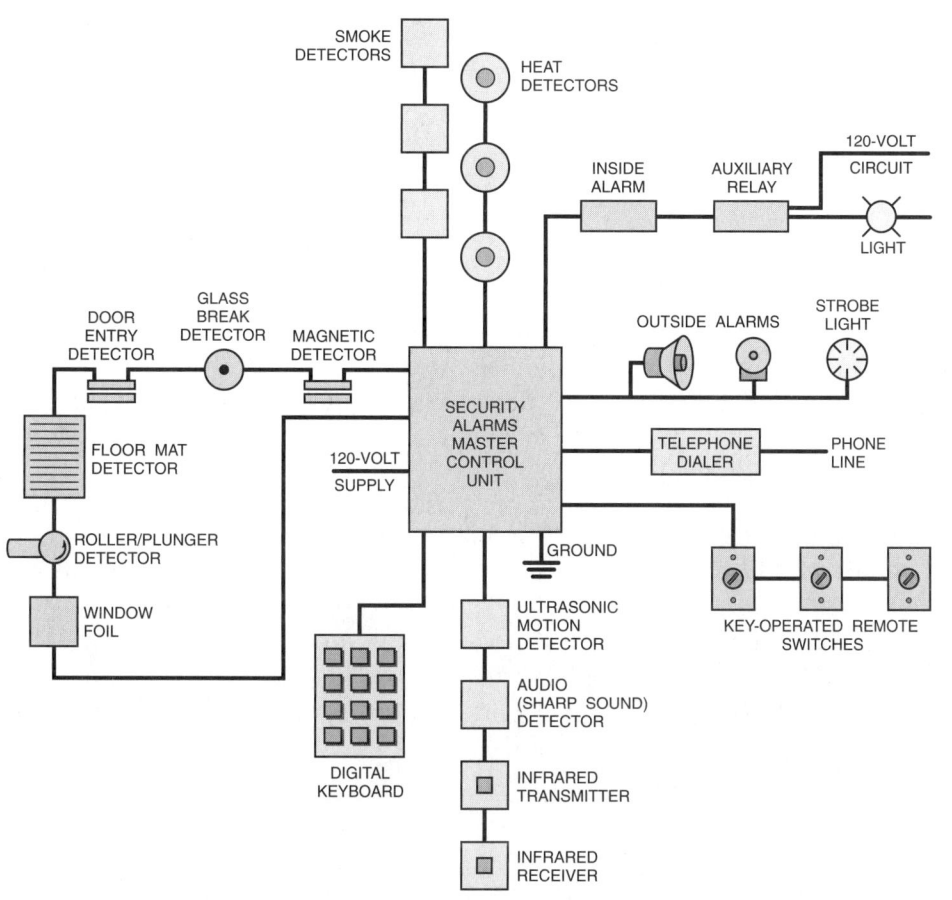

FIGURE 21–11 Diagram of Typical Residential Security System Showing Some of the Devices Available Complete wiring and installation instructions are included with these systems. Check local code requirements in addition to following the detailed instructions furnished with the system. Most of the interconnecting conductors are 18 AWG.

Sprinkling Systems Code

Home sprinkling systems are covered by *NFPA 13D*, which is the *Standard for the Installation of Sprinkler Systems in One- and Two-Family Dwellings and Manufactured Homes*.

Standards Relating to Fire, Smoke, Carbon Monoxide, and Security Devices

UL 217—Standard for Single and Multiple Station Alarms

UL 268—Standard for Smoke Detectors for Fire Protective Signaling Systems

UL 365—Standard for Police Station Connected Burglar Alarm Units

UL 521—Standards for Heat Detectors for Fire Protection Signaling Systems

UL 539—Standard for Single and Multiple Station Heat Detectors

UL 609—Standard for Local Burglar Alarm Units and Systems

UL 827—Standard for Central Station Alarm Services

UL1023—Standard for Household Burglar Alarm System Units

UL 1610—Standard for Central Station Burglar Alarm Units

UL 1641—Standard for Installation and Classification of Residential Burglar Alarm Systems

UL 2034—Standard for Single and Multiple Station Carbon Monoxide Alarms

Summary

- *NFPA 72* presents the minimum requirements for the proper selection, installation, operation, and maintenance of fire alarm warning equipment.
- *NFPA 70, Article 760* details the requirements of wiring fire alarm systems
- The International Code Council (ICC) contains building code requirements for the installation of smoke alarms.
- Smoke, heat, and carbon monoxide alarms are installed in residences to give the occupants early warning of the presence of toxic fumes.
- Heat detectors are not a replacement for smoke detectors or smoke alarms.
- Alarm systems must be powered by 120-volt ac along with a secondary backup battery source, in most cases, 9-volts.
- Once installed, smoke, heat, and carbon monoxide alarms must be tested periodically for proper operation and maintained as recommended by the manufacturer.
- Smoke alarms should be replaced every 10 years.
- Home sprinkler systems are covered by *NFPA 13D*.

Key Terms

Carbon monoxide alarm/detector   A device that detects the presence of the toxic gas.

Heat detector   A device that detects heat and can be either electrical or mechanical in operation. The most common types are the thermocouple and the electropneumatic; both respond to changes in ambient temperature. Typically, if the ambient temperature rises above a predetermined threshold, then an alarm signal is triggered.

International Code Council (ICC)   A membership association dedicated to building safety and fire prevention.

Ionization   A type of smoke alarm that contain a low-level radioactive source which supplies particles that ionize the air in the detector's smoke chamber.

Photoelectric   A type of smoke alarm that has a light sensor that measure the amount of light in a chamber.

Smoke alarm/detector   Safety device that detects airborne smoke and issues an audible alarm, thereby alerting nearby people to the danger of fire.

Review Questions

1. In your own words, explain the terms *alarm* and *detector*.
2. The requirements for household fire alarm systems is found in what *NFPA Code* and in what chapter of this *Code*?
3. Do heat alarms and carbon monoxide alarms take the place of smoke alarms? Explain.
4. Name two basic ways that smoke alarms are powered.
5. Name the two types of smoke alarms.

6. List the absolute minimum level of smoke alarm protection in a new one-family dwelling.

7. Circle the correct answer from the following statements:
   a. Smoke alarms installed in new one- and two-family homes shall be battery operated only so as not to be affected by a power outage.
   b. Smoke alarms installed in new one- and two-family homes shall be dual powered by a 120-volt circuit and a battery.

8. Circle the correct answer from the following statements:
   a. Smoke alarms installed in new one- and two-family homes shall be interconnected so that, if any one of them is triggered, all other alarms will also sound off.
   b. Smoke alarms installed in new one- and two-family homes shall not be interconnected so that each alarm will operate independently of all other alarms in the home.

9. Circle the correct answer from the following statements:
   a. Always install smoke and fire alarms in dead airspaces.
   b. Never install smoke and fire alarms in dead airspaces.

10. Circle the correct answer from the following statements:
    a. Mount wall-mounted smoke alarms so that the top edge is not closer than 4 in. (102 mm) from the ceiling and not more than 12 in. (305 mm) from the ceiling.
    b. Mount wall-mounted smoke alarms anywhere on the wall but not lower than 36 in. (914 mm) from the ceiling.

11. Cooking and baking in the kitchen can produce quite a bit of smoke. The best choice for a smoke alarm in the kitchen is the photoelectric type. The better choice would be to install a (heat alarm) or (carbon monoxide alarm). Circle the correct answer.

12. An important but often overlooked requirement in the *National Fire Alarm Code, NFPA 72* is that alarms must be replaced after being in service for (circle the correct answer):
    a. not more than 5 years
    b. not more than 10 years
    c. not more than 15 years

13. Carbon monoxide is _____, and is _____, _____, and _____.

14. Circle the correct answer from the following statements.
    a. Carbon monoxide is heavier than air.
    b. Carbon monoxide will always rise to the ceiling.
    c. Carbon monoxide will drop to the floor.
    d. Carbon monoxide is about the same weight as air.

15. The more complex household fire alarm systems are covered in what article of the *National Electrical Code®*? Circle the correct answer.
    a. *Article 72*
    b. *Article 310*
    c. *Article 760*

16. Circle the correct answer from the following statements.
    a. Always connect a fire alarm circuit or system to a power supply that has ground-fault circuit interrupter (GFCI) protection.
    b. Never connect a fire alarm circuit or system to a power supply that has ground-fault circuit interrupter (GFCI) protection.
    c. It makes no difference whether or not the branch-circuit is GFCI or AFCI protected because the alarms are required to have battery backup power. In the event of a loss of power for whatever reason, the alarms would still be operative.

17. Circle the correct answer from the following statements.
    a. Conductors for fire alarm systems shall not be installed in the same raceways, cables, or electrical boxes as the light and power conductors.
    b. It is permissible to install fire alarm conductors in the same raceways, cables, or electrical boxes as the light and power conductors because the conductors for fire alarm systems covered in *Article 760* are small and would easily fit into other electrical raceways and boxes.

18. Fire alarm cable generally used in more complex residential fire alarm and security systems is marked (FLP), (low-voltage bell wire), or (THHN). Circle the correct answer.

19. When installing a fire alarm system or a complete security system package in a home, always follow the installation instructions from:

   a. the manufacturer of the product and applicable codes.

   b. your neighbor, because he knows a lot about codes and standards.

   c. the man at the home center who sold you the product.

Home of the Future

After studying this chapter, you should be able to:

- Define the purpose and use of UPNP device.
- List the various types of devices that will be networked and automated in the home of the future.
- List and describe the various devices in the Xanboo product line, and discuss how they are controlled.
- Describe the future of home game consoles.
- Explain how the use of GPS and VPN is changing how individuals will access the Internet as well as their home computer.

OBJECTIVES

Introduction
Universal Plug-and-Play
Home Automation
Video
Home Game Consoles
Wireless Roaming
The Car
Out There
Pen Tablets: Notebooks of the Future?

OUTLINE

Introduction

The home of the future is not that far in the future. In fact, it is here now—you just have to know where to find the equipment to make your home the sci-fi home of your dreams. Essentially, when you think about the electronic devices available for your home and how they integrate with your home network you should consider the following:

- Video and home entertainment systems
- Security, fire, and protection systems
- Telephone and voice communications
- Home automation of such things as lighting, heating, and cooling
- Home office and home education systems
- Transportation such as your car, truck, and/or motorcycle.

Everything from television sets and set-top boxes to your refrigerator can now be connected to your home network. This chapter introduces some of the new, networked home appliances, digital phone systems, security systems, home automation systems, and on-line entertainment systems. Kids and game enthusiasts will be excited to learn about which game systems allow them to play with others across the Internet or over their home network.

Universal Plug-and-Play

To create the interconnected world of the future, a standard was required. Equipment vendors of all types—home security companies, computer manufacturers, consumer electronics companies, and companies that make products such as printers, digital cameras, and home appliances—formed an organization to create such a standard. The standard is known as **Universal Plug-and-Play (UPNP) (www.upnp.org).**

What is UPNP? UPNP is software that has the ability to automatically recognize devices (such as computers, scanners, appliances, etc.) when they are connected to a "residential gateway" so that these devices can communicate with each other.

Cisco and 3Com have postponed their plans to develop the **residential gateways (RGs)** for now. They believe that consumers are more worried about sharing printers and scanners than appliances. Residential gateways will be crucial components for the connectivity of the different appliances and equipment around the house.

New wireless products are now under development using a new chipset that incorporates both Wi-Fi and Bluetooth technologies.

Home Automation

Cisco, a large network equipment manufacturer, has worked with familiar home-appliance companies such as Whirlpool to develop next-generation products and services. Many of the new appliances being developed for the home will have the ability to communicate with each other over a home network and over the Internet to perform tasks such as automatic software and firmware upgrades, service requests and diagnostics, and less "self-centered" tasks such as shopping for food (Figure 22–1) or setting your air-conditioning system based on developing weather patterns.

The Korean company, LG Electronics Inc. **(http://www.lge.com)**, is making news with its new line of Internet-ready home appliances. The following sections describe some of the products.

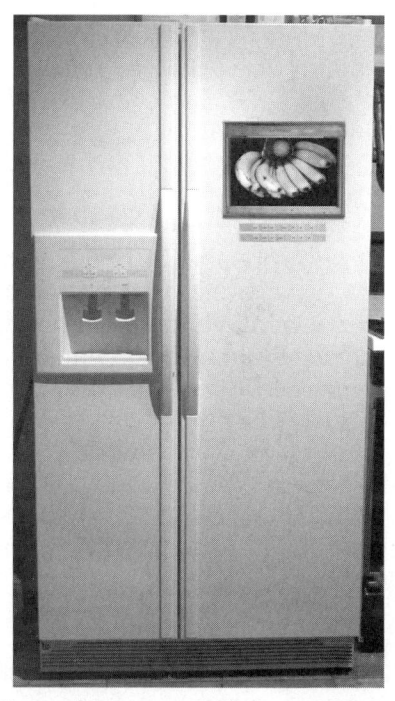

FIGURE 22–1 Next generation refrigerators will do your shopping.

Digital Multimedia Side-By-Side Fridge Freezer with LCD Display

What more can you expect from your refrigerator other than to store your food, keep it fresh, and possibly supply some chilled water and ice? Well the GRD-267DTV fridge from LG Electronics can do that and more. This high-tech refrigerator is not only Internet-ready but is PDA and cell-phone compatible. You can call this unit from the store and ask it to tell you what you need.

This high-tech refrigerator maintains an automated inventory that can tell you what you have without opening the door. No more hunting for expiration dates or moldy cream cheese—it also keeps track of how long food has been stored. Believe it or not this unit will even warn you when you are running low in a product—for example, when you are almost out of milk.

Also, it is not just for keeping things cold! It comes with a 15-inch LCD display, stereo, and digital camera. You can watch your favorite films on this cable and Internet-ready, radio-ready entertainment system and can even play MP3s. These are only some of the fun features. We have not even mentioned the state-of-the-art technology to keep your food fresh.

Air-Conditioning System

You can control the temperature of your house over the Internet or just call on the phone and add a few degrees for those cold winter nights. You have complete remote control of your home's climate-control system by network or cell phone. You can turn the unit on and off or make adjustments remotely. The LG Electronics air conditioner can also contact you when the self-diagnostics software senses a problem. Your home network can expand to allow devices like your air conditioner to be a part of your home network, and your cell phone to act as a network terminal.

Gadgets

The imagination is the only limit to the types of household appliances that either now connect to your home network or will soon. LG Electronics is already working on an Internet-ready microwave and washing machine. Games and home-entertainment systems are already attachable to your home network. It does not take much imagination to see what is possible. Enjoy the following scenario, for example.

You are driving home from work and you call the house to start the dinner you placed in the oven cooking. The thermostat also adjusts a couple of degrees and the refrigerator cuts in to remind you to pick up some butter. You drive into your driveway and the garage-door remote sends your private key to the garage-door opener, which signals the home-security unit to disengage and alerts the network to your arrival.

The home security system knows you are in the garage and senses the door entry into the house. A soft voice asks you if you would like to hear your messages or read them in e-mail. A biosensor notices that your heart rate is elevated and adjusts ambient music and lighting to add a calming effect. You sit down to scan your e-mail and notice that your dog has logged in and sent you an instant message alerting you to the fact that he has not been fed, and that he successfully guarded the house against intruders.

The previous example is largely fiction, but Xanboo Inc. **(http://www.xanboo.com),** a New York-based company, claims to have "created the first smart home management system with real-time video notification and device control that enables users to control and monitor their home or businesses from anywhere in the world, via the Internet." The Xanboo products are used to monitor the systems in your home. This is not a security system, but it may help determine when something in your house is amiss. Some of the products that Xanboo is currently selling are

- **Xanboo System Controller** This is the main unit, which connects to a PC and receives and processes the signals for all of the other Xanboo products.
- **Xanboo Color Video Camera** This video camera has a built-in motion sensor and microphone. It can take video or still pictures every ten seconds.
- **Xanboo Acoustic Sensor** The sensor detects sound between 70 and 100 decibels.
- **Xanboo Door/Window Sensor** This sensor monitors when doors or windows are opened.
- **Xanboo Power On/Off Sensor** This sensor detects when there is a power failure in your appliances or machines.
- **Xanboo Temperature Sensor** This sensor monitors the temperature of the surrounding area where the sensor is situated.
- **Xanboo Water Sensor** This sensor detects if the water level has risen higher than 1/8 of an inch. It is used to detect leaky refrigerators, washing machines, dishwashers, broken pipes, and so on.

All of the Xanboo products are controlled by the system controller, which is connected to a PC via a USB cable. The Xanboo software processes all of the information from the system controller. Information from the sensors is available to the owner over the Internet through the Xanboo Web site. Xanboo requires a small fee to access your sensors remotely through their site. Xanboo is preparing a new line of control products that will be available soon. These products are less passive and will allow remote users to control systems in their home over a network. Some of the new products will be: Power On/Off Control Switch, Temperature Control, Water Control, Garage-Door Control, and Pool Shark (monitors pools PH, temperature, and excessive movement).

Video

The ability to broadcast high-quality home video over your broadband network connection is becoming a reality. DSL providers will soon be competing with cable companies to provide you with quality digital programming. You will soon hear of standards such as MPEG-4 and H.26L, which offer video compression as high as 400 percent greater than current MPEG-2 standards in current use according to a March 26, 2002 article on CNN.com.

Some cars already have DVD players installed. It is only a matter of time until movies can be downloaded via satellite straight to the car.

Home Game Consoles

Just as with other familiar home appliances, video games will also be an integral part of your home network. Home game consoles like Sony's Playstation 2 and Microsoft's Xbox are introducing network capabilities to the console market. You can play against other people on your own network or over the Internet.

Neither of these two major gaming consoles is Internet ready, but both can be used over your home network.

Sony's PS2 Broadband, (not available in the United States at the time of publication) allows Playstation 2 users to download movies, games, and music from the Internet. Using the Playstation 2 USB ports you can connect to a USB network adapter or modem. You will have to refer to the console's technical documentation to determine which network adapters are compatible.

The Xbox Gateway is a third-party Internet service that enables Xbox users to interconnect using special software that acts as an Ethernet gateway between game consoles. This allows games with System Link functionality to be played between users over the Internet. The Xbox has an Ethernet port built-in, ready to use and connect to other computers or Xboxes.

Nintendo's Gamecube will soon enter the market for broadband games. You will be able to buy an adapter for your Gamecube, choosing between an Ethernet adapter and a 56K modem adapter.

As toys become part of the network, you might be able to control them from your computer. Imagine the following:

> *You:* "Kids it's 10 PM, time to turn the game off and go to bed."
>
> *Kids:* "But we're in the last level, it'll only take three more hours to finish the game."
>
> *You:* "Ok, save the game because I am shutting it down in five minutes." Five minutes later you reach into your Control Panel, you see the game connection, and CLICK!—disconnected.
>
> *Kids:* "DAD!!!"
>
> *You:* "Good night."

Wireless Roaming

Imagine you have installed an 802.11 wireless network in your house that allows you to grab your laptop and walk around the house unhindered by network cables, perhaps even into your own backyard—but then what? There is a new system being built that will allow your 802.11 devices to roam while directly connected to the cellular telephone network. This connection is different from the low-speed wireless connection you can achieve with wireless modems. This is a direct wireless network connection exactly like the one you have in home networks.

Currently available only in Europe for the GSM-based phone system, this technology will soon migrate to the United States-based CDMA digital cellular system.

The Car

Your car is an extension of your home, and for those who commute in large cities, it probably feels like home. The automobile has remained relatively unchanged over the last 70 years. You have an engine, some lights, and a radio. Yes, it is true that cars are now computer controlled, and even the braking systems on cars are microprocessor controlled. It is only now, however, that we seeing a *real* revolution in auto technology.

Global Positioning Systems (GPS) guide drivers with maps, even vocally telling a driver when to make a turn to reach a destination. GPS systems are no longer one-way location finders. Systems like OnStar send help and human remote assistance by using the information from a built-in GPS. It will not be long before cars are self-guided by systems similar to GPS.

The car's audio system has changed over the years. It has migrated from AM to FM, from tape to CD. Now, your system can play traditional CDs or music CDs recorded on your home computer or play MP3 music files downloaded from the Internet. Some of the newest models can even play DVDs, displaying the image on small liquid-crystal displays.

Computers in the Car

PCs for the car are now available. When you have that nagging question and wish you were in front of your computer at home so you could log into the Internet, you will have the computer right there in the car. Fitting into the slot where you would find your car radio, these small PCs integrate much of the technology mentioned earlier. The PC will integrate with a GPS to give you driving directions, play CDs and MP3 files, and provide functions like retrieving your e-mail and other features you would expect from a regular computer.

There are ways the computer in your car can be connected to your network at home, even allowing access to the files on your home computer. You can create a **Virtual Private Network (VPN)** that will give you safe and secure access to your home network just as though you were right in the house (or garage).

Out There

The people who brought you Virtual Reality (VR) have not been sucked into a computer somewhere, living out their lives in a bit stream as some movies suggest. Instead they are working on ways to bring education and collaboration remotely

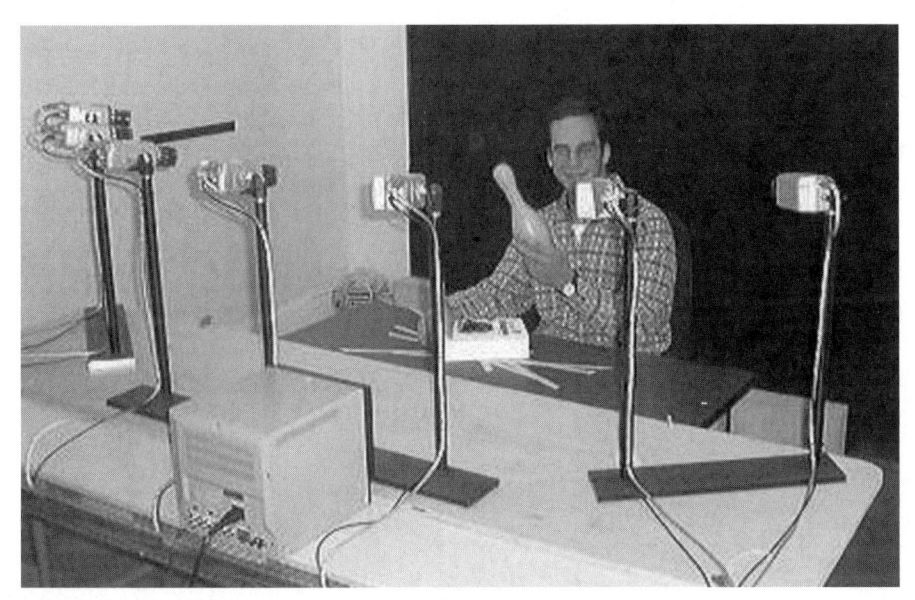

FIGURE 22–2 Immersion Workplace in Real Life

to places like your office or bedroom. The technology, in its current state, requires a level of Internet access only available in universities and research laboratories like the one at NTII (**http://www.advanced.org**).

Figure 22–2 shows the workplace of the future—perhaps it is not too different from what you may have in your home office. The researcher in the photo is holding a vase that will be included as an object in the virtual world shown in Figure 22–3. Notice that the bookcases and filing cabinets shown in the virtual world do not exist in the physical world.

FIGURE 22–3 Immersion Workplace as Seen in a Virtual World

TABLE 22-1
Comparison between Fujitsu Notebooks and Pen Tablets

Component	Pen Tablets		Notebooks	
	Stylistic 3500R	*Stylistec LT P-600F*	*E series*	*P-1000*
Processor	500 MHz Intel Celeron	600 MHz Intel Pentium III	700 MHz Crusoe TM5500	1.2 Ghz-M Mobile Intel Pentium III
Video	10.4" reflective SVGA TFT	8.4" transflective SVGA TFT	8.9" wide-format XGA TFT with touch screen	14.1" XGA TFT
Memory	256 MB SDRAM	256 MB SDRAM	128 MB memory	256 MB memory
Hard drive	15 GB shock mounted	15 GB	20 GB	30 GB
Modem	56K modem	56K modem	56K modem	
Network card	10/100 Mbps Ethernet	10/100 Mbps Ethernet	10/100 Mbps Ethernet	10/100 Mbps Ethernet
Dimension	11.0" (W)	9.6" (W)	9.1" (W)	12.13" (W)
	8.5" (D)	6.3" (D)	6.5" (D)	10.39" (D)
	1.1" (H)	1.1" (H)	1.36" (H)	1.52" (H)
	3.2 lbs.	2.65 lbs.	2.5 lbs.	5.8 lbs.
Other features	Sound Blaster	Sound Blaster	Sigma Tel STAC 9723 16-bit stereo audio	Sigma Tel STAC 9723 16-bit stereo audio
	ATI Rage Mobility	ATI Rage Mobility		
	USB ports	USB ports	Two stereo speakers	Dolby Headphone utility
	Stereo headphone jack	CardBus PC card slot	ATI Rage Mobility	
		Compact Flash slot	USB ports	ATI Mobility Radeon D
	Microphone jack	Microphone jack	Only Type II or Type II card slot	USB ports
	Floppy drive port	Headphone jack		
	Keyboard port	Wireless IR keyboard port	32-bit PC CardBus card slot	Support for external monitor and internal display
	Serial port	Keyboard port	External USB floppy drive	External USB floppy drive
	Mouse port	Mouse port	External PCM-CIA CD-ROM	DVD/CD-RW combo
			Headphone jack	Wireless mouse
			Microphone jack	Full-size keyboard
				Two Type II or Type II card slots
				32-bit PC CardBus card slot
				Embedded Smart Card Reader
Price	$3,659	$3,659	$1,499	$2,399

Pen Tablets: Notebooks of the Future?

Companies such as Fujitsu are developing pen tablets with the same capabilities that notebooks had only a year ago. Pen tablets are starting to become very popular. They are thin, light, and compact, and they are becoming more powerful with each new generation. The latest Fujitsu Stylistic 3500R pen tablet has a 500-MHz Intel Celeron processor, 256 MB of SDRAM and a 15-GB hard drive. The tablet is equipped with a 10.4-inch Reflective SVGA LCD display, a built-in modem, and a 10/100 Ethernet adapter. Also included are all the ports to connect a mouse, keyboard, and USB devices. The only big difference between a tablet and a notebook computer so far is the price (see Table 22–1). As soon as tablets come down in price they may very likely replace notebook computers. The pen tablet is the type of computer you currently see used in the 24th century sci-fi TV show, Star Trek.

A related technology that is improving is handwriting-recognition software. The Microsoft Office *XP*® product offers a handwriting recognition add-in product.

Another holy grail for human interface with the computer is voice recognition. Faster hardware has enabled voice recognition to finally become ready for prime time. It is strange enough to see people walking down the street apparently talking to themselves as they speak into hands-free cell phones. Think how strange it will be when people start talking to their notepad computers!

Summary

- This chapter has discussed some of the latest technologies and some that are still on the near horizon.
- It is smart to consider new technologies that may be available over the next three to five years when designing and installing a new home network.
- Fiber-optic cable may be required in the home in the not too distant future.
- Very soon, almost all home devices will be able to communicate over a standard TCP/IP home network.

Key Terms

Residential gateways (RGs) Devices that enable communication among networks in the residence and between residential networks and service providers' networks.

Universal Plug-and-Play (UPNP) A set of computer network protocols promulgated by the UPnP Forum. The goals of UPnP are to allow devices to connect seamlessly and to simplify the implementation of networks in the home and corporate environments. UPnP achieves this by defining and publishing UPnP device control protocols built upon open, Internet-based communication standards.

Virtual Private Network (VPN) A private communications network usually used within a company, or by several different companies or organizations, to communicate over a public network. VPN message traffic is carried on public networking infrastructure (e.g., the Internet) using standard (often insecure) protocols, or over a service provider's network providing VPN service guarded by a well-defined Service Level Agreement (SLA) between the VPN customer and the VPN service provider.

Review Questions

1. What is Universal Plug-and-Play (UPNP)?
2. Explain the purpose of the residential gateway.
3. Describe how the future of home automation will change the use of many basic appliances.
4. What are the benefits of installing a Xanboo system?
5. Explain how connection to a GPS system is going to change access to the Internet and a home computer.

Glossary

1394 Trade Association A nonprofit industry association dedicated to promoting the proliferation of IEEE 1394 technology in the computer, consumer, peripheral, and industrial markets. The 1394 TA supports several working groups that issue technical specifications and bulletins relating to applications of IEEE 1394.

Absorption That portion of fiber-optic attenuation resulting from conversion of optical power to heat.

Access lines A telecommunications circuit provided by a service provider at the demarcation point.

Access network The facilities that connect the residence to the long-haul networks (telephone and Internet).

Access point (AP) A device that connects wireless communication devices together to form a wireless network.

ACR Attenuation to cross-talk ratio; a measure of how much more signal than noise exists in the link by comparing the attenuated signal from one pair at the receiver to the cross-talk induced in the same pair.

Adapter A mechanical device designed to align and join two optical connectors.

Advanced Mobile Phone Service (AMPS) A high-capacity analog land mobile communication system that employs a cellular concept.

American National Standards Institute (ANSI) Oversees voluntary standards in the United States.

Approved (See *NEC® Article 100*): Referring to equipment, devices, raceways, and other electrical materials that are acceptable to the AHJ for installation and use.

Aspect ratio The ratio of the width to the height of an image—typically expressed as a ratio of whole numbers, such as 16:9.

Attenuation The reduction in optical power as it passes along a fiber, usually expressed in decibels (dB); the reduction of signal strength over distance.

Authority having jurisdiction (AHJ) (See *NEC® Article 100*): The local governmental authority that is charged with the regulation of construction projects. The AHJ may be a city, county, or sometimes a state organization, that is usually answerable to a legislative body that assigns the authority.

Auxiliary Disconnect Outlet (ADO) A connector that allows incoming services to be easily disconnected for troubleshooting or other reasons.

Backbone Cable that connects communications closets, entrance facilities, and buildings.

Balun A device that connects a balanced transmission line (such as UTP) to an unbalanced transmission line (such as coaxial cable). Balun is a contraction of balanced-unbalanced.

Bandwidth The frequency spectrum required or provided by communications networks; the range of single frequencies or bit rate within which a fiber-optic component, link, or network will operate.

Base station Part of a wireless network, consisting of a transmitter, receiver, controller, and antenna system.

Bit error rate (BER) The fraction of data bits transmitted that are received in error.

Bluetooth A uniting technology that allows any sort of electronic equipment from computers to cell phones to make its own connection without any wires.

BNC Bayonet CXC connectors.

Bonding The permanent joining of metallic parts to form an electrically conductive path that will ensure electrical continuity and the capacity to conduct safely any current likely to be imposed.

Boxes Housings in the electrical circuit that contain splices and terminations. Boxes can be metallic or nonmetallic and may house devices or simply contain conductors, but they provide a barrier between the electrical system of the structure and the living or working space of that structure.

Box makeup The act of preparing the conductors contained in a device box for the installation of a device. Makeup is accomplished with the intent of making subsequent installation of the device as easy as possible.

Bridge A network device that is used to connect and disconnect computer clusters operating in two separate domains. The two domains are normally not networked and are only temporarily connected through the bridge when communication is required.

Building permit A document issued by the AHJ, usually for a fee, that allows the construction of, or addition to, a building or other structure. In most areas of the country, it is not legal to begin construction before a permit has been obtained.

Bus topology A method of computer networking that connects each mode of the network to a common line (bus). The bus represents a broadcast environment where all devices on the bus hear all transmission simultaneously.

Capacitance The ability of a conductor to store charge.

Carbon monoxide alarm/detector A device that detects the presence of the toxic gas.

CATV An abbreviation for community antenna television, usually delivered by coax cable or hybrid fiber coax (HFC) networks, or cable TV.

Category 3 (CAT3) The UTP cable specified for signals up to 16 MHz, commonly used for telephones.

Category 5E (CAT5E) The UTP cable specified for signals up to 100 MHz, commonly used for all LANs.

Category 6 (CAT6) The UTP cable specified for up to 250 MHz.

CCTV Closed-circuit television, commonly used for security.

Cell Used in two different contexts: a span of coverage in a wireless network, and a block of data in data transmission.

Cellular Digital Packet Data (CDPD) In a wireless network, it allows for a packet of information to be transmitted in between voice telephone calls.

Cellular network Any mobile communications network with a series of overlapping hexagonal cells in a honeycomb pattern.

Change order Authorization to proceed with a change—an addition or a deletion—to scheduled work. In many cases a change order takes the place of a work order.

Channel The end-to-end transmission path between two points at which application-specific equipment is connected.

Cladding The lower refractive index optical coating over the core of the fiber that "traps" light into the core.

Class 1 circuit Class 1 circuits are defined by *Article 725* of the *NEC*®. There are two types of Class 1 circuits: Class 1 remote-control and signaling, and Class 1 power-limited. Class 1 remote-control and signaling has no power limit and can operate up to 600 volts. Class 1 power-limited is limited to 30 volts and 1000 volt-amps.

Class 2 circuit Class 2 circuits are defined by *Article 725* of the *NEC*®. In general, Class 2 circuits are power limited up to 100 volt-amps. Class 2 circuit conductors must be rated up to 150 volts. Current limitations do change based on voltage levels, see Chapter 9, *Table 11* (*a*) and (*b*) of the *NEC*®, which defines the specific voltage ranges and current limitations of inherently limited and non-inherently limited, AC and DC, Class 2 circuits and power supplies.

Class 3 circuit Class 3 circuits are defined by *Article 725* of the *NEC*®. In general, Class 3 circuits are power limited up to 100 volt-amps. Class 3 circuit conductors must also be rated up to 300 volts and must not be smaller than 18 AWG. Current limitations do change based on voltage levels, see Chapter 9, *Table 11* (*a*) and (*b*) of the *NEC*®, which defines the specific voltage ranges and current limitations of inherently limited and non-inherently limited, AC and DC, Class 3 circuits and power supplies.

Client/Server A client/server network designates one node of the system as the main file server and system resource for other connecting nodes of the network.

Coaxial (coax) cable A cable in which a single center conductor is surrounded by a dielectric material, and a cylindrical shield that is often composed of layers of foil and metallic braid. Coax has excellent high-frequency characteristics and is most commonly used for cable television signals.

Code Division Multiple Access (CDMA) A wireless digital technology that allows multiple users to share a single frequency by encrypting each signal with a different code.

Communications Assistance to Law Enforcement Act (CALEA) A law passed by Congress in 1994 that requires telecommunications carriers to ensure that their digital and wireless networks can comply with authorized wiretaps and electronic surveillance.

Glossary • 431

Communication circuits and systems Covered by *Articles 800, 810, 820,* and *830* of the *NEC®*. They include telecommunication circuits, security and alarm circuits, radio and TV antenna systems, CCTV and radio distribution systems, and network-powered broadband.

Communications infrastructure The cabling systems and their associated pathways that support the distribution of information throughout the home.

Conduit bodies (See *NEC® Article 100*): A type of raceway fitting that allows for a rapid change in direction of wiring, or for the tapping of a raceway, without the use of a box. Conduit bodies have removable covers that allow access to the interior of the fitting to facilitate the installation of the conductors.

Construction drawings A series of drawings, sometimes called *blueprints*, that show the intended design of a building or other structure. There are usually different construction drawings for each of the different trades involved in the construction process.

Construction process Collectively, the procedures followed by the various trades for the successful completion of the construction project.

Convergence The merging of telephony, computing, and television into a single digital network.

Copyright Ownership of intellectual property within limits set by law, which typically provides the copyright holder with exclusive rights to print, distribute, copy, and make derivative works for a limited time period.

Cross-talk Signal coupling from one pair to another.

Current loop Transmission using variable current to carry information as in a simple analog telephone.

Cut notches Sections along the edge of a joist or stud for the installation of cables or raceways. Notches must be cut with a saw, and usually the opening must be covered with a plate to protect the cable or raceway. Notches must meet the requirements of the *NEC®* and the UBC.

Cut sheets (*See* shop drawings.)

Daisy-chain A method of cabling where end points are connected directly together without returning to a common point.

Damper A device, such as a valve or movable plate, that controls the flow of air in a duct.

Delay Scew The maximum difference of propagation time in all pairs of cable.

Demarcation point A point at which operational control or ownership changes.

Device boxes Electrical boxes intended to house and make available devices such as switches and receptacles.

Digital home controller (DHC) A device that can control the operation of DHRs and the transfer of content between DHSs and DHRs.

Digital home renderer (DHR) A device that renders content from the DHS.

Digital home server (DHS) A device that serves as a source for content—either streaming content or for file transfer.

Digital Living Network Alliance (DLNA) An alliance of leading companies in the consumer electronics, mobile and personal computer industries. Its aim is to align the companies and have industry standards, which will allow products from all companies to be compatible with each other and to enable a network of electronic devices in the home.

Digital multimeter (DMM) Modern multimeters are exclusively digital, and identified by the term DMM or digital multimeter. In such as instrument, the signal under test is converted to a digital voltage and an amplifier with an electronically controlled gain preconditions the signal. Since the digital display directly indicates a quantity as a number, there is no risk of *parallax* causing an error when viewing a reading.

Digital Rights Management (DRM) Technology to enable copy protection and secure distribution of copyrighted material.

Digital Subscriber Line (DSL) or **cable modem** A technology that uses the existing copper loop plant to provide broadband access by using frequencies above the voice band.

Digital Transmission Content Protection (DTCP) A user-friendly copy protection scheme used by IEEE 1394 networks.

Digital Visual Interface (DVI) An uncompressed digital video signal standard developed for the interface between a computer and a monitor.

Direct-Sequence Spread Spectrum (DSSS) A spread-spectrum technique that resists interference by mixing in a series of pseudo-random bits with the actual data.

Distribution Device (DD) A facility within the residence that contains the main cross-connect or interconnect where one end of each of the outlet cables terminates.

Domain Name Server (DNS) Converts the numbered Internet address code of a URL to a series of recognizable words.

Drilled holes Holes that are drilled in framing members for the installation of electrical conductors or cables. Holes that are drilled in studs or joists must meet the requirements of the *NEC®* and the UBC.

E911 service Enhanced 911 (E911) service automatically reports the telephone number and location of 911 calls when they are connect to a PSP.

Ethernet The first computer networks, known as Ethernets, were developed in the 1970s by Xerox, which ran at maximum speeds of 2.94 Mb/sec on coaxial cable. Xerox soon joined with Intel and Digital Equipment Corporation to develop a standard for a 10-Mb/sec Ethernet, known as DIX (Digital, Intel, Xerox). Later, the development of the 10Base-T Ethernet in 1987 communicated at speeds of 10 Mbps on a twisted pair line; such networks could reliably transmit over a 100-m run.

Entertainment server A device used to store digital content (audio, video, photos, etc.) that can be accessed by any other device on the home network.

Environmental airspaces Airspaces within a structure that are intended as part of the structure's heating and cooling systems. These areas are used to circulate the air through the furnace or air conditioner and then return the air to the rooms of the building. Wiring within these areas is restricted because of the possibility of the rapid spreading of fire through the air-handling system should a fire occur.

F connector The standard connector for community antenna television (CATV)

Fair use Provisions in the copyright laws that provide some allowed uses for copyrighted works, such as quoting brief passages in a review.

FCC Federal Communications Commission; oversees all communications issues in the United States.

Fiber Distributed Data Interface (FDDI) A fiber distributed data interface (FDDI) uses optical fiber to transport high-speed data over a computer network.

Figure of merit The ratio of a satellite's receiver antenna gain to the system noise temperature.

File Transfer Protocols (FTP) Used to upload or download files to or from network servers.

Fine print note (FPN) A type of entry in the *NEC*® that provides explanatory information but is not formally enforceable as part of the *NEC*®.

Firewall Any of a number of security schemes that prevent unauthorized users from gaining access to a computer network or that monitor transfers of information to and from the network.

FireWire® FireWire®, also known as iLink®, is a high-bandwidth, serial communication standard developed by IEEE (IEEE 1394).

Flag note An icon used on construction drawings that alerts the construction team to the existence of additional information or requirements listed elsewhere on the drawings.

Frequency Division Multiple Access (FDMA) The division of the frequency band into channels allocated for wireless cellular communication.

Frequency-Hopping Spread Spectrum (FHSS) A spread-spectrum technique that resists interference by moving the signal rapidly from frequency to frequency in a pseudo-random way.

Gateways Used to interconnect incompatible computer systems, e-mail systems, or any network devices that would otherwise be unable to communicate directly with each other.

Geostationary orbit (GEO) satellite A satellite that circles the equator without inclination and whose rotational period matches the Earth's own.

Global System for Mobile Communications (GSM) A second-generation global wireless system.

Graded index (GI) A type of multimode fiber that uses a graded profile of refractive index in the core material to correct for dispersion.

Grounding Establishing a conducting connection between an electrical circuit or equipment and the earth.

H.323 An ITU-T standard for transmitting multimedia communications (such as audio and video) over a packet-switched network.

Headend The main distribution point in a CATV system

Heat detector A device that detects heat and can be either electrical or mechanical in operation. The most common types are the thermocouple and the electropneumatic; both respond to changes in ambient temperature. Typically, if the ambient temperature rises above a predetermined threshold, then an alarm signal is triggered.

High-Definition Copy Protection (HDCP) A copy prevention scheme used by DVI.

High Definition Television (HDTV) A digital television system with horizontal and vertical resolution approximately twice that of SDTV and an aspect ratio of 16:9.

Home automation system (HAS) Equipment and infrastructure that support automatic control of home services, such as climate and lighting control, security and fire alarms, and so on.

Home-run A method of cabling where each end point is connected back to a central point.

Host Large computer used with terminals, usually a mainframe.

Hub A hub connects multiple user stations of a computer network through multiple ports. The hub does not make any decision or determination on the data and merely acts as a multiport repeater. This device fits into layer one of the OSI model.

Hypertext Transport Protocol (HTTP) The language that computers use when talking to each other for moving hypertext files across the Internet.

Impedance The AC resistance.

Improved Mobile Telephone System (IMTS) A full-duplex automatic switching system.

Inherently Limited An inherently limited power supply is clamped internally and unable to deliver more than a specific amount of energy to a load. Any attempt to push an inherently limited source past its maximum limit will cause it to either shut down or self-destruct in a safe manner.

Integrated Services Digital Network (ISDN) Integrated Services Digital Network (ISDN) is a dial-up service to the Internet. It does not offer a dedicated or always-on connection to a network like DSL.

Interlaced scan A video component or signal that scans even- and odd-numbered lines in alternating frames. The opposite of progressive scan.

Inspection The process whereby the AHJ enforces established installation and construction standards. The AHJ periodically reviews the progress of construction and either approves or rejects the quality of the construction and ensures compliance with minimum standards. These reviews are accomplished by physically inspecting the building, and the person who performs the inspections is usually referred to as an *inspector*.

Installation instructions The directions provided by the manufacturer concerning the procedures to be followed in preparing electrical equipment, devices, or other materials for use. These instructions must be closely followed in installing electrical materials; otherwise, unsatisfactory operation may result in fire or other safety hazards.

Institute of Electrical and Electronics Engineers (IEEE) Professional society that oversees network standards.

Instrumentation Tray Cable (ITC) is covered by *Article 727* of the *NEC®*. ITC specifications are limited to 150 V and 5 A, for sizes 22 to 12 AWG wire. Size 22 AWG is limited to 150 V and 3 A. ITC can only be used and installed in industrial establishments that are maintained and supervised by qualified personnel; it is not to be used for general-purpose installations or as a substitute for Class 2 wiring (Class 2 wiring is covered by *Article 725* of the *NEC®*).

International Residential Code (IRC) A set of minimum rules for the installation of building systems and the construction of new dwellings. It is the most widely applied standard for construction in the United States; however, it is not recognized in all areas of the country.

International System of Units (SI) The measurement system, sometimes called the *metric system* in the United States, defined by the use of meters and kilograms, that uses base 10 unit divisions (SI units). The SI system is widely employed in almost all areas of the world except in the United States. The *NEC®* uses the SI system as the primary measurement system, with the English system, defined by the use of feet and pounds, used as a secondary system.

International Electrotechnical Committee (IEC) Oversees international communications standards.

Integrated Services Digital Network (ISDN) Digital circuit for carrying voice, video and data simultaneously.

International Code Council (ICC) A membership association dedicated to building safety and fire prevention.

Internet Protocol (IP) A routing protocol that contains the addresses of the sender and the intended network destination.

Internet telephony Another name for Voice over Internet Protocol (VoIP)

Ionization A type of smoke alarm that contain a low-level radioactive source which supplies particles that ionize the air in the detector's smoke chamber.

J Hook A hook shaped like the letter "J" used to suspend cables.

Joist A framing member that makes up part of a flooring system.

Key system A small (often electromechanical) telecommunications switching system that allows a user to pick up or hold a call on any several lines.

Labeled (See *NEC® Article 100*): Referring to materials and equipment that have been found to meet certain requirements for safety and function by a testing agency recognized by the AHJ. An identification label, marking, or decal from the testing agency is attached to the materials to identify them as having been tested.

LAN (Local Area Networks) Refers to the linking of computers and users in a small geographical area, such as an office complex, school, or college campus.

LAN extenders Remote-access, multilayer switches that are used to filter out and forward network traffic coming from a host router.

Lighting outlet boxes Boxes that are designed to supply outlets for luminaires (lighting fixtures). These boxes are usually round in shape, can be metallic or nonmetallic, and have 8-32 threaded holes for the connection of the luminaire (lighting fixture) support hardware.

Listed (See *NEC® Article 100*): Referring to materials and equipment that have been found to meet certain requirements for safety and function by a testing agency recognized by the AHJ. The materials are placed on a list to identify them as having been tested.

Local area networks (LANs) The most widely installed LAN, specified in IEEE 802.3

Low-frequency effects (LFE) LFE is an abbreviation that is commonly used in describing an audio track contained within a 5.1 motion picture sound mix. The signal from this track, ranging from 10 Hz to 120 Hz, is normally sent to a subwoofer.

Mainframe A large computer used to store and process massive amounts of data.

Mandatory rules Describing the rules and procedures listed in the *NEC®* with the terms *shall* and *shall not*. These rules must be followed to comply with the requirements of the *NEC®*.

Messenger cable The aerial cable used to attach communications cable that has no strength member of its own.

Metal oxide varistor (MOV) The most commonly used type of surge suppressor for communication circuits.

Metric designator A dimension corresponding to trade size, as employed for equipment and raceways, measured using the SI system.

MHz Megahertz; millions of cycles per second

Minicomputer A mid-sized computer, usually fitting within a single cabinet about the size of a refrigerator, that has less memory than a mainframe.

Mobile Switching Center (MSC) A control center for wireless services that is connected to the local exchange for wired telephones.

Modem A device that allows a computer to communicate over an analog telephone line.

Modular jack A female connector for wall or panel installation; mates with modular plugs.

Modular plug A standard connector used with wire, with four to ten contacts, to mate cables with modular jacks.

Mulitmode fiber A device that removes optical power in higher-order modes in fiber.

National Electrical Code® **(*NEC®*)** A set of minimum rules for the design and installation of electrical systems and devices published by the National Fire Protection Association as document NFPA 70. It is the most widely applied standard for electrical installations in the United States; however, it is not recognized in all areas of the country.

National Fire Protection Association (NFPA) The organization devoted to fire safety and prevention that publishes, along with many other documents, the *NEC®*.

National Institutes of Standards and Technology (NIST) Establishes primary standards in the United States.

National Television Standards Committee (NTSC) The analog television format used in North America, as well as Central/South America, Japan, Korea, Taiwan, and the Philippines.

Network Refers to the interconnection of computers for the purpose of intercommunication and the sharing of information.

Network address port translation (NAPT) An extension to network address translation (NAT) that allows multiple port identifiers to be multiplexed into a single external IP address. NAPT is described in RFC 2663.

Network address translation (NAT) A method, often implemented in routers, that allows IP addresses to be mapped from one address realm to another. NAT is described in RFC 1631.

Network interface (NI) The point of interconnection between a user terminal and a private or public network.

Network interface device (NID) The point of connection between networks. In a residence, the location of the demarcation point.

NEXT Near end cross-talk; a measure of interference between pairs in UTP cable.

Notch plate A metal plate that is installed over a notch or over a hole that is closer than 11.4 in (32 mm) from the edge of a framing member in order to protect the cables or raceway.

Numerical aperture (NA) A measure of the angular acceptance of an optical fiber.

Occupational Safety and Health Administration (OSHA) A branch of the federal government that is charged with improving workplace safety and encouraging the establishment of safe workplace practices.

Open System Interconnection (OSI) The reference model that represents the primary architectural model for intercomputer communications.

Operating System The OS is the focus of a computing experience. It is the first software you will see when you boot up a computer, the software that will guide you through your applications, and the software that will safely shut down the computer when you are ready to end a session. The OS is referred to as a graphic user interface (GUI), meaning it uses pictures and icons to interact with the user. It is the software that allows programs to be launched and used, together with organizing and controlling the hardware.

Overcurrent protection The purpose of overcurrent protection is to protect cables and circuits from excessive levels of current flow that could result in a potential fire hazard. Examples of overcurrent protection include fuses, circuit breakers, or active electronic feedback circuits.

Parallel communication Sends multiple binary bits, all at one time, between two points in a communication system. In most cases, entire binary words can be sent simultaneously, helping to reduce the total transfer time and making parallel communications much faster than that of serial communication.

Passive link An optical fiber test configuration consisting of a duplex run of fiber and the pair of connectors on each end.

Patch panel A cross-connection using jacks and patchcords to interconnect cables.

PBX Private branch exchange.

Permanent link A test configuration for a link that excludes test cords and patch cords.

Permissive rules Referring to the rules and procedures listed in the *NEC* that are allowed but that are not strictly required. These rules are identified in the *NEC* with the terms *shall be permitted* and *shall not be required*.

Personal protective equipment (PPE) Safety equipment, such as hardhats, gloves, safety glasses, and work boots, that is provided by individual construction workers for their own use.

Personal Video Recorder (PVR) A device that records television signals onto a hard disk. Similar to a VCR, except with a large disk drive instead of removable tapes.

Phase Alternate Line (PAL) The analog television format used in most of Europe, Asia, Australia, and Africa.

Photoelectric A type of smoke alarm that has a light sensor that measure the amount of light in a chamber.

Physical addressing Defines how devices are addressed at the data link layer; it is similar to a zip code as opposed to an actual house address.

Pigtail (as referring to splices) An extra conductor that is added to a splice for connection to a device.

Pixels The smallest individually addressable units in a display or image. Short for "picture element."

Ping A standard troubleshooting utility available on most network OSs. Besides determining whether a remote computer is "alive," ping also indicates something about the speed and reliability of a network connection.

Pipe fill When installing electrical wiring within a raceway, the number of conductors to be run should be limited to a certain percentage of fill. Limiting the number of conductors within a raceway will allow for the easy addition and removal of future wires to avoid damaging those already existing. In addition, not stuffing the raceway to the maximum fill will allow room for the dissipation of heat. To calculate pipe fill, refer to Chapter 9, *Tables 1* through *5* of the *NEC*®, Pipe Fill.

Plaster ring An accessory that is used with square boxes to allow them to be used in flush installations.

Post Office Protocol 3 (POP3) Defines the method and protocol for retrieving e-mail from an e-mail server.

Power-limited circuits Power-limited circuits are limited in output and capacity. Power-limited falls into 3 varieties, Class 1, Class 2, and Class 3. Class 1 power-limited is limited to a maximum output power of 1000 VA and 30 volts. The limitations of Class 2 and Class 3 power-limited circuits are defined by Chapter 9, *Table 11 a* and *b* of the *NEC*®. In general, Class 2 and Class 3 power-limited circuits must not exceed 100 VA. The allowable voltages for Class 2 and Class 3 circuits may vary as high as 150 volts, with maximum current levels calculated based on the level of supply voltage.

Power-Limited Tray Cables (PLTC) A special type of listed and labeled, nonmetallic-sheathed cabling that is intended for use in cable trays of factories or industrial establishments. *Table 725.61* of the *NEC*® also states that PLTC may be used as an approved substitute for Class 2 or 3 wiring in general-purpose or dwelling locations. PLTC is rated for 300 V, 100 VA, and can be purchased in sizes 22 AWG through 12 AWG.

Power Sum NEXT Near end cross-talk tested with all pairs but one energized to find the total amount of cross-talk caused by simultaneous use of all pairs for communication.

Private branch exchange (PBX) A telephone system that can switch calls between internal lines and external lines to a central office.

Progressive scan A video component or signal that scans or displays each line of a video frame in sequence. The opposite of interlaced scan.

Protocol A formal set of rules governing the transfer and exchange of data through the network medium and individual layers of the OSI.

Pulse code modulation (PCM) Digital representation of an analog signal where the magnitude of the signal is sampled regularly at uniform intervals, then quantized to a series of symbols in a digital (usually binary) code. PCM is used in digital telephone systems and is also the standard form for digital audio in computers and various compact disc formats. It is also standard in digital video.

Punchdown block A connection block incorporating insulation displacement connections for interconnecting copper wires with a special insertion tool.

Rafter A framing member that makes up part of a roof support system.

Render To produce a version of the original content—typically on a video screen or audio output device.

Remote-control circuit Circuits that control other circuits. Control voltages often can vary from as high as 600 V to as low as 5 V, depending on the type of system. As an example, motor-control circuits typically will use 120 V on starter coils, but many other types of control circuits will operate from 24 V or less. Coiled relays, transistors, and silicon-controlled rectifiers (SCRs) often will be used to accomplish the task. Common uses for remote-control circuits are motor controls, elevators, conveyor systems, automated processes, and garage door openers.

Repeater Provides the function of passing data or radio transmissions from one system or antenna cell to another.

Residential gateways (RGs) Devices that enable communication among networks in the residence and between residential networks and service providers' networks.

Retrofit The installation of communications infrastructure into an existing residence.

RG Residential gateway

Ring One conductor in a phone line, connected to the "Ring" of the contact on old-fashioned phone plugs; a network where computers are connected in series to form a ring—each computer in turn has an opportunity to use the network.

Ripping The process of uploading the content of an audio CD and converting it to MP3 format.

RJ-11 6 position modular jack/plug.

RJ-45 A modular 8-pin connector, actually referring to a specific telephone application, but usually referring to the connector used in the TIA/EIA 568 standard.

Router A network bridge that is used to link entirely separate networks. A bridge connects clusters of computers within the same network, whereas a router is used to connect isolated networks.

Satellite earth station Establishes and maintains continuous communication links with all other earth stations in the system through a satellite repeater.

Scattering The change of direction of light after striking small particles that causes loss in optical fibers.

Schedule A layout, usually in the form of a table, that is part of the construction drawings and provides detailed information concerning materials, devices, or equipment to be installed. For example, detailed information about the various luminaires (lighting fixtures) to be installed is often conveyed using a schedule.

Scenes A set of predefined HAS actions that are executed in response to a single command. Similar to a macro in computer terminology.

Scope of work A construction document that describes the work that is to be accomplished and the companies or trades that are to complete the work. This document provides a framework for the other construction documents.

Screened twisted-pair (ScTP) UTP cable with an outer shield under the jacket to prevent interference.

SCTE Society of Cable and Telecommunications Engineers.

Sequential Couleur a Memoire (SECAM) An analog television format developed in France.

Serial communication The process of sending data one bit at one time, sequentially, over a communications channel or computer bus.

Session Initiation Protocol (SIP) An IETF standard protocol (in the OSI applications layer) for initiating an interactive user session involving multimedia elements such as voice and video.

Shielding Stops interference and cross-talk by absorbing magnetic fields.

Shielded Twisted-Pair Cable (STP) Character impedance of STP is typically 150 Ω. Although STP can provide more protection against noise and EMF, it is not widely used because of its incompatibility with most of the new modular connectors and terminators currently used in the industry.

Shop drawings/cut sheets Drawings, usually provided by the manufacturer of equipment or materials to be installed in the building or structure, that show details of fabrication, such as dimensions, color, and materials used.

Short conduit bodies A type of conduit body designed for use with smaller sizes of raceways but with limited bending space of the conductors. The length of short conduit bodies is considerably less than the length of a regular conduit body of the same raceway trade size.

Signaling circuit A signaling circuit is defined in *Article 725* of the *NEC®*. Signaling circuits are circuits that activate notification devices. Examples may include lights, doorbells, buzzers, sirens, annunciators, and alarm devices.

Simple Mail Transport Protocol (SMTP) A network protocol that allows for the sending of e-mail and messages from a user station to an e-mail server.

Simple Network Transfer Protocol (SNTP) Exists as a means of monitoring computers, routers, switches, and network usage.

Single-mode fiber A fiber with a small core, only a few times the wavelength of light transmitted, that allows only one mode of light to propagate; commonly used with laser sources for high-speed, long-distance links.

Smart appliance A machine built for a specific purpose (such as a refrigerator or washing machine) that contains a computer and Internet interface. Sometimes referred to as an *Internet appliance* or *network appliance*.

Smoke alarm/detector Safety device that detects airborne smoke and issues an audible alarm, thereby alerting nearby people to the danger of fire.

Soft handoff Allows a handset to communicate with multiple base stations simultaneously in a wireless network.

Specifications A construction document that usually controls the quality of the construction installations. It may list the brand names of the materials to be used during construction, certain procedures to be employed, and directions about communications between the various construction trades and the management team.

Splice The act of connecting two individual conductors together to form one continuous conductor, or the location of that connection. Splices usually occur in boxes and are accomplished using proper methods and materials.

Spread spectrum In a wireless network, users are separated by assigning them digital codes within a broad range of the radio frequency.

Square boxes Electrical boxes that are square in shape and can be metallic or nonmetallic. Square boxes can be used for flush and surface installations by employing plaster rings or industrial covers and are the most common type of box used in commercial electrical work.

Standard-Definition Television (SDTV) A digital television system with picture quality equivalent to CDTV.

Step index fiber A multimode fiber where the core is all the same index of refraction.

Stud A framing member that makes up a part of a wall. Framed walls also usually have a top plate and a bottom plate.

Structured cabling system (SCS) A cabling system that is designed to support a range of applications with standard interfaces and specified transmission performance.

Submittals Cut sheets or shop drawings submitted to an architect or engineer for approval prior to the equipment's installation in the building or structure. Submittals are usually required by the building specifications.

Subwoofer A type of loudspeaker dedicated to the reproduction of bass frequencies, typically from about 20 Hz to about 200 Hz.

Symbols Icons used on construction drawings to represent various design features such as switches, receptacles, and luminaires (lighting fixtures).

Telecommunication closet Obsolete (though still frequently used) terminology, replaced by *telecommunication room*.

Telecommunication outlet A connecting device (usually a modular jack or coaxial connector) in the home on which the outlet cable terminates.

Telecommunication room (TR) An enclosed space for housing telecommunications equipment, cable terminations, and cross-connect cabling; the location of the horizontal cross-connect. Formerly known as the *telecommunication closet*.

Teleworking Doing work outside the traditional workplace. This could be a traditional telecommuting environment where the employee works from home instead of traveling to an office, or it could involve working on the road while traveling, working at a customer's site, and so on.

TIA/EIA A vendor-based group that writes voluntary interoperability standards for communications and electronics.

Time Division Multiple Access (TDMA) A second-generation wireless digital technology that allows users to access the assigned bandwidth on a time-sharing basis.

Time domain reflectometer (TDR) A testing device used for copper cable that operates like radar to find length, shorts or opens, and impedance mismatches.

Tip One conductor in a phone line, connected to the "Tip" of the old-fashioned phone plug.

Token Ring Network (TRN) A type of network connection where each node of the network is connected in series with the next node through a connecting cable.

Topology Defines how computer networks are connected. Examples include token ring, bus, star, and tree.

Trade size A system employed in the *NEC®* to define certain standard sizes of electrical equipment and raceways. For example, 1.2-in. (16-mm) internal diameter trade size conduit can actually measure between .526 in. (13.36 mm) and .660 in. (16.76 mm).

Transport Control Protocol (TCP) One of the core protocols of the Internet protocol suite. Using TCP, applications on networked hosts can create *connections* to one another, over which they can exchange data or packets. The protocol guarantees reliable and in-order delivery of sender-to-receiver data. TCP also distinguishes data for multiple, concurrent applications (e.g. Web server and email server) running on the same host.

Tree topology A type of bus topology that connects nodal points of a computer network to multiple branches.

Uniform Resource Locator (URL) Provides the necessary protocols for finding and accessing Web sites or information files residing on the Internet.

Universal Plug-and-Play (UPNP) A set of computer network protocols promulgated by the UPnP forum. The goals of UPnP are to allow devices to connect seamlessly and to simplify the implementation of networks in the home and corporate environments. UPnP achieves this by defining and publishing UPnP device control protocols built upon open, Internet-based communication standards.

Universal Serial Bus (USB) A high-bandwidth data communication standard that can support multiple devices within an entire system.

Uninterruptible power supply (UPS) Alternating-current power supply with battery backup to keep critical applications running for a short time in the event of a power failure.

Universal Service Fund (USF) A fund, mandated by the FCC, that is designed to offset the higher cost of providing telephone service in some (primarily rural) areas and thus provide universal, affordable telephone service to all households.

Unshielded twisted-pair (UTP) cable Cable consisting of eight insulated copper conductors twisted together into four pairs without any screen or shields.

User Datagram Protocol (UDP) A connectionless network protocol that does not send back any type of acknowledgment to the originating transmitter regarding the status of data packets.

Utilization equipment (See *NEC® Article 100*): Equipment that requires electricity to function. Utilization equipment includes appliances and luminaires (lighting fixtures) but does not include raceways, boxes, or devices.

Virtual Private Network (VPN) A private communicatiosn network usually used within a company, or by several different companies or organizations, to communicate over a public network. VPN message traffic is carried on public networking infrastructure (e.g., the Internet) using standard (often insecure) protocols, or over a service provider's network providing VPN service guarded by a well-defined Service Level Agreement (SLA) between the VPN customer and the VPN service provider.

Voice over Internet Protocol (VoIP) A technology that allows voice calls to be digitized and transmitted over the Internet using the Internet protocol.

Waterproof boxes Electrical boxes designed to be used in wet and damp areas, such as outdoors, where moisture can enter the raceway system or the boxes themselves, thereby causing faulting problems.

Wide area network (WAN) A computer network covering a wide geographical area, involving a vast array of computers. This is different from personal area networks (PANs), metropolitan area networks (MANs), or local area networks (LANs) that are usually limited to a room, building, or campus. The most well-known example of a WAN is the Internet.

Wire connector A fitting allowing for a solderless method of splicing conductors.

WiFi Set of product compatibility standards for wireless local area networks (WLAN) based on the IEEE 802.11 specificiation.

Wireless A communications system in which electromagnetic waves carry a signal through atmospheric space rather than along a wire.

Wireless Application Protocol (WAP) A standard for wireless data delivery and communications over the Internet.

Wireless LANs (WLANs) Transmit and receive data over the air, minimizing the need for wired connections.

Work order An authorization to proceed with scheduled work. Work orders usually involve work that has been priced and contracted or that has been agreed to be completed on a "time and materials" basis.

Index

A
Absorption, 271
Access-control systems, 152–153
Access lines, 254
Access network, 5
Access point (AP), 349
Adapter, 97
Address Resolution Protocol (ARP), 45
Advanced Communications Technology Satellite (ACTS), 78
Advanced Mobile Phone Service (AMPS), 54
Alarm systems, 397–416
Analog access, 57–58
Analog cellular system, 67
Analog signals, 19
Analog telephony, 96–97, 100
Angle pulls, 208
 front-access, 209
Antennas, television, 387, 389–395
Application and process layer, 43–45
Approved, 168
Aspect ratio, 124
Attenuation, 330–331, 361
Attenuation to cross-talk ratio (ACR), 252, 363
Audio and music systems, 115–122
 cabling, 122, 374
 encoding, 116–117
 formats, 117–118
 sampling, 116
 storage capacity, 117
Audio circuits, 292
Authority having jurisdiction (AHJ), 164

B
Balun, 23
Bandwidth, 89
Baseband, 33
Base station, 54
Bending losses, 273
Bluetooth®, 64–66
 block diagram, 65
Blu-ray disks, 117, 137
Binary data packets, 30
Bit error rate (BER), 331
BNC (bayonet-mount connector), 322
Bonding, 262
Boxes, 194–219
 accessibility, 395
 for conductors 4 AWG and larger, 205, 209
 for conductors 6 AWG and smaller, 195–205, 207–208
 functions of, 194–195
 installing, 209, 212
 makeup, 222, 235, 237
 organizing, 239
 sizing, 205–209
 types of, 195–205
Breakout cable, 275
Bridge, 35–36, 257, 308
BridgeCo system, 6
Bridged connections, 98
Broadband, 33, 56
Broadband access, 3
Broadband wireless systems, 72–73
Broadcast, 34
 TV, 90
Building permits, 164
Bundling cables, 228, 232
Bus network, 31
Byte, 19

C
Cable modem, 17
Cabling, audio, 122
 TIA-570B recommendations, 123
Cabling standards, 244–265

Cables,
 bundling, 228, 232
 choosing, 274–278
 in closet or equipment room, 343–344
 environmental and safety issues, 260–264
 flammability classes, 260–262
 grades of, 256–259
 hardware specifications, 260
 mounting hardware, 343
 ratings and markings, 276
 roughing-in, five rules for, 237, 239
 run parallel, 230, 234
 substitutions, types of, 262–263
 support of, 296–297
 testing, 354–355
 types of, 255–256, 274–276
 unshielded twisted-pair categories, 248–253
Cable trays, 294, 342
Capacitance, 330, 331
Carbon monoxide alarms, 399–400, 410–412
 building codes, 409
 installation, 411–412
 maintenance/testing, 408–410
 standards, 415–416
Carrier sense multiple access with collision detection (CSMA/CD), 32
CATV (community antenna) networks, 92–93, 321, 373, 385
Cathode-ray tube display, 131
CCTV (closed-circuit), 321
CD capacity, 119
CEDIA demo system, 7
Cell, 54–55
Cell phone, 4, 100
Cellular, 54

439

440 • Index

Cellular Digital Packet Data (CDPD), 62–63
Cellular network, 54–56
Central office (CO), 96
Change orders, 172. *See also* Work orders
Channel, 247
    measurements, 248
Channel-surfing scenario, 142–143
Characteristic impedance, 330
Chrominance, 127
Circuits
    classification of, 283–284
    comparison of class 1, 2, and 3, 286
    separation of, 284, 291–292
Circuit switching, 38
Cladding, 268
Class 1
    composite cables, 296
    conductor ampacity ratings, 285
    identification and markings, 291
    insulation requirements, 285
    jacketed, 296
    manholes, 294
    overcurrent protection devices, 285
    power limited circuits, 284
    remote-control and signaling circuits, 283
Class 2 and 3
    associated systems, 293
    cable substitutions, 289
    enclosures with single openings, 293
    identification and marking, 291
    installation requirement for multiple circuits, 295
    length, 360–361
    manholes, 294
    power sources, 287–288
    reclassification of, 290–291
    separation requirements, 291, 293
    voltage and insulation requirements, 289
    wiring methods, 288
Classification of circuits, 280–299
Client/server network, 37–38
Climate control, 148–149
Closed-circuit TV (CCTV), 90
Closed-loop power distribution, 294
Closet-to-closet pulls, 341
Coaxial (coax) cable, 2–3, 91, 333
    connectors, 322–323
    construction of, 91
    for television signals, 372–374
    testing, 354
Code. See National Electrical Code®
Code Division Multiple Access (CDMA), 60–61
Collision detection, 32–33
Commercial Building Telecommunications Cabling Standard (TIA/EIA-568-B), 244–245

Communication circuits, 283, 292
    installation requirements for, 295
Communications Assistance to Law Enforcement Act (CALEA), 107
Communications infrastructure design, 367–383
Communications, technology of, 88–93
Composite cables, 296
Computer network addressing, 46–47
Computer
    in the car, 424
    Internet access, 395
    wiring for, 396
Conductors
    calculation of, in a raceway, 297–298
    connecting, 230, 233
    identifying, 235, 237, 239, 240
    interior communications, 307
    support of, 296–297
Conduit bodies, 219
Conduit cable pulls, 341–342
Conference of European Posts and Telecommunications (CEPT), 60
Construction drawings, 169–170
Construction process, 163–191
Consumer Electronic Association (CEA), 1
Content, 2, 113
Convergence, 5
Copyright, 114
Cross-connects, 257, 308
Crossed pairs, 358
Cross-talk, 330, 331, 363
    preventing, 332
Current loop, 89
Cut sheets, 171

D

Daisy-chain, 4, 97
Data applications, 100–110
Data errors, types of, 40–41
Datagram, 40
Data-link layer, 28
Data rate, 66
Delay skew, 252, 364
Demarcation point, 255
Device drivers, 47
Device boxes, 195, 199
    flush-mounted, 215
    surface-mounted, 212, 215
Digital access, 58–63
Digital Control Channel (DCCH), 60
Digital home controller (DHC), 113
Digital home renderer (DHR), 113
Digital home server (DHS), 113
Digital Living Network Alliance (DLNA), 113
Digital multimeter (DMM), 354
Digital rights management (DRM), 114–115

Digital signals, 19
Digital subscriber line (DSL), 3, 11
Digital television
    formats, 126
    standards, 125–126
Digital Transmission Content Protection (DTCP), 115
    proposal, 115
Digital Visual Interface (DVI), 115, 130
Direct Broadcast Satellite (DBS), 17–18
Direct-Sequence Spread Spectrum (DSSS), 62
Distribution cable, 274–275
Distribution device (DD), 5, 95, 316, 368–369
    requirements for location of, 369
Domain Name Server (DNS), 46
Downlink, 75
Drilling or notching studs, rafters, and joists, 222, 224, 226
DSL. *See* Digital subscriber line
DV, 129
DVD-audio (DVD-A), 117
DVD capacity, 119
DVD region codes, 129
DVD-video, 129

E

E911 service, 107
Electrical cabling installation, 221–241
    IRC requirements for, 222, 224, 226
    NEC® requirements for, 222
Electrical circuits
    types of, 281–282
Electromagnetic emissions, 264
Electromagnetic interference (EMI), 18
E-mail, 11
Encapsulation, 28
Encoding, 116
Environmental airspaces, 226, 228
Entertainment applications, 112–144
Entertainment network, 3
Entertainment server, 137–138
Equal-level far-end cross-talk loss (ELFEXT), 251, 364
Error notification, 28
Ethernet, 3, 11–12
    10Base-T, 13

F

Fair use, 114
Far-end cross-talk (FEXT), 364
FCC, 322
Fiber attenuation, 271
Fiber bandwidth, 272–273

Fiber-optic cable, 92, 273
    installation of, 276–278
    overbuild, 93
F connector, 322
Figure of merit, 76
    formula for, 77–78
File sharing, 102–103
File Transfer Protocol (FTP), 43, 44
Fine print note (FPN), 168
Fire alarm systems, 154
    building codes, 410
    exceeding minimum levels
        of protection, 409–410
    post-installation requirements,
        409
    requirements, 399, 412–413
    standards, 415–416
Firestopping, 343
Firewall, 37, 47
Firewire®, 4, 20, 22, 318
    diagram of, 7
Flag note, 170
Flash, 44
Flat panel display, 132
Flow control, 28
Footprint, 75
Format shifting, 115
Four-corner scenario, 141–142
Frequency Division Multiple Access
    (FDMA), 58
Frequency-Hopping Spread
    Spectrum (FHSS), 62
Future applications, 138–144
    air-conditioning system, 421
    automation, 420–423
    car, 424
    digital multimedia fridge
        freezer, 421
    documentation, 376
    flexibility, 375–376
    gadgets, 422–423
    home, 419–427
    home game consoles, 423
    pen tablets, 427
    residential gateways (RGs), 420
    universal plug-and-play, 420
    video, 423
    wireless roaming, 424

G
Gaming, 136–137
Gateway, 37
Geostationary orbit satellite
    (GEO), 78
Geosynchronous satellite (GEO),
    76–79
Global Positioning System (GPS),
    79–81
Global System for Mobile
    Communications (GSM), 60,
    84–85
Glossary of terms, 429–438
Graded index (GI) fiber, 269, 270, 272

Grounding, 262
Grounding electrode, 306

H
H.323, 105
Headend, 92
Heat alarm
    building codes, 410
    maintenance/testing, 408–410
    standards, 415–416
    types of, 402
    wiring, 405–408
Heat detector, 398–399
    installation requirements,
        403–408
Heating, ventilation, and air-
    conditioning systems,
    148–149
Hierarchical star topology, 245–247
    backbone system of, 246
    horizontal subsystem of, 246
High-Definition Copy Protection
    (HDCP), 115
High-Definition Television
    (HDTV), 125
Hoistways, 294
Hollow wall fasteners, 213
Home automation, 147–161
Home automation system (HAS), 5,
    147–148, 158, 374–375
    of the future, 420–423
    remote access to, 159
    scenes, 159
    security cameras, 375
    smart appliances, 375
Home networking, 1
Home Networking and Information
    Technology (HNIT), 1
Home-run, 4
Home systems controller (HSC), 157
Home security devices, 5
Home theatre, 133–138
    components of, 133–134
    special considerations for, 374
    surround sound, 134–136
Horizontal cable pulls, 337–340
Horizontal cable supports, 340–341
Host, 29
Hot, 174
Hub, 31, 35
Hypertext markup language
    (HTML), 43
Hypertext Transport Protocol
    (HTTP), 43–45

I
IEC 61833 Series, 129–130
IEEE 1394 LAN, 101, 115
iLink®, 4, 22
Impedance, 331, 360
Improved Mobile Telephone System
    (MTS), 54

Infrared (IR) waves, 54
    transmission, 334
Infrastructure, communications
        design process, 368–383
    documentation, 376, 381
    entertainment network,
        372–374, 381
    floor plan, example of, 378–382
    future flexibility, 375–376, 381
    home automation systems,
        374–379, 381
    placement of components,
        368–369, 379
    retrofit installations, 376–378
    six stages of, 368, 379–381
    voice and data network,
        370–372, 380
Inherently limited, 287
Insertion loss, 249
Inspection, 164, 172
Installation instructions, 170–171
Institute of Electronic and
    Electronic Engineers (IEEE), 11
Instrumentation tray cable (ITC),
    289–290
Integrated Services Digital Network
    (ISDN), 11
Integrated systems, 157–161
INTELSAT, 73–74
Interconnects, 257
Interlaced scan, 124
International cabling standards,
    264–265
International Code Council
    (ICC), 399
International Electrotechnical
    Commission (IEC), 264
International Mobile
    Telecommunication 2000
    (IMT-2000), 84
International Residential Code
    (IRS), 164
International Standards
    Organization (ISO), 264
International System of Units (SI),
    166, 168
International wireless
    communications, 83–85
Internet
    access to, 13–18, 101–102, 395
    addressing, 46–47
    telephony, 103
Internet Control Message Protocol
    (ICMP), 45
Internet Mail Access Protocol
    (IMAP), 45
Internet Protocol (IP), 19, 42
    layer, 27–28
Intrusion alarms, types of, 152
Ionization smoke alarm, 401–402
IRIDIUM communications,
    82–83
IS-136 specification, 60

Isochronous bandwidth allocation, 133
ITU-R BT.601/709, 128–129

J
Java, 44
J hook, 340–341
Jitter, 40
Job site organization, 165
Job site safety, 174
Joint Photographic Experts Group (JPEG), 134
Junction boxes. *See* pull boxes

K
Ka-band, 78
Key telephone system, 98, 99

L
Labeled, 168
LAN. *See* Local area network
LAN extender, 36–37
Latency, 40
Lighting control systems, 150–152
  benefits of, 150
  communication media for, 151
  technology, 150–151
Lighting outlet boxes, 199–200
Line-of-sight, 70, 71
Listed, 168–169
Load-side applications, 288–289
Local area network (LAN), 3–4, 11
  infrared, 71–72
  microwave, 69–70
  radio, 70–71
Local Multipoint Distribution System (LMDS), 72
Lockout and tag-out, 174
Loose tube cable, 275
Lost packets, 40–41
Low bandwidth, 40
Low-bandwidth bidirectional analog signal, 96
Low earth orbit (LEO) satellite, 81–82
Low frequency effects (LFE) speaker, 134
Luminance, 127

M
Macrocell, 56
Mainframe, 29
Mandatory rules, 166
Media Access Control (MAC)
  address, 19
  layer, 68
Medium earth orbit (MEO) satellite, 81–83
MHz, 90
Metal oxide varistor (MOV), 325
Metric designator, 168

Microcell, 56
Minicomputer, 29
Mobile Switching Center (MSC), 54, 55
Mobile Telephone System (MTS), 54–56
Modem, cable, 17
  dial-up, 11, 14
  high-speed, 100
Modular jack, 304
Modular plug, 304
Modular outlets, 308–309
Modulation, 14, 91
Modulator, 91
Motion Picture Association of America (MPAA), 114
Motion Picture Experts Group (MPEG), 128
Multicast, 34
Multichannel Multipoint Distribution System (MMDS), 72
Multimode fiber, 269
Multiple-player on-line role-playing game (MORG), 136
Multi-Station Access Unit (MSAU), 31

N
Narrowband, 56
National Association of Home-Based Businesses, 4
National Electrical Code® (*NEC*®), 33–34, 164, 166
  Article 100, 191
  Article 110.14(B), 230
  Article 250, 263
  Article 725, 280–299
  Article 300.22, 226
  Article 760, 412
  Article 770, 262
  Article 800, 260
  Article 810, 391–395
  Article 820, 324, 388–389
  Article 830.1, 33
  Article 830.179, 33–34
  Article 830.90, 34
  listing and labeling, 169–170
  measurements and, 166, 168
  NFPA 70, 398
  NFPA 72, 398, 403–408
  organization of, 166, 167
  rules for cable television, 388–389
  rules for installing antennas and lead-in wires, 391–395
  Section 725.51, 288–289
  Section 725.52, 288–289
National Fire Alarm Code®, 398–399
National Fire Protection Association (NFPA), 166
National Television Standards Committee (NTSC), 124

Near-end cross-talk loss (NEXT), 250, 354, 362
Network, 11
  cabling, 23–25, 332–333
  devices, 34–37
  history of, 11–13
Network address translation (NAT), 101
Network address port translation (NAPT), 101
Network cabling, 329–345
  handling, 335–336
  installation, 336–337
  power line carrier, 334–335
  pulling, 337–343
  requirements, 330–332
  types, 332–333
Networking, 10
Network interface (NI), 303
Network interface device (NID), 95, 255
Networking fundamentals, 10–49
Networking software, 29–30
  management, 37–38
Network layer, 27
Network Interface Card (NIC), 18–19
Network Operating System (NOS), 30, 38
NFPA markings for copper cable, 261
NFPA markings for optical fiber cable, 263
Non-power-limited circuits, 292
Nordic Mobile Telephone (NMT), 58
Notch plate, 222
Notes, 170
Numerical aperture (NA), 269

O
Occupational Safety and Health Administration (OSHA), 174
Open DECconnect, 357
Opens, 358
Open System Interconnection (OSI) model, layers of, 25–28
Operating systems (OS), 37–38, 48
Optical fiber, 18
  cable, 333
  circuits, 292
  specifications, 270
  types of, 269–270
Optical fiber and cable, 268–278
OSI. *See* Open system interconnection
Outlet locations, 259
Overhead, 66
  formula for, 67

P
Packets, 60
Packet switching, 39
Parallel communication, 22–23

Parallel port, 22
Pathways, 259
Peer-to-peer network, 37
Peripheral sharing, 102–103
Permanent link, 247
Permissive rules, 166
Permits, 172
Personal Communications Network, 56
Personal Communications System (PCS), 56, 58
Personal Computer Memory Card Industry Association (PCMCIA), 68
Personal protective equipment (PPE), 174
Personal video recorder (PVR), 114
Phase Alternate Line (PAL), 124–125
Photoelectric smoke alarm, 401–402
Physical addressing, 28
Physical star network, 24
Picocell, 56
Pigtail 222, 233, 235
Ping, 47
Pipe fill, 298
Pixels, 126
Plans, 169–170. *See also* Construction Drawings
Plaster rings, 202
Point-to-point links, 38
Post Office Protocol 3 (POP3), 43, 45
Power-limited circuits, 282, 291
Power-limited tray cable (PLTC), 289–290
Power line carrier networks, 334–335
Power sources, interconnection of, 288
Power-sum equal-level far-end cross-talk loss (PSELFEXT), 251
Power-sum near-end cross-talk loss (PSNEXT), 250, 362–363
Presentation layer, 27
Private branch exchange (PBX), 98
Programmed power distribution, 294
Progressive scan, 124
Propagation, 54
Projection display, 132
Propagation delay, 252, 364
Protocols, 26
Pulse code modulation (PCM), 116
Pull boxes
 flush-mounted, 215
 surface-mounted 215

Q
Qualified person, 191
Quantized, 116

R
Raceways, 342
Radio frequency (RF), 54
Receiver/transmitter, 75
Recording Industry Association of America (RIAA), 114
Regional differences, 173
 "Wireman's Guide" for, 173
Remote-control circuits, 282
Repeater, 35
Residential electrical cabling installation, 221–241
Residential gateway, (RG), 5, 420
Residential network, 1–7
 low voltage applications, 48–49
Resistance, 360
Retrofit installations, 376–378
 cables and pathways, 377
 design considerations, 377
 tools and techniques, 378
Return loss, 251, 360
Reversed pairs, 358
Ring, 309
Ring network, 30
Ripping, 117
RJ-11 and RJ-14 connectors, 309–310
RJ-46 pinouts,
 for Ethernet, 101
 for telephony, 97
Riser installation, 342–343
Router, 36
 protocols, 26

S
Sampling, 116
Sampling rate, 116
Satellite antennas, 389–391
 typical wiring for, 391
Satellite communications, 73–83
Satellite earth station, 76
Satellite network connection, 17–18
Scattering, 271
Scenes, 150
 examples of, 160–161
Schedules, 170
Scope of work, 171–172
Screened twisted-pair (ScTP) cables, 332
Security systems, 152–154, 375, 413–415
Sequential Couleur a Memoire (SECAM), 125
Serial communication link, 20, 21
Serial port, 21
Server, 29
Session Initiation Protocol (SIP), 105
Sessions layer, 27
Shielded twisted-pair (STP) cable, 13, 333
Shielding, 332
Shop drawings, 171
Short conduit bodies, 219
Shorts, 358

SI. *See* International System of Units
Signal attenuation, 18
Signaling circuits, 283
Simple Mail Transport Protocol (SMTP), 43, 45
Simplex cable, 274
Single-mode, 89, 270
Smart appliances, 155–157
 cabling for, 375
 connectivity for, 157
Smoke alarms, 398–399
 building codes, 410
 installation requirements, 403–408
 maintenance/testing, 408–410
 minimum level of protection 403–405
 standards, 415–416
 types of, 401–402
Smoke detectors, 398–399
 codes, 400–401
 installation, 399–400
 standards, 415–416
 types of, 400–401
Society of Cable Telecommunications Engineers (SCTE), 260, 322
Soft handoff, 61
Solid wall fasteners, 214
Speaker-level signals, 119
Speakers, types of, 120–122, 134
Specifications, 171
Speed of light, 80
Splice cap, 230
Splices, 222
Split pairs, 358–359
Spread spectrum, 61
Sprinkling system code, 415
Square boxes, 202, 2-4–205
Standard-Definition Television (SDTV), 125
Star network, 30
Step-index fiber, 270
Straight pulls, 208
Structural return loss, 360
Structured cabling system (SCS), 2, 95
Submittals, 171
Subwoofer, 134
Super-audio CD (SACD), 117
Supply-side applications, 288–289
Surround sound, 134–136
 configurations, 135
Switch, 36
Symbols, 170
SyncML, 66

T
1394 Trade Association, 6
TCP/IP
 and the Internet, 41
 hierarchy, 41–42

Telecommunications Industry Association (TIA), 244
Telecommunications outlet (TO), 95, 245, 368, 374
Telecommunications room (TR), 245
Telephone
  cabling and wiring, 89, 302–319
  circuit protection 306–307
  inside wiring, 303–305
  networking, 316–318
  operation of, 89–90
  separation, 312
  wiring components, 307–309
Telephone installation
  code requirements 305–306
  planning, 310–312
  safety, 313–314
  second line, 309–310
  small office, 310
  steps, 314–315
  testing, 315
Teleworking, 4
Telnet, 44–45
Television, 384–396
  antennas, 387, 389–391
  black-and-white, 90
  CATV, 385
  cable television, 385–386
  closed-circuit, 90
  Code rules for, 388–389, 391–395
  digital, 125–126
  installing wiring for, 384–389
  voltage hazards, 386–387
Testing voice, data, and video wiring, 353–365
Throughput, 66
TIA/EIA 568 standards, 24, 245–253, 354–355, 365
  architecture, 245
  fiber specifications, 253
  hierarchical star topology, 245–247
  residential infrastructure, 254–260
  T568A, 356
  T568B, 356
  terminology 245
TIA-570-B security cabling
  recommendations, 153–154
  standards 253–260
Time Division Multiple Access (TDMA), 59–60
Time domain reflectometer (TDR), 358, 360–361
Tip, 309
Token, 30
Token Ring Network (TRN), 12
Topology
  bus, 31
  combinational, 31
  network, 28, 30–31
  tree, 31

Total Access Communications System (TACS), 58
Trade size, 168
Transmission Control Protocol (TCP), 42–43
Transponder, 75
Transport layer, 27
Trunks, 99

U
Underwriters Laboratories, 171
Unicast, 34
Uniform Resource Locator (URL), 46
Uninterruptible power supply (UPS), 108
Universal Mobile Telecommunications System (UMTS), 84
Universal plug-and-play, 420
Universal Service Fund (USF), 107
Universal Service Order Code (USOC), 356–357
Universal signal bus (USB), 20, 21–22
Unlicensed PCS, 56
Unshielded twisted-pair (UTP) cable, 3, 13, 23–24, 332
  categories of, 248–253
  standards, 365
  testing, 354–355
Uplink, 75
User Datagram Protocol (UDP), 43
Utilization equipment, 168

V
Video display interface, 139–140
Video cable installation
  cabling, 321–322, 326–327
  Code requirements, 324
  grounding, 324–325
  indoor circuits, 326
  messenger cables, 326, 342
  substitutions, 327
  surge suppressors, 324
  routing outdoor circuits, 325–326
  termination, 322–323
Video systems, 122–133
  cabling, 133, 326–327
  displays, 130–133
  encoding, 126–130
  formats, 123–126
  installations, 321–327
  monitoring, 153
  standards, 128–130
  storage, 130
Video transmission, 90–91
Virtual Private Network (VPN), 424
Voice applications, 95–100
Voice and data network, 3
  access technologies, 371
  cable selection, 371

telecommunications outlets, 370–371
wireless access points, 371–372
Voice over IP (VoIP), 103–109
  implementation, 105–106
  pros and cons, 107–109
  regulatory issues, 107
  residential network architecture, 106
  technology, 104–107

W
Waterproof boxes, 204–205
Whole-house audio, 118–122
  speakers, 120–122
  zones, 119–120
WiFi, 101, 350
Wire connector, 230
Wire mapping, 355–361
  problems, 358–361
Wire nut, 230
Wired/wireless network integration, 103
Wiring for computers, 395
Wireless access points (WAP), 371–372
Wireless Application Protocol (WAP), 63–64
Wireless applications, 63–66
Wireless communications, 53–85
Wireless, definition of, 53, 349
Wireless LANs (WLANs), 67–73
  planning for, 372
  standards, 68
Wireless Local Loop (WWL), 72
Wireless networks, 53
  cabling for, 348–352
  installation, site survey for, 350–352
  interoperability/interference, 350
  system layout, 349–350
  testing, 353–365
Wireless personal area networks (WPANs), 372
Wireless transmission, 333–334
Work orders, 172. *See also* Change orders

X
Xanboo products, 422–423

Y
YCrCb video format
  advantages to, 127–128
  equations for, 127

Z
Zip cord, 274
Zones, 120–121